Antitumoral Properties of Natural Products

Antitumoral Properties of Natural Products

Volume 1

Editor

Roberto Fabiani

MDPI • Basel • Beijing • Wuhan • Barcelona • Belgrade • Manchester • Tokyo • Cluj • Tianjin

Editor
Roberto Fabiani
Department of Chemistry, Biology and Biotechnology
University of Perugia
Italy

Editorial Office
MDPI
St. Alban-Anlage 66
4052 Basel, Switzerland

This is a reprint of articles from the Special Issue published online in the open access journal *Molecules* (ISSN 1420-3049) (available at: https://www.mdpi.com/journal/molecules/special_issues/molecules_APoNP).

For citation purposes, cite each article independently as indicated on the article page online and as indicated below:

LastName, A.A.; LastName, B.B.; LastName, C.C. Article Title. *Journal Name* **Year**, *Article Number*, Page Range.

Volume 1
ISBN 978-3-03943-098-7 (Hbk)
ISBN 978-3-03943-099-4 (PDF)

Volume 1-2
ISBN 978-3-03943-116-8 (Hbk)
ISBN 978-3-03943-117-5 (PDF)

© 2020 by the authors. Articles in this book are Open Access and distributed under the Creative Commons Attribution (CC BY) license, which allows users to download, copy and build upon published articles, as long as the author and publisher are properly credited, which ensures maximum dissemination and a wider impact of our publications.
The book as a whole is distributed by MDPI under the terms and conditions of the Creative Commons license CC BY-NC-ND.

Contents

About the Editor ix

Roberto Fabiani
Antitumoral Properties of Natural Products
Reprinted from: *Molecules* **2020**, *25*, 650, doi:10.3390/molecules25030650 1

Yollanda E. M. Franco, Marcia Y. Okubo, Adriana D. Torre, Paula P. Paiva, Marcela N. Rosa, Viviane A. O. Silva, Rui M. Reis, Ana L. T. G. Ruiz, Paulo M. Imamura, João E. de Carvalho and Giovanna B. Longato
Coronarin D Induces Apoptotic Cell Death and Cell Cycle Arrest in Human Glioblastoma Cell Line
Reprinted from: *Molecules* **2019**, *24*, 4498, doi:10.3390/molecules24244498 9

Viviane Aline Oliveira Silva, Marcela Nunes Rosa, Aline Tansini, Olga Martinho, Amilcar Tanuri, Adriane Feijó Evangelista, Adriana Cruvinel Carloni, João Paulo Lima, Luiz Francisco Pianowski and Rui Manuel Reis
Semi-Synthetic Ingenol Derivative from *Euphorbia tirucalli* Inhibits Protein Kinase C Isotypes and Promotes Autophagy and S-Phase Arrest on Glioma Cell Lines
Reprinted from: *Molecules* **2019**, *24*, 4265, doi:10.3390/molecules24234265 25

Chanatip Ooppachai, Pornngarm Limtrakul (Dejkriengkraikul) and Supachai Yodkeeree
Dicentrine Potentiates TNF-α-Induced Apoptosis and Suppresses Invasion of A549 Lung Adenocarcinoma Cells via Modulation of NF-κB and AP-1 Activation
Reprinted from: *Molecules* **2019**, *24*, 4100, doi:10.3390/molecules24224100 43

Izabela N. Faria Gomes, Renato J. Silva-Oliveira, Viviane A. Oliveira Silva, Marcela N. Rosa, Patrik S. Vital, Maria Cristina S. Barbosa, Fábio Vieira dos Santos, João Gabriel M. Junqueira, Vanessa G. P. Severino, Bruno G Oliveira, Wanderson Romão, Rui Manuel Reis and Rosy Iara Maciel de Azambuja Ribeiro
Annona coriacea Mart. Fractions Promote Cell Cycle Arrest and Inhibit Autophagic Flux in Human Cervical Cancer Cell Lines
Reprinted from: *Molecules* **2019**, *24*, 3963, doi:10.3390/molecules24213963 59

Chayaporn Subkamkaew, Pornngarm Limtrakul (Dejkriengkraikul) and Supachai Yodkeeree
Proanthocyanidin-Rich Fractions from Red Rice Extract Enhance TNF-α-Induced Cell Death and Suppress Invasion of Human Lung Adenocarcinoma Cell A549
Reprinted from: *Molecules* **2019**, *24*, 3393, doi:10.3390/molecules24183393 75

Jia-Sin Yang, Renn-Chia Lin, Yi-Hsien Hsieh, Heng-Hsiung Wu, Geng-Chung Li, Ya-Chiu Lin, Shun-Fa Yang and Ko-Hsiu Lu
CLEFMA Activates the Extrinsic and Intrinsic Apoptotic Processes through JNK1/2 and p38 Pathways in Human Osteosarcoma Cells
Reprinted from: *Molecules* **2019**, *24*, 3280, doi:10.3390/molecules24183280 91

Jun-Man Hong, Jin-Hee Kim, Hyemin Kim, Wang Jae Lee and Young-il Hwang
SB365, *Pulsatilla* Saponin D Induces Caspase-Independent Cell Death and Augments the Anticancer Effect of Temozolomide in Glioblastoma Multiforme Cells
Reprinted from: *Molecules* **2019**, *24*, 3230, doi:10.3390/molecules24183230 103

Ziqi Yue, Xin Guan, Rui Chao, Cancan Huang, Dongfang Li, Panpan Yang, Shanshan Liu, Tomoka Hasegawa, Jie Guo and Minqi Li
Diallyl Disulfide Induces Apoptosis and Autophagy in Human Osteosarcoma MG-63 Cells through the PI3K/Akt/mTOR Pathway
Reprinted from: *Molecules* **2019**, *24*, 2665, doi:10.3390/molecules24142665 119

Hana Jin, Young Shin Ko, Sang Won Park, Ki Churl Chang and Hye Jung Kim
13-Ethylberberine Induces Apoptosis through the Mitochondria-Related Apoptotic Pathway in Radiotherapy-Resistant Breast Cancer Cells
Reprinted from: *Molecules* **2019**, *24*, 2448, doi:10.3390/molecules24132448 133

Tai-Shan Shen, Yung-Ken Hsu, Yi-Fu Huang, Hsuan-Ying Chen, Cheng-Pu Hsieh and Chiu-Liang Chen
Licochalcone A Suppresses the Proliferation of Osteosarcoma Cells through Autophagy and ATM-Chk2 Activation
Reprinted from: *Molecules* **2019**, *24*, 2435, doi:10.3390/molecules24132435 147

Camila Ramalho Bonturi, Mariana Cristina Cabral Silva, Helena Motaln, Bruno Ramos Salu, Rodrigo da Silva Ferreira, Fabricio Pereira Batista, Maria Tereza dos Santos Correia, Patrícia Maria Guedes Paiva, Tamara Lah Turnšek and Maria Luiza Vilela Oliva
A Bifunctional Molecule with Lectin and Protease Inhibitor Activities Isolated from *Crataeva tapia* Bark Significantly Affects Cocultures of Mesenchymal Stem Cells and Glioblastoma Cells
Reprinted from: *Molecules* **2019**, *24*, 2109, doi:10.3390/molecules24112109 159

Liang-Tzung Lin, Wu-Ching Uen, Chen-Yen Choong, Yeu-Ching Shi, Bao-Hong Lee, Cheng-Jeng Tai and Chen-Jei Tai
Paris Polyphylla Inhibits Colorectal Cancer Cells via Inducing Autophagy and Enhancing the Efficacy of Chemotherapeutic Drug Doxorubicin
Reprinted from: *Molecules* **2019**, *24*, 2102, doi:10.3390/molecules24112102 173

Taotao Ling, Walter H. Lang, Julie Maier, Marizza Quintana Centurion and Fatima Rivas
Cytostatic and Cytotoxic Natural Products against Cancer Cell Models
Reprinted from: *Molecules* **2019**, *24*, 2012, doi:10.3390/molecules24102012 187

Li-Ying Bo, Tie-Jing Li and Xin-Huai Zhao
Effect of Cu/Mn-Fortification on In Vitro Activities of the Peptic Hydrolysate of Bovine Lactoferrin against Human Gastric Cancer BGC-823 Cells
Reprinted from: *Molecules* **2019**, *24*, 1195, doi:10.3390/molecules24071195 203

Li Tian, Fan Cheng, Lei Wang, Wen Qin, Kun Zou and Jianfeng Chen
CLE-10 from *Carpesium abrotanoides* L. Suppresses the Growth of Human Breast Cancer Cells (MDA-MB-231) In Vitro by Inducing Apoptosis and Pro-Death Autophagy Via the PI3K/Akt/mTOR Signaling Pathway
Reprinted from: *Molecules* **2019**, *24*, 1091, doi:10.3390/molecules24061091 217

Hairui Yang, Xu Bai, Henan Zhang, Jingsong Zhang, Yingying Wu, Chuanhong Tang, Yanfang Liu, Yan Yang, Zhendong Liu, Wei Jia and Wenhan Wang
Antrodin C, an NADPH Dependent Metabolism, Encourages Crosstalk between Autophagy and Apoptosis in Lung Carcinoma Cells by Use of an AMPK Inhibition-Independent Blockade of the Akt/mTOR Pathway
Reprinted from: *Molecules* **2019**, *24*, 993, doi:10.3390/molecules24050993 229

Haet Nim Lim, Seung Bae Baek and Hye Jin Jung
Bee Venom and Its Peptide Component Melittin Suppress Growth and Migration of Melanoma Cells via Inhibition of PI3K/AKT/mTOR and MAPK Pathways
Reprinted from: *Molecules* **2019**, 24, 929, doi:10.3390/molecules24050929 . . . . **255**

Xianxian Wei, Lijie Xia, Dilinigeer Ziyayiding, Qiuyan Chen, Runqing Liu, Xiaoyu Xu and Jinyao Li
The Extracts of *Artemisia absinthium* L. Suppress the Growth of Hepatocellular Carcinoma Cells through Induction of Apoptosis via Endoplasmic Reticulum Stress and Mitochondrial-Dependent Pathway
Reprinted from: *Molecules* **2019**, 24, 913, doi:10.3390/molecules24050913 . . . . **269**

Johanna Willer, Karin Jöhrer, Richard Greil, Christian Zidorn and Serhat Sezai Çiçek
Cytotoxic Properties of Damiana (*Turnera diffusa*) Extracts and Constituents and A Validated Quantitative UHPLC-DAD Assay
Reprinted from: *Molecules* **2019**, 24, 855, doi:10.3390/molecules24050855 . . . . **283**

Pengfei Liu, Yuchen Xiang, Xuewen Liu, Te Zhang, Rui Yang, Sen Chen, Li Xu, Qingqing Yu, Huzi Zhao, Liang Zhang, Ying Liu and Yuan Si
Cucurbitacin B Induces the Lysosomal Degradation of EGFR and Suppresses the CIP2A/PP2A/Akt Signaling Axis in Gefitinib-Resistant Non-Small Cell Lung Cancer
Reprinted from: *Molecules* **2019**, 24, 647, doi:10.3390/molecules24030647 . . . . **299**

Selma Ferhi, Sara Santaniello, Sakina Zerizer, Sara Cruciani, Angela Fadda, Daniele Sanna, Antonio Dore, Margherita Maioli and Guy D'hallewin
Total Phenols from Grape Leaves Counteract Cell Proliferation and Modulate Apoptosis-Related Gene Expression in MCF-7 and HepG2 Human Cancer Cell Lines
Reprinted from: *Molecules* **2019**, 24, 612, doi:10.3390/molecules24030612 . . . . **317**

About the Editor

Roberto Fabiani

Working Experience

1995–2002: Technician at the University of Perugia, Department of Cell and Molecular Biology.

2002–2010: Researcher at the Faculty of Sciences, Department of Medical-Surgical Specialties and Public Health, Section of Molecular Epidemiology and Environmental Hygiene -University of Perugia.

2010 to today: Associate Professor in the Scientific Sector (MED/42) General and Applied Hygiene, Department of Chemistry, Biology and Biotechnology – University of Perugia.

Education

1988 -Graduated summa cum laude in Biological Sciences at the University of Perugia.

1989–1990 Holder of a Fellowship for a Research on "Action mechanisms of some disinfectants" at the Department of Microbiology of the University of Perugia.

1991–1994 Holder of a 3 years Fellowship of the Italian "Ministero della Pubblica Istruzione/ Ricerca Scientifica" at the Department of Clinical Chemistry of the University of Uppsala, Sweden.

1994 Ph.D. in Medical Science at the Department of Clinical Chemistry of the University of Uppsala, Sweden.

1997 Holder of a Fellowship of the Italian "CNR" at the Federal Centre for Nutrition, Institute of Nutritional Physiology, Karlsruhe, Germany.

1999 Holder of a Fellowship of the Italian "CNR" at the Federal Centre for Nutrition, Institute of Nutritional Physiology, Karlsruhe, Germany.

Research Interest

The scientific activity of Dr. Roberto Fabiani has been mainly devoted to investigate the relationships between nutrition and cancer. He has studied the cancer risk associated to the consumption of some foods and cellular effects induced by possible risk and protective factors on the carcinogenesis process. He has been interested to study the oxidative DNA damage induced and/or prevented by different environmental pollutants and their metabolites, intestinal substances such as bile acids and short chain fatty acids, and natural compounds present in the different foods, in particular from Mediterranean diet. In addition, he was involved in the study of chemopreventive activities of secoiridoid phenolic compounds isolated from extra virgin olive oil on different cell systems both in vitro and in vivo. Major attention has been paid to the effects these compounds on proliferation, cell cycle, differentiation and apoptosis of tumor cells, and to the molecular mechanisms involved.

Editorial
Antitumoral Properties of Natural Products

Roberto Fabiani

Department of Chemistry, Biology and Biotechnology, University of Perugia, 06126 Perugia, Italy; roberto.fabiani@unipg.it

Received: 27 January 2020; Accepted: 28 January 2020; Published: 3 February 2020

Cancer is one of the major causes of death worldwide. It is a multifactorial heterogeneous disease characterized by a multistep process initiated by genetic alterations of normal cells which become malignant cells. These cells are characterized by uncontrolled cell growth, immortality, invasiveness, and ability to form distant metastasis. Natural bioactive molecules may interfere with the carcinogenesis process by altering the tumor cell behavior and targeting several molecular abnormally activated signaling pathways.

Among different cancer types, glioblastoma is the most frequent and aggressive of all malignant brain tumors. Glioblastoma is highly invasive, and its treatment include surgery, radiation, and chemotherapy with temozolomide (TMZ). Nevertheless, patient prognosis remains poor and associated with a low survival rate. In this Special Issue, Franco et al. have investigated the anticancer properties of coronarin D, a diterpene isolated from a dichloromethane extract of *Hedychium coronarium* in a glioblastoma cell line (U-251) [1]. They found that this compound was able to inhibit proliferation and induce G1 cell cycle arrest and apoptosis in U-251 cells. The authors proposed that coronarin D-induced effects were mediated by an overproduction of reactive oxygen species, which promoted phosphorylation of H2AX and ERK, increased the expression of p21, and activated caspases. Noteworthy is the observation that coronarin D was in some cases even more effective than TMZ. Similarly, Silva et al. have demonstrated that ingenol-3-dodecanoate (IngC), a semi-synthetic ingenol derivative from *Euphorbia tirucalli*, was more effective at inhibiting the growth of different glioma cell lines than TMZ [2]. Overexpression of p21 was also observed in both GAMG and U373 glioblastoma cell lines. On the other hand, IngC promoted S-phase arrest but was not able to induce apoptosis. Worth noting is the observation that IngC promoted glioma cell autophagy as evidenced by both the accumulation of LC3B-II and the development of acidic vesicular organelles. In addition, IngC treatment resulted in potent inhibition of protein kinase C activity, showing differential actions against various PKC isotypes [2]. Hong at al. demonstrated a cytotoxic effect of SB365, a saponin D extracted from the roots of *Pulsatilla koreana*, on U87-MG and T98G glioblastoma multiforme (GBM) cells [3]. This compound induced caspase-independent cell death, inhibited autophagic flux, and deteriorated lysosomal stability and mitochondrial membrane potential (MMP) in U87-MG cells. Very importantly, this paper showed also an additive effect between SB365 and TMZ on glioblastoma cell proliferation both in vitro and in vivo using a mouse U87-MG xenograft model [3]. Furthermore, Bonturi et al. have studied the effect of a plant-derived protein obtained from *Crataeva tapia* tree bark lectin (CrataBL) on U87 glioblastoma cells in co-culture with mesenchymal stem cells [4]. They showed that the mixed cells grown in 1:1 co-culture were more sensitive to the CrataBL than each of the individual cell types with regards to both inhibition of proliferation and induction of death. Corrêa et al. have developed and characterized liposomal nanocapsules loaded with purified tarin, a lectin naturally found in taro corms (*Colocasia esculenta*), which were able to dose-dependently inhibit glioblastoma (U-87 MG) and breast (MDA-MB-231) tumor cells, while free tarin did not affect tumoral cell growth [5]. Finally, a study by Pham et al. has investigated the effects of neferine on another brain tumor, neuroblastoma [6]. They showed that neferine suppressed proliferation and induced both apoptosis and autophagy in human neuroblastoma IMR32 cells [6].

Osteosarcoma (OS) is one of the most frequent bone tumors, with a high prevalence in teenagers and young adults. Due to its high metastatic potential, the prognosis of osteosarcoma patients is poor. Although the main treatment is surgery, anticancer strategies based on development of new agents that target particular intracellular signaling pathways in osteosarcoma cells are needed. The study of Yang et al. has shown that CLEFMA (4-[3,5-bis(2-chlorobenzylidene)-4-oxo-piperidine-1-yl]-4-oxo-2-butenoic acid), a synthetic analog of curcumin, decreased viability and induced apoptosis in human osteosarcoma U2OS and HOS cells [7]. This compound activated both extrinsic and intrinsic apoptotic pathways and, by using specific inhibitors, the authors demonstrated that these effects were mediated by the phosphorylation of c-Jun N-terminal kinases (JNK) and p38, while the ERK pathway was not involved [7]. Diallyl disulfide (DADs), a natural organic compound present in garlic and scallion, inhibited proliferation, induced apoptosis, and arrested the cell cycle at the G2/M phase of MG-63 osteosarcoma cells [8]. In addition, DADs induced the formation of autophagosome as revealed by the increased expression of LC3-II protein which was reduced by the autophagy inhibitor 3-methyladenine. These anticancer effects were mediated by the ability of DADs to inhibit the PI3K/Akt/mTOR signaling pathway [8]. Similar to DADs, Shen et al. showed that licochalcone A, a flavonoid extracted from licorice root, reduced proliferation, caused G2/M phase cell cycle arrest, and induced apoptosis and autophagy in HOS and MG-63 osteosarcoma cells [9]. Furthermore, it was found that licochalcone A induced rapid phosphorylation of Chk2 and ATM, suggesting that the ATM–Chk2 pathway may contribute to its effect on G2/M phase arrest. By using the autophagy inhibitor chloroquine, it was also demonstrated that the observed autophagy was associated with licochalcone A-induced apoptosis [9]. Hsu et al. have reported that coronarin D reduced the proliferation of HOS and MG-63 osteosarcoma cells while had a minor cytotoxic effect in human fibroblasts (MRC-5) [10]. This compound induced apoptosis and mitotic phase arrest in osteosarcoma cells and JNK was found to play a crucial role in coronarin D-induced effects [10].

Lung cancer is the leading cause of cancer incidence and mortality worldwide, with 2.1 million new cases and 1.8 million deaths predicted in 2018. The high mortality and low survival rate, that have remained unchanged in recent years, support the efforts to find new antineoplastic drugs to overcome this malignancy. In the present Special Issue, four papers have described the anticancer activity of different natural products on lung cancer cell in vitro [11–14]. Ooppachai et al. showed that dicentrine, an aporphine alkaloid found in the roots of *Lindera megaphylla* and several other plants, potentiated the TNF-induced apoptosis in A549 lung adenocarcinoma cells [11]. This compound was also able to inhibit the TNF-induced invasion, migration, and expression of metastasis-associated proteins. These effects were due to, at least in part, to the suppression of TAK-1, MAPK, Akt, AP-1, and NF-kB signaling pathways [11]. Similar effects were induced by treatment of A549 lung adenocarcinoma cells with a proanthocyanidin-rich fraction obtained from red rice [12]. Similarly, antrodin C (ADC), a maleimide derivative isolated from mycelium of *Taiwanofungus camphoratus*, exerted its anticancer activities (antiproliferative and pro-apoptotic) by suppressing both the Akt/mTOR and AMPK signaling pathways in human lung adenocarcinoma cell line SPCA-1 [13]. Importantly, this study showed that ADC was not toxic toward normal human lung cells BEAS-2B [13]. Cucurbitacin B (CuB) is a natural tetracyclic triterpenoid compound mainly found in Cucurbitaceae. The study of Liu et al. reported, for the first time, that CuB induced EGFR degradation and inhibited CIP2A/PP2A/Akt activities in different in gefitinib-resistant non-small cell lung cancer cell lines [14].

After lung cancer, breast cancer is the second main cause of death by cancer in women. Triple-negative breast cancer (TNBC), characterized by the absence of expression of three receptors (estrogen, progesterone, and human epidermal growth factor receptor 2), is an aggressive type of breast cancer that is difficult to treat. Jin et al. have reported the antitumoral properties of 13-ethylberberine (13EBR), a derivative of berberine (alkaloid isolated from *Cotridis rhizoma*), on MDA-MB-231 cells, which are a common model for TNBC [15]. 13-EBR reduced proliferation and induced apoptosis in both MDA-MB-231 and radiotherapy-resistant RT-R MDA-MB-231 cells. These effects were mediated by the promotion of reactive oxygen species production and regulation of the apoptosis-related proteins involved in the intrinsic pathway [15]. In the same cell model, Tian et al. showed antitumor activity of 4-epi-isoinuviscolide (CLE-10), a sesquiterpene lactone isolated from *Carpesium abrotanoides* L. [16]. This compound induced pro-death autophagy and apoptosis in MDA-MB-231 cells by upregulating the protein expressions of LC3-II, p-ULK1, Bax, and Bad, and downregulating p-PI3K, p-Akt, p-mTOR, p62, Bcl-2, and Bcl-xl [16]. Tan et al. have used another model of breast cancer (MCF-7 cells) to study the antiproliferative activity of the water soluble natural yellow *Monascus* pigments [17]. These compounds reduced the migration and invasion of MCF-7 cells, and these activities were associated with a downregulation of the expression of matrix metalloproteinases and vascular endothelial growth factor [17].

Gastric cancer is the fourth most common cancer and the second leading cause of cancer death worldwide. Several studies have been performed to find new therapeutic strategies based on bioactive phytochemicals with a lower toxicity. Zeylenone (Zey), a cyclohexene oxide isolated from the leaves of *Uvaria grandiflora*, has shown multiple anticancer activities on different cell lines. In their study, Yang et al. have demonstrated that this compound was able to induce mitochondrial apoptosis and to inhibit migration and invasion in SGC7901 and MGC803 gastric cancer cells in vitro [18]. In addition, Zey downregulated the expression of matrix metalloproteinase 2/9 and inhibited the phosphorylation of AKT and ERK. Of particular interest is the in vivo observation that Zey effectively inhibited tumor growth in nude BALB/c mice bearing BGC823 xenografts without any evident cytotoxic effects [18]. Bo et al. have investigated the anticancer activity of bovine lactoferrin hydrolysate (BLH) with added Cu^{2+} and Mn^{2+} on human gastric cancer BGC-823 cells [19]. They showed that the fortified BLH products had higher activities than BLH alone as evidenced by induction of apoptosis and activation of the classic caspase-3-dependent apoptotic pathway [19]. As opposed to the use of single purified molecules, several studies have highlighted the anticancer effects of plant-derived fractions and/or extracts on different cell models. Gomes at al. have demonstrated a cytotoxic effect of different *Annona coriacea* Mart. derived fractions on cisplatin-resistant cervical cancer cell lines (CaSki, HeLa, and SiHa) and on a normal keratinocyte cell line (HaCaT) [20]. Lin et al. have investigated the tumor-suppressive effects of an ethanol extract from *Paris polyphylla* in DLD-1 human colorectal carcinoma cells [21]. They found that cell death was induced by the upregulation of autophagy markers and treatment in combination with doxorubicin enhanced its cytotoxicity (12). Wei et al. have studied the anticancer activity of an ethanol extract from *Artemisia absinthium* and some subfractions on hepatocellular carcinoma cells [22]. The results showed the inhibition of cells growth and induction of apoptosis which might be mediated by the endoplasmic reticulum stress and mitochondrial-dependent pathway [22]. In addition, it was demonstrated an inhibition of tumor growth in vivo using the H22 tumor mouse model (H22 cells were subcutaneously injected in male Kunming mice and tumor sizes were monitored over time). Interestingly, the extract improved the survival of tumor mice without obvious toxicity and side effects [22]. Willer et al. have assayed extracts and fractions derived from damiana (*Turnera diffusa*) against different myeloma cell lines (NCI-H929, U266, and MM1S) [23]. They identified the flavanone naringenin as the most active compound able to decrease viability in particular in NCI-H929 cells. Furthermore, apigenin 7-O-(4"-O-p-E-coumaroyl)-glucoside was identified as being cytotoxic for the first time. This study also described the first validated UHPLC-DAD method for the quantification of phenolic constituents in *Turnera diffusa* [23]. Huang et al. have found that a *Ganoderma tsugae* ethanol extract, a Chinese natural and herbal product, significantly inhibited expression of SREBP-1 and its downstream genes associated with lipogenesis in prostate cancer cells (LNCaP and C4-2) [24]. These effects were associated to the inhibition of cell growth, migration, and invasion, and induction of apoptosis [24]. Ferhi et al. have shown the antiproliferative effects ethanol and water extracts from grape leaves on HepG2 hepatocarcinoma, MCF-7 human breast cancer cells, and vein human umbilical (HUVEC) cells [25]. In cancer cells, both extracts induced the expression of the pro-apoptotic gene Bax and reduced the expression of the anti-apoptotic gene Bcl-2. Interesting, the extracts did not show toxic effects on vein umbilical HUVEC cells [25]. Elansary et al. have characterized the phenolic profiles of

Catalpa speciosa, *Taxus cuspidata*, and *Magnolia acuminata* bark extracts and studied their antiproliferative activity against different cancer cell lines (MCF-7, HeLa, Jurkat, T24, and HT-29) [26]. Yang et al. have screened 11 different lichen acetone extracts on the stemness potential of colorectal cancer cells and have isolated the most active compound tumidulin from *Niebla* sp. [27]. This compound reduced spheroid formation and the mRNA expression and protein levels of different cancer stem markers (ALDH1, CD133, CD44, Lgr5, and Musashi-1) in CSC221, DLD1, and HT29 cells [27]. Alvarado-Sansininea et al. have isolated quercetagetin and patuletin from *Tagetes erecta* and *Tagetes patula* flower ethanol extracts and tested for their antiproliferative, necrotic, and apoptotic activity on different cancer cell lines (CaSki: cervical, MDA-MB-231: breast, SK-LU-1: lung) [28]. The structure–activity relationship study, including also quercetin for comparison, demonstrated that the presence of a methoxyl group in C6 of the A ring of flavonol patuletin enhanced its anticancer potential [28]. Yu et al. have purified polysaccharides from the stem extract of the medicinal plant *Dendrobium officinale* grown under different planting conditions (in the greenhouse and in the wild) and compared their structure and antitumor properties on HeLa cells [29]. Polysaccharides showed a significant activity only after oxidative degradation to smaller molecular weight species. The fractions from wild plants showed an evident antiproliferative and pro-apoptotic activities while the effects of the fractions from greenhouse plants were not significant [29]. Nguyen et al. have biotransformed three selected anthraquinones into their O-glucoside by a bacteria glycosyltransferase, and tested these products for their antiproliferative affects against various cancer cells (AGS: gastric; HeLa: cervical; Hep-G2: liver) [30]. They found that the glycosylated derivatives were more effective in inhibiting cell growth than their parental aglycones [30].

Kahnt et al. have synthesized 28 new cytotoxic agents starting from the naturally occurring triterpenoids betulinic and ursolic acid [31]. Different ethylenediamine derived carboxamides were tested for cytotoxicity by the sulforhodamine-B colorimetric assay in several tumor cell lines (518A2: melanoma; A2780: ovarian carcinoma; HT29: colon adenocarcinoma; MCF-7: breast adenocarcinoma; 8505C: thyroid carcinoma) and in nonmalignant mouse fibroblasts (NIH 3T3). Two betulinic acid-derived compounds were identified as the most effective with an EC_{50} lower than 1 µM [31]. Unfortunately, these compounds were not selective for tumor cells since they were toxic also toward nonmalignant fibroblasts. Ling et al. have screened a natural product library containing fractions and pure compounds for proliferation inhibition in different cancer cell models [32]. They identified different alkaloid compounds with a potent cytotoxic effect. In particular, homoharringtonine showed an EC_{50} lower that 0.1 µM and together with cephalotaxine, demonstrated potent inhibition of protein synthesis [32]. Lim et al. have demonstrated an antimelanoma effect of bee venom (BV) and that the major active ingredient is melittin, an amphiphilic peptide containing 26 amino acid residues [33]. These effects were mediated by the downregulation of PI3K/AKT/mTOR and MAPK signaling pathways [33].

Three new isochromanes were isolated from *Aspergillus fumigatus* fermentation broth and tested in vitro for their cytotoxic effects by MTT assay of MV4-11 cell line [34]. Only two of them showed a moderate growth inhibition with IC_{50} values of 23.95 and 32.70 µM, respectively [34]. Similarly, four new pentacyclic triterpene were isolated from hexane extract of *Salacia crassifolia* root wood and tested for their cytotoxic activity against human cancer cell lines using the "NCI-60 cell line screen" [35]. Among them, pristimerin showed selective inhibitory activity towards a variety of human tumor cell lines and it was primarily responsible for the cytotoxic activity of the crude extracts [35].

In this Special Issue, six reviews were included aimed to summarize the antitumoral properties of different compounds isolated from several natural sources [36–41]. Liu et al. reviewed the anticancer activities of the compounds porphyran and carrageenan, derived from red seaweed [36]. Possible mechanisms in the anticancer activity of these two polysaccharides were considered along with their possible cooperative actions with other anticancer chemotherapeutics [36]. Wang et al. have reported a mini review on the anticancer activity of the naturally occurring indoloquinoline alkaloids cryptolepine, neocryptolepine, and isocryptolepine, isolated from the roots of *Cryptolepis sanguinolenta* and several of their analogues [37]. They presented an overview of the potential of neocryptolepine and isocryptolepine as scaffolds for the design and development of new anticancer drugs [37]. Yang et al. have reviewed diverse in vitro and in vivo pharmacological properties of capsazepine, a synthetic analogue of capsaicin (the common pungent ingredient of hot chili peppers) [38]. In addition to having an anticancer activity, capsazepine has important anti-inflammatory effects reducing the level of some inflammatory mediators [38]. Liskova et al. provided a comprehensive review of studies focusing on the anticancer effectiveness of dietary phytochemicals, either isolated or as mixtures, which act via targeting cancer stem cells (CSCs) [39]. Among dietary compounds able to target CSCs and some of their abnormally activated signaling pathways, epigallocatechin-3-gallat, resveratrol, genistein, curcumin, isothiocyanates, and diallyl trisulfide have been of particular interest [39]. Girisa et al. have considered and reviewed the potential anticancer activity of zerumbone, a sesquiterpene compound isolated from *Zingiber zerumbet* Smith [40], while Choi has reviewed the anti-inflammatory and anticancer activities of phloretin, a chalcone polyphenol present in apple [41].

Natural products are attractive sources for the development of new medicinal and therapeutic agents. Those with antitumoral potential may be more selective and have weaker adverse effects compared to conventional chemotherapy drugs actually used for cancer treatment. Clinical trials are necessary to demonstrated whether the in vitro and in vivo animal data are reproduced in human, and to allow the application of natural products in cancer prevention and treatment.

Conflicts of Interest: The author declares no conflict of interest.

References

1. Franco, Y.E.M.; Okubo, M.Y.; Torre, A.D.; Paiva, P.P.; Rosa, M.N.; Silva, V.A.O.; Reis, R.M.; Ruiz, A.L.T.G.; Imamura, P.M.; de Carvalho, J.E.; et al. Coronarin D Induces Apoptotic Cell Death and Cell Cycle Arrest in Human Glioblastoma Cell Line. *Molecules* **2019**, *24*, 4498. [CrossRef] [PubMed]
2. Silva, V.A.O.; Rosa, M.N.; Tansini, A.; Martinho, O.; Tanuri, A.; Evangelista, A.F.; Cruvinel Carloni, A.; Lima, J.P.; Pianowski, L.F.; Reis, R.M. Semi-Synthetic Ingenol Derivative from Euphorbia tirucalli Inhibits Protein Kinase C Isotypes and Promotes Autophagy and S-phase Arrest on Glioma Cell Lines. *Molecules* **2019**, *24*, 4265. [CrossRef] [PubMed]
3. Hong, J.M.; Kim, J.H.; Kim, H.; Lee, W.J.; Hwang, Y.I. SB365, Pulsatilla Saponin D Induces Caspase-Independent Cell Death and Augments the Anticancer Effect of Temozolomide in Glioblastoma Multiforme Cells. *Molecules* **2019**, *24*, 3230. [CrossRef] [PubMed]
4. Bonturi, C.R.; Silva, M.C.C.; Motaln, H.; Salu, B.R.; Ferreira, R.D.S.; Batista, F.P.; Correia, M.T.D.S.; Paiva, P.M.G.; Turnšek, T.L.; Oliva, M.L.V. A Bifunctional Molecule with Lectin and Protease Inhibitor Activities Isolated from Crataeva tapia Bark Significantly Affects Cocultures of Mesenchymal Stem Cells and Glioblastoma Cells. *Molecules* **2019**, *24*, 2109. [CrossRef]
5. Corrêa, A.C.N.T.F.; Vericimo, M.A.; Dashevskiy, A.; Pereira, P.R.; Paschoalin, V.M.F. Liposomal Taro Lectin Nanocapsules Control Human Glioblastoma and Mammary Adenocarcinoma Cell Proliferation. *Molecules* **2019**, *24*, 471. [CrossRef]
6. Pham, D.C.; Chang, Y.C.; Lin, S.R.; Fuh, Y.M.; Tsai, M.J.; Weng, C.F. FAK and S6K1 Inhibitor, Neferine, Dually Induces Autophagy and Apoptosis in Human Neuroblastoma Cells. *Molecules* **2019**, *23*, 3110. [CrossRef]
7. Yang, J.S.; Lin, R.C.; Hsieh, Y.H.; Wu, H.H.; Li, G.C.; Lin, Y.C.; Yang, S.F.; Lu, K.H. CLEFMA Activates the Extrinsic and Intrinsic Apoptotic Processes through JNK1/2 and p38 Pathways in Human Osteosarcoma Cells. *Molecules* **2019**, *24*, 3280. [CrossRef]
8. Yue, Z.; Guan, X.; Chao, R.; Huang, C.; Li, D.; Yang, P.; Liu, S.; Hasegawa, T.; Guo, J.; Li, M. Diallyl Disulfide Induces Apoptosis and Autophagy in Human Osteosarcoma MG-63 Cells through the PI3K/Akt/mTOR Pathway. *Molecules* **2019**, *24*, 2665. [CrossRef]
9. Shen, T.S.; Hsu, Y.K.; Huang, Y.F.; Chen, H.Y.; Hsieh, C.P.; Chen, C.L. Licochalcone A Suppresses the Proliferation of Osteosarcoma Cells through Autophagy and ATM-Chk2 Activation. *Molecules* **2019**, *24*, 2435. [CrossRef]
10. Hsu, C.T.; Huang, Y.F.; Hsieh, C.P.; Wu, C.C.; Shen, T.S. JNK Inactivation Induces Polyploidy and Drug-Resistance in Coronarin D-Treated Osteosarcoma Cells. *Molecules* **2018**, *23*, 2121. [CrossRef]

11. Ooppachai, C.; Limtrakul Dejkriengkraikul, P.; Yodkeeree, S. Dicentrine Potentiates TNF-α-Induced Apoptosis and Suppresses Invasion of A549 Lung Adenocarcinoma Cells via Modulation of NF-κB and AP-1 Activation. *Molecules* **2019**, *24*, 4100. [CrossRef] [PubMed]
12. Subkamkaew, C.; Limtrakul Dejkriengkraikul, P.; Yodkeeree, S. Proanthocyanidin-Rich Fractions from Red Rice Extract Enhance TNF-α-Induced Cell Death and Suppress Invasion of Human Lung Adenocarcinoma Cell A549. *Molecules* **2019**, *24*, 3393. [CrossRef] [PubMed]
13. Yang, H.; Bai, X.; Zhang, H.; Zhang, J.; Wu, Y.; Tang, C.; Liu, Y.; Yang, Y.; Liu, Z.; Jia, W.; et al. Antrodin C, an NADPH Dependent Metabolism, Encourages Crosstalk between Autophagy and Apoptosis in Lung Carcinoma Cells by Use of an AMPK Inhibition-Independent Blockade of the Akt/mTOR Pathway. *Molecules* **2019**, *24*, 993. [CrossRef] [PubMed]
14. Liu, P.; Xiang, Y.; Liu, X.; Zhang, T.; Yang, R.; Chen, S.; Xu, L.; Yu, Q.; Zhao, H.; Zhang, L.; et al. Cucurbitacin B Induces the Lysosomal Degradation of EGFR and Suppresses the CIP2A/PP2A/Akt Signaling Axis in Gefitinib-Resistant Non-Small Cell Lung Cancer. *Molecules* **2019**, *24*, 647. [CrossRef]
15. Jin, H.; Ko, Y.S.; Park, S.W.; Chang, K.C.; Kim, H.J. 13-Ethylberberine Induces Apoptosis through the Mitochondria-Related Apoptotic Pathway in Radiotherapy-Resistant Breast Cancer Cells. *Molecules* **2019**, *24*, 2448. [CrossRef]
16. Tian, L.; Cheng, F.; Wang, L.; Qin, W.; Zou, K.; Chen, J. CLE-10 from Carpesium abrotanoides L. Suppresses the Growth of Human Breast Cancer Cells (MDA-MB-231) In Vitro by Inducing Apoptosis and Pro-Death Autophagy Via the PI3K/Akt/mTOR Signaling Pathway. *Molecules* **2019**, *24*, 1091. [CrossRef]
17. Tan, H.; Xing, Z.; Chen, G.; Tian, X.; Wu, Z. Evaluating Antitumor and Antioxidant Activities of Yellow Monascus Pigments from Monascus ruber Fermentation. *Molecules* **2019**, *23*, 3242. [CrossRef]
18. Yang, S.; Liao, Y.; Li, L.; Xu, X.; Cao, L. Zeylenone Induces Mitochondrial Apoptosis and Inhibits Migration and Invasion in Gastric Cancer. *Molecules* **2018**, *23*, 2149. [CrossRef]
19. Bo, L.Y.; Li, T.J.; Zhao, X.H. Effect of Cu/Mn-Fortification on In Vitro Activities of the Peptic Hydrolysate of Bovine Lactoferrin against Human Gastric Cancer BGC-823 Cells. *Molecules* **2019**, *24*, 1195. [CrossRef]
20. Gomes, I.N.F.; Silva-Oliveira, R.J.; Oliveira Silva, V.A.; Rosa, M.N.; Vital, P.S.; Barbosa, M.C.S.; Dos Santos, F.V.; Junqueira, J.G.M.; Severino, V.G.P.; Oliveira, B.G.; et al. Annona coriacea Mart. Fractions Promote Cell Cycle Arrest and Inhibit Autophagic Flux in Human Cervical Cancer Cell Lines. *Molecules* **2019**, *24*, 3963. [CrossRef]
21. Lin, L.T.; Uen, W.C.; Choong, C.Y.; Shi, Y.C.; Lee, B.H.; Tai, C.J.; Tai, C.J. Paris Polyphylla Inhibits Colorectal Cancer Cells via Inducing Autophagy and Enhancing the Efficacy of Chemotherapeutic Drug Doxorubicin. *Molecules* **2019**, *24*, 2102. [CrossRef] [PubMed]
22. Wei, X.; Xia, L.; Ziyayiding, D.; Chen, Q.; Liu, R.; Xu, X.; Li, J. The Extracts of Artemisia absinthium L. Suppress the Growth of Hepatocellular Carcinoma Cells through Induction of Apoptosis via Endoplasmic Reticulum Stress and Mitochondrial-Dependent Pathway. *Molecules* **2019**, *24*, 913. [CrossRef] [PubMed]
23. Willer, J.; Jöhrer, K.; Greil, R.; Zidorn, C.; Çiçek, S.S. Cytotoxic Properties of Damiana (Turnera diffusa) Extracts and Constituents and A Validated Quantitative UHPLC-DAD Assay. *Molecules* **2019**, *24*, 855. [CrossRef]
24. Huang, S.Y.; Huang, G.J.; Wu, H.C.; Kao, M.C.; Huang, W.C. Ganoderma tsugae Inhibits the SREBP-1/AR Axis Leading to Suppression of Cell Growth and Activation of Apoptosis in Prostate Cancer Cells. *Molecules* **2019**, *23*, 2539. [CrossRef]
25. Ferhi, S.; Santaniello, S.; Zerizer, S.; Cruciani, S.; Fadda, A.; Sanna, D.; Dore, A.; Maioli, M.; D'hallewin, G. Total Phenols from Grape Leaves Counteract Cell Proliferation and Modulate Apoptosis-Related Gene Expression in MCF-7 and HepG2 Human Cancer Cell Lines. *Molecules* **2019**, *24*, 612. [CrossRef] [PubMed]
26. Elansary, H.O.; Szopa, A.; Kubica, P.; Al-Mana, F.A.; Mahmoud, E.A.; Zin El-Abedin, T.K.A.; Mattar, M.; Ekiert, H. Phenolic Compounds of Catalpa speciosa, Taxus cuspidate, and Magnolia acuminata have Antioxidant and Anticancer Activity. *Molecules* **2019**, *24*, 412. [CrossRef] [PubMed]
27. Yang, Y.; Bhosle, S.R.; Yu, Y.H.; Park, S.Y.; Zhou, R.; Taş, I.; Gamage, C.D.B.; Kim, K.K.; Pereira, I.; Hur, J.S.; et al. Tumidulin, a Lichen Secondary Metabolite, Decreases the Stemness Potential of Colorectal Cancer Cells. *Molecules* **2018**, *23*, 2968. [CrossRef] [PubMed]
28. Alvarado-Sansininea, J.J.; Sánchez-Sánchez, L.; López-Muñoz, H.; Escobar, M.L.; Flores-Guzmán, F.; Tavera-Hernández, R.; Jiménez-Estrada, M. Quercetagetin and Patuletin: Antiproliferative, Necrotic and Apoptotic Activity in Tumor Cell Lines. *Molecules* **2018**, *23*, 2579. [CrossRef]

29. Yu, W.; Ren, Z.; Zhang, X.; Xing, S.; Tao, S.; Liu, C.; Wei, G.; Yuan, Y.; Lei, Z. Structural Characterization of Polysaccharides from Dendrobium officinale and Their Effects on Apoptosis of HeLa Cell Line. *Molecules* **2018**, *23*, 2484. [CrossRef]
30. Nguyen, T.T.H.; Pandey, R.P.; Parajuli, P.; Han, J.M.; Jung, H.J.; Park, Y.I.; Sohng, J.K. Microbial Synthesis of Non-Natural Anthraquinone Glucosides Displaying Superior Antiproliferative Properties. *Molecules* **2018**, *23*, 2171. [CrossRef]
31. Kahnt, M.; Fischer Née Heller, L.; Al-Harrasi, A.; Csuk, R. Ethylenediamine Derived Carboxamides of Betulinic and Ursolic Acid as Potential Cytotoxic Agents. *Molecules* **2018**, *23*, 2558. [CrossRef] [PubMed]
32. Ling, T.; Lang, W.H.; Maier, J.; Quintana Centurion, M.; Rivas, F. Cytostatic and Cytotoxic Natural Products against Cancer Cell Models. *Molecules* **2019**, *24*, 2012. [CrossRef] [PubMed]
33. Lim, H.N.; Baek, S.B.; Jung, H.J. Bee Venom and Its Peptide Component Melittin Suppress Growth and Migration of Melanoma Cells via Inhibition of PI3K/AKT/mTOR and MAPK Pathways. *Molecules* **2019**, *24*, 929. [CrossRef] [PubMed]
34. Guo, D.L.; Li, X.H.; Feng, D.; Jin, M.Y.; Cao, Y.M.; Cao, Z.X.; Gu, Y.C.; Geng, Z.; Deng, F.; Deng, Y. Novel Polyketides Produced by the Endophytic Fungus Aspergillus Fumigatus from Cordyceps Sinensis. *Molecules* **2018**, *23*, 1709. [CrossRef] [PubMed]
35. Espindola, L.S.; Dusi, R.G.; Demarque, D.P.; Braz-Filho, R.; Yan, P.; Bokesch, H.R.; Gustafson, K.R.; Beutler, J.A. Cytotoxic Triterpenes from Salacia crassifolia and Metabolite Profiling of Celastraceae Species. *Molecules* **2018**, *23*, 1494. [CrossRef]
36. Liu, Z.; Gao, T.; Yang, Y.; Meng, F.; Zhan, F.; Jiang, Q.; Sun, X. Anti-Cancer Activity of Porphyran and Carrageenan from Red Seaweeds. *Molecules* **2019**, *24*, 4286. [CrossRef] [PubMed]
37. Wang, N.; Świtalska, M.; Wang, L.; Shaban, E.; Hossain, M.I.; El Sayed, I.E.; Wietrzyk, J.; Inokuchi, T. Structural Modifications of Nature-Inspired Indoloquinolines: A Mini Review of Their Potential Antiproliferative Activity. *Molecules* **2019**, *24*, 2121. [CrossRef]
38. Yang, M.H.; Jung, S.H.; Sethi, G.; Ahn, K.S. Pleiotropic Pharmacological Actions of Capsazepine, a Synthetic Analogue of Capsaicin, against Various Cancers and Inflammatory Diseases. *Molecules* **2019**, *24*, 995. [CrossRef]
39. Liskova, A.; Kubatka, P.; Samec, M.; Zubor, P.; Mlyncek, M.; Bielik, T.; Samuel, S.M.; Zulli, A.; Kwon, T.K.; Büsselberg, D. Dietary Phytochemicals Targeting Cancer Stem Cells. *Molecules* **2019**, *24*, 899. [CrossRef]
40. Girisa, S.; Shabnam, B.; Monisha, J.; Fan, L.; Halim, C.E.; Arfuso, F.; Ahn, K.S.; Sethi, G.; Kunnumakkara, A.B. Potential of Zerumbone as an Anti-Cancer Agent. *Molecules* **2019**, *24*, 734. [CrossRef]
41. Choi, B.Y. Biochemical Basis of Anti-Cancer-Effects of Phloretin-A Natural Dihydrochalcone. *Molecules* **2019**, *24*, 278. [CrossRef] [PubMed]

© 2020 by the author. Licensee MDPI, Basel, Switzerland. This article is an open access article distributed under the terms and conditions of the Creative Commons Attribution (CC BY) license (http://creativecommons.org/licenses/by/4.0/).

Article

Coronarin D Induces Apoptotic Cell Death and Cell Cycle Arrest in Human Glioblastoma Cell Line

Yollanda E. M. Franco [1,2,*,†], Marcia Y. Okubo [3,4,†], Adriana D. Torre [3], Paula P. Paiva [3], Marcela N. Rosa [5], Viviane A. O. Silva [5], Rui M. Reis [5,6,7], Ana L. T. G. Ruiz [4,8], Paulo M. Imamura [9], João E. de Carvalho [4,8] and Giovanna B. Longato [1,2]

1. Research Laboratory in Molecular Pharmacology of Bioactive Compounds. São Francisco University, Bragança Paulista 12916–900, SP, Brazil; giovanna.longato@usf.edu.br
2. Posgraduate program in Health Science, São Francisco University, Bragança Paulista 12916–900, SP, Brazil
3. Chemical, Biological and Agricultural Pluridisciplinary Research Center (CPQBA), University of Campinas–UNICAMP, Paulínia 13148–218, SP, Brazil; yumiokuboh@gmail.com (M.Y.O.); adriana_biotec@yahoo.com.br (A.D.T.); paula.p21@gmail.com (P.P.P.)
4. Posgraduate program in dentistry, Piracicaba Dental School, University of Campinas, Piracicaba 13 414–903, SP, Brazil; ana.ruiz@fcf.unicamp.br (A.L.T.G.R.); carvalho@fcf.unicamp.br (J.E.d.C.)
5. Molecular Oncology Research Center, Barretos Cancer Hospital, Barretos 14.784–400, SP, Brazil; nr.marcela2@gmail.com (M.N.R.); vivianeaos@gmail.com (V.A.O.S.); ruireis.hcb@gmail.com (R.M.R.)
6. Life and Health Sciences Research Institute (ICVS), School of Medicine, University of Minho, 4710–057 Braga, Portugal
7. ICVS/3B's–PT Government Associate Laboratory, 4710–057 Braga, Portugal
8. Faculty of Pharmaceutical Sciences, University of Campinas, UNICAMP, Campinas 13081–970, SP, Brazil
9. Institute of Chemistry, University of Campinas–UNICAMP, P.O. Box 6154, Campinas 13083–970, SP, Brazil; imam@iqm.unicamp.br
* Correspondence: yollanda.moreira@hotmail.com; Tel.: +55-11-996672895
† These authors contributed equally to this work.

Academic Editor: Roberto Fabiani
Received: 25 October 2019; Accepted: 21 November 2019; Published: 9 December 2019

Abstract: Glioblastoma (GBM) is the most frequent and highest–grade brain tumor in adults. The prognosis is still poor despite the use of combined therapy involving maximal surgical resection, radiotherapy, and chemotherapy. The development of more efficient drugs without noticeable side effects is urgent. Coronarin D is a diterpene obtained from the rhizome extract of *Hedychium coronarium*, classified as a labdane with several biological activities, principally anticancer potential. The aim of the present study was to determine the anti–cancer properties of Coronarin D in the glioblastoma cell line and further elucidate the underlying molecular mechanisms. Coronarin D potently suppressed cell viability in glioblastoma U–251 cell line, and also induced G1 arrest by reducing p21 protein and histone H2AX phosphorylation, leading to DNA damage and apoptosis. Further studies showed that Coronarin D increased the production of reactive oxygen species, lead to mitochondrial membrane potential depolarization, and subsequently activated caspases and ERK phosphorylation, major mechanisms involved in apoptosis. To our knowledge, this is the first analysis referring to this compound on the glioma cell line. These findings highlight the antiproliferative activity of Coronarin D against glioblastoma cell line U–251 and provide a basis for further investigation on its antineoplastic activity on brain cancer.

Keywords: coronarin D; glioblastoma; apoptosis; cell cycle arrest; natural products

1. Introduction

Comprising over 100 diseases, cancer is characterized by disordered cell growth and tissue invasion that can spread to other regions of the body, leading to metastasis. It is a multifactorial heterogeneous disease and one of the major worldwide causes of mortality [1]. Among all types of cancers, brain tumors are one of the less prevalent, accounting for about 2% of all types of malignant tumors, but considered one of the most worrying ones [2,3]. As the most frequent (70–75%), gliomas were originally characterized as lesions originated from glial cells, which play a role of support, protection, and nourishment for neurons in the central nervous system (CNS) [4]. However, stem–like cells within the CNS are now thought to be the cells of origin of several primary brain tumor types, including glioblastomas [5,6]. Gliomas are one of the most fatal tumors, presenting a high mortality rate, 29–35%, of the CNS tumors in adolescents and young adults [7,8]. Among gliomas, glioblastomas (GBM) are the most aggressive and frequent subtype [9,10]. GBM is highly invasive and presents a median survival of only 14.6 months, even after aggressive treatment with surgery, radiation, and chemotherapy. The difficulties in human GBM therapy are due to the pathological characteristic and numerous drug–resistance mechanisms. Temozolomide (TMZ) comprises the standard treatment for glioblastoma, but unlike classic chemotherapeutics, TMZ does not induce DNA damage or misalignment of segregating chromosomes directly. It is a DNA alkylating agent, which leads to base mismatches that initiate futile DNA repair cycles; eventually, DNA strands break, which in turn induces cell death. The addition of TMZ to the standard treatment protocol was hailed as a major breakthrough in GBM therapy. Despite this, patients' prognosis remains dismal with a five–year overall survival below 10% [11]. For this reason, more efficient therapeutic approaches are required for enhancing the treatment effect [12].

One of these approaches is based on cell death induction. There are many cell death morphotypes described in the literature [13]. Searching for new anticancer agents, investigation into cell death mechanisms, such as apoptosis, necrosis, necroptosis or other under–explored forms of cell death, is a significant strategy to afford more selective and efficient drugs. Further, as many clinically–established drugs are based on natural products, there are still many researches focused on finding new compounds from natural products [14].

Native from Asia and the Pacific [15], *Hedychium coronarium* (Zingiberaceae family) is an invasive species in Brazil [16]. Popularly known as white garland–lily, butterfly lily, napoleon, narcissus, Olympia, white ginger or "lírio do brejo" (in Brazil), the *H. coronarium* rhizomes are used as a starch source [17] and in traditional medicine for treatment of inflammation, diabetes, and rheumatic pain, among other uses [15]. Among pharmacological evaluations, the ethanolic extract of *H. coronarium* rhizomes induces apoptosis on HeLa cells by promoting cell cycle arresting at G1 phase, upregulating p53, p21, and Bax expression as well as downregulating cyclin D1, cyclin–dependent kinases CDK–4, CDK–6, and Bcl–2 expression [18]. Moreover, chemical evaluation of these extracts afforded the isolation of several labdane–like diterpenes with anti–inflammatory action [19,20], antiallergic [21], antibacterial [22], and cytotoxic effects over A–549– lung cancer, SK–N–SH– human neuroblastoma, MCF–7 breast cancer, and HeLa cervical cancer cell lines [23].

One of these diterpenes, coronarin D, has been reported as a promising antiproliferative and anti–inflammatory agent. Coronarin D inhibits the β–hexosaminidase release in RBL–2H3 cells [21] in addition to increasing the in vivo inhibition of the acetic acid–induced vascular permeability in mice [19]. Further, Coronarin D exhibits antiproliferative, pro–apoptotic, anti–invasive, antiangiogenic, antiosteoclast, and anti–inflammatory activity by suppressing NF–κB and the gene products regulated by this pathway of osteoclastogenesis [24]. Recently, Coronarin D has been described as inducing apoptosis in human hepatocellular carcinoma (HCC) [25] and in human oral cancer (OSCC) [26] through the c–Jun N–terminal kinases (JNK) pathway while it has induced reactive oxygen species–mediated cell death in human nasopharyngeal cancer cells (NPC) through inhibition of p38 mitogen–activated protein kinase (MAPK) and activation of JNK [27].

Based on these significant activities, the present study sought to further elucidate the Coronarin D mechanism of action on cell death of the human tumor cell line U–251 (glioblastoma). As far as we know, this is the first report concerning the Coronarin D mechanism of action on glioblastoma cancer cell line.

2. Results

2.1. Isolation and Characterization of Coronarin D

Coronarin D (Figure 1) was obtained from the dichloromethane crude extract of *Hedychium coronarium* rhizomes. The rhizomes were collected by Dr. Paulo Matsuo Imamura and identified at the herbarium of the State University of Campinas (UEC 163701). The identification of Coronarin D was done by comparison of experimental ^{1}H– and ^{13}C–NMR data (Figure S1 and Table S1) with those described by Itokawa et al. [28].

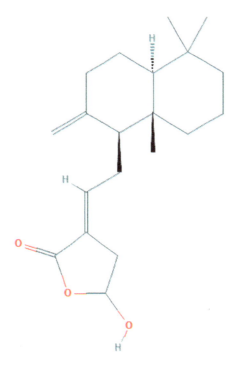

Figure 1. Coronarin D molecular structure, data from [29].

2.2. In Vitro Antiproliferative Activity Assay

Coronarin D presented an interesting antiproliferative activity (Figure 2, Table 1), with U–251 (glioblastoma), 786–0 (kidney), PC–3 (prostate), and OVCAR–3 (ovary) as the most sensitive ones, and total growth inhibition (TGI) values <50 µM.

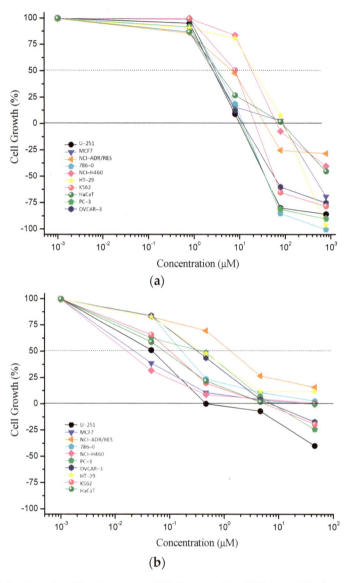

Figure 2. In vitro antiproliferative activity of (**a**) Coronarin D and (**b**) doxorubicin hydrochloride (positive control) after 48 h of treatment. Concentration range: 0.785–785 µM for Coronarin D; 0.043–43.1 µM for doxorubicin hydrochloride. Human tumor cell lines: U–251 (glioblastoma), MCF7 (breast), NCI–ADR/RES (multidrug resistant ovary), 786–0 (kidney), NCI–H460 (lung, non–small cells tumor), PC–3 (prostate), OVCAR–3 (ovary), HT–29 (colon), K562 (chronic myelogenous leukemia). Human non–tumor cell line: HaCaT (keratinocyte).

The glioma cell line (U–251) was chosen to continue the in vitro experimental procedures, taking into account that the treatment for this tumor type is still scarce and requires alternative therapies, as previously mentioned. Considering the TGI value, the concentrations of 2.5, 5, and 10 µM were chosen to proceed with the cell cycle; concentrations of 10, 20, and 40 µM were chosen for the flow cytometry and 40 µM for the Western blot assay.

Table 1. Antiproliferative effect of Coronarin D and doxorubicin hydrochloride expressed as the concentration required for total growth inhibition (TGI, µM) after 48 h of exposition.

Cell Lines	Coronarin D TGI (µM)	Doxorubicin Hydrochloride TGI (µM)
U–251	18.6 ± 0.5	2.4 ± 0.4
MCF7	105.0 ± 5.1	11.6 ± 1.8
NCI-ADR/RES	550.1 ± 79.9	>43.1 *
786–0	36.4 ± 4.2	17.8 ± 3.4
NCI-H460	640.8 ± 11.3	26.7 ± 1.8
PC–3	17.1 ± 0.6	21.4 ± 4.2
OVCAR-3	41.6 ± 2.1	19.2 ± 1.9
HT-29	534.1 ± 31.8	>43.1 *
K562	56.6 ± 2.4	10.0 ± 1.2
HaCaT	12.9 ± 1.7	1.1 ± 0.1

TGI (concentration required for total growth inhibition of each cell line) values expressed as mean ± standard error of two independent experiments. *: TGI values higher than the highest experimental concentration. Human tumor cell lines: U–251 (glioblastoma), MCF7 (breast), NCI-ADR/RES (multidrug resistant ovary), 786–0 (kidney), NCI-H460 (lung, non–small cells tumor), PC–3 (prostate), OVCAR-3 (ovary), HT-29 (colon), K562 (chronic myelogenous leukemia). Human non–tumor cell line: HaCaT (keratinocyte).

2.3. Cell Cycle Assay

Comparing to untreated U–251 cells, Coronarin D induced cell cycle arrest at G1 phase, in a concentration–dependent way and independent of time exposure (Figure 3). The increasing G1 subpopulation was concomitant with a significant reduction on cell subpopulations at S phase (Figure 3a) and G2/M phase (Figure 3a,b), proportionally to Coronarin D concentration. Figure 3c,d reveal the histogram of the most representative concentration (10 µM).

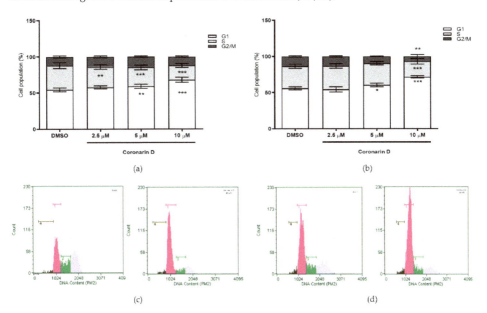

Figure 3. Quantification of U–251 in phases G1, S, and G2 after (**a**) 24 h and (**b**) 48 h of treatment with vehicle (DMSO) and Coronarin D at concentrations of 2.5, 5, and 10 µM. Histograms of the most representative concentration (10 µM) are presented at (**c**) 24 h and (**d**) 48 h. The values were expressed as mean ± standard deviation of two replicates of the same experiment. * $p < 0.05$; ** $p < 0.01$ and *** $p < 0.001$. (Two–way ANOVA: Bonferroni).

2.4. Phosphatidylserine (PS) Externalization Assay

According to Figure 4, after 12 h of U–251 exposition, the concentrations 20 and 40 µM reduced cell viability and increased the number of cells labeled with annexin V–PE (17.88% and 25.88%, consecutively) and doubly labeled with annexin V–PE/7–AAD (7.30 and 13.00%, consecutively). After 24 h of treatment with Coronarin D at 10, 20, and 40 µM, the cell viability was dramatically reduced in comparison with the control (63.4%, 52.00%, 28.88%) and there was an increase of cells labeled with annexin V–PE (26.32%, 23.18%, 22.75%) and doubly labeled with annexin V–PE/7–AAD (9.50%, 19.42%, 42.00%, consecutively). Coronarin D induces cell death through a concentration–dependent effect—the higher the concentration, the more advanced cell death process. The population of non–viable cells labeled only by 7–AAD did not increase significantly, indicating that the treatments with Coronarin D induced cell death characterized by phosphatidylserine exposure, being a type of programmed cell death.

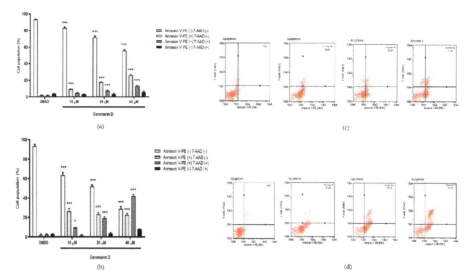

Figure 4. Percentage of U–251 cells stained with annexin V–PE and 7–AAD after (**a**) 12 h and (**b**) 24 h of treatment with vehicle (DMSO) and Coronarin D at 10, 20, and 40 µM concentrations. Dotplots are presented at (**c**) 12 h and (**d**) 24 h. The values are expressed as mean ± standard deviation of two replicates of the same experiment. * $p < 0.05$ and *** $p < 0.001$. (Two–way ANOVA: Bonferroni).

2.5. Detection of Activated Caspases

The results obtained for caspases corroborate with annexin assay. The highest concentrations (20 and 40 µM) led to caspases activation without cell membrane disruption in 14.3% and 12.5% of cells, respectively. The percentage of cells doubly labeled, which means, caspases activation with cell membrane disruption increased for 20 µM concentration and this percentage was even higher for 40 µM concentration (49.7%) (Figure 5).

2.6. Mitochondrial Membrane Potential Assay

The induction of death by intrinsic apoptosis is usually triggered by some stimulus or stress that leads to a mitochondrial response and may result in the depolarization of its outer membrane. Untreated cells showed high intracellular fluorescence intensity indicating that mitochondria were able to sequester a greater amount of rhodamine 123, whereas in cells treated with Coronarin D at 20 µM and 40 µM for 6, 9, and 12 h, there was an intracellular fluorescence signal reduction (Figure 6a–c), being more intense at 12 h of treatment. This result suggests that Coronarin D induces loss of mitochondrial membrane preceding or concomitant with caspase activation and phosphatidylserine externalization.

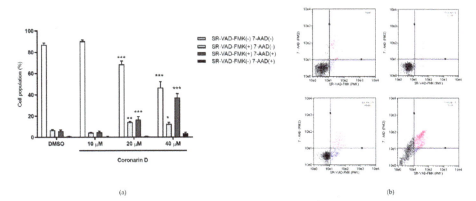

Figure 5. (a) Percentage of U–251 cells stained with SR–VAD–FMK and 7–AAD after 24 h of treatment with vehicle (DMSO) and Coronarin D at 10, 20, and 40 μM concentrations. Dotplots are presented at (b). The values are expressed as mean ± standard deviation of two replicates of the same experiment. * $p < 0.05$, ** $p < 0.01$, and *** $p < 0.001$. (Two–way ANOVA: Bonferroni).

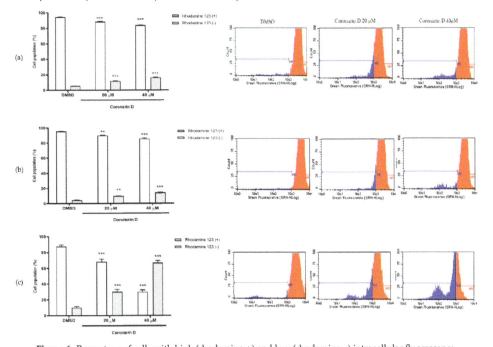

Figure 6. Percentage of cells with high (rhodamine +) and low (rhodamine −) intracellular fluorescence intensity after (a) 6 h, (b) 9 h, and (c) 12 h of treatment with vehicle (DMSO) and Coronarin D at 20 μM and 40 μM. The values were expressed as mean ± standard deviation of two replicates of the same experiment. ** $p < 0.01$ and *** $p < 0.001$. (ANOVA Two-way: Bonferroni).

2.7. Measurement of Hydrogen Peroxide (H_2O_2) Generation

The induction of death by intrinsic apoptosis is usually triggered by a stress that leads to depolarization of the outer membrane of mitochondria and the release of reactive oxygen species (ROS), more specifically hydrogen peroxide. This was measured by examining the fluorescence intensity of DCF. The intensity of fluorescence is proportional to intracellular hydrogen peroxide levels [30]. Data suggest that over 80% of the cell population presented high fluorescence intensity (DCF +) after 90 min of treatment with Coronarin D, even at the lowest concentration (10 µM), indicating the presence of H_2O_2 on these cells (Figure 7).

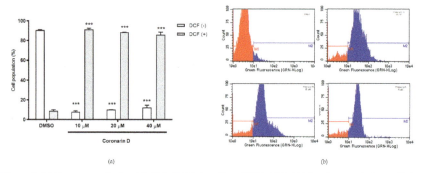

Figure 7. (a) Percentage of cells with high (DCF +) and low (DCF −) 2,7–dichlorofluorescein intracellular fluorescence intensity after 90 min of treatment with vehicle (DMSO) and Coronarin D at 10, 20, and 40 µM. Histograms are presented at (b). The values are expressed as mean ± standard deviation of two replicates of the same experiment. *** $p < 0.001$. (ANOVA Two–way: Bonferroni).

2.8. Western Blotting Assay

In order to confirm the cell signaling pathway of Coronarin D, some proteins involved with proliferation, cell death and cell cycle were evaluated after 24 h of treatment. Figure 8 revealed a decrease of total ERK protein, cleavage of poly (ADP–ribose) polymerases (PARP) and cleavage of caspases 3, 7, and 9 as well as increase of phosphorylation of ERK and p–H2AX histone. The p21 protein, related with cell cycle arrest, was also overexpressed. Of note, Coronarin D activated caspases 7 and 9, as well as PARP and p21 in a more intense way compared with the chemotherapeutic drug TMZ.

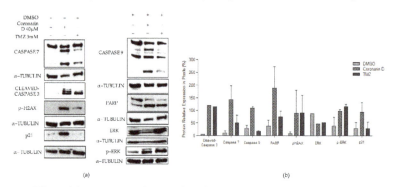

Figure 8. Effects of Coronarin D and temozolomide (TMZ) on expression of caspase 9, caspase 7, cleaved–caspase 3, poly (ADP–ribose) polymerase (PARP), p–H2AX, ERK, and p–ERK proteins in glioblastoma cell line (U–251). The α–tubulin was used as a positive control. The values for caspase 7, caspase 9, PARP, p–H2AX, p–ERK, and p21 were expressed as mean ± standard deviation of three replicates of the same experiment. Values for cleaved–caspase 3 and ERK were obtained from one experiment.

3. Discussion

Since Coronarin D has shown several biological activities as mentioned before and presents a potential clinical application in cancer therapy, it is important to have a clear understanding of its mechanism of action. In this study, we showed that most of the tumor cell lines evaluated were sensitive to the treatment with Coronarin D and, among them, U–251 (glioblastoma) was the cell line chosen to continue the evaluation of the mechanism of action of this compound. We demonstrated that Coronarin D induces cell cycle arrest at G1 phase and apoptosis of glioblastoma cells in a concentration–time dependent manner.

Many natural products can suppress proliferation by arresting cells at phases in the cell cycle [25]. The p21 is a small protein with 165 amino acids and belongs to the CIP/Kip family of CDK inhibitors. The p21 can arrest the cell cycle progression in G1/S and G2/M transitions by inhibiting CDK4,6/cyclin–D and CDK2/cyclin–E, respectively [31,32]. In addition, some studies have shown that H2AX is required for p21–induced cell cycle arrest after replication stalling [33]. The results herein presented indicate that Coronarin D inhibits glioblastoma (U–251) cell growth by inducing cell cycle arrest at G1 phase after increasing expression of p21, likely mediated by the phosphorylation of H2AX.

Coronarin D was also able to trigger cell death with the activation of caspases 9, 3, and 7 and phosphatidylserine exposure, characteristics of apoptosis, and with further rupture of the cell membrane [13]. There was a gradual and time–dependent reduction of the mitochondrial membrane potential (MMP) in U–251 cells treated with Coronarin D, which is a feature of the intrinsic apoptotic pathway that occurs in response to various intracellular stress conditions centered on mitochondria [34]. The ROS production after the cell treatment with Coronarin D suggests that it can act as a second messenger, signaling to the activation of the apoptotic process, since this can activate effector caspases. In addition, ROS can lead to DNA damage that, in turn, activates the p21 pathway and results in cell cycle arrest [35].

Coronarin D led to an increase in the expression of protein kinase ERK, as well as PARP cleavage. ERK is part of the MAPK family and, when activated, can mediate mechanisms of cell proliferation and apoptosis [36,37]. Some studies have reported that the activation of ERK could be a result of DNA damage that subsequently leads to cell cycle arrest and apoptosis [38,39]. In addition, it is known that intracellular ROS lead to the activation of ERK and subsequent apoptosis [40–43]. Poly (ADP–ribose) polymerases (PARPs) are a family of enzymes involved in cellular homeostasis, including DNA transcription, cell–cycle regulation, and DNA repair [44]. This protein is really relevant in the apoptosis pathway, because it has a positive regulation in tumors and when it is inactivated leads to the cleavage of caspase 3 that is involved with the apoptosis process. Studies describe that some natural products are responsible for cleaving PARP as well as activating the caspase cascade as a mechanism of action on the induction of apoptosis [45,46]. Relating all these data reported we can suggest that Coronarin D could induce apoptosis involving ROS generation and ERK activation in U–251 cell line through an intrinsic and caspase–dependent pathway.

It has been reported in the literature that Coronarin D has pro–apoptotic potential, including potentiation of PARP cleavage and a reduction in the expression of anti–apoptotic gene products, such as apoptosis protein–1, TRAF–2 cellular inhibitory proteins, surviving, and Bcl–2 [24]. Consistent with our findings it has been shown that Coronarin D triggers apoptosis by activating caspase–dependent proteins and altering the expression of Bcl–2, Bcl–xL, and Bak in human hepatoma cell lines [25]. Moreover, Dimas et al. and Mahaira et al. demonstrated that compounds that contain labdane–type diterpenes triggered apoptosis in human colon cancer cells and myeloid leukemia cells [47,48].

The results obtained propose that Coronarin D elevates the generation of ROS (H_2O_2), which promotes phosphorylation of H2AX and consequently damage to the DNA. In addition, an increase in the expression of p21 leads cell cycle arrest at the check point between G1 and S. ROS generation also increases ERK phosphorylation and loss of mitochondrial membrane potential that allows the release of cytochrome c (not evaluated in this work), and consequently the cleavage of caspases (9, 3, and 7) and PARP protein. The caspase activation, in turn, leads cell to death through the mechanism of intrinsic apoptosis. In conclusion, as far as we know, this is the first study to elucidate the mechanism of action of Coronarin D in glioblastoma cell line and the results obtained highlight Coronarin D as a promising anticancer compound in Figure 9.

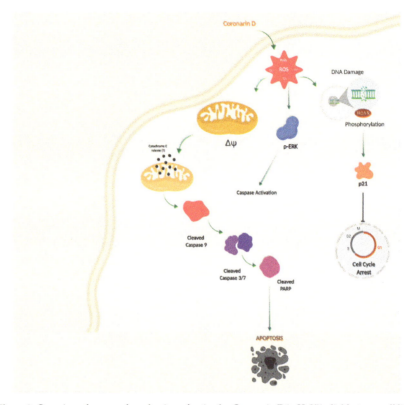

Figure 9. Overview of proposed mechanism of action by Coronarin D in U–251 glioblastoma cell line.

4. Materials and Methods

4.1. Chemicals and Equipments

Column chromatography on silica gel and thin layer chromatography (TLC) were obtained from Merck (Darmstad, Hesse, Germany). Culture medium RPMI 1640, fetal bovine serum, and Hank's balanced salt solution (HBSS) were obtained from Gibco® (Gaithersburg, MD, USA). Penicillin/streptomycin (1000 U/mL:1000 µg/mL) was purchased from LGC Biotechnology (Cotia, SP, BRA) and trypsin–EDTA 0.25% from Vitrocell® (Campinas, SP, BRA). Sulforhodamine B (SRB), trizma base, trichloroacetic acid (TCA), DCFH–DA, and Rhodamine–123 dyes were obtained from Sigma–Aldrich® (Allentown, PA, USA). Doxorubicin hydrochloride was obtained from Eurofarma (Jurubatuba, SP, BRA). Temozolomide (TMZ) was acquired from Sigma–Aldrich. Dimethylsulfoxide (DMSO), silica gel, and dichloromethane were supplied by Merck (Darmstad, Hesse, Germany). Guava Cell Cycle®, Guava Nexin Reagent®, and Guava Multicaspase Kit® were purchased from Millipore® (Burlington, MA, USA). The nitrocellulose membranes were obtained from Hybond–C™Extra, Amersham Biosciences (Piscataway, NJ, USA). Anti–caspase 7, anti–p–H2–AX histone (Ser139), α–tubulin, anti–caspase 9, anti–p44/42 MAPK (ERK1/2), and anti–P21 were obtained from Cell Signaling and Bradford reagent was purchase from Bio–Rad Hybond–C™ (Hercules, CA, USA). The equipments used were rotary evaporator Buchi–Merck (Darmstad, Hesse, Germany), microplate reader (Molecular Devices®, model versaMax), and flow cytometer Guava EasyCyte Mini Flow Cytometry System (Millipore®).

4.2. Isolation of Coronarin D Obtained from Hedychium Coronarium

The rhizomes of *Hedychium coronarium* were collected in January 2011 in the city of Embu, São Paulo, Brazil. Dried–milled rhizomes (480 g) were extracted with dichloromethane (4 L) by static maceration (7 days) at room temperature followed by filtration. The plant residues were re–extracted twice more following the same procedure. After fluidic extract grouping and solvent evaporation under a vacuum, the final crude extract (32.5 g) was purified by column chromatography on silica gel using hexane and hexane/ethyl acetate mixtures (1%, 2%, 5%, 10%, 20%, 50%) as eluent. Coronarin D was isolated after successive column chromatography from fractions obtained by 20% and 50% eluents. The complete identification of Coronarin D was done through experimental spectral data (^{1}H– and ^{13}C–NMR) in comparison to those described in the literature [28].

4.3. In Vitro Anticancer Activity Assay

4.3.1. Cell Culture

The tumor cell lines U–251 (glioblastoma), MCF7 (breast), NCI/ADR–RES (resistant ovary), 786–0 (kidney), NCI–H460 (lung), PC–3 (prostate), OVCAR–3 (ovary), HT–29 (colon), and K–562 (leukemia) were kindly provided from the National Cancer Institute at Frederick, Maryland, USA. The non–tumor cell line HaCaT (keratinocytes) was kindly provided by FOP/Unicamp. Cells were cultured in complete medium (RPMI–1640) supplemented with 5% heat fetal bovine serum and 1% penicillin/streptomycin at 37 °C with 5% CO_2.

4.3.2. Antiproliferative Activity

Antiproliferative activity was assessed by the sulforhodamine B (SRB) colorimetric assay as previously reported [30]. First, the stock solution of Coronarin D (0.3 mM) was prepared aseptically using DMSO followed by serial dilution in complete medium. The cells were seeded in 96–well plates (3×10^4 cells/mL, 100 µL/well), incubated for 24 h and treated with Coronarin D at final concentrations of 0.79, 7.9, 78.5, and 785.0 µM (100 µL/well), in triplicate, and then incubated for 48 h at 37 °C in 5% CO_2. A second plate, named T0, was prepared to infer the absorbance value of untreated cells at the sample addition moment. The antineoplastic agent doxorubicin hydrochloride was used as a positive control, at final concentrations of 0.0431, 0.431, 4.31, and 43.1 µM (100 µL/well), in triplicate. The final concentration of DMSO ($\leq$0.25%) did not affect cell viability [46]. Subsequently, the cells were fixed with 50% trichloroacetic acid and stained with SRB protein dye (0.4%). Determination of protein content was performed using a microplate reader (Molecular Devices®, VersaMax model) at 540 nm. Using the absorbance values, the cell growth (%) for each cell line, at each sample concentration, was calculated considering at 100% of cell growth the difference between the absorbances of untreated cells after 48 h incubation (T_1) and at the sample addition moment (T_0). The curve cell growth vs. sample concentration was plotted and the effective concentration TGI (concentration required for total cell growth inhibition) was calculated by sigmoidal regression using the Origin 8.0 software (OriginLab Corporation, Northampton, MA, USA).

4.4. Cell Cycle Analysis

This experiment was done using Guava® Cell Cycle reagent, following the fabricant's instructions. Briefly, U–251 cells (3×10^4 cells/well) in 12–well plates with complete medium were incubated at 37 °C with 5% CO_2 for 24 h. Then, the complete medium was replaced by RPMI medium without fetal bovine serum (FBS) for cell cycle synchronization and cells were incubated for another 24 h. After that, the cells were treated with Coronarin D (2.5, 5, and 10 µM) during the 24 h. The cells were harvested, collected, centrifuged (5 min, 2500 rpm), and the supernatant discarded. After fixation with 70% cold ethanol (24 h, 4 °C), each cell suspention was centrifuged and washed with PBS, the supernatant was discarded, and then the Guava Cell Cycle reagent (200 µL/cell suspension) was added. After 20 min at room temperature in the dark, each cell suspension was analyzed (5000 events/replicate) by flow cytometry. Using the Guava Cell cycle® software (Austin, TX, USA), the subpopulations at G1, S, and G2/M phases of the cell cycle were quantified in percentage. The analyses were done in biological triplicate of two experiments. Statistical evaluation was done by two–way ANOVA followed by Bonferroni test using GraphPad Prism.

4.5. Phosphatidylserine (PS) Externalization Assay

This experiment was done using Guava® Nexin Reagent, following the fabricant's instructions. Briefly, U–251 cells (3×10^4 cells/well) in 12–well plates with complete medium were incubated at 37 °C with 5% CO_2 for 24 h. Then cells were treated with Coronarin D (10, 20, and 40 µM) for 12 and 24 h. After trypsinization and washing, each cell suspension was stained with 100 µL of Guava Nexin Reagent that consists of annexin–V conjugated with phycoeritrin (PE) and 7–aminoactinomycin D (7–AAD) for 20 min at room temperature in the dark and then analyzed (2000 events/replicate) by flow cytometry. Using the Guava Nexin® software (Austin, TX, USA), each cell population was registered and quantified at four cell subpopulations named as viable cells ((–) annexin–V/PE (–) 7–AAD); only PS externalization ((+) annexin–V/PE (–) 7–AAD); both PS externalization and membrane permeabilization ((+) annexin–V/PE (+) 7–AAD); only membrane permeabilization ((–) annexin–V/PE (+) 7–AAD). The analyses were done in biological triplicate of two experiments. Statistical evaluation was done by two–way ANOVA followed by Bonferroni test using GraphPad Prism software.

4.6. Detection of Activated Multicaspases

This experiment was done using Guava® Caspase Kit, following the fabricant's instructions. Briefly, U–251 cells (3×10^4 cells/well) in 12–well plates with complete medium were incubated at 37 °C with 5% CO_2 for 24 h. Then cells were treated with Coronarin D (10, 20, and 40 µM) for 12 and 24 h. After that, the caspase inhibitor probe (10 µL/well) was added. It comprises a sulforhodamine derivative of valylalanylaspartic acid fluoromethyl ketone (SR-VAD–FMK). After incubation (1 h in the dark) and medium discarding, the cells were washed, trypsinized, and centrifuged. Each resulting cell pellet was resuspended in wash buffer (100 µL/pellet) and co-stained with the caspase–7–AAD working solution (200 µL/pellet) After 10 min at room temperature in the dark, cells were analyzed (2000 events/replicate) by flow cytometry. Using the Guava® MultiCaspase software (Austin, TX, USA), each cell population was registered and quantified at four cell subpopulations named as viable cells (SR-VAD–FMK (–) and 7–AAD (–)); only activated caspases (SR-VAD–FMK (+) and 7–AAD (–)); both activated caspases and membrane permeabilization (SR-VAD–FMK (+) and 7–AAD (+)) and only membrane permeabilization (SR-VAD–FMK (–) and 7–AAD (+)). The analyses were done in biological triplicate of two experiments. Statistical evaluation was done by Two–way ANOVA followed by Bonferroni test using GraphPad Prism software.

4.7. Mitochondrial Membrane Potential Assay

Briefly, U–251 cells (3×10^4 cells/well) in 12–well plates with complete medium were incubated at 37 °C with 5% CO_2 for 24 h. Then cells were treated with Coronarin D (20 and 40 µM) for 6 h. Rhodamine–123 solution (1 µg/mL in medium RPMI 1640 plus 10% FBS, 1 mL/well) was added after medium remoting. After 15 min, the medium was aspirated and all cells were washed with RPMI 1640 plus 10% FBS (1 mL/well, twice), trypsinized, and collected. The cells were analyzed (5000 events/replicate) by flow cytometry. Using the Guava® Express Pro software (Austin, TX, USA) each cell population was registered and quantified at two cell subpopulations named as viable cells (Rhodamine–123(+)); altered mitochondrial membrane potential (Rhodamine–123(–)). The analyses were done in biological duplicate of one experiment. Statistical evaluation was done by one–way ANOVA followed by Tukey's test using GraphPad Prism software.

4.8. Measurement of Hydrogen Peroxide (H_2O_2) Generation

DCFH–DA (Dichlorodihydro–fluorescein diacetate) is a stable, fluorogenic, and non–polar compound, which can readily diffuse into cells and get deacetylated by intracellular esterases to a non–fluorescent 2,7–dichlorodihydrofluorescein (DCFH) which is later oxidized by intracellular hydrogen peroxide into highly fluorescent 2,7–dichlorofluorescein (DCF). The intensity of fluorescence is proportional to intracellular hydrogen peroxide levels [30].

Briefly, U–251 cells (3×10^4 cells/well) in 12–well plates with complete medium were incubated at 37 °C with 5% CO_2 for 24 h. After medium removing, the cells were washed with Hank's buffered salt solution (HBSS, 1 mL/well), treated with DCFH–DA solution (10 µM in HBSS, 1 mL/well), followed by 30 min incubation in the dark. After probe removing and cell washing with HBSS buffer, cells were treated with Coronarin D (20 and 40 µM) for 90 min. After trypsinization, cell suspension in HSBB was analyzed (5000 events/replicate) by flow cytometry. Using the Guava® Express Pro software, each cell population was registered and quantified at two cell subpopulations named as viable cells (DCF(−)) and increased intracellular hydrogen peroxide level (DCF(+)). The analyses were done in biological duplicate of one experiment. Statistical evaluation was done by one–way ANOVA followed by Tukey's test using GraphPad Prism software.

4.9. Western Blotting Assay

U–251 cells (1×10^6 cells/well) in 6–well plates with complete medium were incubated at 37 °C with 5% CO_2 for 24 h. Then cells were treated with a negative control (DMSO), Coronarin D (40 µM), and TMZ (used as positive control) for 24 h. TMZ was dissolved in DMSO to prepare a stock concentration of 1000 mM, which was further diluted in cell culture medium to working concentrations. The cells were exposed at $3 \times IC_{50}$ concentration values of TMZ for 24 h in DMEM (0.5% FBS) [49]. After washing with PBS (1 mL/well, twice), cells were lysed with lysis buffer (70 µL/well); Tris 50 mM pH 7.6–8, NaCl 150 mM, EDTA 5 mM, Na_3VO_4 1 mM, NaF 10 mM, Na pyrophosphate 10 mM, and 1% NP–40 and supplemented with a cocktail of inhibitors (DTT– Dithiothreitol, leupeptin hemisulfate, aprotinin, PSMF– Phenylmethylsulfonyl fluoride, and EDTA) for 1 h, followed by centrifugation (4 °C, 15 min, 13,000 rpm). The protein concentration was determined by Bradford reagent [50]. All samples (untreated and Coronarin D–treated cells) were total cellular protein (20 µg protein/sample) and separated by electrophoresis on 10% gradient gels in SDS–PAGE and blotted onto nitrocellulose membranes by electroblotting in transfer buffer (Trizma base, glycine, distilled water, and methanol). Then the membranes were blocked with skimmed milk powder solution at 5% diluted in tris–buffered saline and Tween 20 (TBST) and incubated overnight with the primary antibodies diluted in bovine serum albumin (BSA) solution (5% in TBST). The primary antibodies were anti–caspase 7, anti–cleaved caspase 3, anti–caspase 9, anti–cleaved PARP, anti–P21 (dilution 1:1000), anti–p44/42 MAPK (ERK1/2), and anti–p–H2-AX histone (Ser139) (dilution 1:500). The antibody α–tubulin was used as the control. After this, the membranes were incubated with peroxidase–conjugated secondary antibody anti–mouse or anti–rabbit (1:5000) and also diluted in BSA solution (5% in TBST). The proteins levels were detected using the enhanced chemiluminescence (ECL–GE) method. The detection of the chemiluminescent signal was performed in the Photo Quant LAS 4000 mini (GE) photo documentation system and the bands were analyzed and quantified using Image J software (obtained from imagej.nih.gov/ij/download/).

4.10. Statistical Analysis

The data are provided as the final result with mean ± standard error. For the statistical analysis of the experiments two–way analysis of variance (ANOVA) was used. All analyses were performed with significance level at $p < 0.05$, using GraphPad Prism version 8.0.0 for Windows, GraphPad Software (San Diego, CA, USA).

Supplementary Materials: The following are available online. Structural identification of Coronarin D was assessed by ^{1}H–NMR and ^{13}C–NMR (Figures S1 and S2). Data obtained from ^{13}C–NMR are presented at Table S1.

Author Contributions: Investigation and writing—original draft, Y.E.M.F. and M.Y.O.; methodology, A.D.T. and P.P.P.; methodology and writing—review and editing, M.N.R. and V.A.O.S.; resources and writing—review and editing, R.M.R.; validation and writing—review and editing, A.L.T.G.R.; resources, P.M.I.; supervision and funding acquisition, J.E.d.C.; conceptualization, funding acquisition, project administration, supervision, and writing—review and editing, G.B.L.

Funding: This research was funded by grants from the Fundação de Amparo à Pesquisa do Estado de São Paulo (FAPESP 2014/06636–7 and 2016/06137–6), financiadora de Estudos e Projetos FINEP (MCTI/FINEP/MS/SCTIE/DECIT–01/2013–FPXII–BIOPLAT).

Conflicts of Interest: The authors declare no conflict of interest.

References

1. Siegel, R.L.; Miller, K.D.; Jemal, A. Cancer statistics 2018. *CA Cancer J. Clin.* **2018**, *68*, 7–30. [CrossRef] [PubMed]
2. Bray, F.; Ferlay, J.; Soerjomataram, I.; Siegel, R.L.; Torre, L.A.; Jemal, A. Global cancer statistics 2018: GLOBOCAN estimates of incidence and mortality worldwide for 36 cancers in 185 countries. *CA Cancer J. Clin.* **2018**, *68*, 394–424. [CrossRef] [PubMed]
3. Ostrom, Q.T.; Gittleman, H.; Truitt, G.; Boscia, A.; Kruchko, C.; Barnholtz–Sloan, J.S. CBTRUS Statistical Report: Primary Brain and Other Central Nervous System Tumors Diagnosed in the United States in 2011–2015. *Neuro. Oncol.* **2018**, *1*, 20–86. [CrossRef] [PubMed]
4. Alcantara, L.S.; Sun, D.; Pedraza, A.M.; Vera, E.; Wang, Z.; Burns, D.K. Cell–of–origin susceptibility to glioblastoma formation declines with neural lineage restriction. *Nat. Neurosci.* **2019**, *18*, 545–555. [CrossRef]
5. Molinaro, A.M.; Taylor, J.W.; Wiencke, J.K.; Wrensch, M.R. Genetic and molecular epidemiology of adult diffuse glioma. *Nat. Rev. Neurol.* **2019**, *21*, 405–417. [CrossRef]
6. Sanai, N.; Alvarez–Buylla, A.; Berger, M.S. Neural Stem Cells and the Origin of Gliomas. *N. Engl. J. Med.* **2005**, *25*, 811–822. [CrossRef]
7. Jones, C.; Perryman, L.; Hargrave, D. Paediatric and adult malignant glioma: Close relatives or distant cousins? *Nat. Rev. Clin. Oncol.* **2012**, *29*, 400–413. [CrossRef]
8. Diwanji, T.P.; Engelman, A.; Snider, J.W.; Mohindra, P. Epidemiology, diagnosis, and optimal management of glioma in adolescents and young adults. *Adolesc. Health Med. Ther.* **2017**, *8*, 99–113. [CrossRef]
9. Kesari, S. Understanding Glioblastoma Tumor Biology: The Potential to Improve Current Diagnosis and Treatments. *Semin. Oncol.* **2011**, *38*, S2–10. [CrossRef]
10. Wen, P.Y.; Kesari, S. Malignant Gliomas in Adults. *N. Engl. J. Med.* **2008**, *31*, 492–507. [CrossRef]
11. Strobel, H.; Baisch, T.; Fitzel, R.; Schilberg, K.; Siegelin, M.D.; Karpel–Massler, G. Temozolomide and Other Alkylating Agents in Glioblastoma Therapy. *Biomedicines* **2019**, *7*, 69. [CrossRef] [PubMed]
12. Panosyan, E.H.; Lasky, J.L.; Lin, H.J.; Lai, A.; Hai, Y.; Guo, X. Clinical aggressiveness of malignant gliomas is linked to augmented metabolism of amino acids. *J. Neurooncol.* **2016**, *128*, 57–66. [CrossRef] [PubMed]
13. Galluzzi, L.; Vitale, I.; Aaronson, S.A.; Abrams, J.M.; Adam, D.; Agostinis, P. Molecular mechanisms of cell death: Recommendations of the Nomenclature Committee on Cell Death 2018. *Cell Death. Differ.* **2018**, *23*, 486–541. [CrossRef] [PubMed]
14. Newman, D.J.; Cragg, G.M. Natural Products as Sources of New Drugs from 1981 to 2014. *J. Nat. Prod.* **2016**, *79*, 629–661. [CrossRef]
15. Chan, E.W.C.; Wong, S.K. Phytochemistry and pharmacology of ornamental gingers, Hedychium coronarium and Alpinia purpurata: A review. *J. Integr. Med.* **2015**, *13*, 368–379. [CrossRef]
16. Zenni, R.D.; Ziller, S.R. An overview of invasive plants in Brazil. *Rev. Bras. Botânica.* **2011**, *34*, 431–446. [CrossRef]
17. Bento, J.A.C.; Ferreira, K.C.; de Oliveira, A.L.M.; Lião, L.M.; Caliari, M.; Júnior, M.S.S. Extraction, characterization and technological properties of white garland–lily starch. *Int. J. Biol. Macromol.* **2019**, *135*, 422–428. [CrossRef]
18. Ray, A.; Jena, S.; Dash, B.; Sahoo, A.; Kar, B.; Patnaik, J. Hedychium coronarium extract arrests cell cycle progression, induces apoptosis, and impairs migration and invasion in HeLa cervical cancer cells. *Cancer Manag. Res.* **2019**, *11*, 483–500. [CrossRef]

19. Matsuda, H.; Morikawa, T.; Sakamoto, Y.; Toguchida, I.; Yoshikawa, M. Labdane–type diterpenes with inhibitory effects on increase in vascular permeability and nitric oxide production from Hedychium coronarium. *Bioorg. Med. Chem.* **2002**, *10*, 2527–2534. [CrossRef]
20. Kiem, P.V.; Anh, H.L.T.; Nhiem, N.X.; Van Minh, C.; Thuy, N.T.K.; Yen, P.H. Labdane–Type Diterpenoids from the Rhizomes of Hedychium coronarium Inhibit Lipopolysaccharide–Stimulated Production of Pro–inflammatory Cytokines in Bone Marrow–Derived Dendritic Cells. *Chem. Pharm. Bull.* **2012**, *60*, 246–250. [CrossRef]
21. Morikawa, T.; Matsuda, H.; Sakamoto, Y.; Ueda, K.; Yoshikawa, M. New farnesane–type sesquiterpenes, hedychiols A and B 8,9–diacetate, and inhibitors of degranulation in RBL–2H3 cells from the rhizome of Hedychium coronarium. *Chem. Pharm. Bull.* **2002**, *50*, 1045–1049. [CrossRef] [PubMed]
22. Reuk–ngam, N.; Chimnoi, N.; Khunnawutmanotham, N.; Techasakul, S. Antimicrobial Activity of Coronarin D and Its Synergistic Potential with Antibiotics. *Biomed. Res. Int.* **2014**, *2014*, 1–8. [CrossRef] [PubMed]
23. Suresh, G.; Prabhakar Reddy, P.; Suresh Babu, K.; Shaik, T.B.; Kalivendi, S.V. Two new cytotoxic labdane diterpenes from the rhizomes of Hedychium coronarium. *Bioorg. Med. Chem. Lett.* **2010**, *20*, 7544–7548. [CrossRef] [PubMed]
24. Kunnumakkara, A.B.; Ichikawa, H.; Anand, P.; Mohankumar, C.J.; Hema, P.S.; Nair, M.S. Coronarin D, a labdane diterpene, inhibits both constitutive and inducible nuclear factor–κB pathway activation, leading to potentiation of apoptosis, inhibition of invasion, and suppression of osteoclastogenesis. *Mol. Cancer Ther.* **2008**, *7*, 3306–3317. [CrossRef]
25. Lin, H.; Hsieh, M.; Hsieh, Y. Coronarin D induces apoptotic cell death through the JNK pathway in human hepatocellular carcinoma. *Environ. Toxico.* **2018**, *33*, 1–9. [CrossRef]
26. Liu, Y.; Hsieh, M.; Lin, J.; Lin, G.C.C.; Chuang, Y.L.Y.; Chen, Y.H.M. Coronarin D induces human oral cancer cell apoptosis though upregulate JNK1/2 signaling pathway. *Environ. Toxicol.* **2019**, *34*, 513–520. [CrossRef]
27. Chen, J.; Hsieh, M.; Lin, S.; Lin, C.; Lo, Y.; Chuang, Y. Coronarin D induces reactive oxygen species–mediated cell death in human nasopharyngeal cancer cells through inhibition of p38 MAPK and activation of JNK. *Oncotarget* **2017**, *8*, 108006–1080019. [CrossRef]
28. Itokawa, H.; Morita, H.; Katou, I.; Takeya, K.; Cavalheiro, A.; de Oliveira, R.; Ishige, M.; Motidome, M. Cytotoxic Diterpenes from the Rhizomes of Hedychium coronarium. *Planta Med.* **1988**, *24*, 311–315. [CrossRef]
29. PubChem Database. Available online: https://pubchem.ncbi.nlm.nih.gov/compound/Coronarin--D (accessed on 30 September 2019).
30. Longato, G.B.; Fiorito, G.F.; Vendramini–Costa, D.B.; de Oliveira Sousa, I.M.; Tinti, S.V.; Ruiz, A.L.T.G. Different cell death responses induced by eupomatenoid–5 in MCF-7 and 786–0 tumor cell lines. *Toxicol. Vitr.* **2015**, *29*, 1026–1033. [CrossRef]
31. Abbas, T.; Dutta, A. p21 in cancer: Intricate networks and multiple activities. *Nat. Rev. Cancer* **2009**, *14*, 400–414. [CrossRef]
32. Karimian, A.; Ahmadi, Y.; Yousefi, B. Multiple functions of p21 in cell cycle, apoptosis and transcriptional regulation after DNA damage. *DNA Repair.* **2016**, *42*, 63–71. [CrossRef] [PubMed]
33. Fragkos, M.; Jurvansuu, J.; Beard, P. H2AX Is Required for Cell Cycle Arrest via the p53/p21 Pathway. *Mol. Cell Biol.* **2009**, *15*, 2828–2840. [CrossRef] [PubMed]
34. Wang, C.; Youle, R.J. The Role of Mitochondria in Apoptosis. *Annu. Rev. Genet.* **2009**, *43*, 95–118. [CrossRef] [PubMed]
35. Circu, M.L.; Aw, T.Y. Reactive oxygen species, cellular redox systems, and apoptosis. *Free Radic. Biol. Med.* **2010**, *15*, 749–762. [CrossRef]
36. Steelman, L.S.; Chappell, W.H.; Abrams, S.L.; Kempf, C.R.; Long, J.; Laidler, P. Roles of the Raf/MEK/ERK and PI3K/PTEN/Akt/mtor pathways in controlling growth and sensitivity to therapy–implications for cancer and aging. *Aging* **2011**, *3*, 192–222. [CrossRef]
37. Cagnol, S.; Chambard, J.C. ERK and cell death: Mechanisms of ERK–induced cell death—apoptosis, autophagy and senescence. *FEBS J.* **2010**, *277*, 2–21. [CrossRef]
38. Bacus, S.S.; Gudkov, A.V.; Lowe, M.; Lyass, L.; Yung, Y.; Komarov, A.P. Taxol–induced apoptosis depends on MAP kinase pathways (ERK and p38) and is independent of p53. *Oncogene* **2001**, *12*, 147–155. [CrossRef]
39. Tang, D.; Wu, D.; Hirao, A.; Lahti, J.M.; Liu, L.; Mazza, B. ERK Activation Mediates Cell Cycle Arrest and Apoptosis after DNA Damage Independently of p53. *J. Biol. Chem.* **2002**, *12*, 12710–12717. [CrossRef]

40. Tan, B.-J.; Chiu, G.N.C. Role of oxidative stress, endoplasmic reticulum stress and ERK activation in triptolide–induced apoptosis. *Int. J. Oncol.* **2013**, *42*, 1605–1612. [CrossRef]
41. Ramachandiran, S.; Huang, Q.; Dong, J.; Lau, S.S.; Monks, T.J. Mitogen–Activated Protein Kinases Contribute to Reactive Oxygen Species–Induced Cell Death in Renal Proximal Tubule Epithelial Cells. *Chem. Res. Toxicol.* **2002**, *15*, 1635–1642. [CrossRef]
42. Zhuang, S.; Yan, Y.; Daubert, R.A.; Han, J.; Schnellmann, R.G. ERK promotes hydrogen peroxide–induced apoptosis through caspase-3 activation and inhibition of Akt in renal epithelial cells. *Am. J. Physiol.* **2007**, *292*, 440–447. [CrossRef] [PubMed]
43. Wang, X.; Martindale, J.L.; Holbrook, N.J. Requirement for ERK Activation in Cisplatin–induced Apoptosis. *J. Biol. Chem.* **2000**, *15*, 39435–39443. [CrossRef] [PubMed]
44. Ricks, T.K.; Chiu, H.-J.; Ison, G.; Kim, G.; McKee, A.E.; Kluetz, P. Successes and Challenges of PARP Inhibitors in Cancer Therapy. *Front. Oncol.* **2015**, *5*, 222. [CrossRef] [PubMed]
45. Wang, S.; Huang, M.; Li, J.; Lai, F.; Lee, H.; Hsu, Y. Punicalagin induces apoptotic and autophagic cell death in human U87MG glioma cells. *Acta. Pharmacol. Sin.* **2013**, *30*, 1411–1419. [CrossRef] [PubMed]
46. Lv, L.; Zheng, L.; Dong, D.; Xu, L.; Yin, L.; Xu, Y. Dioscin, a natural steroid saponin, induces apoptosis and DNA damage through reactive oxygen species: A potential new drug for treatment of glioblastoma multiforme. *Food Chem. Toxicol.* **2013**, *59*, 657–669. [CrossRef] [PubMed]
47. Dimas, K.; Demetzos, C.; Vaos, V.; Ioannidis, P.; Trangas, T. Labdane type diterpenes down–regulate the expression of c–myc protein, but not of bcl–2, in human leukemia T–cells undergoing apoptosis. *Leuk Res.* **2001**, *25*, 449–454. [CrossRef]
48. Mahaira, L.G.; Tsimplouli, C.; Sakellaridis, N.; Alevizopoulos, K.; Demetzos, C.; Han, Z. The labdane diterpene sclareol (labd–14–ene–8, 13–diol) induces apoptosis in human tumor cell lines and suppression of tumor growth in vivo via a p53–independent mechanism of action. *Eur. J. Pharmacol.* **2011**, *666*, 173–182. [CrossRef]
49. Yang, S.H.; Li, S.; Lu, G.; Xue, H.; Kim, D.H.; Zhu, J.-J. Metformin treatment reduces temozolomide resistance of glioblastoma cells. *Oncotarget* **2016**, *7*, 48. [CrossRef]
50. Silva–Oliveira, R.J.; Melendez, M.; Martinho, O.; Zanon, M.F.; de Souza Viana, L.; Carvalho, A.L. AKT can modulate the in vitro response of HNSCC cells to irreversible EGFR inhibitors. *Oncotarget* **2017**, *8*, 53288–53301.

Sample Availability: Samples of the compound Coronarin D are not available from the authors.

© 2019 by the authors. Licensee MDPI, Basel, Switzerland. This article is an open access article distributed under the terms and conditions of the Creative Commons Attribution (CC BY) license (http://creativecommons.org/licenses/by/4.0/).

Article

Semi-Synthetic Ingenol Derivative from *Euphorbia tirucalli* Inhibits Protein Kinase C Isotypes and Promotes Autophagy and S-Phase Arrest on Glioma Cell Lines

Viviane Aline Oliveira Silva [1], Marcela Nunes Rosa [1], Aline Tansini [1], Olga Martinho [1,2,3], Amilcar Tanuri [4], Adriane Feijó Evangelista [1], Adriana Cruvinel Carloni [1], João Paulo Lima [5,6], Luiz Francisco Pianowski [7] and Rui Manuel Reis [1,2,3,*]

1. Molecular Oncology Research Center, Barretos Cancer Hospital, Barretos, São Paulo 14784-400, Brazil; vivianeaos@gmail.com (V.A.O.S.); nr.marcela2@gmail.com (M.N.R.); aline.cpom@hcancerbarretos.com.br (A.T.); olgamartinho@med.uminho.pt (O.M.); adriane.feijo@gmail.com (A.F.E.); drybiomedic@gmail.com (A.C.C.)
2. Life and Health Sciences Research Institute (ICVS), School of Health Sciences, University of Minho, 4710-057 Braga, Portugal
3. ICVS/3B's - PT Government Associate Laboratory, 4806-909 Braga/Guimarães, Portugal
4. Laboratory of Molecular Virology, Departaments of genetics, IB, Federal University of Rio de Janeiro, Rio de Janeiro 21941-902, Brazil; atanuri1@gmail.com
5. Medical Oncology, Barretos Cancer Hospital, Barretos, São Paulo 14784-400, Brazil; jpsnlima@yahoo.com.br
6. Medical Oncology Department, A C Camargo Cancer Center, São Paulo 01509-010, SP, Brazil
7. Kyolab Pesquisas Farmacêuticas, Valinhos, São Paulo 13273-105, Brazil; pianowski@kyolab.com.br
* Correspondence: ruireis.hcb@gmail.com; Tel.: +55-1733216600 (ext. 7090)

Received: 30 October 2019; Accepted: 6 November 2019; Published: 22 November 2019

Abstract: The identification of signaling pathways that are involved in gliomagenesis is crucial for targeted therapy design. In this study we assessed the biological and therapeutic effect of ingenol-3-dodecanoate (IngC) on glioma. IngC exhibited dose-time-dependent cytotoxic effects on large panel of glioma cell lines (adult, pediatric cancer cells, and primary cultures), as well as, effectively reduced colonies formation. Nevertheless, it was not been able to attenuate cell migration, invasion, and promote apoptotic effects when administered alone. IngC exposure promoted S-phase arrest associated with p21CIP/WAF1 overexpression and regulated a broad range of signaling effectors related to survival and cell cycle regulation. Moreover, IngC led glioma cells to autophagy by LC3B-II accumulation and exhibited increased cytotoxic sensitivity when combined to a specific autophagic inhibitor, bafilomycin A1. In comparison with temozolomide, IngC showed a mean increase of 106-fold in efficacy, with no synergistic effect when they were both combined. When compared with a known compound of the same class, namely ingenol-3-angelate (I3A, Picato®), IngC showed a mean 9.46-fold higher efficacy. Furthermore, IngC acted as a potent inhibitor of protein kinase C (PKC) activity, an emerging therapeutic target in glioma cells, showing differential actions against various PKC isotypes. These findings identify IngC as a promising lead compound for the development of new cancer therapy and they may guide the search for additional PKC inhibitors.

Keywords: glioma; cytotoxic activity; semi-synthetic derivative; ingenol; *Euphorbia tirucalli*; protein kinase C; autophagy

1. Introduction

Malignant gliomas are the most common and lethal primary brain tumors in humans [1]. Glioblastoma (WHO grade IV) is the most aggressive and frequent type of glioma [2,3] with dismal prognosis, even when current multidisciplinary treatment is used, with a median survival that has changed little in the last decades [4]. Comprehensive genetic analyses of Glioblastoma (GBMs) have identified a few mutations and pathways as therapeutic targets that contemplate EGFR, PI3K/AKT/mTOR, and Ras/MEK/MAPKinase [5,6]. Less explored, protein kinase C (PKC) proteins have emerged as possible targets due to their hyperactivity or overexpression, concomitant with the decrease of cell proliferation and invasion verified in preclinical glioma models [6–9]. Moreover, some clinical response in patients with refractory high-grade malignant gliomas was reported with PKC-inhibiting drugs [7,10–12]. Nevertheless, despite all biological and clinical advances, it is imperative to identify novel treatment strategies to glioblastoma outcome [4].

Several bioactive products that were derived from plants have been reported to prevent tumorigenesis of different types of tumors [13]. Thus, in the search for new therapies, research with natural products and on their anti-neoplastic mechanisms has emerged as an alternative and successful field [14]. *Euphorbia* species (Euphorbiaceae) have been used in traditional medicine as antimicrobial, antiparasitic, anticancer and other diseases [15]. Several secondary compounds are present in *Euphorbia* species extract and they are responsible for its properties [16,17]. Our group has carried out a bioprospecting program that evaluated the cytotoxicity of *E. tirucalli* compounds in a large panel of human tumor cell lines. We previously showed the cytotoxic effect of euphol, the main constituent of *E. tirucalli* latex, and its antitumor potential in glioma cell lines [18,19].

In addition to euphol, the genus *Euphorbia* also has diterpenes as important bioactive constituents some already approved for pre-cancerous conditions [20–23]. One diterpene that was approved for human use for the treatment of actinic keratosis, ingenol-3-angelate (I3A) (Picato®), from *Euphorbia peplus* demonstrated great antineoplastic potential evaluated in clinical trials for the effective treatment of basal cell carcinoma and squamous cell carcinoma through the modulation of PKCs signaling [24–28]. Some studies have also revealed diterpenes as promising modulators of multidrug resistance (MDR) in tumor cells as well as showing *in vivo* anti-inflammatory activity [29].

Recently, our group reported the cytotoxic potential of three new esters of semi-synthetic ingenol from *E. tirucalli* [20,21]. Among the three derivatives, ingenol-3-dodecanoate (Ingenol C—IngC) effectively promoted cytotoxicity and exhibited antitumoral properties. Besides, IngC showed higher efficacy when compared to I3A and ingenol 3,20-dibenzoate (IDB) from *E. esula* L on esophageal cancer cell lines, two important ingenol diterpenes that can promote PKC activation and anticancer activity [20,27,30]. However, the mechanism underlying IngC-induced antineoplastic effect is not largely understood.

Therefore, in this study, we unravel the antitumor properties of IngC derivative from *E. tirucalli* against glioblastoma-derived cells to provide a comprehensive view of its potential antitumor mechanisms.

2. Results

2.1. IngC promotes Cytotoxic Activity on Glioma Cell Lines More Effectively than Temozolomide but Their Combination Is Not Synergistic

The analyses of antitumor properties of IngC on glioma cells were expanded from our previous study [20]. Thus, the cytotoxicity was assessed by MTS assay in 13 glioma cell lines from commercial (adult and pediatric), primary, and one normal immortalized astrocytic cell line (Table 1). We observed that IngC exhibited dose and time-dependent cytotoxic effects on human glioma cells (Figure S1a). There was a heterogeneous profile to IngC, with each cell line exhibiting a distinct treatment response. The mean IC_{50} values among commercial cells was 6.86 µM, but significantly varied between individual cell lines, with more than a 68-fold difference in the IC_{50} values (IC_{50} range: 0.19–13.09 µM) (Table 1). Primary tumor cell cultures that were derived from glioblastoma surgical biopsies (HCB2 and HCB149) exhibited a more resistant profile to IngC in comparison with commercial cell lines (mean 15.98 µM) (Table 1).

We adopted the criteria of growth inhibition (GI) at a fixed dose of 10 µM, which closely corresponds to the average IC_{50} value of all cell lines at initial screening, to better classify the response to IngC. At this fixed dose, we found that 9.1% (1/11) of cell lines were resistant, 36.4% (4/11) were moderately sensitive, and 54.5% (6/11) were classified as highly sensitive (Figure 1A and Table 1).

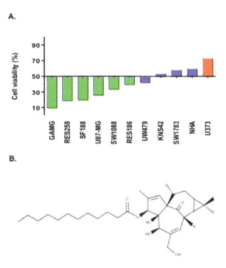

Figure 1. Chemical structures of modified ingenol derivative. (**A**) Cytotoxicity profile of 10 glioma cell lines and one normal human astrocyte exposed to IngC compound. Bars represent the cell viability at 10 µM of IngC. Colors represent the GI score classification: Green (HS = Highly Sensitive); Blue (MS = Moderate Sensitive) and Orange (R = Resistant). (**B**) ingenol-3-dodecanoate (IngC). http://www.chemspider.com/Chemical-Structure.28533061.html.

Furthermore, in comparison with temozolamide (TMZ), IngC showed a median of 106-fold increase in efficacy against glioma cell lines. Additionally, IngC demonstrated a higher selective cytotoxicity index (SI) (0.37–39.05) than TMZ (0.11 to 1.13) (Table 1). However, the combination of IngC and TMZ exposure, promoted antagonistic effects (combination index >1) on 8/9 (88.89%) glioma cell lines (mean CI values: range: 1.13–1.9) (Table 1).

Table 1. Semi-synthetic ingenol derivative (IngC), ingenol-3 angelate (I3A) and temozolomide (TMZ) values of half maximal inhibitory concentration (IC$_{50}$), drug combination studies in glioma cell lines, cell lines origin, and culture conditions.

Cell Line	IngC IC$_{50}$ ±S.D (µM)	I3A IC$_{50}$ ±S.D (µM)	TMZ IC$_{50}$ ±S.D (µM)	Combination Index (CI) *** TMZ+IngC	IngC Growth Inhibition in % at 10 µM *	IngC Growth Inhibition (GI) Score *	S.D	IngC SI **	TMZ SI **	Origin	Culture Conditions	Tumor Type
U87-MG	4.02 ± 2.29	95.15 ± 14.35	746.76 ± 3.15	1.46	74.18	HS	10.46	1.85	0.15	ATCC	DMEM + 10% FBS + 1% P/S	Adult glioma
U373	13.09± 0.84	76.39 ± 19.24	544.75 ± 1.53	0.80	26.88	R	8.19	0.57	0.20	Kindly provided by Dr. Joseph Costello	DMEM + 10% FBS + 1% P/S	
GAMG	0.19 ± 0.05	0.010 ± 0.13	97.00 ± 2.05	UD	90.58	HS	1.32	39.05	1.13	DSMZ	DMEM + 10% FBS + 1% P/S	
SW1088	7.48 ± 0.47	87.62 ± 0.33	979.2 ± 4.00	1.20	66.64	HS	10.18	0.99	0.11	ATCC	DMEM + 10% FBS + 1% P/S	
SW1783	7.4 ± 0.93	91.12 ± 1.59	>1000	1.83	41.39	MS	0.63	1.00	UD	ATCC	DMEM + 10% FBS + 1% P/S	
RES186	10.76 ± 2.6	89.56 ± 3.62	714.75 ± 7.08	1.35	60.4	HS	23.2	0.69	0.15	kindly provided by Dr. Chris Jones	DMEM + 10% FBS + 1% P/S	Pediatric glioma
RES259	5.28 ± 1.54	78.15 ± 23.66	206.05 ± 6.09	1.13	81.4	HS	5.51	1.41	0.53	kindly provided by Dr. Chris Jones	DMEM + 10% FBS + 1% P/S	
KNS42	8.10 ± 1.17	84.84 ± 34.86	>1000	1.9	46.14	MS	5.07	0.92	UD	kindly provided by Dr. Chris Jones	DMEM + 10% FBS + 1% P/S	
UW479	8.89 ± 0.86	72.63 ± 12.45	>1000	1.2	57.57	MS	8.61	0.83	UD	kindly provided by Dr. Chris Jones	DMEM + 10% FBS + 1% P/S	
SF188	3.38 ± 1.24	0.039±0.02	>1000	1.80	80.24	HS	2.8	2.20	UD	kindly provided by Dr. Chris Jones	DMEM + 10% FBS + 1% P/S	
HCB2	11.79± 1.04	ND	ND	ND	59.4	MS	ND	0.63	ND	Barretos Cancer Hospital	DMEM + 10% FBS + 1% P/S	Primary culture
HCB149	20.16± 1.34	ND	ND	ND	20.3	R	ND	0.37	ND	Barretos Cancer Hospital	DMEM + 10% FBS + 1% P/S	
NHA	7.42 ± 2.46	37.59 ± 8.34	110.5 ± 1.05	ND	59.77	MS	ND			ECACC	DMEM + 10% FBS + 1% P/S	Normal Human Astrocyte

* **Growth inhibition** (GI) was calculated as a percentage of untreated controls, and its values were determined at a fixed dose of 10 µM. Samples exhibiting more than 60% growth inhibition in the presence of 10 µM IngC were classified as highly sensitive (HS); as resistant (R) when showing less than 40%; and as moderately sensitive (MS) when showing between 40 and 60% growing inhibition. ** The selectivity index (SI) is the ratio between the IC$_{50}$ values for NHA (IngC IC$_{50}$ = 7.42 ± 2.46 and TMZ IC$_{50}$ = 110.5 ± 1.05 µM) and those for the glioma cell lines. *** The Combination Index (CI) was analyzed using CalcuSyn Software version 2.0. The CI value significantly lower than 1.0, indicates drug synergism; CI value significantly higher than 1.0, drug antagonism; and CI value equal to 1.0, additive effect. UD = undetermined; ND = not determined; * IngC (ingenol-3-dodecanoate); I3A (ingenol-3-angelate); TMZ (temozolomide); FBS (fetal bovine serum). ATCC (American Type Culture Collection); DSMZ (German Collection of Microorganisms and Cell Cultures; ECACC (European Collection of Authenticated Cultures).

2.2. IngC Exhibts Higher Cytotoxic Activity than Other Ingenol-Ester Class on Glioma Cells

We further compared the antitumor activity of IngC with ingenol-3-angelate (I3A), the other compound of the same class adopted in clinical practice. For commercial cell lines, the mean IC_{50} values ranged from 0.19–13.09 µM for IngC, and from 0.01–95.15 µM for I3A, which indicated that IngC displayed a higher ingenol-ester cytotoxicity on glioma (Table 1).

2.3. Biological Properties of IngC in Cancer Cell Lines

2.3.1. IngC Inhibits Proliferation and Induces S-Phase Arrest but Fails to Attenuate Migration and Invasion on Glioma Cells

Two representative cell lines for IngC sensitivity (GAMG line) and resistance response (U373 line) were selected to explore the biological role of IngC in cancer cells (Table 1). We first characterized the cell proliferation capabilities by colony formation assay and BrdU incorporation. IngC was able to reduce or inhibit significantly colony formation of GAMG cells, but not U373 (Figure 2A,B). Besides, IngC exhibited dose-dependent proliferation effects on the GAMG with an increase in BrdU- positive cells after 72 h (Figure S1b). IngC was less active amongst U373 cells, the greatest inhibition was observed at the highest IngC (30 µM) dose applied, being able to inhibit a little more than 30% of proliferation. These results suggest that IngC seems to have cytotoxic effects on the anchorage-dependent growth of both malignant glioma cell lines (Figure S1b).

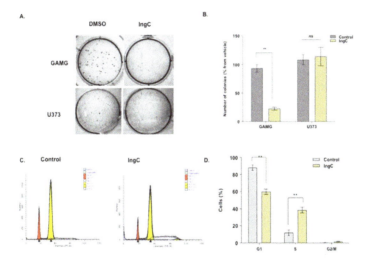

Figure 2. Effect of IngC in colony formation of glioma cells. (**A**) U373 and GAMG cells were seeded and grown in soft-agar medium containing the indicated compounds. (**B**) The number of colonies in each well was counted after 20 days of IngC treatment. The graphs are representative of at least two independent experiments performed in duplicate. n.s. means non-significant. (**C**) U373 cells (untreated and cells treated with IngC) were incubated for 72 h. Next, U373 cells were fixed with ethanol, stained with propidium iodide and cell cycle phase was analyzed by flow cytometry. (**D**) Results shown are the means ± S.D. of three independent experiments. n.s. means non-significant. ** $p < 0.005$.

The effect on the cell cycle profile was characterized by measuring the cellular DNA content. Flow cytometry revealed that IngC exerts significant effects on the cell cycle distribution. U373 cells that were treated with the compound were accumulated in S phase. U373 cells ranged from 11.3% in control to 38.3% when treated with IngC for 72 h (Figure 2C,D).

The impact of IngC on cellular migration and invasion was also evaluated, and no significant effect was observed in the IngC-sensitive or IngC resistant glioma cell lines at the time point and dose investigated (Figure S2a,b).

2.3.2. IngC Induces Cell Death by Other Mechanisms, Not Apoptosis

The effects of IngC on stress, apoptosis, and cell cycle were assessed by human apoptosis and cell stress proteome array. For IngC-sensitive cells (GAMG), the cell stress proteome array assay revealed that IngC exposure at 6 h resulted in a great downregulation of most stress-response proteins, including IDO, PON-3, p–HS27, p–JNKpan, and p-p38 (Figure 3A). In contrast, amongst the IngC-resistant U373 cell line, we observed, at 6h, a marked upregulation of ADAMTS-1, FABP-1, IDO, NF-κB1, and p–HS27 expression and minor changes in (upregulation) other proteins such as DKK4, HIF-2a, p-p38 and PON-1.

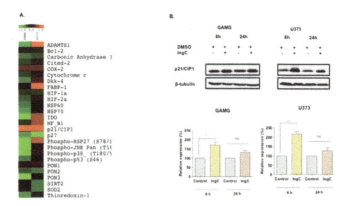

Figure 3. Effect of IngC on cell stress and cell cycle distribution on glioma cell lines. (**A**) A panel of 26 proteins related to cellular stress. Data represented by the heat maps show the proteins modulated after 6 h of IngC treatment (3X IC$_{50}$ value) in glioma cells, GAMG and U373. The quantification and normalization of proteins was performed using the positive controls and untreated controls from the package *Protein Array Analyzer* of Image J software. (**B**) Cells were treated with IC$_{50}$ concentrations of IngC (6 and 24 h) for the indicated time periods. GAMG and U373 cell lysates (20 μg per lane) were analyzed using immunoblotting with anti-p21/cip1. The tubulin was used as an internal control to normalize the amount of proteins applied in each lane. These data are representative of three independent experiments. n.s. means non-significant. * $p < 0.05$ and ** $p < 0.005$.

Remarkably, IngC also increased the levels of p21CIP1/WAF1 and, in a small proportion, increased COX2 expression in both cell lines (Figure 3A). The cell cycle modulation was also validated by immunobloting, which showed the up-regulated expression of cell cycle regulatory proteins p21CIP/WAF1 in both GAMG and U373 cells after 6 and 24 h (Figure 3B), corroborating the previous flow-cytometry analysis of IngC in cell cycle arrest.

Moreover, using the human apoptosis proteome array (R&D systems), containing 35 different proteins related to apoptosis, we assessed the effect of IngC effect after 24 and 72 h on glioma cells (Figure 4A,B). At 24 h mark analysis, the (IngC-sensitive) GAMG cells had an upregulation of catalase, claspin and minor changes in cIAP-2 and FADD. Besides, Clusterin, p-P53 (S392) and at lower levels cIAP-1, H0-1/HM0X1/HSP32, pP53 (S15), and pP53 (S46) were markedly downregulated. At 72 h, IngC treatment promoted the expressive upregulation of p21CIP1/WAF1 and downregulation of cytocrome, p–P53 (S15), SMAC/DIABLO, survivin and TNFRSF1A in GAMG cells. Furthermore, IngC maintained the modulation of catalase, clusterin, pP53 (S392) and pP53 (S46). On the other hand, a greater modulation was observed in the IngC-resistant U373 cells at both times points. IngC downregulated the expression of several proteins at 24 h, especially: claspin, cleaved-caspase-3, clusterin, survivin,

HIF-1, TNFRSF1A, and XIAP. In addition, proteins, such as Bad, BCL-2, cIAP-1, HTRA-2, and p-RAD17, were less intensely downregulated. The treatment also upregulated BcL-x, catalase, cytochrome C, HSP27, HSP60, HSP70, livin, P21, P27, PON2, TRAILR1/DR4, and TRAILR1/DR5. At 72 h, similarly to GAMG cells, p21CIP1/WAF1 was markedly upregulated in U373 cells. Furthermore, IngC kept the modulation of several stress proteins, such as BcL-x, catalase, cytochrome C, HSP27, HSP60, HSP70, cIAP-2, TRAILR1/DR4, and TRAILR1/DR5, and upregulated proteins before decrease as cIPA-1, claspin, cleaved-caspase-3, clusterin, and TNFRSF1A. These results were validated by immunobloting, which confirmed up-regulated expression of DR5 proteins in U373 cells after 24 and 72 h (Figure 4C).

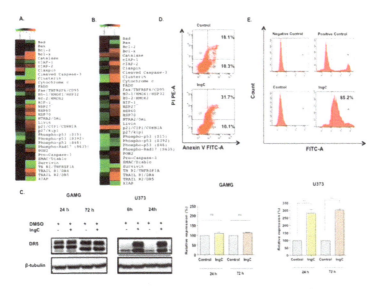

Figure 4. Effect of IngC on apoptosis pathway on glioma cell lines. (**A,B**) Panel of 35 proteins related to apoptosis. The data represented by the heat maps show the proteins modulated after 24 and 72 h of IngC treatment (3X IC_{50} value (13.09 µM) in glioma cells, GAMG and U373. (**C**) Cells were treated with 3X IC_{50} concentrations of IngC (24 and 72 h) for the indicated time periods. GAMG and U373 cell lysates (20 µg per lane) were analyzed using immunoblotting with anti-DR5. The tubulin was used as an internal control to normalize the amount of proteins applied in each lane. This data is representative of three independent experiments. (**D**) After the IngC treatment with 3X IC_{50} for 72 h, U373 cells were fixed, stained with annexin V-FITC/PI and analyzed by FACScan. Data shown are representative of three independent experiments. (**E**) After IngC treatment with 3X IC_{50} for 72 h, DNA fragmentation in U373 cell line was measured with the TUNEL assay using flow cytometry. The graphs are representative of at least three independent experiments performed in duplicate. n.s. means non-significant. ** $p < 0.005$.

The ability of IngC to induce cell death by apoptosis was also analyzed in glioma cells (sensitive or resistant ones) through flow cytometry (Figure 4D). As shown with proteomic profile arrays, no early apoptotic induction was revealed. The percentage of positive annexin V cells, indicative of early apoptosis, was not different from untreated control (10.3% to 10.1%, respectively), while the double positive annexin V/PI cells that are indicative of late apoptosis/necrosis revealed high percentual for IngC when compared to the untreated control (31.7% to 18.1%, respectively; Figure 4D). On the other hand, IngC treatment was able to induce DNA damage in these same lines, as evidenced by the Tunel assay. Close to 85.2% of the U373 cells were Tunel-positive after 72 h of IngC exposure, which suggests DNA fragmentation and a different type of cell death (Figure 4E).

2.3.3. IngC Induces Autophagy in Glioma Cells

Given the importance of autophagy in cell death of gliomas, we wondered whether IngC could interfere in this process. For this, we evaluated autophagy-associated protein LC3-II expression. The GAMG and U373 cells were exposed to IngC for 2 h (IngC at IC_{50} value) and assessed by western blotting. GAMG cells exhibited a marked increase of LC3-II when compared to the untreated control cells after either IngC alone or when combined with Bafilomycin A1 (Baf) (Figure 5A,B). LC3-II levels expression in IngC alone and IngC combined to Baf were especially evident (2 and 2.9–fold increase, respectively) following 2 h of treatment. The presence of acidic vesicular organelles (AVOs), which are a non-specific marker for autophagy, was also analyzed. FACS scanning indicated an increase in the acridine orange positive cells when the cells were treated either with IngC alone or when combined with Baf (GAMG 10 nM). IngC-Baf combination led to greater formation of AVOs in U373 cells (58.0% in IngC versus 15.5% in controls and 97.8% in IngC-Baf versus 25.9% controls), thus indicating the development of AVOs suggestive of the autophagy process (Figure 5C).

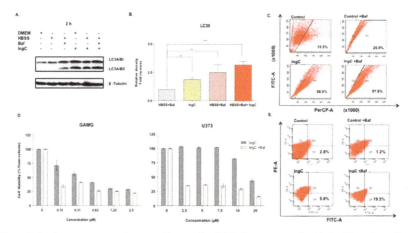

Figure 5. IngC promotes autophagy on glioma cells. (**A**) Cells were treated with the IC_{50} value of IngC for the indicated time periods. GAMG cell lysates (20 µg per lane) were analyzed using immunoblotting with anti-LC3. (**A**,**B**) are representative of three independent experiments with GAMG. Tubulin was used as an internal control to normalize the amount of proteins applied in the treatment without bafilomycin A1 (Baf). (**C**) Development of AVO in IngC-treated cells by detecting green and red fluorescence in acridine orange-stained cells using FACS analysis. U373 cells were treated with IngC (IC_{50} value), and Baf (20 nM) for 72 h. The graphs are representative of at least two independent experiments. FITC indicates green color intensity, while PerCP shows red color intensity. (**D**) Effect of baf on GAMG and U373 cell viability of IngC-treated cells. At 3 h after exposure to IngC, baf was added and cultured until 72 h and evaluated by MTS assay. The viability of the untreated cells was considered as 100%. Results shown are the means ± S.D. of three independent experiments. (**E**) Effect of Baf on IngC-induced apoptosis. After, IngC and bafilomicyn treatment for 72 h, GAMG cells were stained with annexin V-FITC/PI and analyzed by FACScan. Data shown are representative of three independent experiments. ** $p < 0.005$ and *** $p < 0.0001$.

U373 and GAMG were treated with different IngC concentrations for 72 h with the presence of Baf to investigate whether the inhibition of autophagy in late stages could affect the cytotoxicity of IngC. The viability of resistant U373 cells treated with IngC-Baf combination decreased in all concentrations tested as compared to the presence of Baf (20 nM) alone (Figure 5D). Next, we evaluated the double labeling of annexin V/PI in U373 cells to determine the effect of IngC combined with Baf in the apoptosis process. There was no appreciable change of positive cells to annexin V, indicative of early apoptosis, when comparing the control with IngC alone (2.8 to 5.8%) (Figure 5E). On the other hand, when compared to the Baf control, the treatment combining IngC to Baf increased the count of annexin V positive cells (1.2 to 19.5%), which suggesting that the cytotoxicity sensitivity was increased by the apoptosis mechanism (Figure 5E).

2.3.4. IngC Exposure Inhibits Protein Kinase C Isotypes on Glioma Cells

Multiple antitumor effects of diterpenes have been related to the direct modulation of PKCs, important proteins that are involved in cellular signal pathways [31]. We evaluated the PKC isotypes (conventional PKCs (cPKCs) to investigate the possible role of IngC in PKC signaling pathway: PKCα, p-PKCα/βII, p-PKCpanβII, novel PKCs (nPKCs): PKCδ, p-PKCδ, p-PKCδ/θ, p-PKCθ, and atypical PKCs (aPKCs): p-PKCζ/λ, PKCζ as well as PKD1/PKCμ, p-PKC PKDμ (Ser^{916}), and p-PKC PKDμ (Ser^{744})) activation profile using immunoblotting (Figure 6). We also compared the IngC involvement in PKC activity with the known modulator I3A. IngC markedly downregulated the phosphorylation of most PKC isotypes (PKCalpha/beta, PKCpan/betaII, and PKC/PKDμ (Ser^{916})) as compared to I3A in GAMG cell line (Figure 6a). Moreover, similarly to IngC, I3A also decrease PKCδ/θ, PKCδ activation levels and the total expression of PKCα over time. Although less intensely, I3A also decreased the phosphorylation of PKCζ/λ and total levels of PKCζ over time (Figure 6a). In contrast to the general downregulation of PKC isoforms, the PKC/PKDμ (Ser^{744}) isotype had transient phosphorilation by either IngC or I3A. PKC/PKDμ (Ser^{744}) phosphorylation peaked at 6h with IngC and decreased over time, whereas I3A led to a short-lived peak, weaning off before 24 h (Figure 6a).

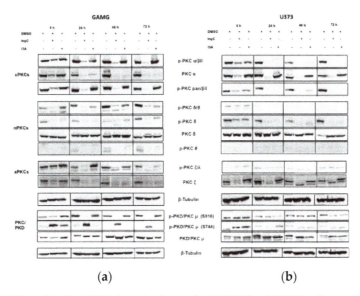

(a)   (b)

Figure 6. Effect of IngC on PKC isoforms in glioma cells. (a) GAMG and (b) U373 cells were incubated with the IC$_{50}$ value for IngC, at 6, 24, 48 and 72 h. Controls were treated with DMSO alone (1%). Whole cell extracts from the same preparation were subjected to western-blotting analysis of PKC isoforms expressions. β-tubulin is shown as an internal control. Results shown are the means ± S.D of two independent experiments.

Further, in the U373 cell line, the PKC activity was modulated in the same way for IngC and I3A treatment, with exception of the total expression of PKCs α and ζ. Unlike GAMG cells, no modulation was found in the PKC/PKDμ (Ser^{744}) isotype in the U373 cell line for both treatments (Figure 6b). These results show that IngC regulates PKCs activity with different responses according to cell sensitiveness.

3. Discussion

Euphorbia tirucalli is widely used as an anticancer drug in Brazilian folk medicine [15]. Our group recently identified *E. tirucalli* derived natural and semi-synthetic compounds, and showed their cytotoxicity effect in a wide-range of human cancer cells [19,20]. Particularly, a diterpene derived, ingenol-3-dodecanoate (IngC), exhibited the highest efficacy in several tumor cell lines, including gliomas [20]. In the present pre-clinical study, we extended and have gained a *comprehensive biological* insight into the underlying molecular mechanisms of IngC in gliomas. We showed that IngC is a major modulator of protein kinase C isotypes and it promotes autophagy and S-phase arrest in gliomas.

The antitumor potential of IngC was studied in a panel of 13 glioma cell lines, including commercial (adult and pediatric), primary cultures, and one normal human astrocyte cell line. IngC exhibited dose-time-dependent cytotoxicity on all glioma cell lines. The different models of glioma cell lines exhibited a heterogeneous profile of response to IngC. At a fixed dose of 10 µM, 9.1% (1/11) of cell lines were resistant, 36.4% (4/11) were moderately sensitive, while 54.5% (6/11) were classified as highly sensitive. This variation in response to IngC seems to be due to the innate differences in the molecular biology underlying adult, pediatric, and especially primary glioma cultures that could better mimic genomic heterogeneity from patients [19,32]. In our study, IngC showed more than twice the cytotoxicity to some cancer cell lines, mainly for GAMG (39.05) and SF188 (2.2) cells, when compared to normal cell lines, an interesting selectivity index preconized [33]. It was not possible to calculate the selectivity index for most of the cancer cell lines that were treated with TMZ, since this standard chemotherapeutic agent was more cytotoxic to the normal astrocyte cell line than to cancer cells.

Our results are in agreement with previous studies, which also demonstrated diterpenes cytotoxicity at micromolar range, such as ingenol 3,20-dibenzoate (IDB) from *E. esula* L in jurkat and breast cancer cells, and ingenol-3-angelate from *E. peplus* (I3A), in human melanoma, cervical cancer, and prostate xenografts [27,30,34,35]. Importantly, we showed that IngC presented higher efficacy when compared to I3A on glioma cells, suggesting this compound as promising against gliomas.

Studies addressing the cytotoxic effect of terpenes/diterpenes in glioma context are scarce. Kaurenoic acid, a bioactive diterpenoid that is present in *Mikania hirsutissima*, was evaluated in U87MG cells [36]. The concentrations used were much higher than the ones used in our study (30 to 70 µM for 24 to 72 h), despite previous reports showing the absence of cytotoxic effects on fibroblasts [37]. We also highlight that some semi-synthetic diterpenes are among the most cytotoxic drugs investigated, with an IC_{50} in the sub-nanomolar range [34].

Herein, we evaluated the intracellular signal pathways that were modulated by short-term exposure to IngC in glioma cells. Interestingly, stress and apoptosis panels of proteome arrays were the most modulated in U373, a drug-resistant cell line. We found that IngC promoted modulation in proteins related to stress and cell cycle, as well as anti-apoptotic and pro-apoptotic protein expression, with special reduction in the levels of pro-apoptotic BAX, BAD, TNRI/TNFRSF1A, and FASTNFR6/CD95, were observed for most of glioma cell lines overtime. In addition, anti-apoptotic proteins, including HSP27, HSP60, HSP70, livin, and PON-2, were upregulated in U373 cells. These results are consistent with flow cytometry studies, showing that changes in p21/CIPI WAf1 and anti-apoptotic factors were more pronounced, indicating that, in the conditions and cells tested, cell cycle arrest, but not apoptosis, contributed to the antiproliferative and cytotoxic effects of IngC in malignant glioma cell lines. Among the diterpenes described, IDB has relevant cell growth inhibition and apoptotic cell induction in jurkat cells and breast cancer cells [30,34]. Additionally, Lizarte and coworkers observed that Kaurenoic acid influences the regulation of several genes that are involved in the apoptotic pathway, including *c-FLIP, caspase 3, caspase 8,* and *miR-21* in U87 cells [26–28,35]. On the other hand, I3A promotes primary necrosis in melanoma, cervical cancer, and prostate xenografts, as well as inducing in vitro and in vivo models of colon cancer, apoptosis senescence, anti-inflammatory, and antitumor immunomodulatory properties in colon cancer. Although our data are discordant from these studies, these findings underscore our limited knowledge regarding the ingenoid pharmacophore and confirm that this paradoxical comportment is still poorly known [34,38].

We assessed the interplay of autophagy and IngC exposure due to the dual role of autophagy in cancer [39]. Temozolomide, the backbone of systemic therapy for glioblastoma, is reported to induce cell death by autophagic mechanisms [40]. Our study suggested that autophagy could play an important role in the antiproliferative mechanism of IngC. IngC induced LC3-II increase and marked formation of AVOs, indicating that this compound might activate an autophagic process. Moreover, the combined treatment of Baf and IngC potentiates the antitumor effect of IngC against malignant glioma cells by the autophagic vacuoles accumulation and apoptosis induction. Furthermore, the cell death that is induced by substances that suppress the autophagy pathway improved the effectiveness of TMZ and natural compounds in glioma cells [41]. These results are particularly important for GBM, since this tumor has been shown to be more resistant to cell death [42,43].

Protein kinase Cs (PKCs) contemplates a family of 14 known isozymes of serine/threonine-specific protein kinases, which are classified into three groups according to their interactions with calcium and diacylglycerol as cofactors; classical PKC (cPKC: α, $β_1$, $β_2$, and γ), novel PKC (nPKC: δ, ε, η, and θ), and atypical PKC (aPKC: ζ ι, ζII, ξ, ν) [6]. PKD1 was initially recognized as a member of the protein kinase C (PKC) family and named PKCμ, however due to some particularities, it was reclassified. PKD1 is now a member of the protein kinase D (PKD) Family [44,45]. The role of PKC and PKD1 in processes that are relevant to neoplastic transformation, proliferation, apoptosis, and tumor cell invasion provides a potential suitable target for anticancer therapy [34,44,46], including glioma [6,9–12]. Provided that most of the biological effects of ingenol esters and derivatives are attributed to protein kinase C (PKC), such as co-carcinogenic and antitumor activity [31,34], we can assume a special interest in this field.

We compared IngC effects with I3A to gain more insight in this issue and found a marked potential inhibitory in the phosphorylation of most of PKC isotypes in sensitive GAMG cells by IngC, and a minor effect for I3A. In the U373 cell line, the PKC activity was modulated in the same way for IngC and I3A treatments, with the exception of total expression of PKCα and ζ. Moreover, IngC treatment in GAMG cells promoted the phosphorylation of PKC/PKDμ (Ser^{744}) isotypes. A substantial PKC/PKDμ (Ser^{744}) activation was also seen with I3A, but rapidly decreased. Of note, the results of I3A found in glioma are surprising, since I3A is a broad range activator of the classical (α, β, γ) and novel (δ, ε, η, θ) protein kinase C isoenzymes inducing direct pro-apoptotic effects in several malignant cells, including melanoma cell lines and primary human acute myelogenous leukemia cells [35,47]. In colon cancer, I3A induced the activation of PKCδ and reduced expression of PKCα, resulting in apoptosis [35]. These divergent biological responses are not completely understood, although it has long been recognized that there is marked heterogeneity in the patterns of biological behaviour induced by these agents' analogs of diacylglycerol (DAG) [48]. The nature and position of the esters structure in the diterpenes ring could explain why analogs of DAG with different side chains induce different biological responses [34,48,49].

PKC isoforms differ, not only in their structure and substrate specificity and mode of activation, but also in their tissue distribution, subcellular localization, and biological functions [47,48]. The activation of PKC isoenzymes results in changes in their subcellular location following translocation. These observations clearly illustrate that the crosstalk between pro- and anti-apoptotic PKC isoforms is important, and the final effect of the PKC-agonist ingenol esters may therefore depend upon the balance between the various isoenzymes that are present within a tumor. In this sense, the drug might be less effective in those tumors with increased levels of the anti-apoptotic isoforms [48]. This issue could also explain why, although IngC treatment had inhibited PKC isoform activities that are involved in migration and invasion in glioma cells, such as PKCα and PKC/PKDμ, it was not able to inhibit cell invasion or migration in either glioma cell tested. These results corroborates with Do Carmo and coworkers, which revealed that, despite that PKCs have clear roles in GBM, the contribution of each isoform depends on residues phosphorylation, oncogenic mutations, cell environment, and type of stimuli/stress [6]. We emphasize that, although the literature and our findings provide the rationale for attempts to exploit PKC as a target for novel forms of treatment of GBM, our incomplete understanding of the cell- and tissue-specificity of the different PKC isoforms may lead to unexpected and/or undesired results in clinical practice becoming important subjects for further studies.

Finally, the combination of different agents with different targets of action might contribute to circumvent the chemoresistance of glioma cells. Recent studies had reported increased cytotoxicity that is induced by combinatory systems while using PKC inhibitors and TMZ [50]. In our study, combinatory therapy with IngC and TMZ promoted an antagonism effect in most of cell lines evaluated, which suggested that, although IngC promotes the inhibition of PKC activities, its administration with standard chemotherapy does not potentiate the effect of each other and could be related to their interaction with different isotypes of PKCs [47]. Further studies could focus on addressing these PKCs mechanisms in combination with other treatment modalities such as radiation and/or chemotherapy, which might help to refine our understanding of the glioma biology, enabling us to develop new therapeutic opportunities against this disease.

4. Materials and Methods

4.1. Cell Culture

Thirteen immortalized glioma cell lines (five adult and five pediatric glioma cell lines, two glioma primary cultures, and one normal human astrocyte) were obtained and cultivated, as indicated in Table 1. The two primary glioma cell lines were derived from surgical glioblastoma biopsies that were obtained at the Neurosurgery Department of the Barretos Cancer Hospital (São Paulo, Brazil). The local ethics committee approved the study protocol and patients signed an informed consent form [51]. The isolated cells were grown in DMEM medium under the same conditions described in Table 1. Besides mycoplasma analysis, all cell lines were authenticated by the Diagnostics Laboratory at the Barretos Cancer Hospital (São Paulo, Brazil) by short tandem repeat (STR) DNA typing, according to the International Reference Standard for Authentication of Human Cell Lines, as previously described [52]. Moreover, the established primary cultures were identified and confirmed by blood that was derived from the same patient.

4.2. Semi-Synthetic Ingenol

The synthesis of semi-synthetic ingenol-3-dodecanoate (IngC) from the sap of *E. tirucalli* was performed by Kyolab Laboratory (Campinas, Brazil) and provided by Amazônia Fitomedicamentos, Brazil (patent). The natural ingenol was altered by the addition of specific ester chains at carbon 3 of the core ring, as previously reported (Figure 1A) [20,21]. The ingenol synthetic derivative was diluted in dimethyl sulfoxide (DMSO) at 10 mM stock. The work dilutions were prepared to obtain a concentration of 1% DMSO. All of the dilutions were stored at −20 °C.

4.3. Cell Viability Analysis and IC_{50} Determination

The cytotoxic effect of IngC, its ingenol-ester analogue, (ingenol-3-angelate) (I3A) (Adipogen (Switzerland)) and temozolomide (TMZ) (Sigma - **T2577**), *was* evaluated while *using* MTS assay (Cell Titer 96 Aqueous cell proliferation assay, Promega, Madison, WI, USA), following the manufacturer's instructions. Cells were treated with increasing concentrations of IngC diluted in DMEM (0.5% fetal bovine serum (FBS)) and incubated for 72 h. Absorbance was measured in the automatic microplate reader Varioskan (Thermo) at 490 nm. The half maximal inhibitory concentration (IC_{50}) was obtained by nonlinear regression while using GraphPad PRISM version 5 (GraphPad Software, La Jolla California USA), as previously described [18,19]. The growth inhibition (GI) was also calculated as a percentage of untreated controls, and its values were determined at a fixed dose of 10 µM (concentration closer to the average IC_{50} value of all cell lines at screening) [53]. Samples exhibiting more than 60% growth inhibition in the presence of 10 µM IngC were classified as highly sensitive (HS), as moderately sensitive (MS) when they were between 40 and 60%, and as resistant (R) when the values were lower than 40% of inhibition, as previous reported [53]. The selectivity index (SI) was obtained by dividing the IC_{50} value of a normal cell line (NHA) by a tumor cell line according to the National Cancer Institute (NCI) [19]. Significant SI values are considered to be greater than or equal to 2.0. [33]. The assays were performed in triplicate and repeated at least three times for each cell line.

4.4. Colony Formation-Assay

Anchorage-independent growth was performed while using a soft-type-agar assay, as reported [18]. Medium was changed every 72 h, and DMEM + 0.5% FBS containing IngC at IC_{50} concentrations values was added on GAMG, and U373 cell lines. The colonies formed were stained with 0.05% crystal violet for 15 min. and photo-documented. The analyses were performed by Image J Software. The assay was performed in two biological replicates and the experiments were done in duplicate.

4.5. Migration and Invasion Assays

Cell migration and invasion effects on GAMG and U373 cell lines were evaluated by wound healing assay and BD Biocoat Matrigel Invasion Chambers (354480, BD Biosciences), as previously described [54]. The shown images are representative of three independent experiments performed in triplicates.

4.6. Cell Cycle and Cell Death Assays

Cell cycle and cell death assays were performed by flow cytometry, as previously described [19]. The cells were plated onto a six-well plate at a density of 1×10^6 cells/well, allowed to adhere for at least 24 h and serum starved for 12 h. Additionally, the cells were exposed to 3X IC_{50} values of IngC for a period of 72 h in DMEM (0.5% FBS). The cell cycle distribution (G1, S, and G2/M) as well as double staining with Annexin V-FITC/PI and tunel assays, were determined with a flow cytometer *BD FACSCanto II* (*BD Biosciences*) and analyzed with the software BD FACSDiva (*BD Biosciences*), following the manufacturer's recommended protocol. Approximately, 2×10^4 cells were evaluated for each sample in both assays. The analyses were performed in experimental and biological triplicates.

4.7. Proteome Arrays

The relative protein expression levels of a panel of 35 proteins related to apoptosis and 26 proteins related to cellular stress were obtained while using the Proteome Profiler Human Apoptosis Array (R&Dsystems- #ARY009) and Proteome Profiler Human Cell Stress Array (R&Dsystems-#ARY018), according to the manufacturer's instructions and as previously reported [54]. The selected cell lines (GAMG and U373) were treated with IngC for 6 and 24 h, for stress array and 24 and 72 h for apoptosis array while using a concentration equivalent to 3 x IC_{50} value of each glioma cell line. In addition, THE expression of some of the proteins was validated by western-blot analysis as described below.

4.8. Western Blotting

Protein expression after IngC treatment was evaluated by western blotting. Cells were plated onto a six-well plate at a density of 1×10^6 cells/well, allowed to adhere at least 24 h, and then serum-starved in DMEM (0.5% FBS). The cells were exposed at IC_{50} values of IngC for several time points of 6, 24, 48, and 72 h in DMEM (0.5% FBS) and total protein was separated, as previously described for western blotting analysis [19]. Antibodies included anti-DR5, anti-p21/Cip1, anti-total PKCs (PKCα, PKCδ and PKCζ), anti-phosphorylated PKCs (p-PKC PKDμ (S916), p-PKC PKDμ (S744), p-PKCα/βII, p-PKCpanβII, p-PKCδ, p-PKCδ/θ, p-PKCθ, PKCζ/λ), and β-tubulin. All of the antibodies were diluted at 1:1000 and purchased from Cell Signaling Technology.

4.9. Autophagy Analysis: LC3 Expression, Acidic Vesicular Organelles (AVOs) and IngC and Autophagy Inhibitor Combination Effect on Glioma Cell Lines

GAMG and U373 cells were plated onto a six-well plate at a density of 5×10^5 cells/well, and allowed to adhere to evaluate the effect of IngC in the autophagy process. The growth medium was replaced by Hanks balanced salt solution (HBSS; Invitrogen) for cells starvation (two rinses in HBSS before being placed in HBSS). Cells were treated with 10 nM (GAMG) and 20 nM (U373) of bafilomycin A1 (Baf), to inhibit autophagy flux [19,41] or with IngC, while using a concentration equivalent to the IC_{50} of each cell line; or Baf and IngC combined; or DMEM alone as control. After 2, 6, or 24 h, the protein extracts was subjected to western blot analysis, as described above. For this, we used the primary polyclonal antibodies LC3A/B (dilution 1:1000; Cell signaling) and β-tubulin (dilution 1:5000; Cell Signaling Technology), as a loading control.

We also detected the acidic vesicular organelles (AVO) in the IngC-treated cells through vital staining with acridine orange, as reported previously [41]. The assay was performed according to the conditions for autophagy analysis, as mentioned above. Subsequently, 72 h after IngC exposure, cells were stained with acridine orange at a final concentration of 1 µg/mL for 15 min, washed twice in PBS 1×, and analyzed with a flow cytometer *BD FACSCanto II* (*BD Biosciences*) and BD FACSDiva software (*BD Biosciences*) following the manufacturer's recommended protocol. These analyses were performed in experimental and biological triplicates.

GAMG and U373 cells were plated onto a 96-well plate at a density of 5×10^3 cells/well, and allowed to adhere, and then, increasing concentrations of IngC were added to determine the effect of the autophagy inhibitor combine with IngC on cell viability. To inhibit autophagy, a fixed dose of Baf (10 nM for GAMG cells and 20 nM for U373 cells) was added to the culture 3 h after IngC treatment, as described [19]. The cell viability assay was evaluated after 72 h while using the Cell Titer 96 Aqueous test One Solution Cell Proliferation Assay (Promega), and measured as described above. The data were obtained and normalized relative to the average survival of untreated samples, or only treated with Baf. The analyses were performed in experimental and biological triplicates.

4.10. Drug Combination Studies

Combination studies from IngC and TMZ were performed with fixed concentrations (determined by the IC_{50} value) of the standard chemotherapeutic agent temozolomide, simultaneously exposed to increasing concentrations of IngC and evaluated by MTS assay as previously described above. Drug interactions were evaluated by the combination index while using CalcuSyn software version 2.0 (Biosoft; Ferguson, MO, USA), as previously described [55,56]. Synergy was defined as CI values that were significantly lower than 1.0; antagonism as CI values significantly higher than 1.0; and, additive as CI values that are equal to 1.0 [55,56] at drug IC_{50} value for each cell line.

4.11. Statistical Analysis

Data were expressed as the mean ± standard deviation (SD) of three independent experiments. We applied the Student's t-test for comparing two different conditions, whereas two-way analysis of variance (ANOVA) was used for assessing the differences between more groups. *p*-values <0.05 were considered to be significant. All of the analyses were performed whle using the aforementioned GraphPad PRISM version 8 (GraphPad Software, La Jolla, CA, USA).

5. Conclusions

Our current findings add a new layer of complexity to understand the diterpene mechanism, including its modulation of the autophagic process and providing a comprehensive view of IngC in glioma. Importantly, this study supports ongoing efforts targeting PKC proteins in cancer therapy with IngC and its indicates as lead semi-synthetic diterpene based PKC inhibitors, which represents a novel and promising antitumor drug to target cancer cells.

Supplementary Materials: The following are available online. Figure S1: Time and concentration cytotoxic effect of IngC exposure on human cancer cell lines. Figure S2: Effect of IngC on migration and invasion of glioma cells.

Author Contributions: Conceptualization, R.M.R. and V.A.O.S.; methodology, V.A.O.S., M.N.R., A.C.C., J.P.L., O.M. and A.T.; software, A.F.E. and A.T.; validation, V.A.O.S. and M.N.R.; formal analysis, V.A.O.S., M.N.R., O.M., A.F.E. and A.T.; investigation, V.A.O.S. and M.N.R.; resources, R.M.R. and L.F.P.; writing—original draft preparation, V.A.O.S.; visualization and revising it critically for important intellectual content, O.M., A.T., J.P.L. and R.M.R.; supervision, R.M.R.; compounds preparation L.F.P.; project administration, R.M.R.; funding acquisition, R.M.R.

Funding: This research received was funded by Amazônia Fitomedicamentos Ltda, and Barretos Cancer Hospital, all from Brazil.

Acknowledgments: Amazônia Fitomedicamentos Ltda provided the ingenol semi-synthetic compounds. The Amazônia Fitomedicamentos Ltda. is the sole and exclusive owner of the respective intellectual property rights.

Conflicts of Interest: The authors confirm that this article content has conflicts of interest. This study was supported by grants from Amazônia Fitomedicamentos Ltda as part of the ingenol pre-clinical studies and Viviane A. O. Silva and Marcela N. Rosa received a scholarship from Amazônia Fitomedicamentos Ltda. to conduct the study.

Abbreviations

AVOs	Acidic Vesicular Organelles
ANOVA	Analysis of variance
ATCC	American Type Culture Collection
Baf	bafilomycin A1
CI	Combination index
DNA	Deoxyribonucleic acid
DSMZ	German Collection of Microorganisms and Cell Cultures
DMEM	Dulbecco's modified Eagle's medium
DMSO	Dimethyl sulfoxide
ECACC	European Collection of Authenticated Cultures
FBS	Fetal bovine serum
FDA	Food and Drug Administration
GBM	Glioblastoma
GI	growth inhibition
g	Gram
IC_{50},	Half maximal inhibitory concentration
IngC,	ingenol-3-dodecanoate; I3A, ingenol-3-angelate
IDB	ingenol 3,20-dibenzoate
mL	Milliliter
MTS	3-(4,5-dimethylthiazol-2-yl)-5-(3-carboxymethoxyphenyl)-2-(4-sulfophenyl)-2H91tetrazolium)
NCI	National Cancer Institute
PKC	Protein kinase C
P/S	Penicillin/streptomycin solution
RPMI-1640	Roswell Park Memorial Institute
SD	Standard deviation
STR	Short tandem repeat
TMZ	temozolamide
WHO	World Health Organization
uM	Micromolar

References

1. Weller, M.; Wick, W.; Aldape, K.; Brada, M.; Berger, M.; Pfister, S.M.; Nishikawa, R.; Rosenthal, M.; Wen, P.Y.; Stupp, R.; et al. Glioma. *Nat. Rev. Dis Primers.* **2015**, *1*, 15017. [CrossRef] [PubMed]
2. Miranda-Filho, A.; Pineros, M.; Soerjomataram, I.; Deltour, I.; Bray, F. Cancers of the brain and CNS: Global patterns and trends in incidence. *Neuro Oncol.* **2017**, *19*, 270–280. [CrossRef] [PubMed]

3. Ostrom, Q.T.; Gittleman, H.; Liao, P.; Vecchione-Koval, T.; Wolinsky, Y.; Kruchko, C.; Barnholtz-Sloan, J.S. CBTRUS Statistical Report: Primary brain and other central nervous system tumors diagnosed in the United States in 2010-2014. *Neuro Oncol.* **2017**. [CrossRef] [PubMed]
4. Weller, M.; van den Bent, M.; Tonn, J.C.; Stupp, R.; Preusser, M.; Cohen-Jonathan-Moyal, E.; Henriksson, R.; Le Rhun, E.; Balana, C.; Chinot, O.; et al. European Association for Neuro-Oncology Task Force on, G., European Association for Neuro-Oncology (EANO) guideline on the diagnosis and treatment of adult astrocytic and oligodendroglial gliomas. *Lancet Oncol.* **2017**, *18*, 315–329. [CrossRef]
5. Gladson, C.L.; Prayson, R.A.; Liu, W.M. The pathobiology of glioma tumors. *Annu Rev. Pathol* **2010**, *5*, 33–50. [CrossRef]
6. Do Carmo, A.; Balca-Silva, J.; Matias, D.; Lopes, M.C. PKC signaling in glioblastoma. *Cancer Biol. Ther.* **2013**, *14*, 287–294. [CrossRef]
7. Da Rocha, A.B.; Mans, D.R.; Regner, A.; Schwartsmann, G. Targeting protein kinase C: New therapeutic opportunities against high-grade malignant gliomas? *Oncologist* **2002**, *7*, 17–33. [CrossRef]
8. Zellner, A.; Fetell, M.R.; Bruce, J.N.; De Vivo, D.C.; O'Driscoll, K.R. Disparity in expression of protein kinase C alpha in human glioma versus glioma-derived primary cell lines: Therapeutic implications. *Clin. Cancer Res.* **1998**, *4*, 1797–1802.
9. Bredel, M.; Pollack, I.F. The role of protein kinase C (PKC) in the evolution and proliferation of malignant gliomas, and the application of PKC inhibition as a novel approach to anti-glioma therapy. *Acta Neurochir.* **1997**, *139*, 1000–1013. [CrossRef]
10. Pollack, I.F.; DaRosso, R.C.; Robertson, P.L.; Jakacki, R.L.; Mirro, J.R., Jr.; Blatt, J.; Nicholson, S.; Packer, R.J.; Allen, J.C.; Cisneros, A.; et al. A phase I study of high-dose tamoxifen for the treatment of refractory malignant gliomas of childhood. *Clin. Cancer Res.* **1997**, *3*, 1109–1115.
11. Muanza, T.; Shenouda, G.; Souhami, L.; Leblanc, R.; Mohr, G.; Corns, R.; Langleben, A. High dose tamoxifen and radiotherapy in patients with glioblastoma multiforme: A phase IB study. *Can. J. Neurol. Sci.* **2000**, *27*, 302–306. [CrossRef] [PubMed]
12. Schwartz, G.K.; Ward, D.; Saltz, L.; Casper, E.S.; Spiess, T.; Mullen, E.; Woodworth, J.; Venuti, R.; Zervos, P.; Storniolo, A.M.; et al. A pilot clinical/pharmacological study of the protein kinase C-specific inhibitor safingol alone and in combination with doxorubicin. *Clin. Cancer Res.* **1997**, *3*, 537–543. [PubMed]
13. Cragg, G.M.; Pezzuto, J.M. Natural Products as a Vital Source for the Discovery of Cancer Chemotherapeutic and Chemopreventive Agents. *Med. Princ. Pract.* **2016**, *25*, 41–59. [CrossRef] [PubMed]
14. Stupp, R.; Mason, W.P.; van den Bent, M.J.; Weller, M.; Fisher, B.; Taphoorn, M.J.; Belanger, K.; Brandes, A.A.; Marosi, C.; Bogdahn, U.; et al. European Organisation for, R.; Treatment of Cancer Brain, T.; Radiotherapy, G.; National Cancer Institute of Canada Clinical Trials, G., Radiotherapy plus concomitant and adjuvant temozolomide for glioblastoma. *N. Engl. J. Med.* **2005**, *352*, 987–996. [CrossRef]
15. Dutra, R.C.; Campos, M.M.; Santos, A.R.; Calixto, J.B. Medicinal plants in Brazil: Pharmacological studies, drug discovery, challenges and perspectives. *Pharmacol. Res.* **2016**, *112*, 4–29. [CrossRef]
16. Dutra, R.C.; Bicca, M.A.; Segat, G.C.; Silva, K.A.; Motta, E.M.; Pianowski, L.F.; Costa, R.; Calixto, J.B. The antinociceptive effects of the tetracyclic triterpene euphol in inflammatory and neuropathic pain models: The potential role of PKCepsilon. *Neuroscience* **2015**, *303*, 126–137. [CrossRef]
17. Passos, G.F.; Medeiros, R.; Marcon, R.; Nascimento, A.F.; Calixto, J.B.; Pianowski, L.F. The role of PKC/ERK1/2 signaling in the anti-inflammatory effect of tetracyclic triterpene euphol on TPA-induced skin inflammation in mice. *Eur. J. Pharmacol.* **2013**, *698*, 413–420. [CrossRef]
18. Silva, V.A.O.; Rosa, M.N.; Tansini, A.; Oliveira, R.J.S.; Martinho, O.; Lima, J.P.; Pianowski, L.F.; Reis, R.M. In vitro screening of cytotoxic activity of euphol from Euphorbia tirucalli on a large panel of human cancer-derived cell lines. *Exp. Ther. Med.* **2018**, *16*, 557–566. [CrossRef]
19. Silva, V.A.O.; Rosa, M.N.; Miranda-Goncalves, V.; Costa, A.M.; Tansini, A.; Evangelista, A.F.; Martinho, O.; Carloni, A.C.; Jones, C.; Lima, J.P.; et al. Euphol, a tetracyclic triterpene, from Euphorbia tirucalli induces autophagy and sensitizes temozolomide cytotoxicity on glioblastoma cells. *Investig. New Drugs.* **2019**, *37*, 223–237. [CrossRef]
20. Silva, V.A.O.; Rosa, M.N.; Martinho, O.; Tanuri, A.; Lima, J.P.; Pianowski, L.F.; Reis, R.M. Modified ingenol semi-synthetic derivatives from Euphorbia tirucalli induce cytotoxicity on a large panel of human cancer cell lines. *Investig. New Drugs* **2019**, *37*, 1029–1035. [CrossRef]

21. Abreu, C.M.; Price, S.L.; Shirk, E.N.; Cunha, R.D.; Pianowski, L.F.; Clements, J.E.; Tanuri, A.; Gama, L. Dual role of novel ingenol derivatives from Euphorbia tirucalli in HIV replication: Inhibition of de novo infection and activation of viral LTR. *PLoS ONE* **2014**, *9*, e97257. [CrossRef] [PubMed]
22. Vasas, A.; Rédei, D.; Csupor, D.; Molnár, J.; Hohmann, J. Diterpenes from European Euphorbia Species Serving as Prototypes for Natural-Product-Based Drug Discovery. *Eur. J. Org. Chem.* **2012**, *2012*, 5115–5130. [CrossRef]
23. Beres, T.; Dragull, K.; Pospisil, J.; Tarkowska, D.; Dancak, M.; Biba, O.; Tarkowski, P.; Dolezal, K.; Strnad, M. Quantitative Analysis of Ingenol in Euphorbia species via Validated Isotope Dilution Ultra-high Performance Liquid Chromatography Tandem Mass Spectrometry. *Phytochem. Anal.* **2018**, *29*, 23–29. [CrossRef] [PubMed]
24. Lebwohl, M.; Swanson, N.; Anderson, L.L.; Melgaard, A.; Xu, Z.; Berman, B. Ingenol mebutate gel for actinic keratosis. *N. Engl. J. Med.* **2012**, *366*, 1010–1019. [CrossRef] [PubMed]
25. Berman, B. New developments in the treatment of actinic keratosis: Focus on ingenol mebutate gel. *Clin. Cosmet. Investig. Dermatol.* **2012**, *5*, 111–122. [CrossRef]
26. Gillespie, S.K.; Zhang, X.D.; Hersey, P. Ingenol 3-angelate induces dual modes of cell death and differentially regulates tumor necrosis factor-related apoptosis-inducing ligand-induced apoptosis in melanoma cells. *Mol. Cancer Ther.* **2004**, *3*, 1651–1658.
27. Hampson, P.; Chahal, H.; Khanim, F.; Hayden, R.; Mulder, A.; Assi, L.K.; Bunce, C.M.; Lord, J.M. PEP005, a selective small-molecule activator of protein kinase C, has potent antileukemic activity mediated via the delta isoform of PKC. *Blood* **2005**, *106*, 1362–1368. [CrossRef]
28. Wang, D.; Liu, P. Ingenol-3-Angelate Suppresses Growth of Melanoma Cells and Skin Tumor Development by Downregulation of NF-kappaB-Cox2 Signaling. *Med. Sci. Monit.* **2018**, *24*, 486–502. [CrossRef]
29. Duarte, N.; Gyemant, N.; Abreu, P.M.; Molnar, J.; Ferreira, M.J. New macrocyclic lathyrane diterpenes, from Euphorbia lagascae, as inhibitors of multidrug resistance of tumour cells. *Planta Med.* **2006**, *72*, 162–168. [CrossRef]
30. Vigone, A.; Tron, G.C.; Surico, D.; Baj, G.; Appendino, G.; Surico, N. Ingenol derivatives inhibit proliferation and induce apoptosis in breast cancer cell lines. *Eur. J. Gynaecol. Oncol.* **2005**, *26*, 526–530.
31. Mochly-Rosen, D.; Das, K.; Grimes, K.V. Protein kinase C, an elusive therapeutic target? *Nat. Rev. Drug Discov.* **2012**, *11*, 937–957. [CrossRef] [PubMed]
32. Paugh, B.S.; Qu, C.; Jones, C.; Liu, Z.; Adamowicz-Brice, M.; Zhang, J.; Bax, D.A.; Coyle, B.; Barrow, J.; Hargrave, D.; et al. Integrated molecular genetic profiling of pediatric high-grade gliomas reveals key differences with the adult disease. *J. Clin. Oncol.* **2010**, *28*, 3061–3068. [CrossRef] [PubMed]
33. Suffness, M. Assays related to cancer drug discovery. *Methods Plant. Biochem* **1990**, *6*, 71–133.
34. Blanco-Molina, M.; Tron, G.C.; Macho, A.; Lucena, C.; Calzado, M.A.; Munoz, E.; Appendino, G. Ingenol esters induce apoptosis in Jurkat cells through an AP-1 and NF-kappaB independent pathway. *Chem. Biol.* **2001**, *8*, 767–778. [CrossRef]
35. Serova, M.; Ghoul, A.; Benhadji, K.A.; Faivre, S.; Le Tourneau, C.; Cvitkovic, E.; Lokiec, F.; Lord, J.; Ogbourne, S.M.; Calvo, F.; et al. Effects of protein kinase C modulation by PEP005, a novel ingenol angelate, on mitogen-activated protein kinase and phosphatidylinositol 3-kinase signaling in cancer cells. *Mol. Cancer Ther.* **2008**, *7*, 915–922. [CrossRef] [PubMed]
36. Lizarte Neto, F.S.; Tirapelli, D.P.; Ambrosio, S.R.; Tirapelli, C.R.; Oliveira, F.M.; Novais, P.C.; Peria, F.M.; Oliveira, H.F.; Carlotti Junior, C.G.; Tirapelli, L.F. Kaurene diterpene induces apoptosis in U87 human malignant glioblastoma cells by suppression of anti-apoptotic signals and activation of cysteine proteases. *Braz. J. Med. Biol. Res.* **2013**, *46*, 71–78. [CrossRef]
37. Cavalcanti, B.C.; Costa-Lotufo, L.V.; Moraes, M.O.; Burbano, R.R.; Silveira, E.R.; Cunha, K.M.; Rao, V.S.; Moura, D.J.; Rosa, R.M.; Henriques, J.A.; et al. Genotoxicity evaluation of kaurenoic acid, a bioactive diterpenoid present in Copaiba oil. *Food Chem. Toxicol.* **2006**, *44*, 388–392. [CrossRef]
38. Hasler, C.M.; Acs, G.; Blumberg, P.M. Specific binding to protein kinase C by ingenol and its induction of biological responses. *Cancer Res.* **1992**, *52*, 202–208.
39. Yun, C.W.; Lee, S.H. The Roles of Autophagy in Cancer. *Int. J. Mol. Sci.* **2018**, *19*, 3466. [CrossRef]
40. Noonan, J.; Zarrer, J.; Murphy, B.M. Targeting Autophagy in Glioblastoma. *Crit. Rev. Oncog.* **2016**, *21*, 241–252. [CrossRef]

41. Kanzawa, T.; Germano, I.; Komata, T.; Ito, H.; Kondo, Y.; Kondo, S. Role of autophagy in temozolomide-induced cytotoxicity for malignant glioma cells. *Cell Death Differ.* **2004**, *11*, 448–457. [CrossRef] [PubMed]
42. Kaza, N.; Kohli, L.; Roth, K.A. Autophagy in brain tumors: A new target for therapeutic intervention. *Brain Pathol.* **2012**, *22*, 89–98. [CrossRef] [PubMed]
43. Lomonaco, S.L.; Finniss, S.; Xiang, C.; Decarvalho, A.; Umansky, F.; Kalkanis, S.N.; Mikkelsen, T.; Brodie, C. The induction of autophagy by gamma-radiation contributes to the radioresistance of glioma stem cells. *Int. J. Cancer* **2009**, *125*, 717–722. [CrossRef] [PubMed]
44. Johannes, F.J.; Prestle, J.; Eis, S.; Oberhagemann, P.; Pfizenmaier, K. PKCu is a novel, atypical member of the protein kinase C family. *J. Biol. Chem.* **1994**, *269*, 6140–6148.
45. Valverde, A.M.; Sinnett-Smith, J.; Van Lint, J.; Rozengurt, E. Molecular cloning and characterization of protein kinase D: A target for diacylglycerol and phorbol esters with a distinctive catalytic domain. *Proc. Natl. Acad. Sci. USA* **1994**, *91*, 8572–8576. [CrossRef]
46. Patel, R.; Win, H.; Desai, S.; Patel, K.; Matthews, J.A.; Acevedo-Duncan, M. Involvement of PKC-iota in glioma proliferation. *Cell Prolif.* **2008**, *41*, 122–135. [CrossRef]
47. Ersvaer, E.; Kittang, A.O.; Hampson, P.; Sand, K.; Gjertsen, B.T.; Lord, J.M.; Bruserud, O. The protein kinase C agonist PEP005 (ingenol 3-angelate) in the treatment of human cancer: A balance between efficacy and toxicity. *Toxins* **2010**, *2*, 174–194. [CrossRef]
48. Kedei, N.; Lundberg, D.J.; Toth, A.; Welburn, P.; Garfield, S.H.; Blumberg, P.M. Characterization of the interaction of ingenol 3-angelate with protein kinase C. *Cancer Res.* **2004**, *64*, 3243–3255. [CrossRef]
49. Kulkosky, J.; Sullivan, J.; Xu, Y.; Souder, E.; Hamer, D.H.; Pomerantz, R.J. Expression of latent HAART-persistent HIV type 1 induced by novel cellular activating agents. *AIDS Res. Hum. Retroviruses* **2004**, *20*, 497–505. [CrossRef]
50. Balca-Silva, J.; Matias, D.; do Carmo, A.; Girao, H.; Moura-Neto, V.; Sarmento-Ribeiro, A.B.; Lopes, M.C. Tamoxifen in combination with temozolomide induce a synergistic inhibition of PKC-pan in GBM cell lines. *Biochim. Biophys. Acta* **2015**, *1850*, 722–732. [CrossRef]
51. Cruvinel-Carloni, A.S.R.; Torrieri, R.; Bidinotto, L.T.; Berardinelli, G.N.; Oliveira-Silva, V.A.; Clara, C.A.; de Almeida, G.C.; Martinho, O.; Squire, J.A.; Reis, R.M. Molecular characterization of short-term primary cultures and comparison with corresponding tumor tissue of Brazilian glioblastoma patients. *Transl. Cancer Res.* **2017**, *6*, 332–345.
52. Dirks, W.G.; Faehnrich, S.; Estella, I.A.; Drexler, H.G. Short tandem repeat DNA typing provides an international reference standard for authentication of human cell lines. *ALTEX* **2005**, *22*, 103–109.
53. Konecny, G.E.; Glas, R.; Dering, J.; Manivong, K.; Qi, J.; Finn, R.S.; Yang, G.R.; Hong, K.L.; Ginther, C.; Winterhoff, B.; et al. Activity of the multikinase inhibitor dasatinib against ovarian cancer cells. *Br. J. Cancer* **2009**, *101*, 1699–1708. [CrossRef] [PubMed]
54. Martinho, O.; Silva-Oliveira, R.; Miranda-Goncalves, V.; Clara, C.; Almeida, J.R.; Carvalho, A.L.; Barata, J.T.; Reis, R.M. In Vitro and In Vivo Analysis of RTK Inhibitor Efficacy and Identification of Its Novel Targets in Glioblastomas. *Transl. Oncol.* **2013**, *6*, 187–196. [CrossRef] [PubMed]
55. Chou, T.C.; Talalay, P. Quantitative analysis of dose-effect relationships: The combined effects of multiple drugs or enzyme inhibitors. *Adv. Enzyme Regul.* **1984**, *22*, 27–55. [CrossRef]
56. Bruzzese, F.; Di Gennaro, E.; Avallone, A.; Pepe, S.; Arra, C.; Caraglia, M.; Tagliaferri, P.; Budillon, A. Synergistic antitumor activity of epidermal growth factor receptor tyrosine kinase inhibitor gefitinib and IFN-alpha in head and neck cancer cells in vitro and in vivo. *Clin. Cancer Res.* **2006**, *12*, 617–625. [CrossRef] [PubMed]

© 2019 by the authors. Licensee MDPI, Basel, Switzerland. This article is an open access article distributed under the terms and conditions of the Creative Commons Attribution (CC BY) license (http://creativecommons.org/licenses/by/4.0/).

Article

Dicentrine Potentiates TNF-α-Induced Apoptosis and Suppresses Invasion of A549 Lung Adenocarcinoma Cells via Modulation of NF-κB and AP-1 Activation

Chanatip Ooppachai [1], Pornngarm Limtrakul (Dejkriengkraikul) [1,2,3] and Supachai Yodkeeree [1,2,3,*]

1. Department of Biochemistry, Faculty of Medicine, Chiang Mai University, Chiang Mai 50200, Thailand; chanatip.fang@gmail.com (C.O.); pornngarm.d@cmu.ac.th (P.L.(D.))
2. Anticarcinogenesis and Apoptosis Research Cluster, Faculty of Medicine, Chiang Mai University, Chiang Mai 50200, Thailand
3. Center for Research and Development of Natural Products for Health, Chiang Mai University, Chiang Mai 50200, Thailand
* Correspondence: yodkeelee@hotmail.com

Academic Editor: Roberto Fabiani
Received: 22 October 2019; Accepted: 11 November 2019; Published: 13 November 2019

Abstract: Numerous studies have indicated that tumor necrosis factor-alpha (TNF-α) could induce cancer cell survival and metastasis via activation of transcriptional activity of NF-κB and AP-1. Therefore, the inhibition of TNF-α-induced NF-κB and AP-1 activity has been considered in the search for drugs that could effectively treat cancer. Dicentrine, an aporphinic alkaloid, exerts anti-inflammatory and anticancer activities. Therefore, we investigated the effects of dicentrine on TNF-α-induced tumor progression in A549 lung adenocarcinoma cells. Our results demonstrated that dicentrine effectively sensitizes TNF-α-induced apoptosis in A549 cells when compared with dicentrine alone. In addition, dicentrine increases caspase-8, -9, -3, and poly (ADP-ribose) polymerase (PARP) activities by upregulating the death-inducing signaling complex and by inhibiting the expression of antiapoptotic proteins including cIAP2, cFLIP, and Bcl-XL. Furthermore, dicentrine inhibits the TNF-α-induced A549 cells invasion and migration. This inhibition is correlated with the suppression of invasive proteins in the presence of dicentrine. Moreover, dicentrine significantly blocks TNF-α-activated TAK1, p38, JNK, and Akt, leading to reduced levels of the transcriptional activity of NF-κB and AP-1. Taken together, our results suggest that dicentrine could enhance TNF-α-induced A549 cell death by inducing apoptosis and reducing cell invasion due to, at least in part, the suppression of TAK-1, MAPK, Akt, AP-1, and NF-κB signaling pathways.

Keywords: TNF-α; dicentrine; apoptosis; metastasis; lung adenocarcinoma

1. Introduction

Lung cancer is one of the leading causes of cancer-associated mortality worldwide. Nonsmall-cell lung cancer (NSCLC) is responsible for more than 80% of all lung cancers. Despite the fact that advancements have been made in cancer therapies, the overall survival rate of patients has remained unchanged in recent years [1]. Several studies have provided evidence to support the hypothesis that the tissue damage caused by inflammation can initiate or promote the development of lung cancer [2,3]. Chronic inflammation has emerged as a key contributor of cancer cell survival, angiogenesis, and metastasis. Inflammatory cytokines, such as interleukin-6 (IL-6), IL-1, and tumor necrosis factor-alpha (TNF-α), can promote cancer progression [4–7]. Among them, TNF-α contributes to the survival and metastasis of lung cancer, while the level of TNF-α in the tumor tissues and serum obtained from patients with NSCLC has increased significantly along with the clinical stage of the tumor [8,9].

Accordingly, even more complicated roles for TNF-α in cancer cases have emerged. The anticancer property of TNF-α is mainly achieved by inducing cancer cell death. This pathway is initiated by TNFR1 internally signaling to form complex II, which consists of TRADD, RIP1, FADD, and caspase-8 [10–12]. Caspase-8 is autoactivated, leading to the initiation of the caspase cascade and the induction of apoptotic cell death [13,14]. On the other hand, TNF-α stimulates proliferation, survival, angiogenesis, and metastasis in most cancer cells that are resistant to TNF-α-induced cell death [15]. After binding to TNFR1, TNF-α can increase the expression of antiapoptotic (cIAPs, XIAP, Bcl-2, Bcl-xl, and cFLIP), angiogenic (VEGF and COX-2), and invasive (MMP-9, MT1-MMP, uPA, urokinase-type plasminogen activator receptor (uPAR), and intercellular adhesion molecule 1 (ICAM-1)) proteins via the TAK-1, MAPKs, Akt, IKK, AP-1, and NF-κB signaling pathways [10,16,17]. Therefore, the identification of sensitizing agents that are capable of suppressing TNF-α-induced survival signaling could be an attractive discovery in facilitating the enhancement of TNF-α-mediated apoptosis and tumor progression.

Dicentrine is an aporphine alkaloid found in the roots of L. megaphylla and several other plants [18]. Previous investigations have shown that dicentrine possesses multiple pharmacological activities, including platelet aggregation inhibition capabilities, antinociceptive and anticancer activities [19–21]. Recently, our previous findings have demonstrated that dicentrine inhibited the inflammation in lipopolysaccharide (LPS)-treated RAW 264.7 cells via the suppression of the AP-1, NF-κB, and MAPKs signaling pathways [22]. However, the effect of dicentrine on TNF-α-induced apoptosis and metastasis in lung cancer cells has not yet been elucidated.

In the current study, we have investigated the mechanism, by which dicentrine inhibits the TNF-α-induced expression and the survival of metastasis proteins. We have also determined the effects of dicentrine on the MAPKs, Akt, NF-κB, and AP-1 signaling pathways in TNF-α-induced A549 cells.

2. Results

2.1. Dicentrine Potentiates TNF-α-Induced Apoptosis in A549 Lung Adenocarcinoma Cells

To examine whether dicentrine enhanced TNF-α-induced cell death, A549 cells were incubated with dicentrine (0–40 μM) and cotreated with or without TNF-α (25 ng/mL) for 24 h. The cell viability was then determined by 3-(4,5-dimethylthiazol-2-yl)-2,5-diphenyltetrazolium bromide (MTT) assays. As is shown in Figure 1B, a treatment of A549 cells with dicentrine alone significantly decreased cell viability in a dose-dependent manner. However, a combined treatment of the cells with TNF-α and dicentrine at 25, 30, and 40 μM reduced the degrees of cell viability to 54.85%, 54.35%, and 45.15%, respectively, which significantly increased the level of cytotoxicity to a greater degree than the treatment with dicentrine alone. Next, we investigated whether dicentrine-potentiated TNF-α-induced cell death was associated with apoptosis by using propidium iodide (PI) staining assays and detecting a SubG1 cell population by flow cytometric analysis. As is shown in Figure 1C,D, the combined treatment of the cells with TNF-α and dicentrine at 25–40 μM significantly increased the number of apoptotic cells in a dose-dependent manner, when compared with the treatment of dicentrine alone.

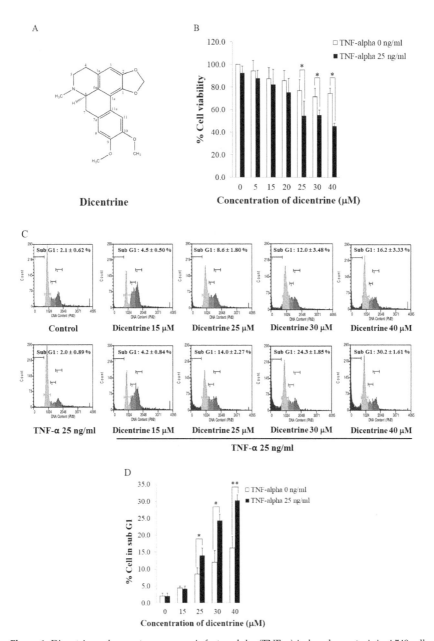

Figure 1. Dicentrine enhances tumor necrosis factor-alpha (TNF-α)-induced apoptosis in A549 cells. (**A**) Structure of dicentrine. A549 cells were pretreated with various concentrations of dicentrine for 4 h and then cotreated with 25 ng/mL of TNF-α for 24 h. (**B**) Cell viability was determined by 3-(4,5-dimethylthiazol-2-yl)-2,5-diphenyltetrazolium bromide (MTT) assays. (**C,D**) Cell cycle distribution was stained with propidium iodide (PI) and analyzed by flow cytometry to measure a SubG1 cell population, which represented the apoptotic cells. The experiments were performed in triplicate. The data are represented as mean ± S.D. * indicates $p < 0.05$, and ** indicates $p < 0.01$, compared with those treated with dicentrine alone.

2.2. Dicentrine Enhances TNF-α-Induced Apoptosis in a Caspase-Dependent Manner and Inhibits the Expression of Antiapoptotic Proteins

Since apoptosis is mainly mediated by caspase enzymes, we investigated whether dicentrine affected the TNF-α-induced proteolytic processing of caspase-8, caspase-9, caspase-3, and a caspase-3 substrate poly(ADP-ribose) polymerase (PARP) cleavage using western blot analysis. Notably, the treatment of A549 cells with TNF-α alone did not induce the proteolytic processing of caspase-8, caspase-3, and PARP, when compared with the vehicle control. However, the combined treatment of TNF-α and dicentrine resulted in an increase in the cleavage of caspase-8, caspase-9, caspase-3, and PARP in a dose-dependent manner (Figure 2A). Upon the stimulation of TNF-α, RIP could interact with the FADD protein, which in turn recruited procaspase-8 to form a death-inducing signaling complex (DISC). This complex then stimulated the caspase-8 activation that subsequently induced apoptosis. Coimmunoprecipitation was performed to determine whether dicentrine enhanced TNF-α-induced apoptosis accompanied by the increased DISC formation. As shown in Figure 2B, an increased interaction between the RIP and procaspase-8 in the combined treatment with dicentrine and TNF-α was observed when compared with that in the control. Overexpression of antiapoptotic proteins, such as cIAP2, c-FLIP, and Bcl-xl, has been linked with the inhibition of TNF-α-induced apoptosis. Therefore, we examined whether dicentrine could modulate the TNF-α-induced expression of these antiapoptotic proteins. As shown in Figure 2C,D, the induction of cFLIPs, cIAP2, and Bcl-XL by TNF-α was reduced by dicentrine in a dose-dependent manner. These results showed that dicentrine effectively enhanced the apoptotic effects of TNF-α due to the downregulation of antiapoptotic proteins.

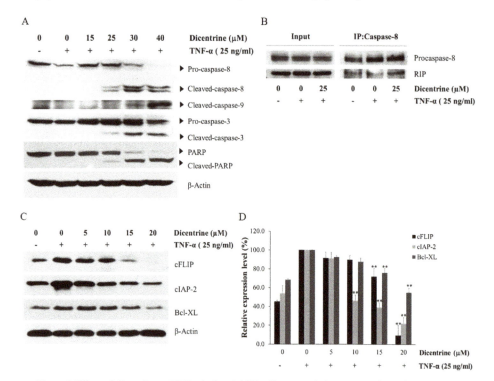

Figure 2. Effects of dicentrine on TNF-α-induced A549 cells apoptosis in a caspase-dependent manner and the expression of antiapoptotic proteins. A549 cells were pretreated with various concentrations of dicentrine for 4 h and then cotreated with 25 ng/mL of TNF-α to 24 h. After the combined treatment

with dicentrine (0–40 µM) and TNF-α, the whole cell extracts were prepared and analyzed by the western blot analysis to detect the expression of caspase-8, -9, -3, and poly(ADP-ribose) polymerase (PARP) (**A**). The complex between procaspase-8 and receptor-interacting protein (RIP) was determined by coimmunoprecipitation after incubating with dicentrine (25 µM) for 4 h and then cotreatment with 25 ng/mL of TNF-α for 12 h (**B**). The levels of antiapoptotic proteins such as cFLIPs, cIAP-2, and Bcl-XL was measure after cotreated with dicentrine (0–20 µM) and TNF-α. (**C,D**). The data are represented as mean ± S.D. ** indicates $p < 0.01$, when compared to those treated with TNF-α alone.

2.3. Dicentrine Inhibits TNF-α-Induced A549 Cells Invasion and Migration

Because TNF-α plays an important role in lung cancer metastasis, the effects of dicentrine on TNF-α-induced A549 cells invasion and migration were investigated. The results revealed that the invasive cells with the TNF-α treatment were increased by 4.52-fold when compared with those treated with the control, whereas dicentrine at 10–20 µM significantly decreased TNF-α-induced A549 cells invasion in a dose-dependent manner (Figure 3A). Moreover, cell migration using the treatment of the cell with TNF-α alone was increased to 2.52-fold when compared with that with the control, while the dicentrine treatment blocked the TNF-α-mediated migration in a dose-dependent manner (Figure 3B). Therefore, dicentrine may suppress the TNF-α-induced migration and the invasion of A549 cells.

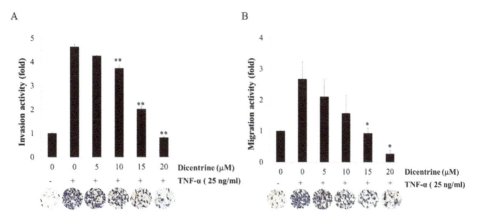

Figure 3. Effects of dicentrine on TNF-α-induced invasion and migration. The matrix gel was coated on membrane filters for invasion assays (**A**), and the gelatin was coated for migration assays (**B**). A549 cells with the number of 1.25×10^5 cells were cultures in the upper chamber and incubated with various concentrations of dicentrine (0–20 µM) in a Dulbecco's modified Eagle's medium (DMEM)-free medium, and the lower chamber was filled with 25 ng/mL of TNF-α. After 24 h of incubation, the cells that actively migrated to the lower surface of the filter were determined. The data are represented as mean ± S.D. of three independent experiments. Sample groups were significantly different from the TNF-α-treated group (* indicates $p < 0.05$, and ** indicates $p < 0.01$).

2.4. Dicentrine Inhibits TNF-α-Induced Expression of Metastasis-Associated Proteins

In the process of cancer metastasis, MT1-MMP, MMP-9, uPAR, ICAM-1, and Cox-2 are responsible for cell migration, invasion, and adhesion. The expression of these proteins was upregulated by TNF-α in various types of cancer cells including lung cancer. Therefore, the effects of dicentrine on TNF-α-induced expression of MT1-MMP, uPAR, ICAM-1, and Cox-2 in A549 cells were detected by the western blot analysis. As shown in Figure 4A,B, dicentrine at 10–20 µM significantly decreased the enhancement of TNF-α-induced MT1-MMP, uPAR, and COX-2 expression. In addition, dicentrine at 15 and 20 µM reduced the expression of ICAM-1. We next used gelatin zymography assays to examine the inhibitory effect of dicentrine on the TNF-α-induced MMP-9 secretion. As is shown in

Figure 4C,D, the TNF-α-induced MMP-9 secretion was significantly inhibited in the presence of 15 and 20 µM of dicentrine.

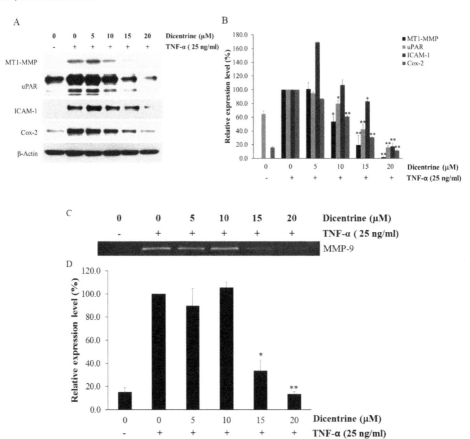

Figure 4. Dicentrine suppresses TNF-α-induced expression of metastatic protein. A549 cells were pretreated with various concentrations of dicentrine (0–20 µM) for 4 h and then cotreatment with 25 ng/mL of TNF-α for 24 h. The whole cell extracts were prepared and analyzed by the western blot analysis using antibodies against cell metastatic proteins (MT1-MMP, uPAR, ICAM-1, and Cox-2) (**A,B**). The culture supernatants of the treated cells were collected, and the secretions of MMP-9 were analyzed by gelatin zymography (**C,D**). The data are represented as mean ± S.D. * indicates $p < 0.05$, and ** indicates $p < 0.01$, when compared to those treated with TNF-α alone.

2.5. Dicentrine Inhibits TNF-α-Induced NF-κB and AP-1 Activation

The activation of NF-κB and AP-1 transcriptional activity in several types of cancer cells can promote tumor progression by regulation of many genes in terms of antiapoptosis, cell proliferation, and cell invasion. To investigate whether dicentrine affected the TNF-α-induced NF-κB and AP-1 activation, the nucleus translocation and the phosphorylation of NF-κB and AP-1 were determined. As shown in Figure 5A,B, dicentrine at 10–20 µM could significantly inhibit the TNF-α-induced p65 phosphorylation and block the TNF-α-induced nuclear translocation of p65 in a dose-dependent manner. We next tested the regulation of dicentrine on the transcription activity of AP-1. The treatment of the cells with TNF-α alone enhanced the nucleus translocation and the phosphorylation of c-Jun (AP-1). Dicentrine could significantly inhibit the TNF-α-induced AP-1 translocation and could effectively block the TNF-α-induced phosphorylation of AP-1 in a dose-dependent manner (Figure 5C,D). These results indicated that dicentrine inhibited the TNF-α-induced activation of NF-κB and AP-1 through the inhibition of the phosphorylation and the nuclear translocation of p65 and AP-1.

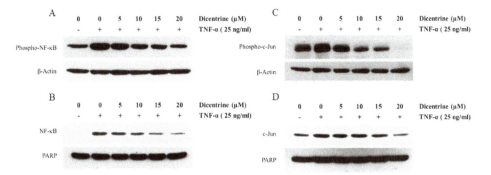

Figure 5. Effects of dicentrine on TNF-α-induced NF-κB and AP-1 activation. A549 cells were pretreated with difference concentrations of dicentrine (0–20 μM/mL) for 12 h and then cotreated with 25 ng/mL of TNF-α for 45 min. The phosphorylation levels of NF-κB and AP-1 in cytoplasmic extracts were detected by the western blot analysis (**A,C**), and the nuclear extracts were prepared to analyze the nuclear translocation (**B,D**). Data from a typical experiment are presented here, while similar results were obtained from three independent experiments.

2.6. Effects of Dicentrine on TNF-α-Induced TAK-1, IκB-α, Akt, and MAPKs Signaling Pathways

It has been reported that TNF-α-induced NF-κB and AP-1 activation is mediated through the sequential interaction of the TNF receptor with TRADD, TRAF2, and TAK-1, which then leads to the phosphorylation of IKK and induces the degradation of IκB-α, along with the phosphorylation of MAPKs and the PI3K/Akt signaling pathway. Therefore, the effects of dicentrine on the TNF-α-induced activation of TAK-1, IKK, IκB-α, Akt, and MAPKs, including Erk1/2, p38, and JNK, were investigated by the western blot analysis. As shown in Figure 6A, the treatment of A549 cells with the TNF-α-induced phosphorylation of TAK-1 and dicentrine inhibited the activation in a dose-dependent manner. The NF-κB activation by TNF-α was mediated via the IKK signaling pathway, resulting in the IκB-α degradation. To examine the effects of dicentrine on IκB-α activity, we determined whether dicentrine affected the TNF-α-induced IκB-α degradation. As is shown in Figure 6B, dicentrine blocked the TNF-α-dependent IκB-α degradation. Since the IKK complex acts as a convergence point for a variety of activating signals for NF-κB and plays a critical role in degradation of IκB-α, we investigated whether dicentrine inhibited the TNF-α-induced phosphorylation of IKK. As shown in Figure 6C, dicentrine at 20 μM suppressed the TNF-α-induced phosphorylation of IKK. Moreover, we investigated the effects of dicentrine on the TNF-α-stimulated phosphorylation of p38, JNK, and Erk1/2. As is shown in Figure 6D, dicentrine at 15 and 20 μM inhibited the TNF-α-induced phosphorylation of JNK and the p38 signaling pathway, whereas dicentrine had no effect on the TNF-α-induced phosphorylation of the ERK1/2 signaling pathway. On the other hand, the treatment with dicentrine alone also inhibited the TNF-α-induced phosphorylation of the Akt signaling pathway (Figure 6E).

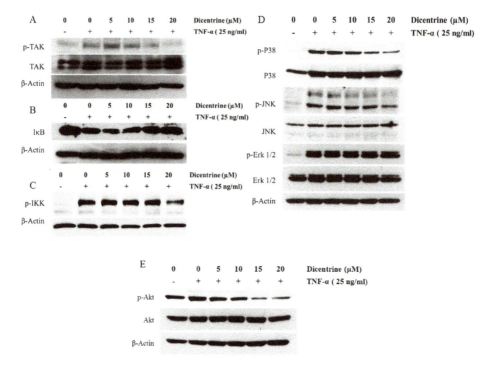

Figure 6. Effects of dicentrine on TNF-α-induced TAK, IKK, Akt, and MAPKs signaling pathways. A549 cells were pretreated with difference concentrations of dicentrine (0–20 µM/mL) for 12 h and then cotreated with 25 ng/mL of TNF-α for 15 min. The whole cell lysate was prepared for the measurements of phosphorylated and nonphosphorylated forms of TAK (**A**), IκB (**B**), IKK (**C**), MAPKs (**D**) and Akt (**E**) by western blot analysis. Data from a typical experiment are depicted here, and similar results were obtained in three independent experiments.

3. Discussion

Dicentrine is an alkaloid that is found in various medicinal plants. It displays activity against many types of cancer by regulating cell cycles, inhibiting topoisomerase II, and inducing apoptosis [18–21]. Despite its various pharmacological activities, the molecular mechanism of dicentrine on TNF-α-induced tumor progression has not been adequately elucidated. The present study was designed to investigate the effect of dicentrine on the enhancement of TNF-α-induced A549 lung adenocarcinoma cell death and to investigate the role of dicentrine as a potent inhibitor of TNF-α-induced cell invasion. Molecular mechanisms of these phenomena have not yet been explored.

Activation of the TNFR1 death receptor by TNF-α can induce either cell survival or cell death, depending on the events that take place downstream of TNFR1 activation [11,17]. A recent study has shown that most cancer cells are resistant to TNF-α-induced cell death, which is known to be involved with the overexpression of antiapoptotic outcomes and the survival of proteins leading to the inhibition of the apoptotic pathway [23,24]. Therefore, the modulation of TNF-α-mediated survival signals may result in the sensitization of cancer cells to the TNF-α-induced apoptosis. Our result from the MTT assays revealed that dicentrine sensitized A549 cells to the TNF-α-induced cell death. TNF-α promotes cancer cell death by inducing apoptosis, necroptosis, and autophagy, depending on the conditions of the cells [25]. The apoptotic DNA fragmentation is being used as one of markers for apoptosis, and fractional DNA content was detected in SubG1 cells. Here, we indicated that dicentrine enhanced TNF-α-induced apoptosis by increasing the SubG1 cell population. TNF-α also induced cell apoptosis via extrinsic and intrinsic pathways by increasing the activation of caspases activity. To confirm the potential effect of dicentrine on TNF-α-mediated apoptosis in A549 cells, we investigated how dicentrine affected the TNF-α-induced activation of caspase-8, -3, -9, and PARP cleavage. The levels of caspase-8, -3, and -9 activation and the accumulation of the cleaved PARP were markedly increased during the combined treatment with TNF-α and dicentrine. During the final phase of apoptosis, caspases cleaved several proteins that are necessary for cell survival and function. Among them, PARP was cleaved by caspase-3 into the fragments. This cleavage is necessary to eliminate PARP activation in response to DNA fragmentation and prevent futile attempts of DNA repair [26]. Thus, PARP play a central role in apoptosis determining the cell fate. Taking the results together, it is possible to suggest that dicentrine potentiates TNF-α-induced apoptosis by activating caspases activity.

The TNF-α-induced apoptosis is initiated by the TNFR1 internalization into endosomes to form complex II or the DISC [11]. The DISC consists of TRADD, RIP-1, FADD, and procaspase-8. The unubiquitinated form of RIP1 can be dissociated from TNFR1, with or without TRADD, to interact with FADD, which in turn recruits procaspase-8 to form the DISC [16,27–29]. The formation of DISC triggers the processing and the activation of caspase-8, leading to the initiation of the caspase cascade and ultimately cell death [13,14]. Therefore, we investigated whether the enhancement effects of dicentrine on TNF-α-induced apoptosis were associated with the formation of DISC. The results from coimmunoprecipitation revealed that the combined treatment of A549 cells with dicentrine and TNF-α increased the complex formation of procaspase-8 and RIP-1. These outcomes indicated that dicentrine potentiated TNF-α-mediated apoptosis by enhancing the DISC formation. The active caspase-8 triggers a caspase cascade by the cleavage of caspase-3. Moreover, caspase-8 cleaves Bid into truncated Bid, which initiates the mitochondrial apoptosis pathway leading to the release of cytochrome c from the mitochondria. Then, cytochrome c is associated with procaspase-9 and Apat-1. This complex processes caspase-9 to an active form. Once activated, caspase-9 continues to cleave caspase-3 and eventually leads to apoptosis [13]. Our result demonstrated that dicentrine enhanced the TNF-α-induced activation of capase-9, which is one of key markers in the intrinsic pathway of apoptosis. Several studies have shown that the overexpression of antiapoptotic proteins including cIAP-2, cFLIPs, Bcl-2, and Bcl-XL in tumor cells has been associated with the inhibition of TNF-α-induced apoptosis [30–32]. Notably, cIAP-2 mediates ubiquitylation by controlling RIP-1 activity, thereby preventing TNF-α-induced cell death [33,34]. Moreover, c-FLIP has been identified as an inhibitor of TNF-α-mediated apoptosis by preventing the homodimerization of caspase-8 in the DISC [35,36]. Therefore, to explain the mechanism, by which dicentrine enhances TNF-α-induced apoptosis, we have examined the modulatory effects of dicentrine on the TNF-α-induced expression of the antiapoptotic capability. We found that dicentrine inhibited the TNF-α-induced expression of cIAP-2, cFLIP, and Bcl-XL. Together, these results suggested that dicentrine enhanced the TNF-α induced cell death, at least in part, by downregulating cIAP-2, cFLIP, and Bcl-XL, leading to the induction of the cell apoptosis in intrinsic and extrinsic pathways.

Tumor metastasis consists of multiple steps, including those involving loose cell–cell junctions, extracellular matrix (ECM) degradation, adherence to the surrounding ECM, and eventually migration through the ECM to the distance sites. The disruption of the basement membranes and the ECM proteins by proteolytic enzymes is an important step. MMPs and uPA are the key of protease in the ECM degradation. High levels of MMPs and uPA are closely linked to the promotion of cancer metastasis [8,15,37]. Moreover, uPAR is involved in lung cancer invasion and could enhance cell proliferation, angiogenesis, and migration. Adhesion molecules such as ICAM-1 are essential for the adhesion of tumor cells to endothelial cells and thus mediate tumor cell metastasis [38,39]. Our results demonstrated that dicentrine suppressed the TNF-α-induced A549 cell invasion and migration. Moreover, dicentrine reduced the degree of TNF-α-induced expression of invasive genes, including MMP-9, MT1-MMP, ICAM-1, and uPAR in A549 cells. In addition, Cox-2 overexpression plays a key role in tumorigenesis by stimulating angiogenesis, cell proliferation, and invasion. Cox-2 is a rate-limiting enzyme catalyzing the synthesis of prostaglandin E2 (PGE2), which induces MMPs and uPA expression [40,41]. Therefore, the modulation of Cox-2 activity is considered to be antitumor compounds. Here, we demonstrated that dicentrine reduced the TNF-α-induced expression of Cox-2 in a dose-dependent manner. This result is similar to that of our previous findings, which have stated that dicentrine reduced the levels of Cox-2 and PGE2 expression in LPS-induced macrophage cells [22].

NF-κB and AP-1 are important transcription factors that play an essential role in the progression process [41]. NF-κB controls the expression of the cell survival gene products *cIAP-2*, *Bcl-2*, *Bcl-xl*, and *cFLIP* and the invasive gene products *MMP-9*, *MT1-MMP*, *ICAM-1*, *uPA*, *uPAR*, and *Cox-2* [42,43]. In most resting cells, NF-κB exists in the cytosol in an inactive form that is associated with IκB-α. Once the cells are exposed to TNF-α, IκB-α is phosphorylated by IKK and subsequently degraded, leading to the phosphorylation and the nuclear translocation of NF-κB. The activated NF-κB then binds to the consensus sequences, activating the expression of the target genes [15,43]. Therefore, we investigated the inhibitory activity of dicentrine on the TNF-α-induced IκB-α degradation, the phosphorylation of IKK, and the NF-κB activity. The data presented herein indicate that dicentrine prevented the degradation of IκB-α and the phosphorylation of IKK and reduced the NF-κB activity by inhibiting the TNF-α-induced NF-κB phosphorylation and the nucleus translocation. This result is in accordance with the outcomes of previous studies, which also demonstrated that the inhibition of NF-κB activity can potentiate the TNF-α-induced cell death and inhibit the cancer cell invasion. AP-1 has been shown to play a role in regulating cancer cell proliferation and survival. AP-1 also controls the genes expression of *MMP-9*, *MT1-MMP*, *Cox-2*, *uPA*, and *uPAR*. The activity of AP-1 is regulated at an expression level and is involved in post-translation modification via phosphorylation. The activated AP-1 is translocated to the nucleus and is bound to the DNA consensus sequence. The inhibition of AP-1 activity have been shown to reduce cancer cell invasion and survival [44,45]. In this study, we investigated the activation of AP-1 by observing the phosphorylation and the translocation of c-Jun in the TNF-α-induced A549 cells. Our results found that dicentrine reduced the TNF-α-induced c-Jun phosphorylation and translocation to the nucleus of A549 cells. These results implied that dicentrine could reduce the level of survival and metastasis proteins by the inhibition of AP-1 and NF-κB activation.

Next, we investigated how dicentrine regulated the activity of NF-κB and AP-1 by examining its effects on several forms of kinase that function upstream of NF-κB and AP-1, such as MAPKs, Akt, and TAK1. Recent studies have indicated that TAK1 is essential for the TNF-α-induced activation of NF-κB via the activation of IKK activity. In addition, TAK1 has been implicated in TNF-α-induced MAPK activation [12,15,44]. Indeed, our study showed, for the first time, that the TNF-α-induced TAK-1 phosphorylation was inhibited by dicentrine. Notably, TNF-α can activate Akt and three MAPK cascades, including ERK1/2, p38, and JNK, which modulate the transcriptional activity of AP-1 and NF-κB. Accumulating evidence indicates that the Akt and MAPKs signaling pathways have been involved with cancer cell survival and metastasis by upregulating the expression of survival and invasive proteins [46]. On the other hand, it has been reported that high levels of reactive oxygen species promoted cancer cells apoptosis by activating the MAPKs signaling pathway [47]. Thus, the activation of MAPKs signaling pathway can induce either cancer cell death or cancer cell survival, depending on the condition of induction. In lung cancer cells, TNF-α-induced inflammation, survival, and metastasis via the MAPKs signaling pathway have been reported [48,49]. Therefore, these experiments were performed to determine whether dicentrine regulated TNF-α to stimulate the activities of Akt and MAPKs. Our results have shown that dicentrine prevented the phosphorylation of p38, JNK, and Akt. These results are consistent with those of previous reports that found the inhibitors of the PI3K/Akt and MAPKs signaling pathways can cause the cell death that is associated with apoptosis and can then reduce the invasion of cancer cells.

4. Materials and Methods

4.1. Materials

Dulbecco's modified Eagle's medium (DMEM), penicilin-streptomycin, and trypsin-ethylenediaminetetra acetic acid (EDTA) were purchased from GIBCO-BRL (Grand Island, NY, USA). Fetal bovine serum (FBS) was purchased from Hyclone (Logan, UT, USA). Gelatin, PI, and MTT were purchased from Sigma-Aldrich (St. Louis, MO, USA). Antibodies specific to caspase-3, caspase-8, caspase-9, cFILP, cIAP-2, Bcl-XL, COX-2, phospho-NF-κB, phospho-c-Jun, c-Jun, phospho-IKK, phospho-IκB, phospho-p38, p38, phospho-JNK, JNK, phospho-ERK1/2, ERK1/2, phospho-AKT, AKT, phospho-TAK, TAK, and RIP were purchased from Cell Signaling Technology Inc. (Beverly, MA, USA). NF-κB, PARP, MT1-MMP, uPAR, ICAM-1, and VEGF were purchased from Santa Cruz Biotechnology (Santa Cruz, CA, USA). Cyclin D1 was purchased from Milipore. The Can Get Signal® Immunoreaction Enhancer Solution was purchased from Toyobo (Osaka, Japan). Matrigel was purchased from Becton Dickinson (Bedford, MA, USA). Dicentrine was ordered from Chengdu Biopurify Phytochemicals Ltd. (Sichuan, China).

4.2. Cell Culture

A549 lung adenocarcinoma cells (American Type Culture Collection (ATCC), CCL-185) were cultured in DMEM supplemented with 10% FBS, 100 U/mL of penicillin, and 100 μg/mL of streptomycin. The cell cultures were maintained in a humidified incubator with an atmosphere comprised of 95% air and 5% CO_2 at 37 °C. For the dicentrine treatment, dicentrine was dissolved in dimethyl sulfoxide (DMSO) and diluted with the culture medium, and the final concentration of DMSO was less than 0.1% (v/v).

4.3. Cell Viability

The cytotoxic effect of dicrntrine to A549 lung adenocarcinoma cells were determined using MTT assays. Briefly, A549 cells with 5.5×10^3 cells per well were placed into 96-well plates containing DMEM with 10% FBS for 24 h. After that, the cells were treated with various dicentrine concentrations (0–40 μM) with or without 25 ng/mL of TNF-α and incubated for 24 h. Then, 15 μL of the MTT solution (5 mg/mL) were added to each well and incubated at 37 °C for 4 h. After the incubation, the formazan crystals in each well were dissolved in 200 μL of DMSO. The absorbance was measured using a microplate reader at 570 nm with a reference wavelength of 630 nm.

4.4. Cell Cycle Arrest Assay

A549 cells were treated with various concentrations of dicentrine (0–40 μM) with or without 25 ng/mL of TNF-α for 24 h. After that, the cells were fixed with ice-cold 70% (v/v) ethanol for 15 min. The staining of the nuclear DNA content was conducted by adding PI (50 μg/mL), 0.05% triton X-100, and RNAse A (25 μg/mL) in PBS, followed by incubation at 37 °C for 30 min in the dark. Cells were washed with cold PBS and resuspended in 500 μL of PBS. The samples were analyzed by an FACScan flow cytometer, and the DNA content in the SubG1 was representative of the apoptotic cells.

4.5. Cell Invasion and Migration Assay

The invasion and migration were tested using the modified Boyden chamber assay as described previously. Briefly, polyvinylpyrrolidone-free polycarbonate filters (Millipore, Carrigtwohill, Tullagreen) with a pore size of 8 μm were coated with gelatin (0.01%, w/v) for cell migration or with Matrigel (12.5 μg/50 μL) for the invasion assays. The A549 cells with the number of 1.25×10^5 were treated with various concentrations of dicentrine (0–20 μM/mL) and placed into the upper chamber for 24 h at 37 °C in 5% CO_2. The medium in the lower chamber contained a serum-free culture medium of NIH3T3 cells with or without TNF-α (25 ng/mL). The cells that had invaded the lower surface of the membrane were fixed with methanol and stained with 1% (w/v) toluidine blue. The cells were migrated to the lower surface of the filter measure by count of the cell.

4.6. Gelatin Zymography

The secretions of MMP-9 from the cells were analyzed by gelatin zymography as described before. The culture supernatant of the treated cell was collected and separated by 10% polyacrylamide gels containing 0.1% (w/v) of gelatin in nonreducing conditions. After electrophoresis, gels were washed twice with 2.5% Triton X-100 for 30 min at room temperature to remove sodium dodecyl sulfate (SDS). The gel was then incubated at 37 °C and kept for 18 h in an activating buffer (50 mM Tris-HCl, 200 mM NaCl, and 10 mM $CaCl_2$, pH 7.4). Gels were stained with Coomassie Brilliant Blue R (0.1%, w/v) and destained in 30% methanol with 10% acetic acid. The MMP-9 activity appeared as a clear band against a blue background

4.7. Coimmunoprecipitation and Western Blot

A549 cells with the number of 1×10^7 were treated with 25 μM/mL of dicentrine for 4 h and 25 ng/mL of TNF-α for 12 h. The cells were lysed in a lysis buffer (150 mM NaCl, 50mM Tris-Hcl, 1% NP-40, 2 mM EDTA, and a proteases inhibitor mixture; Sigma-Aldrich) for 20 min on ice. Cell lysates were cleared by centrifugation at 4 °C for 10 min at 12,000 rpm. After that, incubations with Dynabeads Protein A (Thermo Fisher Scientific, Norway) and polyclonal antiRIP1 occurred overnight at 4 °C. The target protein and its complex were washed twice with the lysis buffer. For the western blot experiments, the proteins were separated by the SDS-PAGE electrophoresis and transferred to a nitrocellulose membrane (GE Healthcare Ltd., UK). Immunoblot analyses were performed using specific primary antibodies overnight at 4 °C. After being washed with TBS containing 0.3% (v/w) Tween-20 (TBST buffer), the membrane was incubated with a secondary antibody for 2 h. The blots were detected with chemiluminescence (Thermo Fisher Scientific Inc., MA, USA).

4.8. Extraction of Nuclear and Whole Cell Lysate

The whole cell lysate were used to determine the expression levels of apoptotic, antiapoptotic, invasive, and signaling proteins, such as caspase-3, caspase-8, PARP-1, cFLIP, cIAP-2, cyclin D1, Bcl-XL, MT1-MMP, uPAR, ICAM-1, Cox-2, phospho-TAK, TAK, phospho-IKK, phospho-IκB, phospho-p38, p38, phospho-JNK, JNK, phospho-ERK1/2, ERK1/2, phospho-AKT, and AKT in A549 cells. Briefly, the cells were pretreated with various concentrations of dicentrine for 4 h and cotreated with or without 25 ng/mL of TNF-α for 24 h. The treated cells were washed with ice-cold PBS and extracted by the RIPA buffer for 20 min on ice. The collected supernatant fraction by centrifugation at 12,000 rpm for 10 min, and the protein concentration was determined with the Bradford protein assay.

Nuclear extraction was used to determine the levels of NF-κB and AP-1 proteins. The treated cells were collected and washed twice with the ice-cold PBS. The cell pellet was extracted by a hypotonic buffer (10 mM HEPES, pH 7.9, 10 mM KCL, 0.1 mM EDTA, 0.1 mM EGTA, 1 mM dithiothreitol, 0.5 mM phenylmethylsulfonyl fluoride, 1 μg/mL leupeptin, and 1 μg/mL aprotinin) for 20 min on ice, after which 5 μL of 10% NP-40 was added. The tube were agitated on a vortex for 15 s and centrifuged at 12,000 rpm for 5 min. The supernatant was collected and was representative of the cytoplasm extract. The nuclear pellets were suspended in an ice-cold nuclear extraction buffer (20 mM HEPES, pH 7.9, 0.4 M NaCl, 1 mM EDTA, 1 mM EGTA, 1 mM dithiothreitol, 1 mM phenylmethylsulfonylfluoride, 2.0 μg/mL leupeptin, and 2.0 μg/mL aprotinin) for 25 min on ice and centrifuged at 12,000 rpm for 10 min. The supernatant was collected and was representative of the nuclear fraction.

To determine the expression level of protein in the whole cell lysate, cytoplasm and nuclear faction were separated by the SDS-PAGE electrophoresis and transferred to the nitrocellulose membrane by electroblotting. The membrane was blocked with 5% skim milk in TBS containing 0.3% (v/v) Tween-20 for 1 h and then incubated with primary antibodies at 4 °C overnight. The membrane was incubated with a secondary antibody for 2 h and detected by chemiluminescence.

4.9. Statistical Analysis

All data are presented as mean ± S.D. Statistical analysis was analyzed with SPSS Software using one-way ANOVA with Dunnett's test. Statistical significance was determined at * $p < 0.05$ and ** $p < 0.01$.

5. Conclusions

In summary, our results suggested that dicentrine enhances TNF-α-induced A549 cell death by inducing apoptosis and by reducing cell invasion due to, at least in part, the suppression of the TAK-1, MAPKs, Akt, AP-1, and NF-κB signaling pathways. These results underscore the potential of dicentrine as a therapeutic agent in treatments of certain types of cancer, such as lung cancer, where TNF-α plays a major role in tumor progression.

Author Contributions: Conceptualization, S.Y. and P.L.(D.); methodology, C.O.; software, C.O.; validation, S.Y.; formal analysis, S.Y. and C.O.; data curation, C.O.; writing of original draft preparation, C.O. and S.Y.; writing of review and editing, P.L.(D.); visualization, S.Y.; supervision, S.Y. and P.L.(D.); project administration, S.Y. and P.L.(D.); funding acquisition, S.Y.

Funding: This work was supported by Faculty of Medicine Research Found, Chiang Mai University. (grant No. 014-2560).

Acknowledgments: This research study was supported by Chiang Mai University and the Center for Research and Development of Natural Products for Health, Chiang Mai University.

Conflicts of Interest: The authors declare that they hold no conflicts of interest.

References

1. Malhotra, J.; Malvezzi, M.; Negri, E.; La Vecchia, C.; Boffetta, P. Risk factors for lung cancer worldwide. *Eur. Respir. J.* **2016**, *48*, 889–902. [CrossRef] [PubMed]
2. Lee, G.; Walser, T.C.; Dubinett, S.M. Chronic inflammation, chronic obstructive pulmonary disease, and lung cancer. *Curr. Opin. Pulm. Med.* **2009**, *15*, 303–307. [PubMed]
3. Gomes, M.; Teixeira, A.L.; Coelho, A.; Araujo, A.; Medeiros, R. The role of inflammation in lung cancer. *Adv. Exp. Med. Biol.* **2014**, *816*, 1–23. [PubMed]
4. Aggarwal, B.B.; Shishodia, S.; Sandur, S.K.; Pandey, M.K.; Sethi, G. Inflammation and cancer: How hot is the link? *Biochem. Pharmacol.* **2006**, *72*, 1605–1621. [CrossRef]
5. Grivennikov, S.I.; Greten, F.R.; Karin, M. Immunity, inflammation, and cancer. *Cell* **2010**, *140*, 883–899. [CrossRef]
6. Mantovani, A. Molecular pathways linking inflammation and cancer. *Curr. Mol. Med.* **2010**, *10*, 369–373. [CrossRef]
7. Landskron, G.; De la Fuente, M.; Thuwajit, P.; Thuwajit, C.; Hermoso, M.A. Chronic inflammation and cytokines in the tumor microenvironment. *J. Immunol. Res.* **2014**, *2014*, 149185. [CrossRef]
8. Shang, G.S.; Liu, L.; Qin, Y.W. IL-6 and TNF-alpha promote metastasis of lung cancer by inducing epithelial-mesenchymal transition. *Oncol. Lett.* **2017**, *13*, 4657–4660. [CrossRef]
9. Derin, D.; Soydinc, H.O.; Guney, N.; Tas, F.; Camlica, H.; Duranyildiz, D.; Yasasever, V.; Topuz, E. Serum levels of apoptosis biomarkers, survivin and TNF-alpha in nonsmall cell lung cancer. *Lung Cancer* **2008**, *59*, 240–245. [CrossRef]
10. Idriss, H.T.; Naismith, J.H. TNF alpha and the TNF receptor superfamily: Structure-function relationship(s). *Microsc. Res. Tech.* **2000**, *50*, 184–195.
11. Chen, G.; Goeddel, D.V. TNF-R1 signaling: A beautiful pathway. *Science* **2002**, *296*, 1634–1635. [CrossRef]
12. Zelova, H.; Hosek, J. TNF-alpha signalling and inflammation: Interactions between old acquaintances. *Inflamm. Res.* **2013**, *62*, 641–651. [CrossRef]
13. Wang, L.; Du, F.; Wang, X. TNF-alpha induces two distinct caspase-8 activation pathways. *Cell* **2008**, *133*, 693–703. [CrossRef] [PubMed]
14. Lin, Y.; Devin, A.; Rodriguez, Y.; Liu, Z.G. Cleavage of the death domain kinase RIP by caspase-8 prompts TNF-induced apoptosis. *Genes Dev.* **1999**, *13*, 2514–2526. [CrossRef] [PubMed]

15. Balkwill, F. TNF-alpha in promotion and progression of cancer. *Cancer Metast. Rev.* **2006**, *25*, 409–416. [CrossRef] [PubMed]
16. Gaur, U.; Aggarwal, B.B. Regulation of proliferation, survival and apoptosis by members of the TNF superfamily. *Biochem. Pharmacol.* **2003**, *66*, 1403–1408. [CrossRef]
17. Wajant, H.; Pfizenmaier, K.; Scheurich, P. Tumor necrosis factor signaling. *Cell Death Differ.* **2003**, *10*, 45–65.
18. Teng, C.M.; Yu, S.M.; Ko, F.N.; Chen, C.C.; Huang, Y.L.; Huang, T.F. Dicentrine, a natural vascular alpha 1-adrenoceptor antagonist, isolated from Lindera megaphylla. *Br. J. Pharmacol.* **1991**, *104*, 651–656. [CrossRef]
19. Lin, H.F.; Huang, H.L.; Liao, J.F.; Shen, C.C.; Huang, R.L. Dicentrine Analogue-Induced G2/M Arrest and Apoptosis through Inhibition of Topoisomerase II Activity in Human Cancer Cells. *Planta Med.* **2015**, *81*, 830–837. [CrossRef]
20. Montrucchio, D.P.; Cordova, M.M.; Santos, A.R. Plant derived aporphinic alkaloid S-(+)-dicentrine induces antinociceptive effect in both acute and chronic inflammatory pain models: Evidence for a role of TRPA1 channels. *PLoS ONE* **2013**, *8*, e67730. [CrossRef]
21. Huang, R.L.; Chen, C.C.; Huang, Y.L.; Ou, J.C.; Hu, C.P.; Chen, C.F.; Chang, C. Anti-tumor effects of d-dicentrine from the root of Lindera megaphylla. *Planta Med.* **1998**, *64*, 212–215. [CrossRef] [PubMed]
22. Yodkeeree, S.; Ooppachai, C.; Pompimon, W.; Limtrakul (Dejkriengkraikul), P. O-Methylbulbocapnine and Dicentrine Suppress LPS-Induced Inflammatory Response by Blocking NF-kappaB and AP-1 Activation through Inhibiting MAPKs and Akt Signaling in RAW264.7 Macrophages. *Biol. Pharm. Bull.* **2018**, *41*, 1219–1227. [CrossRef] [PubMed]
23. Wang, C.Y.; Mayo, M.W.; Korneluk, R.G.; Goeddel, D.V.; Baldwin, A.S., Jr. NF-kappaB antiapoptosis: Induction of TRAF1 and TRAF2 and c-IAP1 and c-IAP2 to suppress caspase-8 activation. *Science* **1998**, *281*, 1680–1683. [CrossRef] [PubMed]
24. Calabrese, F.; Carturan, E.; Chimenti, C.; Pieroni, M.; Agostini, C.; Angelini, A.; Crosato, M.; Valente, M.; Boffa, G.M.; Frustaci, A.; et al. Overexpression of tumor necrosis factor (TNF)alpha and TNFalpha receptor I in human viral myocarditis: Clinicopathologic correlations. *Mod. Pathol.* **2004**, *17*, 1108–1118. [CrossRef] [PubMed]
25. Edinger, A.L.; Thompson, C.B. Death by design: Apoptosis, necrosis and autophagy. *Curr. Opin. Cell Biol.* **2004**, *16*, 663–669. [PubMed]
26. Munoz-Gamez, J.A.; Rodriguez-Vargas, J.M.; Quiles-Perez, R.; Aguilar-Quesada, R.; Martin-Oliva, D.; de Murcia, G.; de Murcia, J.M.; Almendros, A.; de Almodovar, M.R.; Oliver, F.J. PARP-1 is involved in autophagy induced by DNA damage. *Autophagy* **2009**, *5*, 61–74.
27. Lee, E.W.; Seo, J.; Jeong, M.; Lee, S.; Song, J. The roles of FADD in extrinsic apoptosis and necroptosis. *BMB Rep.* **2012**, *45*, 496–508. [CrossRef]
28. Ofengeim, D.; Yuan, J. Regulation of RIP1 kinase signalling at the crossroads of inflammation and cell death. *Nat. Rev. Mol. Cell Biol.* **2013**, *14*, 727–736.
29. Brenner, D.; Blaser, H.; Mak, T.W. Regulation of tumour necrosis factor signalling: Live or let die. *Nat. Rev. Immunol.* **2015**, *15*, 362–374. [CrossRef]
30. Grethe, S.; Ares, M.P.; Andersson, T.; Porn-Ares, M.I. p38 MAPK mediates TNF-induced apoptosis in endothelial cells via phosphorylation and downregulation of Bcl-x(L). *Exp. Cell Res.* **2004**, *298*, 632–642. [CrossRef]
31. Tamatani, M.; Che, Y.H.; Matsuzaki, H.; Ogawa, S.; Okado, H.; Miyake, S.; Mizuno, T.; Tohyama, M. Tumor necrosis factor induces Bcl-2 and Bcl-x expression through NFkappaB activation in primary hippocampal neurons. *J. Biol. Chem.* **1999**, *274*, 8531–8538. [PubMed]
32. Herrmann, J.L.; Beham, A.W.; Sarkiss, M.; Chiao, P.J.; Rands, M.T.; Bruckheimer, E.M.; Brisbay, S.; McDonnell, T.J. Bcl-2 suppresses apoptosis resulting from disruption of the NF-kappa B survival pathway. *Exp. Cell Res.* **1997**, *237*, 101–109. [CrossRef] [PubMed]
33. Mahoney, D.J.; Cheung, H.H.; Mrad, R.L.; Plenchette, S.; Simard, C.; Enwere, E.; Arora, V.; Mak, T.W.; Lacasse, E.C.; Waring, J.; et al. Both cIAP1 and cIAP2 regulate TNFalpha-mediated NF-kappaB activation. *Proc. Natl. Acad. Sci. USA* **2008**, *105*, 11778–11783.
34. Bertrand, M.J.M.; Milutinovic, S.; Dickson, K.M.; Ho, W.C.; Boudreault, A.; Durkin, J.; Gillard, J.W.; Jaquith, J.B.; Morris, S.J.; Barker, P.A. cIAP1 and cIAP2 facilitate cancer cell survival by functioning as E3 ligases that promote RIP1 ubiquitination. *Mol. Cell* **2008**, *30*, 689–700. [PubMed]
35. Safa, A.R. c-FLIP, a master anti-apoptotic regulator. *Exp. Oncol.* **2012**, *34*, 176–184.

36. Zhang, X.; Jin, T.; Yang, H.; DeWolf, W.C.; Khosravi, R.; Olumi, A.F. Persistent c-FLIP(L) Expression Is Necessary and Sufficient to Maintain Resistance to Tumor Necrosis Factor-Related Apoptosis-Inducing Ligand–Mediated Apoptosis in Prostate Cancer. *Cancer Res.* **2004**, *64*, 7086–7091.
37. Ham, B.; Fernandez, M.C.; D'Costa, Z.; Brodt, P. The diverse roles of the TNF axis in cancer progression and metastasis. *Trends Cancer Res.* **2016**, *11*, 1–27.
38. Rao, J.S.; Gondi, C.; Chetty, C.; Chittivelu, S.; Joseph, P.A.; Lakka, S.S. Inhibition of invasion, angiogenesis, tumor growth, and metastasis by adenovirus-mediated transfer of antisense uPAR and MMP-9 in non-small cell lung cancer cells. *Mol. Cancer Ther.* **2005**, *4*, 1399–1408.
39. Lin, Y.C.; Shun, C.T.; Wu, M.S.; Chen, C.C. A novel anticancer effect of thalidomide: Inhibition of intercellular adhesion molecule-1-mediated cell invasion and metastasis through suppression of nuclear factor-kappaB. *Clin. Cancer Res.* **2006**, *12*, 7165–7173.
40. Castelao, J.E.; Bart, R.D., 3rd; DiPerna, C.A.; Sievers, E.M.; Bremner, R.M. Lung cancer and cyclooxygenase-2. *Ann. Thorac. Surg.* **2003**, *76*, 1327–1335.
41. Bu, X.; Zhao, C.; Dai, X. Involvement of COX-2/PGE2 pathway in the upregulation of MMP-9 expression in pancreatic cancer. *Gastroent. Res. Pract.* **2011**, *2011*. [CrossRef]
42. Dempsey, P.W.; Doyle, S.E.; He, J.Q.; Cheng, G. The signaling adaptors and pathways activated by TNF superfamily. *Cytokine Growth Factor Rev.* **2003**, *14*, 193–209. [CrossRef]
43. Jackson-Bernitsas, D.G.; Ichikawa, H.; Takada, Y.; Myers, J.N.; Lin, X.L.; Darnay, B.G.; Chaturvedi, M.M.; Aggarwal, B.B. Evidence that TNF-TNFR1-TRADD-TRAF2-RIP-TAK1-IKK pathway mediates constitutive NF-kappaB activation and proliferation in human head and neck squamous cell carcinoma. *Oncogene* **2007**, *26*, 1385–1397. [CrossRef] [PubMed]
44. Kim, S.; Choi, J.H.; Kim, J.B.; Nam, S.J.; Yang, J.H.; Kim, J.H.; Lee, J.E. Berberine suppresses TNF-alpha-induced MMP-9 and cell invasion through inhibition of AP-1 activity in MDA-MB-231 human breast cancer cells. *Molecules* **2008**, *13*, 2975–2985. [CrossRef]
45. Jung, Y.S.; Lee, S.O. Apomorphine suppresses TNF-alpha-induced MMP-9 expression and cell invasion through inhibition of ERK/AP-1 signaling pathway in MCF-7 cells. *Biochem. Biophys. Res. Commun.* **2017**, *487*, 903–909. [CrossRef]
46. Sun, Y.; Liu, W.Z.; Liu, T.; Feng, X.; Yang, N.; Zhou, H.F. Signaling pathway of MAPK/ERK in cell proliferation, differentiation, migration, senescence and apoptosis. *J. Recept. Signal Transduct. Res.* **2015**, *35*, 600–604.
47. Liu, L.; Zhu, H.; Wu, W.; Shen, Y.; Lin, X.; Wu, Y.; Liu, L.; Tang, J.; Zhou, Y.; Sun, F. Neoantimycin F, a Streptomyces-Derived Natural Product Induces Mitochondria-Related Apoptotic Death in Human Non-Small Cell Lung Cancer Cells. *Front. Pharmacol.* **2019**, *10*, 1042. [CrossRef]
48. Wen, J.; Fu, J.H.; Zhang, W.; Guo, M. Lung carcinoma signaling pathways activated by smoking. *Chin. J. Cancer* **2011**, *30*, 551–558. [CrossRef]
49. Kim, E.K.; Choi, E.J. Pathological roles of MAPK signaling pathways in human diseases. *Biochim. Biophys. ACTA* **2010**, *1802*, 396–405.

Sample Availability: Samples of the compounds are not available from the authors.

© 2019 by the authors. Licensee MDPI, Basel, Switzerland. This article is an open access article distributed under the terms and conditions of the Creative Commons Attribution (CC BY) license (http://creativecommons.org/licenses/by/4.0/).

Article

Annona coriacea Mart. Fractions Promote Cell Cycle Arrest and Inhibit Autophagic Flux in Human Cervical Cancer Cell Lines

Izabela N. Faria Gomes [1,2], Renato J. Silva-Oliveira [2], Viviane A. Oliveira Silva [2], Marcela N. Rosa [2], Patrik S. Vital [1], Maria Cristina S. Barbosa [3], Fábio Vieira dos Santos [3], João Gabriel M. Junqueira [4], Vanessa G. P. Severino [4], Bruno G Oliveira [5], Wanderson Romão [5], Rui Manuel Reis [2,6,7,*] and Rosy Iara Maciel de Azambuja Ribeiro [1,*]

1. Experimental Pathology Laboratory, Federal University of São João del Rei—CCO/UFSJ, Divinópolis 35501-296, Brazil; izabela.faria.tk@hotmail.com (I.N.F.G.); patrikdasilvavital@gmail.com (P.S.V.)
2. Molecular Oncology Research Center, Barretos Cancer Hospital, Barretos 14784-400, Brazil; renatokjso@gmail.com (R.J.S.-O.); vivianeaos@gmail.com (V.A.O.S.); nr.marcela@gmail.com (M.N.R.)
3. Laboratory of Cell Biology and Mutagenesis, Federal University of São João del Rei—CCO/UFSJ, Divinópolis 35501-296, Brazil; mariacristina@ufsj.edu.br (M.C.S.B.); fabiosantos@ufsj.edu.br (F.V.d.S.)
4. Special Academic Unit of Physics and Chemistry, Federal University of Goiás, Catalão 75704-020, Brazil; jgmjunqueira@gmail.com (J.G.M.J.); vanessa.pasqualotto@gmail.com (V.G.P.S.)
5. Petroleomic and forensic chemistry laboratory, Department of Chemistry, Federal Institute of Espirito Santo, Vitória, ES 29075-910, Brazil; brunoliveir_ra20@msn.com (B.G.O.); wandersonromao@gmail.com (W.R.)
6. Life and Health Sciences Research Institute (ICVS), Medical School, University of Minho, 4710-057 Braga, Portugal
7. 3ICVS/3B's-PT Government Associate Laboratory, 4710-057 Braga, Portugal
* Correspondence: ruireis.hcb@gmail.com (R.M.R.); rosy@ufsj.edu.br (R.I.M.d.A.R.); Tel.: +55-173-321-6600 (R.M.R.); +55-3736-904-484 or +55-3799-1619-155 (R.I.M.d.A.R.)

Received: 25 September 2019; Accepted: 29 October 2019; Published: 1 November 2019

Abstract: Plant-based compounds are an option to explore and perhaps overcome the limitations of current antitumor treatments. *Annona coriacea* Mart. is a plant with a broad spectrum of biological activities, but its antitumor activity is still unclear. The purpose of our study was to determine the effects of *A. coriacea* fractions on a panel of cervical cancer cell lines and a normal keratinocyte cell line. The antitumor effect was investigated in vitro by viability assays, cell cycle, apoptosis, migration, and invasion assays. Intracellular signaling was assessed by Western blot, and major compounds were identified by mass spectrometry. All fractions exhibited a cytotoxic effect on cisplatin-resistant cell lines, SiHa and HeLa. C3 and C5 were significantly more cytotoxic and selective than cisplatin in SiHa and Hela cells. However, in CaSki, a cisplatin-sensitive cell line, the compounds did not demonstrate higher cytotoxicity when compared with cisplatin. Alkaloids and acetogenins were the main compounds identified in the fractions. These fractions also markedly decreased cell proliferation with p21 increase and cell cycle arrest in G2/M. These effects were accompanied by an increase of H2AX phosphorylation levels and DNA damage index. In addition, fractions C3 and C5 promoted p62 accumulation and decrease of LC3II, as well as acid vesicle levels, indicating the inhibition of autophagic flow. These findings suggest that *A. coriacea* fractions may become effective antineoplastic drugs and highlight the autophagy inhibition properties of these fractions in sensitizing cervical cancer cells to treatment.

Keywords: natural compounds; cervical cancer; autophagy; cell cycle arrest

1. Introduction

Cervical cancer is the fourth most common cancer among women worldwide, accounting for 7.5% of all female cancer deaths [1]. The number of diagnosed cases is about twice as high in developing countries, such as Brazil, where cervical cancer corresponds to the third most common type among Brazilian women [2]. Despite the high incidence, cervical cancer is one of the tumor types that present significant potential for prevention [3]. Nevertheless, many cases are still diagnosed at an advanced stage, and the therapeutic options are limited [4]. So far, platinum-based chemotherapy remains the only anticancer approach that has improved the results in recurrence and metastatic cervical cancer [5]. However, cisplatin demonstrated extensive side effects, such as myelosuppression and nephrotoxicity, as well as problems related to the resistance and relapse of the disease [5]. Currently, results with molecular-targeted therapies constitute potential alternatives, but the clinical outcomes are still in progress [6]. In this context, natural compounds offer an exciting option to explore and maybe overcome the treatment limitations for cervical cancer.

Autophagy is a homeostatic biological process that maintains cell survival by recycling organelles and molecules, but its role in cancer is still unclear [7]. Autophagic process is activated in many tumors and, when inhibited, can lead to cell death or survival, depending on the tissue type, tumor grade, and therapy [8]. In cervical cancer, autophagy activation is reported as a target of paclitaxel (Taxol), a relevant natural product in cancer chemotherapy resistance [9]. Autophagy targeting has been recognized as a novel therapeutic approach. So far, the establishment of news autophagy modulators is required for cancer treatment [10].

Annonaceae is a common Brazilian plant family, with 29 genera and approximately 386 species [11]. Some species of Annonaceae, a family of plants widely distributed in Brazil, have been related by their biological activity as an anticancer, analgesic, and antimicrobial [12,13]. *Annona coriacea* Mart., a member of the Annonaceae family, is one of the endemic species of the Brazilian Cerrado. It is popularly known as "araticum-liso", "marola", or "araticum do campo" [14]. Among the biological activities already reported for the species are analgesic, anti-inflammatory, carminative, and anthelmintic activity [15]. Recently, methanolic extract of *A. coriacea* seeds exhibited cytotoxicity activity against some cancer cell lines [16]. Although the advantage of obtaining and developing a therapy from leaves rather than other plant parts is clear, potential cytotoxicity activity from *A.coriacea* leaves remains unknown.

The goal of the current study was to evaluate the antineoplastic activity of seven fractions of leaves of *A. coriacea* in human cervical cancer cell lines. We analyzed several biological effects, such as cytotoxicity, proliferation, cell death by apoptosis and autophagy, cell migration, and tumorigenesis, to explore their potential in cervical cancer treatment.

2. Results

2.1. Anonna coriacea Mart. Fractions Contain Acetogenins and Alkaloids in Their Constitution

Analysis of the Electrospray Ionization Fourier Transform Ion Cyclotron Resonance Mass Spectrometry (ESI (-) FT-ICR MS) profile of *A. coriacea* fractions suggests the presence of acetogenins as bulatacin, annonacin, annohexocin, anomuricin E, and coriaheptocinin magnification of 500 to 700 m/z regions in both fractions (C3 and C5). The m/z values of the main molecules found in C3 and C5 are shown in Table 1. Supplementary Table S1 summarizes the major features of the seven fractions isolated.

Table 1. Proposed structures by ESI (-) FT-ICR MS for the main molecules in C3 and C5 fractions from *Annona coriacea*.

Measured m/z	Theoretical m/z	Error (ppm)	DBE	[M-H]⁻	Proposed Compound	Reference
255.2332	255.23324	−1.12	1	$[C_{16}H_{32}O_2-H^+]^-$	palmitic acid	Chen et al., 2016
281.24881	281.24886	0.92	2	$[C_{18}H_{34}O_2-H^+]^-$	oleic acid	Chen et al., 2016
595.45815	595.45822	−0.49	4	$[C_{35}H_{64}O_7-H^+]^-$	asitrocinone	Adewole e Ojewole et al., 2008
595.45838	595.45845	−0.87	4	$[C_{35}H_{64}O_7-H^+]^-$	annonacin	Alkofahi et al., 1988
609.43885	609.43893	−2.85	5	$[C_{35}H_{62}O_8-H^+]^-$	trilobalicin	He et al., 1997
611.45312	611.45314	−0.49	4	$[C_{35}H_{64}O_8-H^+]^-$	annomuricin E	Kim et al., 1998
621.4742	621.4743	−1.17	5	$[C_{37}H_{66}O_7-H^+]^-$	asimicin	Ye et al., 1996
621.47413	621.47418	−0.96	5	$[C_{37}H_{66}O_7-H^+]^-$	bullatacin	Morre et al., 1995
627.4483	627.44832	−0.9	4	$[C_{35}H_{64}O_9-H^+]^-$	annohexocin	Moghadamtousi et al., 2015
627.44823	627.44828	−0.83	4	$[C_{35}H_{64}O_9-H^+]^-$	murihexocin	Kim et al., 1998
635.4540	635.4542	−2.14	6	$[C_{37}H_{64}O_8-H^+]^-$	goniotriocin	Alali et al., 1999
637.46921	637.46927	−1.22	5	$[C_{37}H_{66}O_8-H^+]^-$	bullatalicinone	Hui et al., 1991
637.46905	637.46914	−1.02	5	$[C_{37}H_{66}O_8-H^+]^-$	annoglaucin	Bermejo et al., 2005
641.42889	641.42895	−1.34	4	$[C_{35}H_{64}O_{10}-H^+]^-$	coriaheptocin B/A	Formagio et al., 2015
651.44943	651.44949	−2.65	6	$[C_{35}H_{64}O_{10}-H^+]^-$	ginsenoside Rh5	Vamanu, 2014
653.46442	653.46444	−1.58	5	$[C_{37}H_{66}O_9-H^+]^-$	salzmanolin	Queiroz et al., 2003
669.46005	669.4601	−1.22	6	$[C_{37}H_{68}O_{10}-H^+]^-$	annoheptocin A	Meneses Da Silva et al., 1998
671.47569	671.47575	−1	6	$[C_{37}H_{68}O_{10}-H^+]^-$	annoheptocin B	Meneses Da Silva et al., 1998
763.47932	763.47939	−0.83	12	$[C_{39}H_{70}O_5-H]^-$	squamocin glycosilated	Jamkhande e Wattamwar, 2015

DBE: Double bond equivalent; m/z: mass-to-charge ratio.

2.2. A. coriacea Fractions Promote Cytotoxicity in a Dose- and Time-Dependent Manner in Cervical Cancer Cells Lines

In order to determine the cytotoxicity effects of *A. coriacea* fractions on human cervical cancer cell lines, the cells were cultured and treated with various concentrations of *Annona coriacea* fractions or cisplatin (CIS), respectively, for 72 h, followed by the use of an MTS assay to analyze the cell viability. As shown in Table 2, of the seven fractions used, five reached the IC_{50} (half maximal inhibitory concentration) for the three tested cell lines, and fractions C2 and C4 did not affect cell viability. The IC_{50} values decreased as the concentration of fraction increased, suggesting a dose-dependent manner. The IC_{50} values for the CaSki cell line ranged from 3.6 to 21.4 µg/mL, from 4.1 to 12.9 µg/mL in HeLa, and from 5.1 to 16.1 µg/mL in the SiHa cell line (Table 2). Notably, for the HeLa and SiHa cell lines, the cisplatin-resistant cell lines, all fractions showed a lower IC_{50} than cisplatin (Table 2). However, for CaSki cells, a cisplatin-sensitive cell line, the compounds did not demonstrate higher cytotoxicity as compared with cisplatin.

Table 2. IC_{50} values for *A. coriacea* compounds and cisplatin in cervical cancer cell lines.

	IC_{50} Value (Mean ± SD) µg/mL							
Cell Line	C1	C2	C3	C4	C5	C6	C7	Cisplatin
CaSki	17.8 ± 2.8	ND	6.5 ± 1.8	ND	3.6 ± 0.9	11.7 ± 2.2	21.4 ± 3.3	1.05 ± 1.2
HeLa	12.2 ± 1.5	ND	6.6 ± 1.2	ND	4.1 ± 0.4	12.9 ± 1.9	12.3 ± 0.83	13.6 ± 0.44
SiHa	16.1 ± 2.7	ND	8.7 ± 1.3	ND	5.1 ± 0.6	12.6 ± 1.6	12.7 ± 1.3	15.5 ± 0.93

ND: Not determined; C1: Ethanolic extract; C2: Hexane fraction; C3: Ethyl acetate fraction; C4: Hidroalcoholic fraction; C5: Fraction enriched in acetogenin; C6: Neutral dichloromethane fraction obtained from acid-base extraction; C7: Dichloromethane fraction enriched in alkaloids.

Thus, based on these results, we continued the studies with the two most cytotoxic and selective fractions, C3 and C5 (Table 3). For both fractions, the IC_{50} in HaCaT cells was greater than those observed for cisplatin. Regarding the selectivity indexes, C5 was more selective than C3 in all cell lines tested. Moreover, the fractions in the cisplatin-resistant cell lines (HeLa and SiHa) showed better selectivity than cisplatin. However, for the cisplatin-sensitive cell line, the compounds did not show better selectivity indexes when compared with cisplatin. The same effect was observed over time (from the 3 until 72 h) for C3 and C5 (see Table S2 in the Supplementary Materials).

Table 3. IC$_{50}$ values and selectivity index for the C3 and C5 fractions of cisplatin to tumor cells as compared with HaCaT.

	IC50 Value (Mean ± SD) µg/mL and SI [a]					
Cell Line	C3	C5	Cisplatin	SIC3	SIC5	SI Cisplatin
CaSki	6.5 ± 1.8	3.6 ± 0.9	1.05 ± 1.2	1.57	3.72	4.57
HeLa	6.6 ± 1.2	4.1 ± 0.4	13.6 ± 0.44	1.55	3.27	0.35
SiHa	8.7 ± 1.3	5.1 ± 0.6	15.5 ±0.93	1.17	2.63	0.31
HaCat	10.2 ± 2.4	13.4 ± 1.0	4.8 ± 1.3	R	R	R

[a] Selectivity index is the ratio of the IC$_{50}$ values of the treatments on HaCaT cells to those in the cancer cell lines. SI: Selectivity index; C3: Ethyl acetate fraction; C5: Fraction enriched in acetogenin; R: Reference cell line.

We selected SiHa cells to evaluate the mechanism of action of the C3 and C5 fractions since they are resistant to cisplatin and because these treatments showed greater cytotoxicity and selectivity to this cell line (Figure 1A,B).

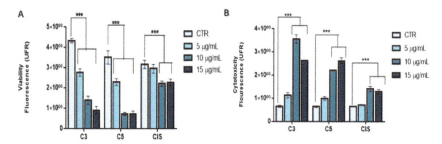

Figure 1. Cytotoxicity in SiHa cells. (**A**) Cell viability measured after 24 h of exposure in SiHa cells. (**B**) Cell cytotoxicity measured after 24 h of exposure in SiHa cells. There was an increase in cytotoxicity and a decrease in viability in a dose-dependent manner ($p < 0.0001$). C3: Ethyl acetate fraction; C5: Fraction enriched in acetogenin; Cis: cisplatin. *** Indicates a statistical difference between groups. UFR: Relative unit of fluorescence.

2.3. A. coriacea Fractions Inhibited Cell Proliferation and Invasion, and Induced Cell Cycle Arrest in Cervical Cancer Cell Lines

We analyzed the effect of C3 and C5 fractions on cell proliferation. The C3 and C5 fractions reduced AKT phosphorylation (Figure 2A,D) and also promoted a reduction in more than 90% of the number of colonies in anchorage-independent growth in comparison to the control (Figure 2B). Moreover, C5 was able to reduce BrdU incorporation significantly (Figure 2C). Using the matrigel invasion assay, we observed that the C5 fraction significantly inhibited cell invasion in SiHa cells. Also, C5 demonstrated higher invasion inhibition when compared with cisplatin (Figure 2E,F).

We further analyzed the expression of proteins involved in the cell cycle. The C3 and C5 treatments increased p21 expression (Figure 3A,B). Regarding the cell cycle, we observed that both C3 and C5 promoted cell cycle arrest in the G2/M phase in SiHa cells after treatment with 5 and 10 µg/mL of each fraction (Figure 3C,D).

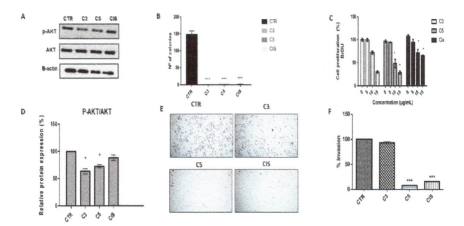

Figure 2. Cell proliferation and invasion upon C3 and C5 treatment (5μg/mL) in SiHa cells (**A**) Western blotting of phospho-AKT (protein kinase B) upon C3 and C5 treatment. (**B**) Number of colonies in the soft agar assay performed for 45 days. (**C**) BrdU incorporation after C3 and C5 treatment ($p < 0.0001$) in SiHa cells. (**D**) Densitometry of p-AKT. (**E**) Invasion inhibition through C3 and C5 treatments in SiHa cells. (**F**) Percentage of invasion cells in SiHa cells (*** $p < 0.0001$; * $p < 0.05$). * Indicates statistical difference between the treatments). C3: Ethyl acetate fraction; C5: Fraction enriched in acetogenin; Cis: cisplatin.

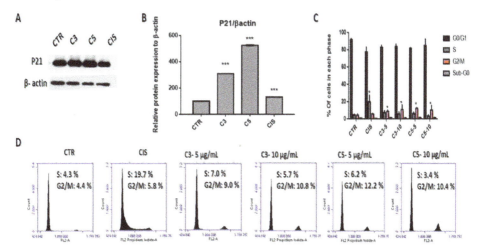

Figure 3. Cell cycle alterations in SiHa cells after exposure to C3 and C5 compounds (**A**) Western blot of p21 in SiHa cells upon C3, C5, and cisplatin treatments. (**B**) Densitometry of p21. (**C**) Cell cycle profile in SiHa cells. (**D**) Cell cycle phase distribution after treatment with C3 and C5. (*** $p < 0.0001$; * $p < 0.05$). C3: Ethyl acetate fraction; C5: Fraction enriched in acetogenin; Cis: cisplatin; DMSO: dimethylsulfoxide.

2.4. Annona coriacea Fractions Promote Cytotoxic Effects by DNA Damage but Do Not Induce Apoptosis

We also analyzed the protein expression of poly (ADP-ribose) polymerase (PARP) and caspase 3 by Western blot. We observed that C3 and C5 treatment increased cleavage of PARP but not caspase 3 (Figure 4A) As opposed to the action of cisplatin, exposure to C3 and C5 fractions did not induce apoptosis (Figure 4B,C). Additionally, the mitochondrial membrane potential was analyzed, and we verified that at higher concentrations, C3 and C5 induce mitochondrial membrane depolarization similar to cisplatin (Figure 4D). Taken together, these results suggest that apoptosis is not the main mechanism of cell death induced by *A. coriacea* Mart. fractions.

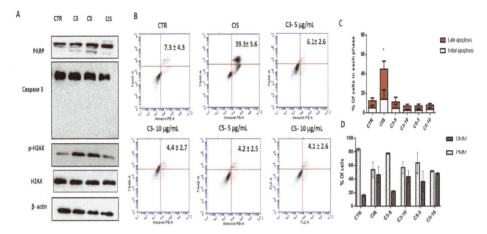

Figure 4. Apoptosis evaluation in SiHa cells upon C3 and C5 compounds. (**A**) Western blot of PARP (Poly (ADP-ribose) polymerase), caspase 3, and H2AX (H2A histone family member X) proteins (**B**) Flow cytometry for SiHa cells. (**C**) Comparison of apoptotic cells upon C3 and C5 treatment. There was a significant increase for cells in apoptosis only for cisplatin (CIS) * p = 0.0282 (**D**) Depolarization of the mitochondrial membrane after treatment with C3 and C5 and cisplatin in the SiHa cell line. C3: Ethyl acetate fraction; C5: Fraction enriched in acetogenin; Cis: cisplatin; DMM: Depolarized mitochondrial membrane; PMM: Polarized mitochondrial membrane. $p < 0.05$). C3: Ethyl acetate fraction; C5: Fraction enriched in acetogenin; Cis: cisplatin.

To better understand the cytotoxic effects of *A. coricea* fractions, we performed the comet assay on HeLa cells, which has a response profile to both fractions that is similar to SiHa. The results showed that *A. coriacea* fractions significantly increased on average the damage score in comparison with the control (Figure 5A,B). Moreover, in agreement with these results, both fractions increased the expression of H2AX phosphorylation, suggesting their role in DNA damage (Figure 4A).

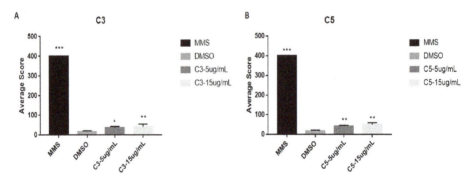

Figure 5. DNA damage evaluation in HeLa cells upon C3 and C5 compounds. (**A**) Genotoxic damage induced by C3. (**B**) Genotoxic damage induced by C5. Data are representative of three experiments. Error bars represent SD. C3: Ethyl acetate fraction; C5: Fraction enriched in acetogenin; Cis: cisplatin; C3: Ethyl acetate fraction; C5: Fraction enriched in acetogenin; Cis: cisplatin; MMS: Methyl methane sulphonate, DMSO: dimethylsulfoxide. *** $p < 0.0001$; ** $p < 0.01$, * $p < 0.05$).

2.5. A. coriacea Fractions Promote Autophagy Flux Inhibition in Cervical Cancer Cell Lines

To explore the role of C3 and C5 in autophagic flux, we analyzed by Western blot the expression of critical proteins for the autophagic pathway. Interestingly, we found these fractions induced an increase in p62 (Figure 6A–C). Furthermore, by acridine orange staining, we verified that C3 and C5 fractions produced a reduction of acidic vesicles (Figure 6D), suggesting that less autophagosome formation could be associated with inhibited autophagy initiation. These findings suggest that A. coriacea fractions may inhibit the initiation steps of autophagy.

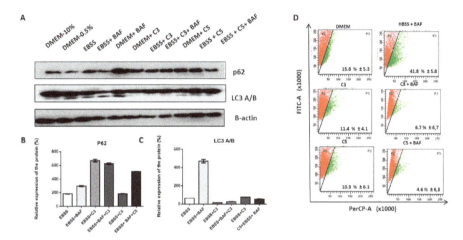

Figure 6. Analysis of the involvement of A. coriacea fractions in autophagy. (**A**) Analysis of the expression of proteins involved in the autophagic flux in SiHa cells. (**B**) Densitometry of p62. (**C**) Densitometry of LC3 B/A (Microtubule-associated protein 1A/1B-light chain 3). (**D**) Acridine orange staining in SiHa cells. There was a reduction in the percentage of formation of acid vesicles, evidenced by a reduction of the fluorescent green signal after treatment with the fractions ($p < 0.05$). HBSS: Hank's balanced salt solution; EBSS: Earle's balanced salt solution; C3: Ethyl acetate fraction; C5: Fraction enriched in acetogenin; Cis: cisplatin. BAF: Bafilomycin.

3. Discussion

Cisplatin is the most frequent chemotherapy agent used for metastatic and refractory cervical tumors; yet, it has demonstrated high recurrence rates due to its high toxicity and resistance [5]. In this context, it is necessary to develop new and less toxic therapeutic approaches. In the present study, we observed that *Annona coriacea* fractions promoted a cytotoxic effect, cell cycle arrest, and inhibit autophagy as well as invasion.

Annona coriacea compounds were able to promote cytotoxicity in cervical cancer cell lines in a dose- and time-dependent-manner. It can also be inferred that these compounds are considered pharmacologically active according to the recommendation of the Institute National Cancer Institute (NCI) for IC_{50} values less than 30 µg/mL [17]. The results are in accordance with previous studies that reported the antiproliferative activity of *Annonaceae* [12,18]. Our results showed that C3 and C5 are selective for tumor cells when compared with normal skin keratinocyte. One limitation of our study is the lack of an ideal normal cervix cell line counterpart to evaluate the selectivity index. Previous studies have considered that a value greater than or equal to 2.0 is an interesting selectivity index, which means that the compound is more than twice more cytotoxic to the tumor cell line as compared with the normal cell line [19].

Among the compounds tested, C3 and C5 were those that showed higher cytotoxicity concerning the other compounds. Moreover, C3 and C5 showed more selectivity when compared with cisplatin in the cisplatin-resistant cell lines. These fractions are rich in acetogenins and alkaloids, as previously described in the *Annonaceae* family [11,14,20–42]. Chemical studies on *Annonaceae* species have identified a large number of acetogenins and alkaloids that possess great biological and pharmacological potential due to their antitumor, cytotoxic, and apoptosis-inducing activities [43]. Many of these metabolites have already been described as acting on crucial enzymes for cell division, such as topoisomerases and the deregulation of the phosphorylative chain by inhibition of the mitochondrial I complex [44]. Taken together, these results suggest that the phytochemical constituents of C3 and C5, among them acetogenins and alkaloids, might contribute substantially to the antineoplastic effect of *A. coriacea* fractions in cervical cancer.

Proliferation inhibition is another characteristic attributed to natural compounds [45,46]. In accordance, we also observed that C3 and C5 treatment promoted a significant increase in the p21 levels, a key regulator of the cell cycle, correlating with G2/M arrest as well as a decrease in BrdU incorporation and p-AKT. Cell cycle arrest in the G2/M transition can be attributed to cytoskeletal disorganization by inhibition of the mitotic spindle [47]. Also, as previously reported, cisplatin treatment induced p53 accumulation and upregulated P21 expression as well as cell cycle arrest in the G1/S phase [48]. Many alkaloids are reported as both mitotic spindle inhibition promoters, such as vinblastine, and as promotors of the inhibition of topoisomerases, such as liriodenine [49,50]. Moreover, cell cycle disruption and repair enzyme overexpression can be attributed to DNA damage [51].

The invasion and clonogenic activity of C3 and C5 were also evaluated in cervical cancer cell lines. The results demonstrated that C3 and C5 treatment significantly reduced the number and size of colonies, as well as the number of invasive cells. A nanoformulation based on curcumin, a natural acetogenin as founded in C3 and C5, has also shown promising results in reducing invasion rates and colonies formed for the SiHa cell line [52]. In this way, it can be inferred that *A. coriacea* fractions could play an inhibitory role in invasion and metastasis processes.

Apoptosis has been described as a key mechanism for the antitumor activity of natural products [53]. Our findings suggest that *A. coriacea* fractions do not induce apoptosis, although we found alterations of the PARP cleavage and H2AX activity involved in DNA repair. The comet assay showed that C3 and C5 promoted an increase in DNA damage. Thus, these data together provide evidence that the cytotoxicity of *A. coricea* fractions can be related to DNA damage directly.

Regarding the autophagic flux, we found decreased expression of LC3 cleavage, increased p62 levels, and negative labeling of acidic vesicles by acridine orange. In accordance with our results, AKT downregulation has already reported as being involved in autophagy and apoptosis through the beclin-1 block, and the PI3K/AKT/mTOR pathway has been implicated as one of the principals of autophagy pathways in gynecological cancers [54]. Moreover, some studies have shown that G2/M cell cycle arrest is a target of autophagy inhibition [55]. Thus, these results indicate that C3 and C5 could inhibit the initiation of autophagy or even the vesicular traffic in cervical cancer.

4. Materials and Methods

4.1. Plant Material

The leaves of *A. coriacea* Mart. were collected in May 2010, at the Federal University of Goias, Catalão, GO, Brazil (18°09′16.4″ S; 47°55′43.2″ W). Dr. Helder N. Consolaro from the Academic Unit of Biotechnology, Federal University of Goias, Catalão, GO, Brazil carried out the identification, and a voucher specimen (no. 47919) was deposited at the Herbarium of Integrated Laboratory of Zoology, Ecology, and Botany in the same university. Registration was made in National System for Management of Genetic Heritage and Associated Traditional Knowledge (SISGEN): A11AE20.

4.2. Preparation of Extracts

Leaves (619 g) from *A. coriacea* were subjected to exhaustive maceration in EtOH (Sigma Aldrich, # 459836) at room temperature. The filtered material was concentrated in a rotary evaporator under reduced pressure at 40 °C to yield the ethanolic extract of the leaves (57.5 g; C1). The ethanolic extract of the leaves was solubilized in MeOH/H2O (3:7, v/v) and subjected to liquid-liquid extraction with n-hexane (Sigma Aldrich # 650552) and ethyl acetate (Sigma-Aldrich, San Luis, EUA #270989; EtOAc). After the evaporation of the solvent under reduced pressure, fractions were obtained: Hexane (12.3 g; C2), EtOAc (20.5 g; C3), and hydroalcoholic (5.0 g; C4). From the separation of C3, fraction C3.3.4.2 (3.3 g; C5) resulted. The ethanolic extract of leaves was subjected to an acid-base extraction resulting in neutral (5.2 g; C6) and alkaloidal (0.32 g; C7) fractions.

4.3. Electrospray Ionization Fourier Transform Ion Cyclotron Resonance Mass Spectrometry (ESI (−) FT-ICR MS)

To identify the main chemical compounds present in *A. coriacea* Mart. fractions, negative electrospray ionization coupled to Fourier transform ion cyclotron resonance mass spectrometry (ESI (−)-FT-ICR MS) analysis was performed as described in the literature [56]. Briefly, 10 µL of each fraction was dissolved in 1000 µL of methanol/toluene (Sigma-Aldrich, # 1.06018; #650579; 50% v/v). Afterward, the solution was basified with 4 µL of NH_4OH (Vetec Fine Chemicals Ltda, Brazil, # 60REAQMO002448). Samples were directly infused at a flow rate of 4.0 mL/min into the ESI (−) source.

The mass spectrometer (model 9.4 T Solarix, Bruker Daltonics, Bremen, Germany) was set to negative ion mode, ESI (−), over a mass range of m/z 150–1500. The ESI source conditions were as follows: A nebulizer gas pressure of 1.5 bar, a capillary voltage of 3.8 kV, and a transfer capillary temperature of 200 °C. The ions' time accumulation was 2 s. ESI (−) FT-ICR mass spectra were acquired by accumulating 32 scans of time-domain transient signals in four mega-point time-domain data sets. All mass spectra were externally calibrated using NaTFA (m/z from 200 to 1200). A resolving power, $m/\Delta m 50\% = 500,000$ (in which $m/\Delta m 50\%$ is the full-peak width at the half-maximum peak height of m/z 400), and mass accuracy of <1 ppm provided the unambiguous molecular formula assignments for singly charged molecular ions.

The mass spectra were acquired and processed using data analysis software (Bruker Daltonics, Bremen, Germany). Elemental compositions of the fractions were determined by measuring the m/z values. The proposed structures for each formula were assigned using the Chemspider (www.chemspider.com) database. The degree of unsaturation for each molecule can be deduced directly from its DBE (double bond equivalent) value according to the equation, $DBE = c - h/2 + n/2 + 1$, where c, h, and n are the numbers of carbon, hydrogen, and nitrogen atoms, respectively, in the molecular formula.

4.4. Cell Lines and Cell Culture

CaSki, HeLa, SiHa cells (ATCC catalog number CRL-1550, CCL-2, and HTB-35, respectively), and one normal keratinocytes cell line (HaCaT) were kindly provided by Dr. Luisa Villa. All the cell lines were maintained in Dulbecco's modified eagle's medium (DMEM1X, high glucose; Gibco, Invitrogen, Grand Island, NY, USA) supplemented with 10% fetal bovine serum (FBS, Gibco, Invitrogen, #26140079) and 1% penicillin/streptomycin solution (P/S, Gibco, #15140122), at 37 °C and 5% $CO_2$2. Authentication of cell lines was performed by the Department of Molecular Diagnostics, Barretos Cancer Hospital. Genotyping confirmed the identity of all cell lines, as previously reported [57]. Moreover, all cell lines were tested for mycoplasma through the MycoAlert™ Mycoplasma Detection Kit (Lonza), following the manufacturer's instructions.

4.5. Drugs

Cisplatin was obtained from Sigma Aldrich (#479306), and its stock solution was prepared in NaCl 0.9%. The stocks solutions of all the fractions were prepared in dimethyl sulfoxide (Sigma-Aldrich, #472301, DMSO). All solutions were stored at −20 °C. Cisplatin was subsequently prepared as intermediate dilutions in DMSO to obtain an equal quantity of DMSO (1% final concentration) in each of the conditions studied. In all experimental conditions, the drugs were diluted in 0.5% FBS culture medium (DMEM-0.5% FBS). Vehicle control (1% DMSO, final concentration) was also used in all experiments.

4.6. Cell Viability and Selectivity Assay

The cell viability and selectivity were performed by MTS (-(4,5-dimethylthiazol-2-yl)-5-(3-carboxymethoxyphenyl)-2-(4-sulfophenyl)-2H-tetrazolium, Promega, Madison, WI, # G3581) as previously described [57]. To determine the IC_{50} values, the cells were seeded into 96-well plates at a density of 5×10^3 cells per well and allowed to adhere overnight in DMEM-10% FBS. Subsequently, the cells were treated with increasing concentrations of the *A. coriacea* fractions (0, 1, 2.5, 5, 7.5, 10, 20, and 25 µg/mL) and cisplatin (0, 1, 3,6, 9, 12, 15, and 18 µg/mL) diluted in DMEM-0.5% FBS for 72 h and analyzed over time (3, 6, 12, 16, 24, and 36 h) [12]. The selectivity index (SI) of *A. coriacea* compounds were determined as previously reported [19]. The SI of the more cytotoxic fractions (C3 and C5) was calculated by the ratio of the IC_{50} values of the treatments in a normal cell line (HaCaT) to those in the cancer cell lines.

The cytotoxicity and viability were also assessed by ApoTox-Glo (Promega, Madison, WI, # G6320). The results are expressed as the mean viable cells relative to DMSO alone (considered as 100% viability) ± SD. For the kinetics assay, the results were calibrated to the starting viability (time 0 h, considered as 100% of viability) and are expressed as the means ± SD. The IC_{50} concentration was calculated by nonlinear regression analysis using GraphPad Prism software. Both assays were done in triplicate at least three times.

4.7. Proliferation Assay

The ELISA-BrdU assay was performed as previously described [58]. Cells were seeded at 5×10^3 densities per well and treated with increasing doses (5, 10, and 15 µg/mL) of the fractions and cisplatin. After 24 h, cell proliferation was detected by the ELISA-BrdU Kit (Roche, Basel, Switzerland #11647229001), following the manufacturer's specifications.

4.8. Cell Cycle Analysis

Cell cycle distribution was analyzed by flow cytometry using propidium iodide (PI) DNA staining [59]. The cells were plated in a six-well plate at a density of 2×10^5 cells per well, and the next day, the cells were treated with fixed concentrations of the fractions and cisplatin. After 24 h, the cells were disrupted and incubated with 40 µg mL^{-1} of PI (Cycle Test Plus BD solution, # 340242) for 10 min at 37 °C, 5% CO_2, as instructed by the manufacturer. Analysis of the PI-labeled cells was performed by a flow cytometer (ACCURIBD Biosciences, San Jose, CA, USA) and the cell cycle phases' distribution was determined as at least 20,000 cells.

4.9. Matrigel Invasion Assay

Cell invasion was measured using BD BioCoat Matrigel invasion chambers (BD Biosciences, San Jose, CA, USA, # 354480), as previously described [6]. Briefly, 2.5×10^4 cells were plated in the matrigel-coated 24-well transwell inserts in DMEM-0.5% FBS containing fractions at a fixed concentration. DMEM-10% FBS was used as a chemoattractant. The cells were allowed to invade for 24 h. The invasive cells, attached to the insert membrane, were fixed with methanol and stained with hematoxylin. Then, images were obtained using a × 10 magnification microscope Eclipse 2220 (Nikon) and the cells were counted in all the fields of the membrane. The results are expressed in relation to the DMSO control (considered as 100% of invasion) as the mean percentage of invasion ± SD.

4.10. Soft Agar Colony Assay

Cell growth and proliferation of SiHa cell lines under anchorage-independent conditions using the soft agar assay were evaluated as described previously [60]. Briefly, 1×10^4 cells were mixed with an equal volume of 0.6% agar and applied into 6-well plates that had been pre-coated with 0.5 mL of 1.2% agar mixed with the same volume of DMEM-20% FBS. The next day, 5 µg/mL of *A. coriacea* fractions diluted in 0.5 mL of serum-free DMEM were added into the wells, and these treatments were exchanged every two days. Cisplatin (15 µg/mL) was used as a positive control. The cells were allowed to form colonies for 45 days before being fixed with methanol and stained with 0.125% crystal violet. Colonies with more than 50 cells were photographed under the light microscope Eclipse 2200 (Nikon) and the number of colonies was analyzed by open CFU (Plos One—http://opencfu.sourceforge.net/) [61]. The results represent the mean of at least three independent experiments.

4.11. Annexin-V-7AAD Assay

This assay was performed using a PE Annexin V Apoptosis Detection Kit (BD Pharmingen, San Diego, CA, USA, #556547) following the manufacturer's specifications. SiHa cells (2×10^5 cells/mL) were seeded into a six-well plate and treated with 5 and 10 µg/mL of C3 and C5 and cisplatin (15 µg/mL). After the treatment, the cells were harvested and washed with phosphate-buffered saline. Next, 100 µL of each sample were taken and placed into a tube containing 5 µL of FITC Annexin V and 5 µL of 7AAD stain. The suspension was mixed, and 400 µL of 1X Assay buffer were added per tube. All samples were analyzed using a flow cytometer (BD ACCURI™, San Jose, CA, USA).

4.12. Analysis of Autophagy Flux

SiHa cells were plated into a six-well plate at a density of 1×10^6 cells/well and allowed to adhere for at least 24 h. Then, the growth medium was replaced with fresh growth medium for control cells, with Hank's balanced salt solution or Earle's balanced salts (EBSS, HBSS; Invitrogen, Carlsbad, Califórnia, USA, EUA, # 14155063, # 14025076) for starved cells (two rinses in HBSS or EBSS before being placed in HBSS or EBSS). The cells were then incubated in HBSS and/or C3 and C5 fractions for 24 h using an equivalent concentration to IC_{50} of the evaluated cell line. Then, 20 nM of bafilomycin A1 (Sigma-Aldrich, #B1793) were added to the fraction treatment with EBSS or HBSS as a control condition. Afterward, the cells were scraped into PBS cold and subjected to Western blot analysis as described below.

4.13. Acridine Orange Staining

Acidic vacuolar organelles (AVOs) were stained by acridine orange (Sigma-Aldrich, San Luis, MO, USA, EUA, # 235,474 AO) as previously described [62]. Concisely, SiHa cells (2×10^5 cells/mL) were seeded into a six-well plate. After exposure to fractions and bafilomycin (BAF; 10 nM) for 24 h, the cells were trypsinized, harvested, and washed with phosphate-buffered saline. After, the cells were stained with fluorescent dye comprising 10 µL of AO (10 µg/mL). The analysis for AVOs was performed by flow cytometry (BD FACSCanto™ II, San Jose, CA, USA). A minimum of 20,000 cells within the gated region was analyzed.

4.14. Detection of Mitochondrial Membrane Potential

MitoStatus Red (BD, Biosences, San Jose, CA, USA, #564697) was used for the analysis of mitochondrial membrane integrity. SiHa cells (2×10^5) were seeded in 6-well plates. After the cells were treated, fixed doses of the fractions (5, 10 µg/mL) and cisplatin (15 µg/mL) were diluted in culture medium for 24 h at 37 °C. In the end, MitoStatus Red (1 µL/mL) was added, following the manufacturer's instructions. After incubation, the cells were disaggregated and analyzed by flow cytometry (BD, ACCURI).

4.15. Comet Assay

The alkaline comet assay (single-cell gel electrophoresis assay) was performed according to Olive and Banáth [63] with adaptations [64]. Briefly, the HeLa cells were seeded in 24-well plates (2×10^5 cells/well) in complete medium. After 24 h, cells were washed twice with PBS 1X and incubated with the different treatments for 3 h in culture medium without serum. The negative control group was treated with PBS and the positive control group was exposed to methyl methane sulphonate (MMS,

120 µM, Sigma-Aldrich, San Luis, MO, USA, EUA, #129925). The quantification of DNA damage was achieved by visual scoring, with the comets being classified from 0 (no damage) to 4 (maximum damage) [65]. For each treatment, 100 comets were analyzed, and the score of damage was calculated employing the equation: Score = 0(C0) + 1 (C1) + 2(C2) + 3(C3) + 4(C4), where C0–C4 are the numbers of comets in each classification of damage. Three independent experiments were performed and the mean of the scores was calculated for each treatment.

4.16. Western Blot

To assess the effect of the drugs on the inhibition of intracellular signaling pathways, the cells were plated at the density of 2×10^5 cells/mL in DMEM-10% FBS into 6-well plates, allowed to grow to 85% of confluence and then serum starved for 2 h, and incubated with IC_{50} values of fractions, diluted in DMEM-0.5% FBS, by 24 h. At the end time, the cells were washed in PBS and lysed with lysis buffer (50 mM Tris (pH 7.6–8), 150 mM NaCl (Sigma-Aldrich, San Luis, MO, USA, EUA, # S9888), 5 mM EDTA (Sigma-Aldrich, San Luis, MO, USA, EUA, # E6758), 1 mM Na_3VO_4 (Sigma-Aldrich, San Luis, MO, USA, EUA, # 450243), 10 mM NaF (Sigma-Aldrich, # 201154), 10 mM sodium pyrophosphate (Sigma-Aldrich,# P8010), 1% NP-40 (Sigma-Aldrich, San Luis, MO, USA, EUA, #74385), and 1/7 of protease cocktail inhibitors (Roche, Amadora, Portugal,# 11697498001). Western blot analysis was done using a standard 10% and 15% sodium dodecyl sulfate-polyacrylamide gel electrophoresis, loading 20 µg of protein per lane. All antibodies were provided by cell signaling and used as recommended by the manufacturer (see Table S3 of the Supplementary Materials). Blot detection was done by chemiluminescence (ECL Western Blotting Detection Reagents, #RPN2109; GE Healthcare, Piscataway, NJ, USA) in Image Quant LAS 4000 mini (GE Healthcare).

4.17. Statistical Analysis

Single comparisons between the conditions studied were made using Student's *t*-test, and the differences between the groups were tested using analysis of variance. The statistical analysis was performed using GraphPad Prism version 5. The level of significance in all statistical analyses was set as $p < 0.05$.

5. Conclusions

In conclusion, our results showed a comprehensive characterization of antitumor mechanisms associated with *Annona coriacea* Mart. fractions that are cytotoxic in cervical cancer cell lines. Also, we highlight the ability of these fractions to inhibit invasion, clonogenic potential, and autophagy as well as an increase of p21 and subsequent cell cycle arrest in G2/M. All these biological activities observed could be attributed to the alkaloids and acetogenins present in these fractions. Further studies are needed to identify the active substances and to characterize their action using in vitro and in vivo models. Nevertheless, the present findings suggest these compounds are a potential candidate for new drug development for cervical cancer.

Supplementary Materials: The following are available online, Supplementary Table S1. Antibodies used in Western Blot analysis; Supplementary Table S2. -Summary all *Annona coriacea* Mart fractions used in the present study; Supplementary Table S3. -IC_{50} values for the *A. coriacea* fractions in kinetics assay on cervical cell lines.

Author Contributions: I.N.F.G. carried out the studies of cell line and cell culture, viability assay, Apotoxiglo Assay, Combination Assay, Invasion, Colony formation Assay, western blot. She also participated of data acquisition and its interpretation and performed the statistical analysis. R.J.S.-O. carried out the studies of cell line and cell culture, western blot, Apotoxiglo Assay, Proliferation Assay, Colony formation Assay. V.A.O.S. and M.N.R. carried out the flow cytometry assay, apoptosis, cell cycle and autophagy and have been involved in revising critically the manuscript. P.S.V. participated in analysis of mass spectrometry and identification of compounds. M.C.S.B. and F.V.d.S. carried out the Comet assay and interpretation of data. J.G.M.J. and V.G.P.S. have been responsible for the preparation of extracts and have been involved in revising critically the manuscript. B.G.O. and W.R. participated in acquisition of mass spectrometry data of *Annona coriacea*, analysis and interpretation of these data, and have been involved in revising it critically for important intellectual content. R.I.M.d.A.R. and R.M.R. conceived the study, participated in its design and coordination, and has made substantial contributions to analysis and interpretation of data and drafted the manuscript. All authors read and approved the final manuscript.

Funding: This study was partially supported by grants from the FINEP (MCTI/FINEP/MS/SCTIE/DECIT-01/2013—FP XII-BIOPLAT), Barretos Cancer Hospital, CAPES, CNPq, FAPEMIG, UFSJ. RMR is a recipient of CNPq Productivity Grant.

Acknowledgments: In this section you can acknowledge any support given which is not covered by the author contribution or funding sections. This may include administrative and technical support, or donations in kind (e.g., materials used for experiments).

Conflicts of Interest: The authors declare that they have no conflict of interests associated with this publication.

References

1. Bray, F.; Ferlay, J.; Soerjomataram, I.; Siegel, R.L.; Torre, L.A.; Jemal, A. Global cancer statistics 2018: GLOBOCAN estimates of incidence and mortality worldwide for 36 cancers in 185 countries. *CA Cancer J. Clin.* **2018**, *68*, 394–424. [CrossRef]
2. Goss, P.E.; Lee, B.L.; Badovinac-Crnjevic, T.; Strasser-Weippl, K.; Chávarri-Guerra, Y.; Louis, J.S.; Villarreal-Garza, C.; Unger-Saldaña, K.; Ferreyra, M.; DeBiasi, M.; et al. Planning cancer control in Latin America and the Caribbean. *Lancet Oncol.* **2013**, *14*, 391–436. [CrossRef]
3. Tota, J.; Ramana–Kumar, A.; El-Khatib, Z.; Franco, E. The road ahead for cervical cancer prevention and control. *Curr. Oncol.* **2014**, *21*, e255–e264. [CrossRef] [PubMed]
4. Quinn, M.A.; Benedet, J.L.; Odicino, F.; Maisonneuve, P.; Beller, U.; Creasman, W.T.; Heintz, A.P.M.; Ngan, H.Y.S.; Pecorelli, S. Carcinoma of the Cervix Uteri. *Int. J. Gynecol. Obstet.* **2006**, *95*, S43–S103. [CrossRef]
5. Chen, J.; Solomides, C.; Parekh, H.; Simpkins, F.; Simpkins, H. Cisplatin resistance in human cervical, ovarian and lung cancer cells. *Cancer Chemother. Pharmacol.* **2015**, *75*, 1217–1227. [CrossRef]
6. Martinho, O.; Silva-Oliveira, R.; Cury, F.P.; Barbosa, A.M.; Granja, S.; Evangelista, A.F.; Marques, F.; Miranda-Gonçalves, V.; Cardoso-Carneiro, D.; De Paula, F.E.; et al. HER Family Receptors are Important Theranostic Biomarkers for Cervical Cancer: Blocking Glucose Metabolism Enhances the Therapeutic Effect of HER Inhibitors. *Theranostics* **2017**, *7*, 717–732. [CrossRef]
7. Orfanelli, T.; Jeong, J.M.; Doulaveris, G.; Holcomb, K.; Witkin, S.S. Involvement of autophagy in cervical, endometrial and ovarian cancer. *Int. J. Cancer* **2014**, *135*, 519–528. [CrossRef]
8. Singh, S.S.; Vats, S.; Chia, A.Y.; Tan, T.Z.; Deng, S.; Ong, M.S.; Arfuso, F.; Yap, C.T.; Goh, B.C.; Sethi, G.; et al. Dual role of autophagy in hallmarks of cancer. *Oncogene* **2018**, *37*, 1142–1158. [CrossRef]
9. Peng, X.; Gong, F.; Chen, Y.; Jiang, Y.; Liu, J.; Yu, M.; Zhang, S.; Wang, M.; Xiao, G.; Liao, H. Autophagy promotes paclitaxel resistance of cervical cancer cells: Involvement of Warburg effect activated hypoxia-induced factor 1-alpha-mediated signaling. *Cell Death Dis.* **2014**, *5*, e1367. [CrossRef]
10. Wang, Y.; Chen, Y.; Chen, X.; Liang, Y.; Yang, D.; Dong, J.; Yang, N.; Liang, Z. Angelicin inhibits the malignant behaviours of human cervical cancer potentially via inhibiting autophagy. *Exp. Ther. Med.* **2019**, *18*, 3365–3374. [CrossRef]
11. Moghadamtousi, S.Z.; Fadaeinasab, M.; Nikzad, S.; Mohan, G.; Ali, H.M.; Kadir, H.A. Annona muricata (Annonaceae): A Review of Its Traditional Uses, Isolated Acetogenins and Biological Activities. *Int. J. Mol. Sci.* **2015**, *16*, 15625–15658. [CrossRef] [PubMed]
12. Silva, V.A.O.; Alves, A.L.V.; Rosa, M.N.; Silva, L.R.V.; Melendez, M.E.; Cury, F.P.; Gomes, I.N.F.; Tansini, A.; Longato, G.B.; Martinho, O.; et al. Hexane partition from *Annona crassiflora* Mart. promotes cytotoxicity and apoptosis on human cervical cancer cell lines. *Investig. New Drugs* **2019**, *37*, 602–615. [CrossRef] [PubMed]
13. Liu, Y.; Liu, D.; Wan, W.; Zhang, H. In vitro mitochondria-mediated anticancer and antiproliferative effects of Annona glabra leaf extract against human leukemia cells. *J. Photochem. Photobiol. B Boil.* **2018**, *189*, 29–35. [CrossRef] [PubMed]
14. Bermejo, A.; Figadere, B.; Zafra-Polo, M.C.; Barrachina, I.; Estornell, E.; Cortes, D. Acetogenins from Annonaceae: Recent progress in isolation, synthesis and mechanisms of action. *Nat. Prod. Rep.* **2005**, *22*, 269–303. [CrossRef] [PubMed]
15. Sousa, O.V.; Del-Vechio-Vieira, G.; Kaplan, M.A.C. Propriedades Analgésica e Antiinflamatória do Extrato Metanólico de Folhas de *Annona coriacea* Mart. (Annonaceae). *Lat. Am. J. Pharm.* **2007**, *26*, 872–877.

16. Formagio, A.; Vieira, M.; Volobuff, C.; Silva, M.; Matos, A.; Cardoso, C.; Foglio, M.; Carvalho, J. In vitro biological screening of the anticholinesterase and antiproliferative activities of medicinal plants belonging to Annonaceae. *Braz. J. Med. Boil. Res.* **2015**, *48*, 308–315. [CrossRef]
17. Suffness, M.; Pezzuto, J.M. Assays related to cancer drug discovery. In *Methods in Plant Biochemistry: Assays for Bioactivity*; Hostettmann, K., Ed.; Academic Press: London, UK, 1990; pp. 71–133.
18. Ma, C.; Wang, Q.; Shi, Y.; Li, Y.; Wang, X.; Li, X.; Chen, Y.; Chen, J. Three new antitumor annonaceous acetogenins from the seeds of Annona squamosa. *Nat. Prod. Res.* **2017**, *31*, 2085–2090. [CrossRef]
19. De Oliveira, P.F.; Alves, J.M.; Damasceno, J.L.; Oliveira, R.A.M.; Dias Júnior, H.; Crotti, A.E.M.; Tavares, D.C. Cytotoxicity screening of essential oils in cancer cell lines. *Rev. Bras. Farmacogn.* **2015**, *25*, 183–188. [CrossRef]
20. Yu, J.G.; Li, T.M.; Sun, L.; Luo, X.Z.; Ding, W.; Li, D.Y. [Studies on the chemical constituents of the seeds from Artabostrys hexapetalus (Annonaceae)]. *Yao Xue Xue Bao Acta Pharm. Sin.* **2001**, *36*, 281–286.
21. Pinheiro, M.L.B.; Xavier, C.M.; De Souza, A.D.L.; Rabelo, D.D.M.; Batista, C.L.; Batista, R.L.; Campos, F.R.; Barison, A.; Valdez, R.H.; Ueda-Nakamura, T.; et al. Acanthoic acid and other constituents from the stem of Annona amazonica (Annonaceae). *J. Braz. Chem. Soc.* **2009**, *20*, 1095–1102. [CrossRef]
22. Kim, G.-S.; Zeng, L.; Alali, F.; Rogers, L.L.; Wu, F.-E.; McLaughlin, J.L.; Sastrodihardjo, S. Two New Mono-Tetrahydrofuran Ring Acetogenins, Annomuricin E and Muricapentocin, from the Leaves of Annona muricata. *J. Nat. Prod.* **1998**, *61*, 432–436. [CrossRef] [PubMed]
23. Ye, Q.; He, K.; Oberlies, N.H.; Zeng, L.; Shi, G.; Evert, D.; McLaughlin, J.L. Longimicins A–D: Novel Bioactive Acetogenins fromAsimina longifolia(Annonaceae) and Structure−Activity Relationships of Asimicin Type of Annonaceous Acetogenins. *J. Med. Chem.* **1996**, *39*, 1790–1796. [CrossRef] [PubMed]
24. Zeng, L.; Wu, F.-E.; McLaughlin, J.L. Annohexocin, a novel mono-THF acetogenin with six hydroxyls, from Annona muricata (Annonaceae). *Bioorg. Med. Chem. Lett.* **1995**, *5*, 1865–1868. [CrossRef]
25. Morré, D.J.; De Cabo, R.; Farley, C.; Oberlies, N.H.; McLaughlin, J.L. Mode of action of bullatacin, a potent antitumor acetogenin: Inhibition of NADH oxidase activity of HeLa and HL-60, but not liver, plasma membranes. *Life Sci.* **1995**, *56*, 343–348. [CrossRef]
26. Kim, G.S.; Zeng, L.; Alali, F.; Rogers, L.L.; Wu, F.E.; Sastrodihardjo, S.; McLaughlin, J.L. Muricoreacin and murihexocin C, mono-tetrahydrofuran acetogenins, from the leaves of Annona muricata. *Phytochemistry* **1998**, *49*, 565–571. [CrossRef]
27. Alali, F.Q.; Rogers, L.; Zhang, Y.; McLaughlin, J.L. Goniotriocin and (2,4-cis- and -trans)-xylomaticinones, bioactive annonaceous acetogenins from Goniothalamus giganteus. *J. Nat. Prod.* **1999**, *62*, 31–34. [CrossRef]
28. Hui, Y.-H.; Rupprecht, J.K.; Anderson, J.E.; Wood, K.V.; McLaughlin, J.L. Bullatalicinone, a new potent bioactive acetogenin, and squamocin from annona bullata (Annonaceae). *Phytother. Res.* **1991**, *5*, 124–129. [CrossRef]
29. Vamanu, E. Antioxidant Properties of Mushroom Mycelia Obtained by Batch Cultivation and Tocopherol Content Affected by Extraction Procedures. *BioMed Res. Int.* **2014**, *2014*, 1–8. [CrossRef]
30. Queiroz, E.F.; Roblot, F.; Laprevote, O.; Paulo Mde, Q.; Hocquemiller, R. Two unusual acetogenins from the roots of Annona salzmanii. *J. Nat. Prod.* **2003**, *66*, 755–758. [CrossRef]
31. Da Silva, E.L.M.; Roblot, F.; Hocquemiller, R.; Serani, L.; Laprevote, O. Structure elucidation of annoheptocins, two new heptahydroxylated C37 acetogenins by high-energy collision-induced dissociation tandem mass spectrometry. *Rapid Commun. Mass Spectrom.* **1998**, *12*, 1936–1944. [CrossRef]
32. Costa, M.S.; Cossolin, J.F.S.; Pereira, M.J.B.; Sant'Ana, A.E.G.; Lima, M.D.; Zanuncio, J.C.; Serrão, J.E. Larvicidal and Cytotoxic Potential of Squamocin on the Midgut of Aedes aegypti (Diptera: Culicidae). *Toxins* **2014**, *6*, 1169–1176. [CrossRef] [PubMed]
33. Jamkhande, P.G.; Wattamwar, A.S. Annona reticulata Linn. (Bullock's heart): Plant profile, phytochemistry and pharmacological properties. *J. Tradit. Complement. Med.* **2015**, *5*, 144–152. [CrossRef] [PubMed]
34. Chen, Y.; Chen, Y.; Shi, Y.; Ma, C.; Wang, X.; Li, Y.; Miao, Y.; Chen, J.; Li, X. Antitumor activity of Annona squamosa seed oil. *J. Ethnopharmacol.* **2016**, *193*, 362–367. [CrossRef] [PubMed]
35. De Pedro, N.; Cautain, B.; Melguizo, A.; Cortes, D.; Vicente, F.; Genilloud, O.; Tormo, J.R.; Peláez, F. Analysis of cytotoxic activity at short incubation times reveals profound differences among Annonaceus acetogenins, inhibitors of mitochondrial Complex I. *J. Bioenerg. Biomembr.* **2013**, *45*, 145–152. [CrossRef]
36. Yu, J.G.; Gui, H.Q.; Luo, X.Z.; Sun, L. Murihexol, a linear acetogenin from Annona muricata. *Phytochemistry* **1998**, *49*, 1689–1692. [CrossRef]

37. Da Silva, E.L.M.; Roblot, F.; Mahuteau, J.; Cavé, A. Coriadienin, the First Annonaceous Acetogenin with Two Double Bonds Isolated from Annona coriaceae. *J. Nat. Prod.* **1996**, *59*, 528–530. [CrossRef]
38. Qin, G.-W.; Li, C.-J.; Wang, L.-Q.; Li, Y.; Min, B.-S.; Nakamura, N.; Hattori, M. Cytotoxic Mono-Tetrahydrofuran Ring Acetogenins from Leaves of Annona montana. *Planta Medica* **2001**, *67*, 847–852.
39. Zhong, J.; Ying, C.; Ruo-Yun, C.; De-Quan, Y. Linear acetogenins from Goniothalamus donnaiensis. *Phytochemistry* **1998**, *49*, 769–775. [CrossRef]
40. Lin, C.-Y.; Chou, C.-J.; Wu, Y.-C.; Chang, F.-R.; Liaw, C.-C.; Chiu, H.-F. New Adjacent Bis-Tetrahydrofuran Annonaceous Acetogenins from Annona muricata. *Planta Medica* **2003**, *69*, 241–246.
41. Liaw, C.-C.; Chang, F.-R.; Lin, C.-Y.; Chou, C.-J.; Chiu, H.-F.; Wu, M.-J.; Wu, Y.-C. New cytotoxic monotetrahydrofuran annonaceous acetogenins from Annona muricata. *J. Nat. Prod.* **2002**, *65*, 470–475. [CrossRef]
42. Da Silva, E.L.M.; Roblot, F.; Laprévote, O.; Sérani, L.; Cavé, A. Coriaheptocins A and B, the First Heptahydroxylated Acetogenins, Isolated from the Roots of Annona coriacea. *J. Nat. Prod.* **1997**, *60*, 162–167. [CrossRef]
43. Matsushige, A.; Kotake, Y.; Matsunami, K.; Otsuka, H.; Ohta, S.; Takeda, Y. Annonamine, a new aporphine alkaloid from the leaves of Annona muricata. *Chem. Pharm. Bull.* **2012**, *60*, 257–259. [CrossRef] [PubMed]
44. Yao, Q.; Lin, M.; Wang, Y.; Lai, Y.; Hu, J.; Fu, T.; Wang, L.; Lin, S.; Chen, L.; Guo, Y. Curcumin induces the apoptosis of A549 cells via oxidative stress and MAPK signaling pathways. *Int. J. Mol. Med.* **2015**, *36*, 1118–1126. [CrossRef] [PubMed]
45. Han, B.; Yao, Z.-J.; Wang, L.-S. Effect of annonaceous acetogenin mimic AA005 on proliferative inhibition of leukemia cells in vitro and its possible mechanisms. *Zhongguo Shi Yan Xue Ye Xue Za Zhi* **2012**, *20*, 549–553. [PubMed]
46. Nordin, N.; Majid, N.A.; Hashim, N.M.; Rahman, M.A.; Hassan, Z.; Ali, H.M. Liriodenine, an aporphine alkaloid from Enicosanthellum pulchrum, inhibits proliferation of human ovarian cancer cells through induction of apoptosis via the mitochondrial signaling pathway and blocking cell cycle progression. *Drug Des. Dev. Ther.* **2015**, *9*, 1437–1448.
47. Hung, D.T.; Jamison, T.F.; Schreiber, S.L. Understanding and controlling the cell cycle with natural products. *Chem. Boil.* **1996**, *3*, 623–639. [CrossRef]
48. Kielbik, M.; Krzyżanowski, D.; Pawlik, B.; Klink, M. Cisplatin-induced ERK1/2 activity promotes G1 to S phase progression which leads to chemoresistance of ovarian cancer cells. *Oncotarget* **2018**, *9*, 19847–19860. [CrossRef]
49. Chen, K.; Shou, L.-M.; Lin, F.; Duan, W.-M.; Wu, M.-Y.; Xie, X.; Xie, Y.-F.; Li, W.; Tao, M. Artesunate induces G2/M cell cycle arrest through autophagy induction in breast cancer cells. *Anti-Cancer Drugs* **2014**, *25*, 652–662. [CrossRef]
50. Newman, D.J.; Cragg, G.M. Natural Products as Sources of New Drugs from 1981 to 2014. *J. Nat. Prod.* **2016**, *79*, 629–661. [CrossRef]
51. Broustas, C.G.; Lieberman, H.B. DNA Damage Response Genes and the Development of Cancer Metastasis. *Radiat. Res.* **2014**, *181*, 111–130. [CrossRef]
52. Zaman, M.S.; Chauhan, N.; Yallapu, M.M.; Gara, R.K.; Maher, D.M.; Kumari, S.; Sikander, M.; Khan, S.; Zafar, N.; Jaggi, M.; et al. Curcumin Nanoformulation for Cervical Cancer Treatment. *Sci. Rep.* **2016**, *6*, 20051. [CrossRef] [PubMed]
53. Millimouno, F.M.; Dong, J.; Yang, L.; Li, J.; Li, X. Targeting Apoptosis Pathways in Cancer and Perspectives with Natural Compounds from Mother Nature. *Cancer Prev. Res.* **2014**, *7*, 1081–1107. [CrossRef] [PubMed]
54. Rashmi, R.; DeSelm, C.; Helms, C.; Bowcock, A.; Rogers, B.E.; Rader, J.; Grigsby, P.W.; Schwarz, J.K. AKT Inhibitors Promote Cell Death in Cervical Cancer through Disruption of mTOR Signaling and Glucose Uptake. *PLoS ONE* **2014**, *9*, e92948. [CrossRef] [PubMed]
55. Han, L.; Zhang, Y.; Liu, S.; Zhao, Q.; Liang, X.; Ma, Z.; Gupta, P.K.; Zhao, M.; Wang, A. Autophagy flux inhibition, G2/M cell cycle arrest and apoptosis induction by ubenimex in glioma cell lines. *Oncotarget* **2017**, *8*, 107730–107743. [CrossRef] [PubMed]
56. Baliano, A.P.; Pimentel, E.F.; Buzina, A.R.; Vieira, T.Z.; Romão, W.; Tose, L.V.; Lenz, D.; de Andrade, T.U.; Fronza, M.; Kondratyuk, T.P.; et al. Brown seaweed Padina gymnospora is a prominent natural wound-care product. *Rev. Bras. Farmacogn.* **2016**, *26*, 714–719. [CrossRef]

57. Silva-Oliveira, R.J.; Silva, V.A.O.; Martinho, O.; Cruvinel-Carloni, A.; Melendez, M.E.; Rosa, M.N.; De Paula, F.E.; Viana, L.D.S.; Carvalho, A.L.; Reis, R.M. Cytotoxicity of allitinib, an irreversible anti-EGFR agent, in a large panel of human cancer-derived cell lines: KRAS mutation status as a predictive biomarker. *Cell. Oncol.* **2016**, *39*, 253–263. [CrossRef]
58. Mead, T.J.; Lefebvre, V. Proliferation assays (BrdU and EdU) on skeletal tissue sections. *Breast Cancer* **2014**, *1130*, 233–243.
59. Visagie, M.H.; Joubert, A.M. In vitro effects of 2-methoxyestradiol-bis-sulphamate on reactive oxygen species and possible apoptosis induction in a breast adenocarcinoma cell line. *Cancer Cell Int.* **2011**, *11*, 43. [CrossRef]
60. Silva-Oliveira, R.J.; Melendez, M.; Martinho, O.; Zanon, M.F.; Viana, L.D.S.; Carvalho, A.L.; Reis, R.M. AKT can modulate the in vitro response of HNSCC cells to irreversible EGFR inhibitors. *Oncotarget* **2017**, *8*, 53288–53301. [CrossRef]
61. Geissmann, Q. OpenCFU, a New Free and Open-Source Software to Count Cell Colonies and Other Circular Objects. *PLoS ONE* **2013**, *8*, 54072. [CrossRef]
62. Wilson, E.N.; Bristol, M.L.; Di, X.; Maltese, W.A.; Koterba, K.; Beckman, M.J.; Gewirtz, D.A. A switch between cytoprotective and cytotoxic autophagy in the radiosensitization of breast tumor cells by chloroquine and vitamin D. *Horm. Cancer* **2011**, *2*, 272–285. [CrossRef] [PubMed]
63. Olive, P.L.; Banáth, J.P. The comet assay: A method to measure DNA damage in individual cells. *Nat. Protoc.* **2006**, *1*, 23–29. [CrossRef] [PubMed]
64. De Oliveira, J.T.; Barbosa, M.C.D.S.; De Camargos, L.F.; Da Silva, I.V.G.; Varotti, F.D.P.; Da Silva, L.M.; Moreira, L.M.; Lyon, J.P.; Santos, V.J.D.S.V.D.; Dos Santos, F.V. Digoxin reduces the mutagenic effects of Mitomycin C in human and rodent cell lines. *Cytotechnology* **2017**, *69*, 699–710. [CrossRef] [PubMed]
65. Collins, A.R. The Comet Assay for DNA Damage and Repair: Principles, Applications, and Limitations. *Mol. Biotechnol.* **2004**, *26*, 249–261. [CrossRef]

Sample Availability: Samples of the compounds are available from the authors.

© 2019 by the authors. Licensee MDPI, Basel, Switzerland. This article is an open access article distributed under the terms and conditions of the Creative Commons Attribution (CC BY) license (http://creativecommons.org/licenses/by/4.0/).

Article

Proanthocyanidin-Rich Fractions from Red Rice Extract Enhance TNF-α-Induced Cell Death and Suppress Invasion of Human Lung Adenocarcinoma Cell A549

Chayaporn Subkamkaew [1], Pornngarm Limtrakul (Dejkriengkraikul) [1,2,3] and Supachai Yodkeeree [1,2,3,*]

[1] Department of Biochemistry, Faculty of Medicine, Chiang Mai University, Chiang Mai 50200, Thailand; subkamkaew@gmail.com (C.S.); pornngarm.d@cmu.ac.th (P.L.(D.))
[2] Anticarcinogenesis and Apoptosis Research Cluster, Faculty of Medicine, Chiang Mai University, Chiang Mai 50200, Thailand
[3] Center for Research and Development of Natural Products for Health, Chiang Mai University, Chiang Mai 50200, Thailand
* Correspondence: yodkeelee@hotmail.com

Academic Editor: Roberto Fabiani
Received: 23 August 2019; Accepted: 17 September 2019; Published: 18 September 2019

Abstract: Tumor necrosis factor-alpha (TNF-α) plays a key role in promoting tumor progression, such as stimulation of cell proliferation and metastasis via activation of NF-κB and AP-1. The proanthocyanidin-rich fraction obtained from red rice (PRFR) has been reported for its anti-tumor effects in cancer cells. This study investigated the molecular mechanisms associated with PRFR on cell survival and metastasis of TNF-α-induced A549 human lung adenocarcinoma. Notably, PRFR enhanced TNF-α-induced A549 cell death when compared with PRFP alone and caused a G0-G1 cell cycle arrest. Although, PRFR alone enhanced cell apoptosis, the combination treatment induced the cells that had been enhanced with PRFR and TNF-α to apoptosis that was less than PRFR alone and displayed a partial effect on caspase-8 activation and PARP cleavage. By using the autophagy inhibitor; 3-MA attenuated the effect of how PRFR enhanced TNF-α-induced cell death. This indicates that PRFR not only enhanced TNF-α-induced A549 cell death by apoptotic pathway, but also by induction autophagy. Moreover, PRFR also inhibited TNF-α-induced A549 cell invasion. This effect was associated with PRFR suppressed the TNF-α-induced level of expression for survival, proliferation, and invasive proteins. This was due to reduce of MAPKs, Akt, NF-κB, and AP-1 activation. Taken together, our results suggest that TNF-α-induced A549 cell survival and invasion are attenuated by PRFR through the suppression of the MAPKs, Akt, AP-1, and NF-κB signaling pathways.

Keywords: proanthocyanins; TNF-α; autophagy; invasion; lung adenocarcinoma

1. Introduction

Lung cancer is the most commonly diagnosed form of cancer and the primary cause of cancer-related mortality for males worldwide. It is also the second leading cause of cancer-related deaths among women globally [1]. Lung cancer is aggressive, and its treatment remains a difficult and challenging task for physicians [2]. Several research studies have indicated that long-term exposure to inhaled carcinogens has the greatest impact on increased risks of lung cancer [1,3–5]. Inhalation of such toxic air pollutants and microorganisms can cause lung injuries and chronic inflammation [6]. Chronic inflammation has been associated with cancer development. Many proinflammatory mediators, especially cytokines, chemokines, and prostaglandins, have been found to promote cancer proliferation, invasion, angiogenesis, and drug resistance [7].

A large number of studies have indicated that TNF-α displays a degree of potential in linking the molecules associated with inflammation and cancer. The data obtained from clinical studies have revealed that the expression level of TNF-α in the tumor tissues and serum samples obtained from patients with non-small cell lung cancer increased along with the clinical stage of the tumor [8,9]. TNF-α plays an important role in the process by binding itself to the tumor necrosis factor R-1 (TNFR-1). After the binding of TNF-α and TNFR-1, the receptor interacts with TRADD (TNFR1-associated death domain) to initiate the recruitment of receptor-interacting protein 1 (RIP1) and TNFR-associated factor 2 (TRAF2) [10]. These complex signaling lead to induce cancer cell survival, proliferation, and metastasis via upregulation of antiapoptotic (Survivin, XIAP, Bcl-xl, Bcl-2, and cFLIP), proliferative (cyclin D and cyclin B1), invasive (MMP-9, MT1-MMP, uPA, and uPAR), and angiogenic (VEGF, and COX-2) proteins by activating NF-κB, activator protein-1 (AP-1), and the mitogen-activated protein kinase (MAPKs) signaling pathway [11,12]. On the other hand, the binding of TNF-α and TNFR-1 can induce the program cell death that is involved with apoptosis by recruiting TRADD-FADD and caspases-8. In fact, TRADD and caspase-8 complex are assembled over a delayed period of time when compared with TRAF2 and RIP1, which results in a sufficient amount of time needed to activate survival signaling. Therefore, the expression of anti-apoptotic proteins and caspase inhibitors, including Bcl-2, Bcl-xl, xIAP, and cFLIP, would be elevated prior to caspase 8 activation [13,14]. Thus, the blockade of TNF-α-induced survival signaling can lead to an increase in the sensitivity of TNF-α-induced cell death. Moreover, many studies have shown that TNF-α expression results in the induction of multiple autophagy markers in breast cancer cells, lung cancer cells, and Ewing's sarcoma cells [15]. A novel function of anti-apoptotic proteins, such as cFLIP, survivin, Bcl-2, and Bcl-xl, that serve as autophagy inhibitors, have been reported in various cells [16,17]. Downregulation of these antiapoptotic proteins could enhance TNF-α-induced cancer cell death via autophagy and apoptosis. Accordingly, the efficient agents that can suppress TNF-α-induced cancer cell progression could be an important part of an attractive and alternative form of cancer therapy.

Proanthocyanidins, also known as condensed tannins, are a class of polymeric phenolic compounds that consist mainly of catechin, epicatechin, gallocatechin, and epigallocatechin units [18]. Recently, our previous findings have demonstrated that proanthocyanidin-rich fractions derived from red rice (PRFR) inhibited inflammation in LPS-treated Raw 264.7 cells via suppression of the AP-1, NF-κB, and MAPKs signaling pathways [19]. Moreover, PRFR reduced human fibrosarcoma, HT1080 cells and breast adenocarcinoma, MDA-MB-231 cells invasion via inhibition of the expression of invasive proteins [20]. Furthermore, PRFR suppressed cell proliferation in human hepatocellular carcinoma, HepG2 cells via the downregulation of survival proteins and induced cell apoptosis by enhancing active apoptotic proteins [21].

However, the effect of PRFR on TNF-α-induced cancer progression has not yet been clarified. Therefore, the purpose of this study was to investigate whether PRFR exerts anticancer effects through suppression of the TNF-α-induced expression of the survival and metastasis proteins by inhibiting the MAPKs, NF-κB, and AP-1 signaling pathways in A549 human lung adenocarcinoma cells.

2. Results

2.1. PRFR Enhanced TNF-α-Induced Cytotoxicity in A549 Lung Adenocarcinoma Cells

The cytotoxicity of PRFR was examined by using trypan blue staining assay. Treatment of A549 cells with PRFR (0–50 µg/mL) for 24 h significantly reduced the viability of the cells in a dose-dependent manner. In particular, treatment of the cells with 40 and 50 µg/mL of PRFR decreased cell viability to 63.0% and 54.6%, respectively. However, a combination treatment of the cells with TNF-α (25 ng/mL), PRFR 40, and 50 µg/mL reduced cell viability to 42.3% and 36.5%, which significantly increased cytotoxicity to greater levels than in the treatment with PRFR alone (Figure 1a). Next, we investigated whether the enhancement activity of PRFR on TNF-α-induced cell death was associated with apoptosis by employing Annexin V-PI staining assay. The results indicate that treatment with PRFR alone at 40 and 50 µg/mL induced the apoptotic population from 3% to 16% and 18%, respectively (Figure 1b). However, co-treatment of PRFR at 40 and 50 µg/mL and TNF-α significantly induced the apoptotic population to a degree that was less than with the treatment of PRFP alone. To confirm whether or not apoptosis is the main cause of PRFR enhanced TNF-α induced cell death, the level of the apoptotic signaling pathway proteins was investigated by including cleaved caspase-8 and PARP-1. As shown in Figure 1c, the levels of caspase-8 and PARP-1 in a combination treatment were lower than for PRFP alone. These results indicate that PRFR could enhance the cytotoxicity effect of TNF-α; however, this result was not limited to the apoptotic pathway.

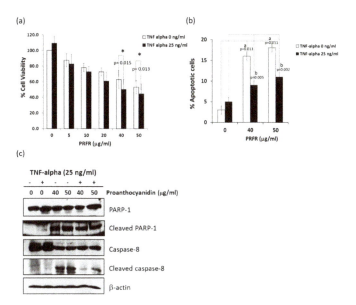

Figure 1. PRFR-enhanced tumor necrosis factor-alpha (TNF-α)-induced cytotoxicity in A549 lung adenocarcinoma cells. (**a**) A549 cells were preincubated with different concentrations of PRFR for 4 h and then co-treated with 25 ng/mL of TNF-α for 24 h. (**b**) Cell apoptosis was determined by Guava Nexin and analyzed by Guava® easyCyte Flow Cytometer to detect the apoptotic cell population. (**c**) The apoptotic proteins were detected by western blotting using the antibodies to caspase-3, caspase-8, and PARP-1. The data are presented as mean ± S.D. with * $p < 0.05$, and ** $p < 0.01$ when compared with the PRFR alone, [a] $p < 0.05$ when compared with the control group, and [b] $p < 0.01$ when compared with the TNF-α alone.

2.2. PRFR Potentiates TNF-α-Induced Autophagy

TNF-α-induced cell death occurred via the apoptosis pathway, but also stimulated autophagy cell death. Therefore, we investigated whether the enhancement activity of PRFR on TNF-α-induced cell death was involved with autophagy. The autophagy vacuoles were labeled by Monodansylcadaverin (MDC) fluorescent staining and analyzed them with a fluorescent microscope. Co-treatment of PRFR and TNF-α significantly increased the number of autophagy vacuoles in A549 cells when compared with TNF-α alone. However, PRFR alone did not induce autophagy vacuoles (Figure 2a,b). To further confirm PRFR mediated autophagy cell death in TNF-α-induced A549 cells, the expression level of LC3B-II, a credible marker of the autophagosome [22,23], was assayed by western blot analysis. Combination treatment with PRFR and TNF-α increased the expression levels of LC3B-II when compared with TNF-α alone, whereas PRFR alone had no effect (Figure 2c). To verify that autophagy plays a major role in the process of PRFR enhancement of TNF-α-induced cell death, the cells were co-treated with 3-MA (autophagy inhibitor), TNF-α, and PRFR for 24 h, and the cell viability was then analyzed. As shown in Figure 2d, combination treatment with 3-MA, PRFR, and TNF-α did not significantly reduce the cell viability when compared with PRFR alone. This results indicated that 3-MA attenuated the enhancement effect of PRFR on TNF-α-induced cell death by reversing the percentage of cell viability to the same level of treatment with PRFR alone (Figure 2d). In addition, the modulation effect of PRFR on the autophagy regulated proteins was determined. The results presented in Figure 2e. show that the induction of survivin, cFLIPs, and Bcl-xl by TNF-α were reduced by PRFR in a dose-dependent manner. Taken together, these results indicate that PRFR could enhance TNF-α-induced A549 cell death via the autophagy and apoptosis pathways.

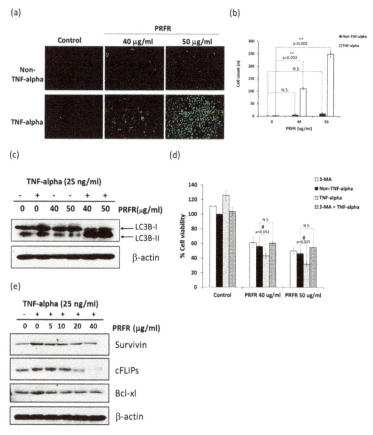

Figure 2. PRFR enhanced TNF-α-induced autophagic cell death in A549 cells. (**a**,**b**) A549 cells were stained with monodansylcadaverin (MDC) after being preincubated with 40 and 50 µg/mL PRFR and then co-treated with 25 ng/mL of TNF-α for 24 h. The data are presented in bar graphs (**b**). (**c**) The expression of autophagosome proteins (LC3B) was detected by western blot analysis using antibodies against LC3B. (**d**) A549 cells were preincubated with 1.5 mM of 3-MA for 1 h and then treated with 40 and 50 µg/mL PRFR and 25 ng/mL of TNF-α for 24 h, and the cell viability was determined using trypan blue assay. (**e**) The expression of survival proteins was detected by western blot analysis using the antibodies against survivin, cFLIPs, and Bcl-xl. Data from a typical experiment are depicted here, while similar results were obtained from three independent experiments. The data are presented as mean ± S.D. with ** $p < 0.01$ when compared with the TNF-α alone, and # $p < 0.05$ when compared with control group (N.S., not significant).

2.3. Effect of PRFR on TNF-α-Induced Cell Proliferation

TNF-α plays an important role in cancer cell proliferation by inducing the expression of proliferative proteins. The effect of PRFR on TNF-α-induced cell proliferation was examined by using PI staining. To determine the anti-proliferative effects of PRFR, A549 cells were pretreated with PRFR (10–40 µg/mL) and then treated with 25 ng/mL of TNF-α. As is shown in Figure 3a,b, the percentages of the G0/G1 phase of the cells receiving the combination treatment with TNF-α and PRFR at 10, 20, and 40 µg/mL, significantly increased from 76.4% to 83.1%, 85.1%, 88.9%, respectively when compared with those of the TNF-α treatment alone. The manner in which TNF-α induced was examined the expression levels of cyclin D1, which are G0/G1 cell cycle regulatory proteins. As is shown in Figure 3b, TNF-α induced the expression levels of cyclin D1 was decreased when the cells were treated with PRFR at 20 and 40 µg/mL.

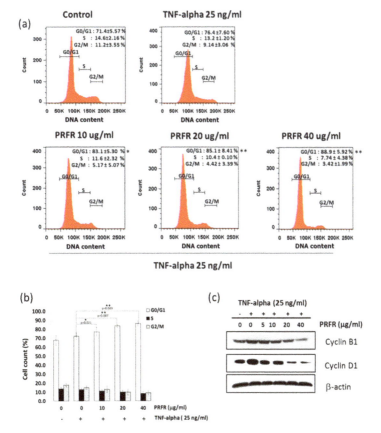

Figure 3. Effect of PRFR on TNF-α-induced cell proliferation. A549 cells were preincubated with various concentrations of PRFR for 4 h and then co-treated with 25 ng/mL of TNF-α for up to 24 h (**a,b**) cell cycle was determined by PI staining and analyzed by flow cytometry to detect the cell cycle arrest. The data present in a bar graph (**b**). (**c**) The expression of proliferative proteins was detected by western blot analysis using the antibodies against cell proliferation proteins (cyclin D1). Data from a typical experiment are depicted here, while similar results were obtained from three independent experiments. The data are presented as mean ± S.D. with * $p < 0.05$ and ** $p < 0.01$ when compared with the TNF-α alone.

2.4. PRFR Inhibited TNF-α-Induced A549 Cell Invasion and Migration

TNF-α plays a crucial role in lung cancer cell invasion. Therefore, the effect of PRFR on TNF-α-induced A549 cell invasion and migration was evaluated. The Figure 4a showed TNF-α efficiently induced A549 cell invasion through the basement membrane by 2.3-fold when compared with the control. However, in the presence of PRFR, TNF-α-induced invasion of A549 cells was significantly inhibited. Moreover, TNF-α also stimulated A549 cell migration by almost two-fold, while PRFR suppressed this activity. TNF-α promoted cancer cell metastasis by upregulating invasive proteins. Therefore, the effect of PRFR on TNF-α-induced proteins was examined that are involved in cancer cell invasion including MT1-MMP, uPA, uPAR, Cox-2, and MMP-9. As is shown in Figure 4c, TNF-α dramatically induced the expression levels of MT1-MMP, uPA, uPAR, and Cox-2 proteins after 24 h, while treatment of the cells with PRFR (0–15 μg/mL) prevented the TNF-α induced expression of these proteins in a dose-dependent manner. Next, the gelatin zymography assay was used to examine the inhibitory effects of PRFR on TNF-α-induced MMP-9 secretions. As is shown in Figure 4d, TNF-α-induced MMP-9 secretions were significantly inhibited in the presence of 10–15 μg/mL of PRFR.

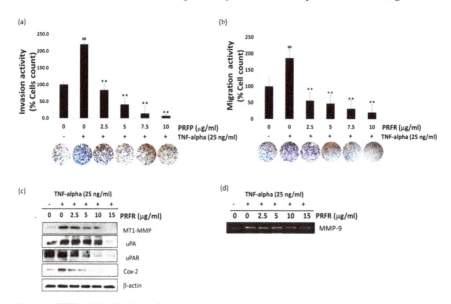

Figure 4. PRFR inhibits TNF-α-induced A549 cell invasion and migration. The matrix gel was coated on the upper surfaces of the membrane filters for invasion assay (**a**) and the gelatin was then coated for migration assay (**b**). Different concentrations of PRFR (0–10 μg/mL) with 1.25×10^5 cells of the A549 cells were seeded into the upper chamber in DMEM serum-free medium, and the lower chamber was filled with 25 ng/mL of TNF-α. After 24 h of incubation, the migrated cells on the lower surface of the filter were determined. After co-treatment with PRFR and TNF-α for 24 h, whole-cell extracts were prepared and analyzed by western blot analysis using antibodies against metastatic proteins (MT1-MMP, uPA, uPAR, and COX-2) (**c**). The culture supernatants of the treated cells were collected, and the secretions of MMP-9 were analyzed by gelatin zymography (**d**). The data are presented as the mean ± S.D. of three independent experiments. Notably, (**a**,**b**) sample groups were found to be significantly different from the TNF-α-treated group (** $p < 0.01$) and the TNF-α alone compared with the control group (## $p < 0.01$).

2.5. Effect of PRFR on TNF-α-Induced NF-κB and AP-1 Activation

NF-κB and AP-1 transcription factors are involved in cancer cell progression. The expression of survival, anti-apoptotic, autophagy, invasive, and angiogenesis genes are controlled by NF-κB and AP-1 transcriptional activity. To investigate whether PRFR affected TNF-α-induced NF-κB and AP-1 activation, nucleus translocation and phosphorylation of NF-κB and AP-1 were determined. As is shown in Figure 5a,b, TNF-α enhanced the nucleus translocation and phosphorylation of c-Jun (AP-1). PRFR could inhibit TNF-α-induced AP-1 translocation and blocked the TNF-α-induced phosphorylation of AP-1 in a dose-dependent manner. Next, the regulation of PRFR on the transcription activity of NF-κB was tested. The co-treatment with PRFR and TNF-α decreased TNF-α-induced nuclear translocation of p-65 (Figure 5c) and the phosphorylation of p65 at ser536 (Figure 5d) in a dose-dependent manner. The data indicate that PRFR can suppress the TNF-α-induced transcriptional activity of AP-1 and NF-κB.

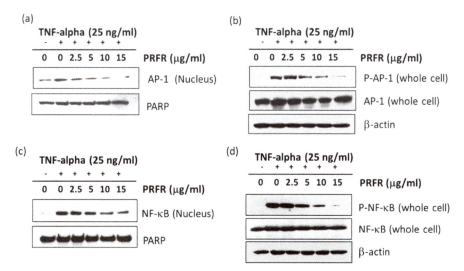

Figure 5. Effect of PRFR on TNF-α-induced NF-κB and activator protein-1 (AP-1) activation. A549 cells were pretreated with PRFR and induced with TNF-α 25 ng/mL for 1 h. The nucleus-extracted fraction was prepared in order to detect c-Jun (AP-1) (**a**) and p65 (NF-κB) levels (**c**) by western blot analysis. A549 cells were pretreated with PRFR for 12 h and induced with TNF-α 25 ng/mL for 15 min. The whole cell lysate was prepared for measurement of the phosphorylated and non-phosphorylated forms of AP-1 and NF-κB (**b,d**). β-Actin and PARP were used as internal loading control proteins in the cytoplasm and nucleus, respectively. Data from a typical experiment are depicted here, while similar results were obtained from three independent experiments.

2.6. Effect of PRFR on TNF-α-Induced MAPK, Akt, and IκB-α Signaling Pathways

TNF-α-activated MAPKs, Akt, and IκB-α signaling pathways have been involved in tumor progression via AP-1 or NF-κB transcriptional activities. Therefore, the effects of PRFR on the TNF-α-induced activation of IκB-α, Akt and MAPKs, including Erk1/2, p38, and JNK, were investigated by western blot analysis. As is shown in Figure 6a,b, TNF-α stimulated the phosphorylation of p38, JNK, and Erk1/2. Additionally, PRFR at 10 and 15 μg/mL inhibited the TNF-α-induced phosphorylation of the JNK and Erk1/2 signaling pathways, whereas, PRFR at 15 μg/mL reduced TNF-α induced phosphorylation of the p38 signaling pathway. On the other hand, the levels of the phosphorylated forms of Akt were induced by TNF-α. The level of TNF-α-induced phosphorylation of Akt was suppressed by PRFR in a dose-dependent manner, while the non-phosphorylation of Akt had no effect. Moreover, NF-κB activation by TNF-α was mediated via NIK and IKK, resulting in IκB-α degradation. In order to examine the effects of PRFR on IκB-α activity, PRFR affected the degree of TNF-α induced IκB-α degradation was determined. As is shown in Figure 6c, PRFR effectively blocked TNF-α-dependent IκB-α degradation.

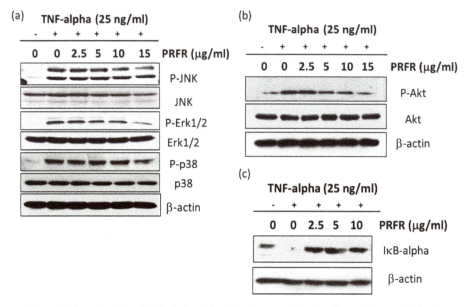

Figure 6. Effect of PRFR on TNF-α-induced MAPK, Akt, and IκB-α signaling pathways. A549 cells were pretreated with PRFR and induced with TNF-α 25 ng/mL for 15 min. The whole cell lysate was prepared for measurement of phosphorylated and non-phosphorylated forms of JNK, Erk1/2, p-38 (**a**), and Akt (**b**) by western blot analysis. The whole cell lysate was also used to determine TNF-α-induced IκB-α degradation (**c**) by western blot analysis. Data from a typical experiment are presented here, while similar results were obtained from three independent experiments.

3. Discussion

Proanthocyanidins are oligomers and polymers of flavanol-3-ol which are found in various fruits, vegetables, and cereals. Notably, they are present in foods such as grape seeds, blackberries, and red rice [18,24]. Our previous study revealed that PRFR exhibited anti-cancer activities by inducing HepG2 hepatocarcinoma cell apoptosis and inhibiting MDA-MB-231 breast cancer cell invasion [20,21]. Despite its various pharmacological activities, the molecular mechanism of PRFR on the anti-tumor effects in A549 lung adenocarcinoma cells has not been elucidated. In this study, we investigated whether PRFR could sensitize TNF-α-induced cell death in lung A549 cancer cells and then act as a potent inhibitor of TNF-α-induced A549 cell metastasis. Molecular mechanisms of this phenomena have been elucidated.

The interaction between TNF-α and TRFR-1 can trigger survival or death signaling pathways. It has been reported that most cancer cells are resistant to TNF-α-induced cell death via increased expressions of survival proteins through the induction transcriptional activity of NF-κB. Thus, a blockade of the activation of the survival signaling pathways may lead to an increase in sensitivity in TNF-α-induced cell death [12]. This study was the first to report that PRFR sensitized A549 lung adenocarcinoma cells to TNF-α-induced cell death. TNF-α promotes cancer cell death by induction apoptosis, necroptosis, and autophagy depending on the condition of the cells. TNF-α induced cell apoptosis via the intrinsic and extrinsic pathways. The results of this study demonstrate that PRFR alone enhanced A549 cell apoptosis with increased caspase-8 activation while inducing PARP cleavage. However, a combination of PRFR and TNF-α induced A549 to apoptosis to a lesser degree than PRFR alone and also revealed a partial effect on caspase-8 activation and the level of the cleaved PARP. This would suggest that apoptosis is more than a mechanism for PRFR to enhance TNF-α-induced A549 cell death.

Autophagy has been extensively reported to play a critical role in the control of cell proliferation, differentiation, and cell death. Autophagy is a highly regulated and fundamental cellular homeostatic process, in which cytoplasmic material is delivered and organelles convert to lysosomes via double membrane vesicles called autophagosomes for degradation. Autophagy is activated in response to various forms of cellular stress, including starvation, hypoxia, radiation, and inflammation [25,26]. Many studies have shown that TNF-α-induced autophagic cell death occurs in various cancer cell types including breast cancer, hepatoma and ovarian cancer [27,28]. Therefore, autophagy is considered a potential pathway in the treatment of cancer. Many natural drug molecules, such as curcumin, celastrol, and bufalin, play important roles in tumor inhibition by inducing autophagy. Thus, PRFR was examined whether it can enhance TNF-α-induced A549 cell death via autophagy cell death. Co-treatment of PRFR and TNF-α led A549 cells to autophagy by accumulating autophagosomes and upregulating the expression of LC3B-II proteins. Whereas, PRFR alone did not induce autophagy, LC3s proteins were found to be a structural protein of autophagosome membranes. The conversion of a soluble form of LC3B-I to LC3B-II is often used to demonstrate active autophagy. To confirm that autophagy is a major process of PRFR in the enhancement of TNF-α induced cell death by using 3-MA as an autophagy inhibitor. The obtained results indicate that using 3-MA could reverse the enhancement effect of PRFR on TNF-α-induced cell death, which would indicate that the way in which PRFR enhanced the cytotoxicity effect of TNF-α was due to autophagy cell death. Recent findings have revealed a novel function of anti-apoptotic proteins, such as FLIP, survivin, Bcl-2, and Bcl-xl, as negative regulators of autophagy. FLIP has been shown to inhibit LC3 lipidation by competitive interaction with ATG3, which in turn blocks autophagy [16]. Moreover, Zhu J., et al. have shown that the inhibition of survivin through the use of siRNA enhanced autophagy by upregulating Beclin-1 [29]. Therefore, in order to explain the mechanism by which PRFR sensitizes TNF-α-induced autophagy, the modulatory effect of PRFR on TNF-α induced FLIP, Bcl-xl, and survivin expression levels was examined. It was found that the levels of FLIP, Bcl-xl and survivin were reduced by PRFR. Together, these results suggest that the manner in which PRFR enhanced TNF-α induced cell death was at least in part accomplished by down regulating FLIP, Bcl-xl and survivin, which then led to autophagic cell death.

Moreover, the results of this study indicate that PRFR can suppress cell proliferation by blocking cell cycle progression in the G1 phase. TNF-α is known to stimulate transcriptions of Cyclin D1, Cyclin E, and Cyclin B1 in order to accelerate the progression of the cell cycle. Cyclin D1 is a key regulator of the G1 checkpoint control [30]. This finding is consistent with our observation that PRFR suppressed TNF-α-induced expression of cyclin D1. This result suggests that PRFR reduced TNF-α induced cell proliferation by inhibiting the expression of cyclin proteins.

The degradation of ECM and the components of the basement membrane through which proteases are the key steps of cancer cell invasion and metastasis. Of these proteases, MMPs such as MMP-9, MT1-MMP, and uPA are thought to play an important role in cancer invasion. Furthermore, COX-2 has been implicated in metastasis, and its overexpression can enhance cellular invasion, proliferation, and induce angiogenesis [31–33]. Previous observations have indicated that TNF-α is an inducer for the invasion and metastasis of A549 cells. These results clearly demonstrate that PRFR inhibits the TNF-α-induced invasion and migration of A549. Moreover, PRER reduced the levels of TNF-α-induced expression of invasive genes, including MMP-9, MT1-MMP, uPA, uPAR, and Cox-2, in the A549 cells.

NF-κB and AP-1 are major key players in TNF-α-mediated tumor progression. NF-κB regulates the expression of the survival gene products cIAP, Bcl-2, Bcl-xl, and FLIP, along with the proliferation of gene products cyclin B1 and cyclin D1, and the invasion of gene products uPA, COX-2, MMP-9, and MT1-MMP, which are known to be induced by TNF-α [31,34,35]. Furthermore, it has been reported that TNF-α could induce autophagy in cancer cells when NF-κB signaling is inhibited [36,37]. A common form of NF-κB is a heterodimer consisting of p50/p65. NF-κB is normally retained in the cytoplasm through interaction with its inhibitor IκB. Upon TNF-α stimulation, IκB-α is catalyzed for phosphorylation by IκB kinase (IKK) leading to IκB-α degradation and allowing for the nuclear translocation of NF-κB, which promotes the transcription of the corresponding genes. Therefore, we have determined the activity of PRFR on TNF-α can induce the degradation of IκB-α and the translocation of NF-κB activity. Our results demonstrated that PRFR prevented degradation of IκB-α and reduced NF-κB activity by inhibiting TNF-α-induced p65 phosphorylation and translocation to the nucleus of the cells. AP-1 has been implicated in regulating cancer cell survival and proliferation. AP-1 also controls the gene expression values of MMP-9, MT1-MMP, Cox-2, uPA, and uPAR. Here, the activation of AP-1 was investigated by observing the phosphorylation and translocation of c-Jun in TNF-α treated cells. In this study, PRFR inhibited TNF-α induced c-Jun phosphorylation and translocation to the nucleus of the A549 cells. This result was in accordance with the findings of an investigation conducted by Qiao Y et al., which also demonstrated that the suppression of AP-1 signaling can potentiate TNFα-induced cell death and inhibit cancer cell invasion [34]. Based on the above-mentioned results, we suggest that PRFR could decrease the level of expression of survival and metastasis proteins by the inhibition of AP-1 and NF-κB activation and are also in agreement with inhibition of AP-1 and NF-κB by epigallocatechin gallate reduced cancer cells survival and metastasis [38,39].

It is accepted that the activation of the MAPKs or Akt signaling pathways is important for regulating survival and metastasis in a variety of cancer cells. TNF-α is bound to the TNF-α receptor-1 which induces NF-κB activation by activating the MAPKs, Akt, and IKK signaling pathways. Moreover, the activation of MAPKs and Akt are important for regulating AP-1 activity. MAPKs are known to be serine/threonine kinase and are composed of several subgroups, such as ERK1/2, JNK, and p38 [35]. It is generally demonstrated that MAPKs signaling pathways regulate metastasis and survival in a variety of cancer cells. Accumulated evidence indicates that the Akt and MAPKs signaling pathways are involved with autophagy. The AKT/mTOR signaling pathway is one of the survival regulatory pathways in both normal and cancer cells, and it can negatively regulate autophagy [40]. Therefore, the experiments were performed to determine whether PRFR regulates TNF-α in order to stimulate the activity of MAPKs and Akt. Our results show that PRFR prevented the phosphorylation of p38, ERK, JNK, and Akt. These results are consistent with those of previous reports which have found that using the inhibitors of the PI3K/Akt and MAPK signaling pathways causes cell death and is associated with autophagy, apoptosis and a reduction in the invasive properties of cancer cells.

4. Materials and Methods

4.1. Chemicals and Reagents

Dulbecco's Modified Eagle Medium (DMEM), trypsin and penicillin-streptomycin were supplied from Gibco (Grand Island, NY, USA). Fetal bovine serum (FBS), RIPA buffer, protease inhibitors and Coomassie Plus™ Protein Assay Reagent were obtained from Thermo Scientific Company (Waltham, MA, USA). Guava Cell Nexin Reagent was purchased from Guava Technologies (Darmstadt, Germany). Nitrocellulose membrane and ECL reagent were supplied from GE Healthcare (Little Chalfont, UK). Gelatin, propidium iodide (PI) and 3-Methyladenine (3-MA) were obtained from Sigma (St. Louis, MO, USA). Antibodies specific to COX-2, and cyclin -D1 were purchased from Millipore (Darmstadt, Germany). Antibodies specific to β-actin, uPA, urokinase-type plasminogen activator receptor (uPAR), poly (ADP-ribose) polymerase (PARP), and p65 were purchased from Santa Cruz Biotechnology (Santa Cruz, CA, USA). Antibodies for the detection of ERK1/2, p38, JNK, c-Jun, and p65 were purchased from Cell Signaling Technology (Danvers, MA, USA). Matrigel was purchased from Becton Dickinson (Bedford, MA, USA).

4.2. Preparation of Proanthocyanidin-Rich Fraction from Red Rice Extract

Whole grains of red rice (Oryza sativa L.) collected from Doi Saket District (Chiang Mai, Thailand) were dehulled and polished in order to obtain the rice germ and bran using a rice de-husker and a rice milling machine (Kinetic (Hubei) Energy Equipment Engineering Co., Ltd., Wuhan, Hubei, China). A voucher specimen was certified by the herbarium at the Flora of Thailand, Faculty of Pharmacy, Chiang Mai University (voucher specimen no. 023148).

Proanthocyanidin-rich fraction (PRFR) was prepared by following the previously reported protocol [20]. Briefly, 440 g red rice bran were soaked in 50% ethanol for 24 h. After that, the mixture was filtered to separate the ethanolic fractions. The ethanolic fractions were evaporated and partitioned with saturated butanol. The saturated butanol fractions were collected and evaporated to obtain the medium polar fractions. Next, the PRFR was prepared form the medium polar fractions by using Sephadex LH20 (GE Healthcare) chromatography (GE Healthcare Ltd., Little Chalfont, UK). The medium polar fractions (3.5 g) were dissolved in methanol and loaded onto a Sephadex LH-20 column. The fractions were sequentially eluted with solutions of 70% methanol, 30% methanol, and 70% acetone, respectively. Total contents of proanthocyanidins in each fraction were determined by vanillin assay. The fractions containing high concentrations of proanthocyanidins were pooled together and freeze-dried in order to obtain PRFR powder. The total amount of proanthocyanidins in the PRFR was 177.22 ± 16.66 mg catechin/g extract.

4.3. Cell Cultures

A549 lung adenocarcinoma cells were supplied by ATCC. The cells were cultured in DMEM supplemented with 100 U/mL penicillin, and 100 µg/mL streptomycin plus 10% FBS. The cultures were maintained in a humidified incubator with an atmosphere comprised of 95% air and 5% CO_2 at 37 °C. For the PRFR treatment, PRFR was dissolved in DMSO and diluted with culture medium, for which the final concentration of DMSO was less than 0.1% (v/v).

4.4. Cell Viability Assay

The cell viability assay of PRFR against A549 lung adenocarcinoma cells was evaluated using trypan blue staining. Briefly, 2×10^4 cells/well were seeded in a 24-well plate and incubated at 37 °C, 5% CO_2 for 24 h in DMEM containing 10% FBS. After that, the cells were treated with or without various concentrations (0–200 µg/mL) of PRFR in DMEM containing 10% FBS for 24 h. At the end of the treatment, the percent of cell viability was also determined from counts of the cells suspended in the medium and counts of those cells removed from the plates by trypsinization. Equal parts of 0.4% trypan blue dye were added to the cell suspension in order to obtain a 1 to 2 ratio. The cell viability in each well was determined using Trypan blue dye and the values were compared with the controls.

4.5. Cell Cycle Arrest Assay

A549 cells were incubated with or without various concentrations of the proanthocyanidin-rich fraction (0–50 µg/mL) for 24 h. Then, the cell suspension was prepared on ice and stained with propidium iodide (PI) for 30 min in the dark. Cells were washed with cold PBS and resuspended in 500 µL. For cell cycle analysis, 1×10^4 events were recorded and then analyzed with the BD FACScan™ flow cytometer (BD Biosciences, San Jose, CA, USA).

4.6. Apoptosis Assay

A549 cells were incubated with or without various concentrations of PRFR (0–50 µg/mL) for 24 h. The cell suspension was then prepared and stained with annexin V and 7-amino actinomycin D (Guava Cell Nexin Reagent; Guava Technologies) for 20 min according to the Guava Nexin Assay protocol. Apoptosis was determined on a Guava PCA Instrument using Guava®Viacount™ Software (Merck Ltd., Darmstadt, Germany).

4.7. Extraction of Nuclear and Whole-Cell Lysate

Whole-cell extraction was done to determine the expression levels of the invasive, apoptotic and survival proteins in the A549 cells. The cells were pretreated with various concentrations of PRFR for 4 h and treated with 25 ng/mL of TNF-α for 24 in DMEM medium to determine the levels of uPA, uPAR, COX-2, Survivin, cFLIPs, cyclin D, LC3B, caspase-8, and PARP-1 proteins. The levels of the

MAPKs and Akt pathway proteins were determined from the cells treated with PRFR. After that, the cells were treated with TNF-α (25 ng/mL) for 15 min. The treated cells were then extracted using a RIPA lysis buffer containing protease inhibitors (1 mM PMSF, 10 µg/mL leupeptin, 10 µg/mL aprotinin) for 20 min on ice. The insoluble matter was removed by centrifugation at 12,000 rpm for 15 min at 4 °C, and the supernatant fraction (whole cell lysate) was collected and protein concentration was determined using Bradford protein assay.

For the preparation of the nuclear extract fractions, after the A549 cells were treated with PRFR (0–15 µg/mL), TNF-α (25 ng/mL) was added to the cells and they were incubated for 1 h at 37 °C. The treated cells were then collected and the cell pellets were suspended with 50 µL of lysis buffer (10 mM HEPES, pH 7.9, 10 mM KC1, mM EDTA, 0.1 mM EGTA, 1 mM dithiothreitol, 0.5 mM phenylmethylsulfonyl -fluoride, 0.1 µg/mL leupeptin, 1 µg/mL aprotinin). The cells were allowed to swell on ice for 20 min, after which, 15 µL of 10% of Nonidet P-40 was added. The tubes were agitated on a vortex and centrifuged at 12,000 rpm for 5 min. The supernatant was collected and was representative of the cytoplasm extract. The nuclear pellets were suspended in ice-cold nuclear extraction buffer (20 mM HEPES, pH 7.9, 0.4 M NaCl, 1 mM EDTA, 1 mM EGTA, 1 mM dithiothreitol, 1 mM phenylmethylsulfonylfluoride, 2.0 µg/mL leupeptin, 2.0 µg/mL aprotinin) with an intermittent vortex for 30 min. The nuclear extract was centrifuged at 12,000 rpm for 10 min, and the supernatant was collected and used to determine the resulting yield of nuclear proteins.

4.8. Western Blotting Analysis

The whole cell lysate or nuclear extractions were subjected to 10–12% SDS-PAGE. The proteins were transferred onto nitrocellulose membranes. The membranes were blocked with 5% non-fat dried milk protein in 0.5% TBS-tween. Thereafter, the membranes were further incubated overnight with the desired primary antibody at 4 °C followed by incubation with horseradish peroxidase conjugated secondary antibody. Bound antibodies were detected using the chemiluminescent detection system and then exposed to the X-ray film (GE Healthcare Ltd., Little Chalfont, U.K.). Equal values of protein loading were confirmed as each membrane was stripped and re-probed with an anti-β-actin antibody.

4.9. Monodansylcadaverine Staining

The treated A549 cells were stained with 0.05 mM Monodansylcadaverin (MDC) in PBS for 30 min at 37 °C. The cells were washed three times with PBS to remove excess MDC. The visualization step employed a Carl Zeiss Microscopy GmbH (Carl Zeiss AG, Jena, Germany) with an excitation wavelength of 460–500 nm and an emission wavelength of 512–542 nm.

4.10. Statistical Analysis

All data are presented as mean ± standard deviation (S.D.) values. Statistical analysis was analyzed with Prism version 6.0 software GmbH (GraphPad Software, Inc. , San Diego, CA, USA) using one-way ANOVA with Dunnett's test. Statistical significance was determined at * $p < 0.05$, ** $p < 0.01$, *** $p < 0.001$, or **** $p < 0.0001$.

5. Conclusions

PRFR was determined that could enhance TNF-α-induced A549 cell death by inducing autophagy and inhibiting cell invasion. PRFR suppressed TNF-α-induced the expression of survival, proliferation and invasive proteins. This was, at least in part, due to the reduced values of the MAPKs, Akt, NF-κB and AP-1 signaling pathways. Therefore, these findings provide important new evidence that can assist researchers to gain a better understanding of the anti-cancer activity of PRFR, which can facilitate further investigations into its potential for use in anti-cancer therapy.

Author Contributions: Conceptualization, S.Y. and P.L.(D.); methodology, S.Y.; software, C.S.; validation, S.Y.; formal analysis, S.Y. and C.S.; data curation, C.S.; writing—original draft preparation, C.S. and S.Y.; writing—review and editing, P.L.(D.); visualization, S.Y.; supervision, S.Y. and P.L.(D.); project administration, S.Y. and P.L.(D.); funding acquisition, S.Y. and P.L.(D.).

Funding: This research study was granted financial support by the Faculty of Medicine, Chiang Mai University. (Grant No. 096/2560).

Acknowledgments: This research study was supported by Chiang Mai University and the Center for Research and Development of Natural Products for Health, Chiang Mai University.

Conflicts of Interest: The authors declare that they hold no conflicts of interest.

References

1. Bray, F.; Ferlay, J.; Soerjomataram, I.; Siegel, R.L.; Torre, L.A.; Jemal, A. Global cancer statistics 2018: GLOBOCAN estimates of incidence and mortality worldwide for 36 cancers in 185 countries. *CA A Cancer J. Clin.* **2018**, *68*, 394–424. [CrossRef] [PubMed]
2. Farbicka, P.; Nowicki, A. Palliative care in patients with lung cancer. *Contemp. Oncol. (Pozn)* **2013**, *17*, 238–245. [CrossRef] [PubMed]
3. Pope, C.A., III; Burnett, R.T.; Thun, M.J.; Calle, E.E.; Krewski, D.; Ito, K.; Thurston, G.D. Lung cancer, cardiopulmonary mortality, and long-term exposure to fine particulate air pollution. *JAMA* **2002**, *287*, 1132–1141. [CrossRef] [PubMed]
4. Beelen, R.; Hoek, G.; van den Brandt, P.A.; Goldbohm, R.A.; Fischer, P.; Schouten, L.J.; Armstrong, B.; Brunekreef, B. Long-Term Exposure to Traffic-Related Air Pollution and Lung Cancer Risk. *Epidemiology* **2008**, *19*, 702–710. [CrossRef] [PubMed]
5. Kim, H.-B.; Shim, J.-Y.; Park, B.; Lee, Y.-J. Long-Term Exposure to Air Pollutants and Cancer Mortality: A Meta-Analysis of Cohort Studies. *Int. J. Environ. Res. Public Health* **2018**, *15*, 2608. [CrossRef] [PubMed]
6. Araújo, A. Inflammation and Lung Cancer Oxidative Stress, ROS, and DNA Damage. *React. Oxyg. Spec. Biol. Hum. Health* **2016**, *1*, 215–223. [CrossRef]
7. Multhoff, G.; Molls, M.; Radons, J. Chronic inflammation in cancer development. *Front. Immunol.* **2012**, *2*. [CrossRef] [PubMed]
8. Shang, G.S.; Liu, L.; Qin, Y.W. IL-6 and TNF-alpha promote metastasis of lung cancer by inducing epithelial-mesenchymal transition. *Oncol. Lett.* **2017**, *13*, 4657–4660. [CrossRef]
9. Perez-Gracia, J.L.; Prior, C.; Guillén-Grima, F.; Segura, V.; Gonzalez, A.; Panizo, A.; Melero, I.; Grande-Pulido, E.; Gurpide, A.; Gil-Bazo, I. Identification of TNF-α and MMP-9 as potential baseline predictive serum markers of sunitinib activity in patients with renal cell carcinoma using a human cytokine array. *Br. J. Cancer* **2009**, *101*, 1876. [CrossRef] [PubMed]
10. Devin, A.; Lin, Y.; Liu, Z.G. The role of the death-domain kinase RIP in tumour-necrosis-factor-induced activation of mitogen-activated protein kinases. *EMBO Rep.* **2003**, *4*, 623–627. [CrossRef] [PubMed]
11. Sethi, G.; Sung, B.; Aggarwal, B.B. TNF: A master switch for inflammation to cancer. *Front. Biosci.* **2008**, *13*, 5094–5107. [CrossRef] [PubMed]
12. Wang, X.; Lin, Y. Tumor necrosis factor and cancer, buddies or foes? *Acta Pharm. Sin.* **2008**, *29*, 1275–1288. [CrossRef] [PubMed]
13. Guicciardi, M.E.; Gores, G.J. Life and death by death receptors. *FASEB J.* **2009**, *23*, 1625–1637. [CrossRef] [PubMed]
14. Van Herreweghe, F.; Festjens, N.; Declercq, W.; Vandenabeele, P. Tumor necrosis factor-mediated cell death: To break or to burst, that's the question. *Cell. Mol. Life Sci.* **2010**, *67*, 1567–1579. [CrossRef] [PubMed]
15. Harris, J. Autophagy and cytokines. *Cytokine* **2011**, *56*, 140–144. [CrossRef] [PubMed]
16. Safa, A.R. Roles of c-FLIP in Apoptosis, Necroptosis, and Autophagy. *J. Carcinog. Mutagen.* **2013**. [CrossRef]
17. Sahni, S.; Merlot, A.M.; Krishan, S.; Jansson, P.J.; Richardson, D.R. Gene of the month: BECN1. *J. Clin. Pathol.* **2014**, *67*, 656–660. [CrossRef]
18. Gabetta, B.; Fuzzati, N.; Griffini, A.; Lolla, E.; Pace, R.; Ruffilli, T.; Peterlongo, F. Characterization of proanthocyanidins from grape seeds. *Fitoterapia* **2000**, *71*, 162–175. [CrossRef]

19. Limtrakul, P.; Yodkeeree, S.; Pitchakarn, P.; Punfa, W. Anti-inflammatory effects of proanthocyanidin-rich red rice extract via suppression of MAPK, AP-1 and NF-κB pathways in Raw 264.7 macrophages. *Nutr. Res. Pract.* **2016**, *10*, 251–258. [CrossRef]
20. Pintha, K.; Yodkeeree, S.; Limtrakul, P. Proanthocyanidin in red rice inhibits MDA-MB-231 breast cancer cell invasion via the expression control of invasive proteins. *Biol. Pharm. Bull.* **2015**, *38*, 571–581. [CrossRef]
21. Upanan, S.; Yodkeeree, S.; Thippraphan, P.; Punfa, W.; Wongpoomchai, R.; Limtrakul Dejkriengkraikul, P. The Proanthocyanidin-Rich Fraction Obtained from Red Rice Germ and Bran Extract Induces HepG2 Hepatocellular Carcinoma Cell Apoptosis. *Molecules* **2019**, *24*, 813. [CrossRef] [PubMed]
22. Tanida, I.; Ueno, T.; Kominami, E. LC3 and Autophagy. In *Autophagosome and Phagosome*; Springer: Berlin/Heidelberg, Germany, 2008; pp. 77–88.
23. Koukourakis, M.I.; Kalamida, D.; Giatromanolaki, A.; Zois, C.E.; Sivridis, E.; Pouliliou, S.; Mitrakas, A.; Gatter, K.C.; Harris, A.L. Autophagosome Proteins LC3A, LC3B and LC3C Have Distinct Subcellular Distribution Kinetics and Expression in Cancer Cell Lines. *PLoS ONE* **2015**, *10*, e0137675. [CrossRef] [PubMed]
24. Gu, L.; Kelm, M.A.; Hammerstone, J.F.; Beecher, G.; Holden, J.; Haytowitz, D.; Gebhardt, S.; Prior, R.L. Concentrations of proanthocyanidins in common foods and estimations of normal consumption. *J. Nutr.* **2004**, *134*, 613–617. [CrossRef] [PubMed]
25. Levine, B.; Yuan, J. Autophagy in cell death: An innocent convict? *J. Clin. Investig.* **2005**, *115*, 2679–2688. [CrossRef] [PubMed]
26. Gozuacik, D.; Kimchi, A. Autophagy as a cell death and tumor suppressor mechanism. *Oncogene* **2004**, *23*, 2891. [CrossRef] [PubMed]
27. Sivaprasad, U.; Basu, A. Inhibition of ERK attenuates autophagy and potentiates tumour necrosis factor-α-induced cell death in MCF-7 cells. *J. Cell. Mol. Med.* **2008**, *12*, 1265–1271. [CrossRef] [PubMed]
28. FBauvy, D.-M.M.M.J. NF-kappaB activation represses tumor necrosis factor-alpha-induced autophagy. *J. Biol. Chem.* **2006**, *281*, 30373–30382.
29. Zhu, J.; Cai, Y.; Xu, K.; Ren, X.; Sun, J.; Lu, S.; Chen, J.; Xu, P. Beclin1 overexpression suppresses tumor cell proliferation and survival via an autophagy-dependent pathway in human synovial sarcoma cells. *Oncol. Rep.* **2018**, *40*, 1927–1936. [CrossRef]
30. Radeff-Huang, J.; Seasholtz, T.M.; Chang, J.W.; Smith, J.M.; Walsh, C.T.; Brown, J.H. Tumor necrosis factor-α-stimulated cell proliferation is mediated through sphingosine kinase-dependent Akt activation and cyclin D expression. *J. Biol. Chem.* **2007**, *282*, 863–870. [CrossRef]
31. Gupta, S.C.; Kim, J.H.; Prasad, S.; Aggarwal, B.B. Regulation of survival, proliferation, invasion, angiogenesis, and metastasis of tumor cells through modulation of inflammatory pathways by nutraceuticals. *Cancer Metastasis Rev.* **2010**, *29*, 405–434. [CrossRef]
32. Liu, B.; Qu, L.; Yan, S. Cyclooxygenase-2 promotes tumor growth and suppresses tumor immunity. *Cancer Cell Int.* **2015**, *15*, 106. [CrossRef] [PubMed]
33. Pang, L.Y.; Hurst, E.A.; Argyle, D.J. Cyclooxygenase-2: A role in cancer stem cell survival and repopulation of cancer cells during therapy. *Stem Cells Int.* **2016**, *2016*. [CrossRef]
34. Qiao, Y.; He, H.; Jonsson, P.; Sinha, I.; Zhao, C.; Dahlman-Wright, K. AP-1 is a key regulator of proinflammatory cytokine TNFα-mediated triple-negative breast cancer progression. *J. Biol. Chem.* **2016**, *291*, 5068–5079. [CrossRef] [PubMed]
35. Balkwill, F. TNF-α in promotion and progression of cancer. *Cancer Metastasis Rev.* **2006**, *25*, 409. [CrossRef] [PubMed]
36. Djavaheri-Mergny, M.; Amelotti, M.; Mathieu, J.; Besançon, F.; Bauvy, C.; Codogno, P. Regulation of autophagy by NF-kappaB transcription factor and reactives oxygen species. *Autophagy* **2007**, *3*, 390–392. [CrossRef] [PubMed]
37. Djavaheri-Mergny, M.; Amelotti, M.; Mathieu, J.; Besançon, F.; Bauvy, C.; Souquère, S.; Pierron, G.; Codogno, P. NF-κB activation represses tumor necrosis factor-α-induced autophagy. *J. Biol. Chem.* **2006**, *281*, 30373–30382. [CrossRef] [PubMed]
38. Jang, J.-Y.; Lee, J.-K.; Jeon, Y.-K.; Kim, C.-W. Exosome derived from epigallocatechin gallate treated breast cancer cells suppresses tumor growth by inhibiting tumor-associated macrophage infiltration and M2 polarization. *BMC Cancer* **2013**, *13*, 421. [CrossRef] [PubMed]

39. Pratheeshkumar, P.; Sreekala, C.; Zhang, Z.; Budhraja, A.; Ding, S.; Son, Y.-O.; Wang, X.; Hitron, A.; Hyun-Jung, K.; Wang, L. Cancer prevention with promising natural products: Mechanisms of action and molecular targets. *Anti-Cancer Agents Med. Chem. (Former. Curr. Med. Chem. Anti-Cancer Agents)* **2012**, *12*, 1159–1184. [CrossRef]
40. Noguchi, M.; Hirata, N.; Suizu, F. The links between AKT and two intracellular proteolytic cascades: Ubiquitination and autophagy. *Biochim. Biophys. Acta (Bba) Rev. Cancer* **2014**, *1846*, 342–352. [CrossRef]

Sample Availability: Samples of the compounds are not available from the authors.

© 2019 by the authors. Licensee MDPI, Basel, Switzerland. This article is an open access article distributed under the terms and conditions of the Creative Commons Attribution (CC BY) license (http://creativecommons.org/licenses/by/4.0/).

Article

CLEFMA Activates the Extrinsic and Intrinsic Apoptotic Processes through JNK1/2 and p38 Pathways in Human Osteosarcoma Cells

Jia-Sin Yang [1,2], Renn-Chia Lin [2,3,4,5], Yi-Hsien Hsieh [6], Heng-Hsiung Wu [7,8,9], Geng-Chung Li [6], Ya-Chiu Lin [2], Shun-Fa Yang [1,2,*] and Ko-Hsiu Lu [3,4,*]

1. Department of Medical Research, Chung Shan Medical University Hospital, Taichung 402, Taiwan
2. Institute of Medicine, Chung Shan Medical University, Taichung 402, Taiwan
3. Department of Orthopedics, Chung Shan Medical University Hospital, Taichung 402, Taiwan
4. School of Medicine, Chung Shan Medical University, Taichung 402, Taiwan
5. Division of Hyperbaric Oxygen Therapy and Wound Medicine, Chung Shan Medical University Hospital, Taichung 402, Taiwan
6. Institute of Biochemistry, Microbiology and Immunology, Chung Shan Medical University, Taichung 402, Taiwan
7. Graduate Institute of Biomedical Science, China Medical University, Taichung 404, Taiwan
8. Research Center of Tumor Medical Science, China Medical University, Taichung 404, Taiwan
9. Center for Molecular Medicine, China Medical University Hospital, Taichung 404, Taiwan
* Correspondence: ysf@csmu.edu.tw (S.-F.Y.); cshy307@csh.org.tw (K.-H.L.); Tel.: +886-4-24739595-34253 (S.-F.Y.)

Academic Editor: Roberto Fabiani
Received: 26 July 2019; Accepted: 5 September 2019; Published: 9 September 2019

Abstract: Due to the poor prognosis of metastatic osteosarcoma, chemotherapy is usually employed in the adjuvant situation to improve the prognosis and the chances of long-term survival. 4-[3,5-Bis(2-chlorobenzylidene)-4-oxo-piperidine-1-yl]-4-oxo-2-butenoic acid (CLEFMA) is a synthetic analog of curcumin and possesses anti-inflammatory and anticancer properties. To further obtain information regarding the apoptotic pathway induced by CLEFMA in osteosarcoma cells, microculture tetrazolium assay, annexin V-FITC/PI apoptosis staining assay, human apoptosis array, and Western blotting were employed. CLEFMA dose-dependently decreased the cell viabilities of human osteosarcoma U2OS and HOS cells and significantly induced apoptosis in human osteosarcoma cells. In addition to the effector caspase 3, CLEFMA significantly activated both extrinsic caspase 8 and intrinsic caspase 9 initiators. Moreover, CLEFMA increased the phosphorylation of extracellular signal-regulated protein kinases (ERK)1/2, c-Jun N-terminal kinases (JNK)1/2 and p38. Using inhibitors of JNK (JNK-in-8) and p38 (SB203580), CLEFMA's increases of cleaved caspases 3, 8, and 9 could be expectedly suppressed, but they could not be affected by co-treatment with the ERK inhibitor (U0126). Conclusively, CLEFMA activates both extrinsic and intrinsic apoptotic pathways in human osteosarcoma cells through JNK and p38 signaling. These findings contribute to a better understanding of the mechanisms responsible for CLEFMA's apoptotic effects on human osteosarcoma cells.

Keywords: apoptosis; CLEFMA; JNK; osteosarcoma; p38

1. Introduction

Osteosarcoma, the most common histological form of primary bone cancer, is most prevalent in teenagers and young adults [1,2]. Surgical *en bloc* resection of the cancer to achieve a complete radical excision has been the treatment of choice for osteosarcoma [2], but its prognosis is poor because of its highly metastatic potential. To decrease its high treatment failure and mortality rates, the combination of surgery and chemotherapy for osteosarcoma has increased long-term survival chances to approximately 68% through limb-sparing surgeries based on radiological staging, surgical techniques, and new chemotherapy protocols [2,3]. Nevertheless, potent metastatic lung diseases are still responsible for one of the most lethal pediatric malignancies to date. Because of this, novel agents that target particular intracellular signaling pathways related to the distinctive properties of osteosarcoma cells need to be developed.

Apoptosis, or programmed cell death, a key regulator of physiological growth control and regulation of tissue homeostasis, is characterized by typical morphological and biochemical hallmarks, including cell shrinkage, nuclear DNA fragmentation and membrane blebbing [4]. Multiple stress-inducible molecules, such as mitogen-activated protein kinase (MAPK)/extracellular signal-regulated protein kinase (ERK), c-Jun N-terminal kinase (JNK), and nuclear factor kappa B (NF-κB), have been implied in transmitting the apoptotic pathway [5,6]. To undergo apoptosis, the activation of important initiator and effector caspases would be initiated through the activation of the extrinsic (receptor) pathway or the stimulation of the intrinsic (mitochondria) pathway [7–9]. Currently, most anticancer strategies in clinical oncology focus on triggering apoptosis in cancer cells. On the contrary, failure to undergo apoptosis may result in treatment resistance. Thereby, understanding the molecular events that regulate apoptosis in response to chemotherapy provides novel opportunities to develop molecular-targeted therapy through the intrinsic and/or extrinsic pathways for osteosarcoma, which is very difficult to cure.

Curcumin (diferuloylmethane), a bright yellow chemical produced by Curcuma longa plants, has been shown to exhibit antioxidant, anti-inflammatory, antibacterial, antiviral, antifungal, and anticancer activities through the modulation of multiple cell signaling pathways [10]. The potent cytotoxic activity of curcumin on osteosarcoma cells has been reported to be mediated by the induction of multiple apoptotic processes [11–15]. However, even though curcumin is safe at high doses (12 g/day) for humans, many reasons, such as its poor absorption, rapid metabolism, and rapid systemic elimination, contribute to the low plasma and tissue levels of curcumin [16]. To improve the poor bioavailability of curcumin, numerous approaches have been undertaken, including the use of adjuvants and structural analogues of curcumin (e.g., EF24 [3,5-bis(2-fluorobenzylidene) piperidin-4-one]).

4-[3,5-Bis(2-chlorobenzylidene)-4-oxo-piperidine-1-yl]-4-oxo-2-butenoic acid (CLEFMA) is a synthetic analog of EF 24 and possesses anti-inflammatory and anticancer properties [17,18]. Using a reverse-phase high-performance liquid chromatography (HPLC) method to analyze the stability of the new drug, CLEFMA has been validated as a potential active anticancer drug-product [19]. In fact, various signaling pathways involved in diverse antitumor properties all depend on different specific tumor types and cell lines. Despite the absence of apoptosis, the curcuminoid CLEFMA has an anti-proliferative activity to induce autophagic cell death via oxidative stress in human lung adenocarcinoma H441 cells, offering an alternative mode of cell death in apoptosis-resistant cancers [17]. Moreover, CLEFMA-induced cell death and tumor growth suppression has been reported to be associated with the cleavage of caspases 3/9 and NF-κB-regulated anti-inflammatory and anti-metastatic effects [20]. As a potent diphenyldihaloketone analogue, CLEFMA has been developed over the past years as an anticancer agent [17]; nonetheless, the effect of CLEFMA on human osteosarcoma cell death remains unclear. Thus, we investigated whether CLEFMA affects the apoptosis of osteosarcoma and attempted to define its underlying mechanisms.

2. Results

2.1. Cytotoxicity of CLEFMA in Osteosarcoma U2OS and HOS Cells

To assess the cytotoxicity of CLEFMA on osteosarcoma U2OS and HOS cells, the [3-(4,5-dimethylthiazol-2-yl)-2,5-diphenyltetrazolium bromide] (MTT) assay was utilized. After 24 h of treatment, the viabilities of U2OS and HOS cells in the presence of concentrations of 5, 10, 20, 40 and 80 μM of CLEFMA were significantly different to that of the controls (0 μM) (Figure 1A,B), and both of the relationships were dose-dependent ($p < 0.001$ and $p < 0.001$). Moreover, a 24 h treatment with

20 µM of CLEFMA showed about a 50% reduction, while a 24 h treatment with 80 µM of CLEFMA decreased the cell viability of U2OS cells by about 90%. In HOS cells, there were reductions of about 70% in 20 µM and about 90% in 80 µM of CLEFMA.

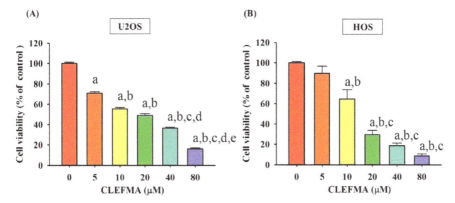

Figure 1. Effects of 4-[3,5-Bis(2-chlorobenzylidene)-4-oxo-piperidine-1-yl]-4-oxo-2-butenoic acid (CLEFMA) on the cell viability of U2OS and HOS cells. Using an [3-(4,5-dimethylthiazol-2-yl)-2,5-diphenyltetrazolium bromide] (MTT) assay, the viability of U2OS and HOS cells treated with CLEFMA (5, 10, 20, 40 and 80 µM) for 24 h was detected, and the effects are illustrated after quantitative analysis. Results are shown as mean ± S.D. (**A**) $n \geq 4$. ANOVA analysis with Scheffe's posteriori comparison was used. F = 386.619, $p < 0.001$. (**B**) $n \geq 4$. ANOVA analysis with Turkey's posteriori comparison was used. F = 53.288, $p < 0.001$. a: Significantly different, $p < 0.05$, when compared to control. b: Significantly different, $p < 0.05$, when compared to 5 µM. c: Significantly different, $p < 0.05$, when compared to 10 µM. d: Significantly different, $p < 0.05$, when compared to 20 µM. e: Significantly different, $p < 0.05$, when compared to 40 µM.

2.2. CLEFMA Induces the Apoptosis of U2OS and HOS Cells

To further examine the mechanism of CLEFMA inhibition of osteosarcoma cell proliferation, the annexin V-FITC/PI apoptosis assay was performed to test the viability of U2OS and HOS cells after a treatment of 5, 10, and 20 µM of CLEFMA for 24 h. The results revealed that the percentage of apoptotic cells was significantly increased in a dose-dependent manner (Figure 2A,B). These findings suggest that CLEFMA induced the apoptosis of osteosarcoma cells.

2.3. CLEFMA Increases the Expression of Cleaved Caspase 3 in U2OS Cells

To identify the underlying mechanism of apoptosis induced by CLEFMA in U2OS cells, we first employed the human apoptosis array to determine apoptosis-related proteins in U2OS cells. Consequently, obvious increases in the expression of cleaved caspase 3, HIF-1α, HO-1, HSP60, survivin and clusterin in U2OS cells were observed after treatment with 20 µM CLEFMA for 24 h. (Figure 3) Among them, the protein that increased the most in quantity was cleaved caspase 3, which was seven-fold that of the original, suggesting that the effector caspase 3 is responsible for the actual dismantling of the U2OS cell.

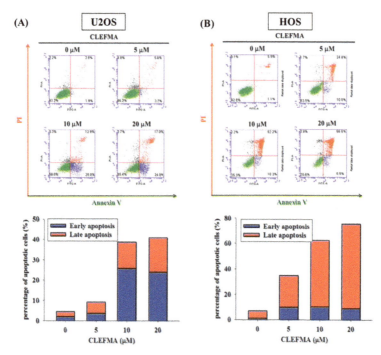

Figure 2. Effects of CLEFMA on the apoptosis of U2OS and HOS cells. (**A**) U2OS and (**B**) HOS cells were treated with CLEFMA (5, 10 and 20 μM) for 24 h and then subjected to flow cytometry after annexin V-FITC/PI staining. Cells that were considered viable were FITC annexin V and PI negative, cells that were in early apoptosis were FITC annexin V positive and PI negative, and cells that were in late apoptosis or already dead were both FITC annexin V and PI positive. Thus, the quantitative analysis of early apoptosis and late apoptosis was summarized to differentiate apoptosis from necrosis.

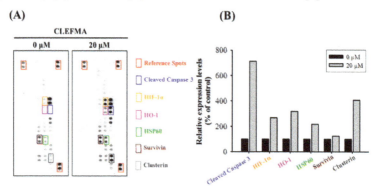

Figure 3. Effects of CLEFMA on the human apoptosis array in U2OS cells. (**A**) After treatment with 20 μM CLEFMA for 24 h in U2OS cells, the human apoptosis array, with 35 apoptosis-related proteins included, was employed as described in the Materials and Methods. (**B**) The five increased proteins were subjected to quantitative analysis.

2.4. CLEFMA Triggers Activation of the Caspase Cascade in U2OS Cells

To investigate the effect of CLEFMA on the caspase cascade in the apoptotic signaling pathway, the effector caspase 3 and its upstream initiators, caspases 8 and 9, as well as their cleaved forms were determined with Western blotting. After treatment with different concentrations of CLEFMA in U2OS cells for 24 h, the higher concentrations of CLEFMA corresponded to higher expressions of the cleaved forms of caspases 3, 8, and 9, in a dose-dependent manner ($p < 0.001$, $p < 0.001$ and $p < 0.001$, respectively), combined with the lesser expressions of caspases 3, 8, and 9, dose-dependently ($p < 0.001$, $p < 0.001$ and $p < 0.001$, respectively). (Figure 4A–C) Thus, we found that CLEFMA induces U2OS cell apoptosis by activating both extrinsic caspase 8- and intrinsic caspase 9-mediated pathways and their downstream effector caspase 3.

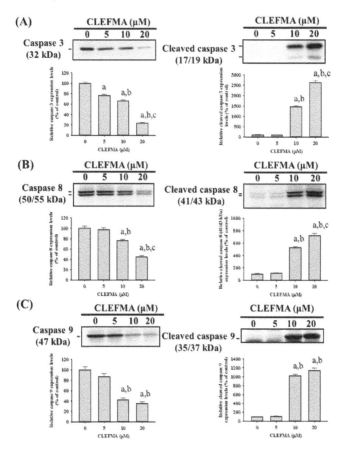

Figure 4. Effects of CLEFMA on the activation of caspases 3, 8 and 9 in U2OS cells. Western blot analysis for caspases 3, 8 and 9 and their active forms after various concentrations (5, 10 and 20 μM) of CLEFMA treatment for 24 h in U2OS cells were measured as described in the Materials and Methods. Subsequently, (**A**) caspase 3 and cleaved caspase 3, (**B**) caspase 8 and cleaved caspase 8, and (**C**) caspase 9 and cleaved caspase 9 were subjected to quantitative analysis. Results are shown as mean ± S.D.; n = 3. ANOVA analysis with Turkey's posteriori comparison was used. Caspase 3: F = 196.205, $p < 0.001$; cleaved caspase 3: F = 478.594, $p < 0.001$. Caspase 8: F = 51.604, $p < 0.001$; cleaved caspase 8: F = 205.373, $p < 0.001$. Caspase 9: F = 37.754, $p < 0.001$; cleaved caspase 9: F = 294.964, $p < 0.001$. a: Significantly different, $p < 0.05$, when compared to control. B: Significantly different, $p < 0.05$, when compared to 5 μM. c: Significantly different, $p < 0.05$, when compared to 10 μM.

2.5. CLEFMA Activates Extrinsic and Intrinsic Apoptotic Processes via JNK and p38 Pathways in U2OS Cells

Since MAPK pathways have been implicated as playing an important role in the action of chemotherapeutic drugs in the regulation of apoptosis and may be part of the signaling pathways that directly affect caspases 3, 8, and 9, the Western blot analysis was employed to further investigate the underlying molecular mechanisms. As shown in Figure 5A–C, CLEFMA increased the phosphorylation of ERK1/2, JNK1/2 and p38, dose-dependently, in U2OS cells ($p < 0.001$, $p < 0.001$ and $p < 0.001$, respectively), indicating that CLEFMA activates the phosphorylation of ERK1/2, JNK1/2 and p38 in U2OS cells. Furthermore, to identify whether the activation of ERK1/2, JNK1/2 and p38 phosphorylation by CLEFMA interferes with the actions of caspases 3, 8, and 9 of the extrinsic and intrinsic apoptotic processes in U2OS cells, we used inhibitors of ERK1/2 (U0126), JNK1/2 (JNK-in-8), and p38 (SB203580) with or without treatment with 20 µM CLEFMA in Western blotting. Cleaved caspases 3, 8, and 9 were activated by 20 µM of CLEFMA ($p < 0.001$, $p < 0.001$ and $p = 0.001$), as expected. (Figure 6) Intriguingly, inhibitors of JNK1/2 (JNK-in-8) and p38 (SB203580) significantly repressed CLEFMA's increase of cleaved caspases 3, 8 and 9 in U2OS cells (JNK-in-8: $p < 0.001$, $p < 0.001$ and $p = 0.013$; SB203580: $p < 0.001$, $p < 0.001$ and $p = 0.003$), but the inhibitor of ERK1/2 (U0126) did not suppress CLEFMA's increase of cleaved caspases 3, 8 and 9 (U0126: $p = 0.088$, $p = 0.568$ and $p = 0.990$). Overall, these findings indicated that JNK1/2 and p38 pathways play a critical upstream role in CLEFMA-mediated apoptosis of extrinsic caspase 8- and intrinsic caspase 9-mediated pathways and their downstream effector caspase 3 in U2OS cells.

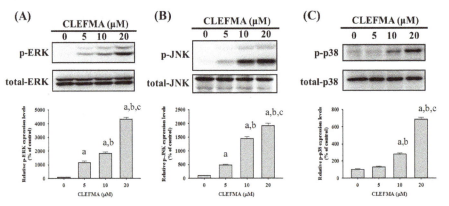

Figure 5. Effects of CLEFMA on the phosphorylation of ERK, c-Jun N-terminal kinases (JNK) and p38 in U2OS cells. Expressions of ERK1/2, JNK 1/2 and p38, as well as their phosphorylation after various concentrations (5, 10 and 20 µM) of CLEFMA treatment for 24 h in U2OS cells, were measured through Western blot analysis. Next, they were subjected to quantitative analysis. Results are shown as mean ± S.D.; $n = 3$. ANOVA analysis with Turkey's posteriori comparison was used. (**A**) p-ERK: F = 275.513, $p < 0.001$; (**B**) p-JNK: F = 205.474, $p < 0.001$; and (**C**) p = p38: F = 292.128, $p < 0.001$. a: Significantly different, $p < 0.05$, when compared to control. B: Significantly different, $p < 0.05$, when compared to 5 µM. c: Significantly different, $p < 0.05$, when compared to 10 µM.

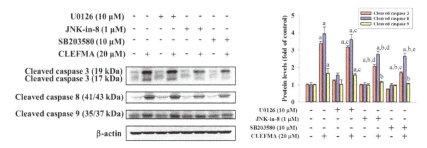

Figure 6. Effects of CLEFMA and inhibitors of ERK1/2 (U0126), JNK1/2 (JNK-in-8), and p38 (SB203580) on cleaved caspases 3, 8 and 9 expression of U2OS cells. Expressions of cleaved caspases 3, 8 and 9 after pretreatment with or without 10 µM of U0126, 1 µM of JNK-in-8, and 10 µM of SB203580 for 1 h followed by 20 µM or without CLEFMA treatment for an additional 24 h in U2OS cells were measured through Western blot analysis. Next, they were subjected to quantitative analysis. Results are shown as mean ± S.D.; $n = 3$. ANOVA analysis with Turkey's posteriori comparison was used. Cleaved caspase 3: $F = 502.398$, $p < 0.001$; Cleaved caspase 8: $F = 95.967$, $p < 0.001$; and cleaved caspase 9: $F = 10.543$, $p < 0.001$. a: Significantly different, $p < 0.05$, when compared to control. b: Significantly different, $p < 0.05$, when compared to 20 µM CLEFMA. c: Significantly different, $p < 0.05$, when compared to U0126. d: Significantly different, $p < 0.05$, when compared to JNK-in-8. e: Significantly different, $p < 0.05$, when compared to SB203580.

3. Discussion

In previous studies, curcumin has been reported to induce the apoptosis of human leukemia THP-1 cells through the activation of JNK/ERK pathways [21] and SHI-1 cells, possibly via both intrinsic and extrinsic pathways triggered by MAPKs (ERK, JNK and p38) signaling [22]. Also, curcumin exerts antitumor effects in retinoblastoma cells by regulating the JNK and p38 pathways [23], while this occurs through ERK1/2 and p38 signaling in malignant mesothelioma cells [24]. In human osteoclastoma cells, curcumin inhibits cell proliferation and promotes apoptosis through JNK, NF-κB and MMP-9 signaling pathways [25]. In spite of its efficacy and safety, curcumin has severely limited bioavailability because of its poor absorption and rapid metabolism [16].

After using the adjuvant to improve the poor bioavailability of curcumin, natural borneol and curcumin synergistically induce the apoptosis of human melanoma A375 cells with the involvement of the downregulation of Akt and ERK1/2 phosphorylation and the upregulation of phosphorylated JNK [26]. Similarly, the JNK/Bcl-2/Beclin1 pathway is thought to play a key role in the induction of apoptosis and autophagic cell death in breast cancer cells by the co-treatment of curcumin and berberine [27]. Additionally, synergistic inhibitory effects of cetuximab and curcumin on human cisplatin-resistant oral cancer CAR cells have been observed through the MAPK pathway and the intrinsic apoptotic process [28]. Moreover, curcumin-based photodynamic therapy induces breast cancer apoptosis through the activation of the ROS-mediated JNK/caspase-3 signaling pathway [29].

In managing patients diagnosed with any form of osteosarcoma, powerful chemotherapeutic drugs are the mainstay. Apart from adjuvants, structural analogues of curcumin (e.g., EF-24 and CLEFMA) have been undertaken to improve the bioavailability of curcumin for chemotherapy [16]. Although the synthetic curcuminoid CLEFMA developed over the past years has focused on anticancer effects against lung cancer cells [17,18,20], no research has been reported on the apoptotic process of CLEFMA in osteosarcoma cells. Here, we intriguingly found that CLEFMA decreases cell viabilities and induces cell apoptosis in human osteosarcoma U2OS and HOS cells.

Currently, the process of apoptosis is triggered by two different signaling pathways. The extrinsic apoptotic signal, which responded mainly to extracellular stimuli, involves death receptors, and the intrinsic apoptotic process, activated by modulators within the cell itself, involves the mitochondria [30,31]. The action of the cascade of caspases is required to conduct apoptosis signal transduction and execution. As in other reports, we discovered that effector caspase 3 plays a critical role in the underlying programs of apoptosis and relies on the activation of its upstream initiators including extrinsic caspase 8 and intrinsic caspase 9 [8,32].

By collecting information from various aspects of signal transduction cascades and cellular metabolism, both pathways continuously process this signaling, and eventually decide on the fate of cells. While CLEFMA's phosphorylation of ERK1/2, JNK1/2 and p38 in U2OS cells was observed in the study, we supposed that CLEFMA's induction of the extrinsic and intrinsic apoptotic pathways was achieved through these three MAPK pathways. Unexpectedly, CLEFMA's increases of cleaved caspases 3, 8, and 9 could be effectively inhibited by co-treatment with inhibitors of JNK (JNK-in-8) and p38 (SB203580), but co-treatment with the ERK inhibitor (U0126) had no effect on the increased effect. Therefore, these findings suggested that CLEFMA activates both extrinsic and intrinsic apoptotic pathways in U2OS cells through JNK and p38 signaling, but the ERK pathway is not involved. CLEFMA's increases of cleaved caspases 3, 8, and 9 could be effectively inhibited with the co-treatment of the ERK inhibitor (U0126), implying that the cleaved caspases 3, 8, and 9 are not the downstream of the CLEFMA's phosphorylation of ERK1/2.

4. Materials and Methods

4.1. Materials

Cell culture materials including Dulbecco's modified Eagle medium (DMEM) and fetal bovine serum (FBS) were purchased from Gibco-BRL (Gaithersburg, MD, USA) and Hyclone Laboratories, Inc. (Logan, UT, USA), respectively. Antibodies specific for p38, phosphorylated p38, β-actin, caspases 3 and 8, and FITC (fluorescein isothiocyanate-labeled) Annexin V Apoptosis Detection Kit I were obtained from BD Biosciences (San Jose, CA, USA). Human Apoptosis Array Kit was purchased from R&D Systems (Minneapolis, MN, USA). Additionally, antibodies specific for ERK1/2, JNK1/2, phosphorylated ERK1/2 and JNK1/2, caspases 9, and cleaved caspases 3, 8 and 9 were purchased from Cell Signaling Technology (Danvers, MA, USA). Unless otherwise specified, all chemicals used in this study were purchased from Sigma-Aldrich (St. Louis, MO, USA).

4.2. Cell Culture and CLEFMA Treatment

Obtained from the Food Industry Research and Development Institute (Hsinchu, Taiwan), the human osteosarcoma U2OS (15-year-old female) cells and HOS (13 year-old female) cells were supplemented with 10% FBS, 1% penicillin/streptomycin, and 5 mL glutamine while being cultured in DMEM and Eagle's MEM, respectively. The cell cultures were maintained at 37 °C in a humidified atmosphere of a 5% CO2 incubator. CLEFMA was purchased from Sigma-Aldrich (St. Louis, MO, USA).

4.3. Microculture Tetrazolium Colorimetric (MTT) Assay

To obtain information regarding the effect of apoptosis induced by CLEFMA, we subjected 8.5×10^4/well U2OS cells and 7.5×10^4/well HOS cells in 24-well plates for 16 h and treated them with different concentrations (5, 10, 20, 40 and 80 µM) of CLEFMA to assay cell viability via MTT [3-(4,5-dimethylthiazol-2-yl)-2,5-diphenyltetrazolium bromide] assay. After the 24 h exposure period, the media were removed and the U2OS and HOS cells were washed with phosphate-buffered saline. Afterwards, the medium was changed and the cells were incubated with MTT (0.5 mg/mL) for 4 h [33,34].

4.4. Annexin V-FITC Apoptosis Staining Assay

About 8.5×10^5 U2OS and HOS cells in one 6 cm plate were cultured and treated with different concentrations (0, 5, 10 and 20 µM) of CLEFMA for 24 h. Subsequently, U2OS cells were harvested with trypsinization together with floating non-viable cells. The FITC Annexin V Apoptosis Detection Kit I was used according to the manufacturer's protocols (BD Biosciences, San Jose, CA, USA); thereafter, the cell cycle analysis was measured by flow cytometry. Combined with PI staining, annexin V-FITC apoptosis staining was performed to differentiate apoptosis from necrosis.

4.5. Human Apoptosis Array

To explore the underlying mechanism of induced apoptosis, a Human Apoptosis Array Kit was used to evaluate protein lysates from vehicle- or 20 μM CLEFMA-treated cells for 24 h according to the manufacturer's protocols (R&D Systems, Minneapolis, MN). The kit detected 35 human apoptosis-related proteins simultaneously. Captured proteins were presented on the nitrocellulose membrane, detected with biotinylated detection antibodies, then finally visualized using chemiluminescent detection reagents.

4.6. Protein Extraction and Western Blot Analysis

To investigate the molecular mechanism further, the initiator and effector caspases and signaling pathways were detected using Western blot analysis. We plated 8.5×10^5 U2OS cells in 6 cm plates for 16 h and treated them with different concentrations (0, 5, 10 and 20 μM) of CLEFMA for 24 h, and the total cell lysates of U2OS cells were prepared as described previously [33–35]. Western blot analysis was performed using specific primary antibodies against caspases 3, 8 and 9, cleaved caspases 3, 8 and 9, and the specific antibodies for unphosphorylated or phosphorylated forms of the three corresponding MAPKs (ERK1/2, JNK1/2, and p38). As described previously, blots were then incubated with a horseradish peroxidase goat anti-rabbit or anti-mouse IgG for 1 h, and the intensity of each band was measured via densitometry [33–35].

4.7. Statistical Analysis

Statistical calculations of the data were performed using one-way analysis of variance (ANOVA) with post hoc Scheffe's and Turkey's tests for more than two groups with unequal and equal sample sizes per group, respectively. Each experiment was performed in triplicate, and three independent experiments were performed. Statistical significance was at $p < 0.05$.

5. Conclusions

Overall, these results demonstrated that CLEFMA decreases cell viabilities and induces the apoptosis of human osteosarcoma U2OS and HOS cells. By activating JNK and p38 pathways, but not via the ERK, both the extrinsic and intrinsic caspase cascades are triggered to induce the apoptosis of U2OS cells. Thus, CLEFMA may be a potential therapeutic agent against human osteosarcoma, whereas the therapeutic potential of CLEFMA combined with chemotherapy in osteosarcoma treatment should warrant evaluation in future research. Further tests are needed to investigate the detailed effects and possible mechanism of CLEFMA on the cell cycle progression and regulatory molecules of human osteosarcoma cells; however, animal studies are needed to justify CLEFMA as a promising candidate as a cytotoxic agent against osteosarcoma in vivo.

Author Contributions: Conceptualization, J.-S.Y., S.-F.Y. and K.-H.L.; methodology, R.-C.L., Y.-H.H., H.-H.W., G.-C.L., and Y.-C.L.; validation, J.-S.Y., S.-F.Y. and K.-H.L.; resources, S.-F.Y.; writing—original draft preparation, J.-S.Y., S.-F.Y. and K.-H.L.; writing—review and editing, J.-S.Y., S.-F.Y. and K.-H.L.

Funding: This research was funded by Chung Shan Medical University Hospital, Taiwan, grant number CSH-2017-D-003. This research was also funded by China Medical University, Taiwan (CMU 106-N-013).

Acknowledgments: The authors would like to express sincere thanks to Eric Wun-Hao Lu of American School in Taichung for proofreading.

Conflicts of Interest: The authors declare no conflict of interest.

References

1. Mirabello, L.; Troisi, R.J.; Savage, S.A. Osteosarcoma incidence and survival rates from 1973 to 2004: Data from the surveillance, epidemiology, and end results program. *Cancer* **2009**, *115*, 1531–1543. [CrossRef] [PubMed]
2. Ottaviani, G.; Jaffe, N. The epidemiology of osteosarcoma. *Cancer Treat. Res.* **2009**, *152*, 3–13. [PubMed]
3. Oertel, S.; Blattmann, C.; Rieken, S.; Jensen, A.; Combs, S.E.; Huber, P.E.; Bischof, M.; Kulozik, A.; Debus, J.; Schulz-Ertner, D. Radiotherapy in the treatment of primary osteosarcoma—A single center experience. *Tumori* **2010**, *96*, 582–588. [CrossRef]
4. Hengartner, M.O. The biochemistry of apoptosis. *Nature* **2000**, *407*, 770–776. [CrossRef]
5. Davis, R.J. Signal transduction by the jnk group of map kinases. *Cell* **2000**, *103*, 239–252. [CrossRef]
6. Karin, M.; Cao, Y.; Greten, F.R.; Li, Z.W. Nf-kappab in cancer: From innocent bystander to major culprit. *Nat. Rev. Cancer* **2002**, *2*, 301–310. [CrossRef]
7. Fulda, S.; Debatin, K.M. Targeting apoptosis pathways in cancer therapy. *Curr. Cancer Drug Targets* **2004**, *4*, 569–576. [CrossRef] [PubMed]
8. Lu, K.H.; Chen, P.N.; Lue, K.H.; Lai, M.T.; Lin, M.S.; Hsieh, Y.S.; Chu, S.C. 2'-hydroxyflavanone induces apoptosis of human osteosarcoma 143 b cells by activating the extrinsic trail- and intrinsic mitochondria-mediated pathways. *Nutr. Cancer* **2014**, *66*, 625–635. [CrossRef] [PubMed]
9. Degterev, A.; Boyce, M.; Yuan, J. A decade of caspases. *Oncogene* **2003**, *22*, 8543–8567. [CrossRef] [PubMed]
10. Kunnumakkara, A.B.; Bordoloi, D.; Harsha, C.; Banik, K.; Gupta, S.C.; Aggarwal, B.B. Curcumin mediates anticancer effects by modulating multiple cell signaling pathways. *Clin. Sci. (Lond.)* **2017**, *131*, 1781–1799. [CrossRef]
11. Chang, Z.; Xing, J.; Yu, X. Curcumin induces osteosarcoma mg63 cells apoptosis via ros/cyto-c/caspase-3 pathway. *Tumour. Biol.* **2014**, *35*, 753–758. [CrossRef] [PubMed]
12. Jin, S.; Xu, H.G.; Shen, J.N.; Chen, X.W.; Wang, H.; Zhou, J.G. Apoptotic effects of curcumin on human osteosarcoma u2os cells. *Orthop. Surg.* **2009**, *1*, 144–152. [CrossRef] [PubMed]
13. Lee, D.S.; Lee, M.K.; Kim, J.H. Curcumin induces cell cycle arrest and apoptosis in human osteosarcoma (hos) cells. *Anticancer Res.* **2009**, *29*, 5039–5044. [PubMed]
14. Singh, M.; Pandey, A.; Karikari, C.A.; Singh, G.; Rakheja, D. Cell cycle inhibition and apoptosis induced by curcumin in ewing sarcoma cell line sk-nep-1. *Med. Oncol.* **2010**, *27*, 1096–1101. [CrossRef] [PubMed]
15. Walters, D.K.; Muff, R.; Langsam, B.; Born, W.; Fuchs, B. Cytotoxic effects of curcumin on osteosarcoma cell lines. *Invest. New Drugs* **2008**, *26*, 289–297. [CrossRef]
16. Anand, P.; Kunnumakkara, A.B.; Newman, R.A.; Aggarwal, B.B. Bioavailability of curcumin: Problems and promises. *Mol. Pharm.* **2007**, *4*, 807–818. [CrossRef] [PubMed]
17. Lagisetty, P.; Vilekar, P.; Sahoo, K.; Anant, S.; Awasthi, V. Clefma-an anti-proliferative curcuminoid from structure-activity relationship studies on 3,5-bis(benzylidene)-4-piperidones. *Bioorg. Med. Chem.* **2010**, *18*, 6109–6120. [CrossRef]
18. Sahoo, K.; Dozmorov, M.G.; Anant, S.; Awasthi, V. The curcuminoid clefma selectively induces cell death in h441 lung adenocarcinoma cells via oxidative stress. *Investig. New Drugs* **2012**, *30*, 558–567. [CrossRef]
19. Raghuvanshi, D.; Nkepang, G.; Hussain, A.; Yari, H.; Awasthi, V. Stability study on an anti-cancer drug 4-(3,5-bis(2-chlorobenzylidene)-4-oxo-piperidine-1-yl)-4-oxo-2-butenoic acid (clefma) using a stability-indicating hplc method. *J. Pharm. Anal.* **2017**, *7*, 1–9. [CrossRef]
20. Yadav, V.R.; Sahoo, K.; Awasthi, V. Preclinical evaluation of 4-[3,5-bis(2-chlorobenzylidene)-4-oxo-piperidine-1-yl]-4-oxo-2-butenoic acid, in a mouse model of lung cancer xenograft. *Br. J. Pharmacol.* **2013**, *170*, 1436–1448. [CrossRef]
21. Yang, C.W.; Chang, C.L.; Lee, H.C.; Chi, C.W.; Pan, J.P.; Yang, W.C. Curcumin induces the apoptosis of human monocytic leukemia thp-1 cells via the activation of jnk/erk pathways. *BMC Complement. Altern. Med.* **2012**, *12*, 22. [CrossRef] [PubMed]
22. Zhu, G.H.; Dai, H.P.; Shen, Q.; Ji, O.; Zhang, Q.; Zhai, Y.L. Curcumin induces apoptosis and suppresses invasion through mapk and mmp signaling in human monocytic leukemia shi-1 cells. *Pharm. Biol.* **2016**, *54*, 1303–1311. [CrossRef] [PubMed]
23. Yu, X.; Zhong, J.; Yan, L.; Li, J.; Wang, H.; Wen, Y.; Zhao, Y. Curcumin exerts antitumor effects in retinoblastoma cells by regulating the jnk and p38 mapk pathways. *Int. J. Mol. Med.* **2016**, *38*, 861–868. [CrossRef] [PubMed]

24. Masuelli, L.; Benvenuto, M.; Di Stefano, E.; Mattera, R.; Fantini, M.; De Feudis, G.; De Smaele, E.; Tresoldi, I.; Giganti, M.G.; Modesti, A.; et al. Curcumin blocks autophagy and activates apoptosis of malignant mesothelioma cell lines and increases the survival of mice intraperitoneally transplanted with a malignant mesothelioma cell line. *Oncotarget* **2017**, *8*, 34405–34422. [CrossRef] [PubMed]
25. Cao, F.; Liu, T.; Xu, Y.; Xu, D.; Feng, S. Curcumin inhibits cell proliferation and promotes apoptosis in human osteoclastoma cell through mmp-9, nf-kappab and jnk signaling pathways. *Int. J. Clin. Exp. Pathol.* **2015**, *8*, 6037–6045. [PubMed]
26. Chen, J.; Li, L.; Su, J.; Li, B.; Chen, T.; Wong, Y.S. Synergistic apoptosis-inducing effects on a375 human melanoma cells of natural borneol and curcumin. *PLoS ONE* **2014**, *9*, e101277. [CrossRef] [PubMed]
27. Wang, K.; Zhang, C.; Bao, J.; Jia, X.; Liang, Y.; Wang, X.; Chen, M.; Su, H.; Li, P.; Wan, J.B.; et al. Synergistic chemopreventive effects of curcumin and berberine on human breast cancer cells through induction of apoptosis and autophagic cell death. *Sci. Rep.* **2016**, *6*, 26064. [CrossRef] [PubMed]
28. Chen, C.F.; Lu, C.C.; Chiang, J.H.; Chiu, H.Y.; Yang, J.S.; Lee, C.Y.; Way, T.D.; Huang, H.J. Synergistic inhibitory effects of cetuximab and curcumin on human cisplatin-resistant oral cancer car cells through intrinsic apoptotic process. *Oncol. Lett.* **2018**, *16*, 6323–6330. [CrossRef]
29. Sun, M.; Zhang, Y.; He, Y.; Xiong, M.; Huang, H.; Pei, S.; Liao, J.; Wang, Y.; Shao, D. Green synthesis of carrier-free curcumin nanodrugs for light-activated breast cancer photodynamic therapy. *Colloids Surf. B Biointerfaces* **2019**, *180*, 313–318. [CrossRef]
30. Gazitt, Y.; Kolaparthi, V.; Moncada, K.; Thomas, C.; Freeman, J. Targeted therapy of human osteosarcoma with 17aag or rapamycin: Characterization of induced apoptosis and inhibition of mtor and akt/mapk/wnt pathways. *Int. J. Oncol.* **2009**, *34*, 551–561. [CrossRef]
31. Park, H.; Bergeron, E.; Senta, H.; Guillemette, K.; Beauvais, S.; Blouin, R.; Sirois, J.; Faucheux, N. Sanguinarine induces apoptosis of human osteosarcoma cells through the extrinsic and intrinsic pathways. *Biochem. Biophys. Res. Commun.* **2010**, *399*, 446–451. [CrossRef] [PubMed]
32. Fulda, S.; Debatin, K.M. Extrinsic versus intrinsic apoptosis pathways in anticancer chemotherapy. *Oncogene* **2006**, *25*, 4798–4811. [CrossRef] [PubMed]
33. Hsieh, Y.S.; Chu, S.C.; Yang, S.F.; Chen, P.N.; Liu, Y.C.; Lu, K.H. Silibinin suppresses human osteosarcoma mg-63 cell invasion by inhibiting the erk-dependent c-jun/ap-1 induction of mmp-2. *Carcinogenesis* **2007**, *28*, 977–987. [CrossRef] [PubMed]
34. Lu, K.H.; Yang, H.W.; Su, C.W.; Lue, K.H.; Yang, S.F.; Hsieh, Y.S. Phyllanthus urinaria suppresses human osteosarcoma cell invasion and migration by transcriptionally inhibiting u-pa via erk and akt signaling pathways. *Food Chem. Toxicol.* **2013**, *52*, 193–199. [CrossRef] [PubMed]
35. Lu, K.H.; Chen, P.N.; Hsieh, Y.H.; Lin, C.Y.; Cheng, F.Y.; Chiu, P.C.; Chu, S.C.; Hsieh, Y.S. 3-hydroxyflavone inhibits human osteosarcoma u2os and 143b cells metastasis by affecting emt and repressing u-pa/mmp-2 via fak-src to mek/erk and rhoa/mlc2 pathways and reduces 143b tumor growth in vivo. *Food Chem. Toxicol.* **2016**, *97*, 177–186. [CrossRef] [PubMed]

Sample Availability: Not available.

© 2019 by the authors. Licensee MDPI, Basel, Switzerland. This article is an open access article distributed under the terms and conditions of the Creative Commons Attribution (CC BY) license (http://creativecommons.org/licenses/by/4.0/).

Article

SB365, *Pulsatilla* Saponin D Induces Caspase-Independent Cell Death and Augments the Anticancer Effect of Temozolomide in Glioblastoma Multiforme Cells

Jun-Man Hong [1], Jin-Hee Kim [2], Hyemin Kim [3], Wang Jae Lee [1] and Young-il Hwang [1],*

[1] Department of Anatomy and Cell Biology, Seoul National University College of Medicine, Seoul 03080, Korea
[2] Department of Biomedical Laboratory Science, Cheongju University, Cheongju 28503, Korea
[3] Research Institute for Future Medicine, Samsung Medical Center, Seoul 06351, Korea
* Correspondence: hyi830@snu.ac.kr; Tel.: +822-740-8209; Fax: +822-745-9528

Academic Editor: Roberto Fabiani
Received: 8 August 2019; Accepted: 4 September 2019; Published: 5 September 2019

Abstract: SB365, a saponin D extracted from the roots of *Pulsatilla koreana*, has been reported to show cytotoxicity in several cancer cell lines. We investigated the effects of SB365 on U87-MG and T98G glioblastoma multiforme (GBM) cells, and its efficacy in combination with temozolomide for treating GBM. SB365 exerted a cytotoxic effect on GBM cells not by inducing apoptosis, as in other cancer cell lines, but by triggering caspase-independent cell death. Inhibition of autophagic flux and neutralization of the lysosomal pH occurred rapidly after application of SB365, followed by deterioration of mitochondrial membrane potential. A cathepsin B inhibitor and N-acetyl cysteine, an antioxidant, partially recovered cell death induced by SB365. SB365 in combination with temozolomide exerted an additive cytotoxic effect in vitro and in vivo. In conclusion, SB365 inhibits autophagic flux and induces caspase-independent cell death in GBM cells in a manner involving cathepsin B and mainly reactive oxygen species, and its use in combination with temozolomide shows promise for the treatment of GBM.

Keywords: *Pulsatilla* saponin D; SB365; glioblastoma multiforme; temozolomide; autophagic flux inhibition; lysosomal membrane permeabilization; mitochondrial membrane potential

1. Introduction

Glioblastoma multiforme (GBM) is the most frequent and most malignant brain tumor, with a mean survival of GBM patients of less than 2 years [1]. Although several therapeutic modalities including immunotherapies are under development [2], the standard therapy for newly diagnosed GBM is surgical resection within a maximum range followed by concomitant chemotherapy and radiotherapy [2,3]. For chemotherapy, temozolomide (TMZ) is the drug of choice [4]. TMZ is an oral alkylating agent that induces DNA methylation at the O^6 position of guanine. The resultant O^6-methylguanine is abnormally paired with thymine, leading to cleavage of DNA strands by the mismatch-repair system, which triggers apoptosis [5]. TMZ is suitable for treating GBM because it can pass the blood–brain barrier [6]. However, resistance to TMZ can be induced in GBM cells by expression of p53, p21, or O^6-methylguanine-DNA methyltransferase (MGMT) [7]. Furthermore, TMZ has side effects such as genotoxicity, fetal toxicity, and lymphocytopenia of T cells and NK cells [8].

Combinations of drugs are typically used to reduce the likelihood of toxicity and side effects [9]. In patients with GBM, combinations of TMZ with inhibitors of autophagic flux (e.g., hydroxychloroquine) have been developed, on the basis that blocking autophagy should enhance the effects of TMZ because autophagy protects against the toxicity of radiotherapy and TMZ [10]. However, such combinations can cause side effects such as anemia, maculopapular rash, hemolysis, and decreased platelet and immune cell counts [10].

SB365 is a saponin D, hederagenin 3-O-α-L-rhamnopyranosyl(1→2)-(β-D-glucopyranosyl(1→4))-α-L-arabinopyranoside, which is extracted from the roots of *Pulsatilla koreana* [11]. Among eight lupane- and nine oleanane-type saponins extracted from *P. koreana*, SB365 showed the greatest antitumor activity in vitro against A-549 (lung cancer), SK-OV-3 (ovarian cancer), SK-MEL-2 (melanoma), and HCT-15 (colon cancer) cells. Indeed, its effect was superior to those of Taxol and doxorubicin [12]. In immunocompromised mice, SB365 suppressed the proliferation of human Huh-7 (liver cancer), MKN-45 (gastric cancer), PANC-1 (pancreatic cancer), and HT-29 (colon cancer) cells, without weight loss or toxicity to normal tissue [13–16]. In a clinical trial involving patients with stage IV pancreatic cancer, SB365 increased the survival rate without inducing side effects [17].

SB365 is reported to induce apoptosis of cancer cells in vitro [13–16,18] and to inhibit the autophagic flux in HeLa (cervical cancer), K562 (leukemia), B16-F10 (melanoma), A549 (lung cancer), and MCF-7 (breast cancer) cells. Moreover, SB365 additively enhanced the anticancer activity of the chemotherapeutic agents 5-fluorouracil, camptothecin, and etoposide in HeLa cells in vitro [19].

The effects of SB365 on GBM cells have, to our knowledge, not yet been investigated. Furthermore, if it inhibits autophagic flux in GBM cells, SB365 in combination with TMZ could be used for the treatment of GBM, replacing chloroquine or hydroxychloroquine.

The aim of this study was to investigate the effects of SB365 alone and in combination with TMZ on GBM cells in vitro and in vivo. To this end, we selected two GBM cell lines, U87-MG and T98G, among dozens of them. These are of human grade IV glioma cells [20]. We selected them because they are the most extensively employed ones in related studies [21], and especially they possess opposite characteristics to the susceptibility to TMZ. U87-MG cells are susceptible to TMZ, while T98G cells are not. T98G cells express O^6-methylguanine-DNA methyltransferase (MGMT), which removes the methyl group at the O^6 position of guanine added by TMZ [22], rendering them resistant to this drug. The survival duration of patients with MGMT-expressing GBM is approximately two years less than that of patients with non-functional methylated MGMT genes [23].

2. Results

2.1. SB365 Inhibited the Proliferation of GBM Cells In Vitro

The proliferation of U87-MG cells treated with SB365 was assayed after 24, 48, and 72 h (Figure 1). At 24 h, cell proliferation was comparable to that of the control group (Figure 1A), irrespective of SB365 concentration. However, after 48 h, 20 μM SB365 reduced cell proliferation by ~30% compared to the control (Figure 1B). After 72 h, 2.5 and 20 μM SB365 reduced cell proliferation by 25% and 80%, respectively, compared to the control ($p < 0.001$) (Figure 1C). Similar results were obtained using TMZ-resistant T98G cells (Supplementary Materials, Figure S1). Calculated half maximal inhibitory concentration (IC50) for 72 h treatment was 8.9 μM.

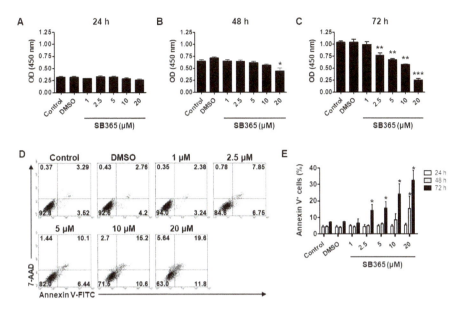

Figure 1. SB365 exerted a cytotoxic effect on U87-MG cells. (**A–C**) SB365 inhibited the proliferation of U87-MG cells. The cells in 96-well plates were treated with SB365 at the indicated concentrations for (**A**) 24, (**B**) 48, or (**C**) 72 h in quadruplicate, and subjected to CCK-8 assay. (**D,E**) SB365 increased the frequency of the annexin V-positive cells. U87-MG cells in six-well plates were treated as above, stained with annexin V and 7-AAD, and subjected to FACS analysis. (**D**) A representative FACS profile after 72 h and (**E**) the frequency of annexin V-positive cells. Experiments were performed independently in triplicate. * $p < 0.05$, ** $p < 0.01$, and *** $p < 0.001$ vs the control.

Moreover, after 24 h, flow cytometry showed that SB365 did not significantly increase the frequency of annexin V-positive cells (Figure 1E and Supplementary Materials Figure S2A). After 48 h, 20 μM SB365 resulted in a significant increase in the frequency of annexin V-positive cells (Supplementary Materials Figure S2B). After 72 h, the frequency of annexin V-positive cells increased by 2.5–20 μM SB365 in a dose-dependent manner (Figure 1D,E). Similar results were obtained using TMZ-resistant T98G cells (Supplementary Materials Figure S3).

2.2. SB365 Induced the Death of GBM Cells in a Caspase-Independent Manner

The cytotoxic effect of SB365 in cancer cells is mediated by apoptosis [13–16,18]. Since FACS showed the presence of few cells in the early stage of the apoptotic process, which are 7-AAD-negative and annexin V-positive [24], we furthered explored SB365-induced apoptosis of U87-MG cells.

The level of cleaved caspase-3, the final caspase of the intrinsic and extrinsic apoptosis pathways [25], in cells treated with 10 μM SB365 for 72 h was evaluated by western blotting (Figure 2A,B). SB365 triggered cleavage of caspase-3 in HT-29 and Huh-7 cells, as reported previously [13,14], but not in U87-MG cells. When the cells were stained with DAPI, SB365-treated HT-29 and Huh-7 cells showed nuclear blebbing and/or fragmentation with a frequency of 1–4 nuclei per a high-power field. However, SB365-treated U87-MG cells showed round or oval nuclei without blebbing and fragmentation (Figure 2C). Thus, SB365 induced caspase-independent cell death (CICD) rather than caspase-dependent apoptosis in U87-MG cells. Similar results were obtained using T98G cells (Supplementary Materials Figure S4).

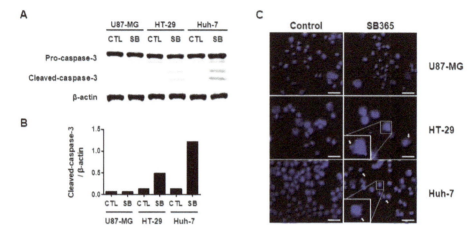

Figure 2. SB365 induced caspase-independent death in U87-MG cells. U87-MG, HT-29 (1 × 10^5/well), and Huh-7 cells (1 × 10^5/well) in six-well plates were treated with 10, 5, and 15 µM SB365, respectively. The calculated IC50 values of SB365 on each cell line were 8.9, 5.1, and 13.2 µM, respectively. (**A**) Cell lysates were subjected to western blotting of caspase-3 cleavage, (**B**) followed by densitometry. (**C**) SB365 induced nuclear fragmentation in HT-29 and Huh-7 cells, but not in U87-MG cells. Cells were treated with 10 µM SB365 for 72 h, adhered to an eight-well multispot slide, and stained with DAPI (blue). Arrows indicate fragmented nuclei. Images were acquired using a fluorescence microscope (× 400). The scale bar represents 50 µm. CTL, control group; SB, SB365-treated group.

2.3. SB365 Induced Autophagic Flux Inhibition in GBM Cells

SB365 reportedly inhibits autophagic flux in HeLa, K562, A549, and MCF-7 cells [19]. Given that autophagy protects against cell damage [26], its inhibition could be involved in SB365-induced death in GBM cells. Thus, we evaluated whether SB365 inhibited autophagic flux in U87-MG cells.

The cells were treated with 10 µM SB365, and the expression of microtubule-associated protein light chain 3 (LC3)-I, II, and p62 was evaluated by western blotting within 24 h. When autophagy is induced, LC3-I is converted to LC3-II in combination with phosphatidylethanolamine in the cytosol to produce autophagosomes, and *p62* binds to ubiquitinated proteins and pulls them into autophagosomes to be decomposed due to subsequent autophagic flux [27]. When the autophagic flux is inhibited, LC3-II and p62 accumulate in the cell [28]. Thus, the LC3-II/I ratio and *p62* were regarded as indicators of autophagic flux inhibition.

The *p62* level and LC3-II/I ratio (Figure 3A,B) increased in a time-dependent manner, indicating that SB365 inhibits autophagic flux. The *p62* level and LC3-II/I ratio in U87-MG and T98G cells remained high until 72 h (Supplementary Materials Figure S5), but the expression of beclin-1 did not change significantly (Figure 3 and Supplementary Materials Figure S5).

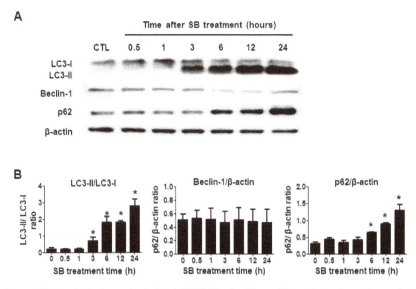

Figure 3. SB365 inhibited autophagic flux in U87-MG cells. Western blot analysis of autophagy-related proteins within 24 h of treatment with SB365. U87-MG cells in a six-well plate were treated with 10 μM SB365 for the indicated times. (**A**) Cell lysates were subjected to western blotting for LC3-I, II, beclin-1, and *p*62, and (**B**) the LC3-II/I, beclin-1/β-actin, and *p*62/β-actin ratios were calculated. The experiment was performed independently in triplicate. * $p < 0.05$ vs the control.

2.4. Inhibition of Autophagic Flux by SB365 is Linked to Lysosomal Neutralization and Reduction of MMP

Since inhibition of autophagic flux is associated with lysosomal dysfunction such as neutralization and permeabilization [29], we performed a lysosomal stability test. Cells were stained with acridine orange and analyzed by flow cytometry. The frequency of cells emitting red fluorescence decreased by 65% at 6 h after SB365 treatment compared to the control and decreased steadily thereafter ($p = 0.05$) (Figure 4A,C).

Next, we measured alterations in mitochondrial membrane potential (MMP), which typically occur after lysosomal dysfunction [30]. Cells were treated with 10 μM SB365 as above, stained with JC-1 for 20 min, and analyzed by flow cytometry. The frequencies of cells with altered MMP were 5.8% and 8.6% higher at 36 and 48 h after SB365 treatment, respectively, compared to the control ($p = 0.01$) (Figure 4B,C).

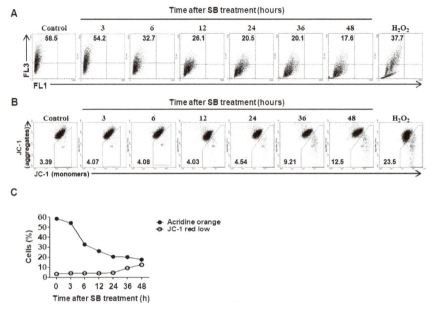

Figure 4. SB365 deteriorated lysosomal stability and mitochondrial membrane potential (MMP) in U87-MG cells. (**A**) SB365 induced lysosomal pH neutralization in U87-MG cells. Cells were treated with 10 µM SB365 for the indicated times, stained with (**A**) 3 µg/mL acridine orange for lysosomal stability measurement. (**B**) SB365 induced mitochondrial depolarization in U87-MG cells. Cells were stained with 2.5 µM JC-1 for 20 min for MMP measurement, harvested, and analyzed by flow cytometry. Cells treated with 0.5 mM H_2O_2 for 2 h constituted the positive control. (**C**) Combination of (**A**) and (**B**). The experiment was performed independently in triplicate.

2.5. Cathepsin B and Reactive Oxygen Species Contribute to SB365-Induced Cell Death

Since lysosomal membrane permeabilization (LMP) is a frequent cause of lysosomal dysfunction, and leads to leakage of cathepsin B and/or cathepsin D from the lysosome into the cytoplasm, resulting in cell death [31–33], we determined whether SB365-induced cell death was due to leakage of cathepsins. To this end, cell proliferation was evaluated 72 h after SB365 treatment in the presence or absence of cathepsin inhibitors. A cathepsin B inhibitor II at 5 µM recovered the cell proliferation inhibited by SB365 by ≥40% ($p = 0.05$) (Figure 5A) and reduced the frequency of cells with altered MMP (Figure 5B). However, a cathepsin D inhibitor (pepstatin A) exerted no such effects (data not shown).

Next, we evaluated whether reactive oxygen species (ROS) were related to SB365-induced cell death, because autophagic flux inhibition [34,35] and MMP deterioration [36] increase intracellular ROS levels, leading to cell death. Cells were treated with the indicated concentrations of the antioxidant *N*-acetyl cysteine (NAC) 1 h after SB365 exposure. After 72 h, NAC recovered the suppression of proliferation caused by SB365 (by ~30% at 0.625 mM and 50% at 2.5 mM) (Figure 5C). However, NAC at 5 mM did not recover the inhibition of cell proliferation. NAC exerted a similar effect in T98G cells, albeit to a lesser degree (Supplementary Materials Figure S6). Considering that MMP deterioration started late during the experiment time (Figure 4B,C) and that NAC could decompose in culture media, we performed the same experiment with 2.5 mM NAC, which at this time was added 24 and 48 h after SB365 treatment, instead of 1 h after (Figure 5D). As a result, NAC recovered the cytotoxicity by SB365 up to over 70% when added at 48 h.

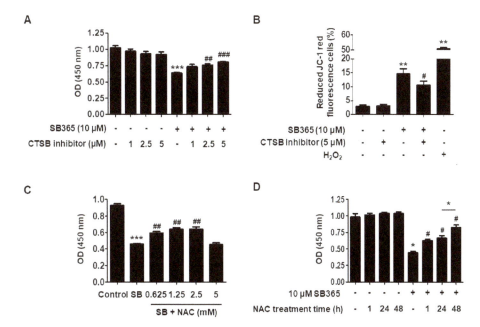

Figure 5. SB365 induced cell death via cathepsin B and ROS in U87-MG cells. (**A**) A cathepsin B inhibitor partially restored inhibited proliferation of U87-MG cells induced by SB365. Cells were cultured in 96-well plates, treated with 10 μM SB365 for 72 h in the presence of the indicated concentrations of cathepsin B inhibitor, and subjected to CCK-8 assay. (**B**) A cathepsin B inhibitor partially recovered SB365 induced MMP deterioration. U87-MG cells were treated with 10 μM SB365 for 72 h in the presence of 5 μM cathepsin B inhibitor, stained with JC-1, and MMP was analyzed by FACS. (**C**) NAC partially reduced the anti-proliferative effect of SB365 in U87-MG cells. Cells were cultured in 96-well plates, treated with 10 μM SB365 for 72 h in the presence of the indicated concentrations of NAC, and subjected to CCK-8 assay. NAC was added to the culture medium 1 h after SB365 treatment. (**D**) The same experiments were performed as in (**C**) with 2.5 mM NAC. However, NAC was treated 24 and 48 h after SB365 treatment, in addition to 1 h treatment. Quadruplicate samples were analyzed independently in triplicate. * $p < 0.05$, ** $p < 0.01$, *** $p < 0.001$ vs the control; # $p < 0.05$, ## $p < 0.01$, and ### $p < 0.001$ vs the SB365 group. CTSB, cathepsin B; NAC, N-acetyl cysteine.

2.6. SB365 and TMZ Additively Inhibited the Proliferation of GBM Cells In Vitro

Since SB365 inhibited autophagic flux in GBM cells, we evaluated its influence on the anticancer activity of TMZ, like other autophagic flux inhibitors such as hydroxychloroquine [10].

U87MG cells were treated with TMZ in the presence or absence of 10 μM SB365 for 72 h, and their proliferation was determined by CCK-8 assay. TMZ alone at 25 and 50 μM inhibited cell proliferation by 37% and 46%, respectively, compared to the control ($p < 0.001$) (Figure 6A). Lower concentrations of TMZ (6.25 and 12.5 μM) also inhibited cell proliferation, albeit not significantly. The combination of TMZ (6.25, 12.5, 25, and 50 μM) and SB365 inhibited cell proliferation by 46%, 48%, 56%, and 63%, respectively ($p = 0.016$) (Figure 6B). Similar results were obtained using T98G cells (Supplementary Materials Figure S7). At low TMZ concentrations, the combination exerted an additive effect on cell proliferation. That is, the combination of 10 μM SB365 with 6.25 and 12.5 μM TMZ increased the inhibition of cell proliferation from 6% to 46%, and from 10% to 48%, respectively (Supplementary Materials Table S1).

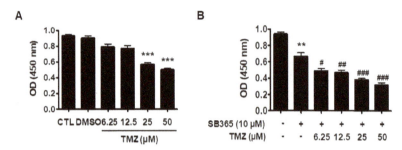

Figure 6. SB365 augmented the cytotoxic effect of TMZ on U87-MG cells. Cells were cultured in 96-well plates, treated with the indicated concentrations of TMZ in the (**A**) absence or (**B**) presence of 10 μM SB365, and subjected to CCK-8 assay. Quadruplicate samples were analyzed independently in triplicate. ** $p < 0.01$, and *** $p < 0.001$ vs the control; # $p < 0.05$, ## $p < 0.01$, and ### $p < 0.001$ vs the SB365 group. CTL, control; TMZ, temozolomide.

2.7. SB365 Inhibited Tumor Growth in the Mouse U87-MG Xenograft Model

Based on the above in vitro results, the effects of SB365 and/or TMZ on tumor growth in vivo were investigated. U87-MG cells were inoculated into both flanks of nude mice. When the tumor volume reached 100–200 mm^3, SB365 (5 mg/kg/every other day, intratumoral) and/or TMZ (2.5 mg/kg/day, intraperitoneal) were administered until day 22. The doses were determined based on previous reports and the results of a pilot study (data not shown). No marked change in body weight was detected (Figure 7A).

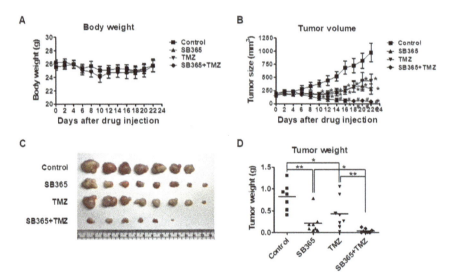

Figure 7. Combination of SB365 with TMZ additively suppressed the growth of U87-MG tumors in a mouse xenograft model. U87-MG cells were subcutaneously inoculated into both flanks of nude mice. When the tumor reached a volume of ~100–200 mm^3, mice were intratumorally administered with SB365 every other day and/or with TMZ intraperitoneally daily for 22 days. The control received vehicle (<3% DMSO). (**A**) Body weight and (**B**) tumor size were measured every other day. The mice were euthanized, and (**C**) the tumors were extracted and (**D**) weighed. $n = 8$ per group. * $p < 0.05$ and ** $p < 0.01$ vs the control. SB365, SB365-treated group; TMZ, temozolomide-treated group.

Tumor growth was significantly inhibited by injection of SB365 or TMZ only compared to the control (p = 0.011) (Figure 7B). In addition, the combination of SB365 and TMZ resulted in significantly greater inhibition of tumor growth compared to TMZ or SB365 only (p = 0.046) (Figure 7B). The tumor weights at the end of the experiment were in agreement with these results (Figure 7C,D).

3. Discussion

In this experiment, SB365 exerted a cytotoxic effect on these cells in a dose-dependent manner. However, this effect was mediated by induction of, not apoptosis, as in other cancer cells, but CICD. The cytotoxic impact of SB365 proceeded as follows: neutralization of the lysosomal pH and inhibition of autophagic flux occurred rapidly, followed by alteration of MMP, and finally, cell death. SB365-induced cell death was partially recovered by treatment with a cathepsin B inhibitor and NAC. Moreover, the combination of SB365 and TMZ exerted an additive effect both in vitro and in vivo.

SB365 is administered intratumorally via direct percutaneous injection to patients with pancreatic cancer [17]. To mimic this, we injected the agent directly into the tumor mass in mice, rather than administering intraperitoneally or orally, as in prior studies [13–16].

The dose-dependency of the cytotoxic effect (Figure 1) of SB365 is in agreement with prior findings in liver, lung, colon, and pancreatic cancer cells [13–16,18]. SB365 induced caspase-3 cleavage and nuclear fragmentation in colon cancer and hepatocarcinoma, but not in GBM cells (Figure 2). Activation of caspase-3 is a converging step of both the intrinsic and extrinsic pathways of caspase-dependent apoptosis [37]. In addition, SB365 did not affect Bcl-2 and Bax expression in U87-MG cells (data not shown) the expression of which decreases and increases, respectively, during initiation of apoptosis [38]. Thus, we assumed that SB365 induced CICD in GBM cells.

To evaluate the mechanism underlying SB365-induced death in GBM cells, we explored its effect on autophagic flux, because CICD in GBM cells by chloroquine [33] and thymoquinone [31] is associated with inhibition of autophagic flux, and SB365 inhibits autophagic flux in other cancer cell lines [19]. The levels of LC3-II and p62 increased at 6 h after SB365 treatment (Figure 3A,B), and remained high up to 72 h (Supplementary Materials Figure S5), which implies that the SB365-induced death of GBM cells may be associated with inhibition of autophagic flux.

SB365 induces autophagy in HeLa cells by increasing ERK phosphorylation and decreasing mTOR activation, though it inhibits subsequent autophagic flux [19]. In hepatocarcinoma [13] and gastric cancer [14] cells, SB365 suppressed the PI3K/Akt/mTOR pathway, which negatively regulates autophagy [39]. However, in U87-MG cells, the p-Akt and p-mTOR levels were unchanged after 24 h of treatment with SB365 (data not shown). Furthermore, the cytotoxic effect of SB365 on U87-MG cells was augmented by pretreatment with a non-toxic concentration of the autophagy inducer rapamycin [40] (Supplementary Materials Figure S8). These results imply that the accumulation of autophagosomes due to inhibition of the autophagic flux caused cell death. Critically, SB365 did not increase the expression of beclin-1 (Figure 3 and Supplementary Materials Figure S5), which is associated with autophagy induction [41]. Therefore, SB365 does not induce autophagy, but inhibits autophagic flux, in U87-MG cells.

Inhibition of autophagic flux can result from lysosomal neutralization [29]. SB365 treatment resulted in simultaneous inhibition of autophagic flux (Figure 3) and lysosomal neutralization (Figure 4A,C). Thus, the SB365-induced inhibition of autophagic flux may be mediated by lysosomal deterioration. Indeed, saponins, in particular oleanane-type saponins such as SB365 [12], reportedly permeabilize the cell membrane [42] and the lysosomal membrane [43]. In addition, a cathepsin B inhibitor partially restored the SB365-induced reduction in cell proliferation (Figure 5A), suggesting that cathepsin B was released from lysosomes and that SB365 induced permeabilization of the lysosomal membrane.

In our results, a cathepsin B inhibitor restored cell death but a cathepsin D inhibitor did not (data not shown). Given that the molecular weights of cathepsins B and D are similar [44], and thus the two molecules would have been released simultaneously, the contradictive effect of each inhibitor would be somewhat unexpected. However, the same results have been reported in paclitaxel-, epothilone B-, and discodermolide-treated human non-small cell lung cancer cells [45] and supraoptimally activated T cells [46]. Possibly, only cathepsin B had been released [44]. Alternatively, these results suggest the varying role of cathepsins depending on the type of cells [30]. The exact mechanisms remain to be determined.

The frequency of the cells with MMP deterioration was only 5.8% at 36 h and 8.6% at 48 h after SB365 treatment (Figure 4B,C). These are low values considering that MMP deterioration directly led to the SB365-induced cell death. Indeed, the phenomena caused by various factors secreted from the mitochondria when MMP deterioration occurs, such as the activation of caspase-3/9 leading to apoptosis by cytochrome c, degrading DNA by endonuclease G, and chromatin condensation by AIF [47] were not observed in this experiment. Another substance that is released from deteriorated mitochondria is ROS. Autophagic flux inhibition, which was induced by SB365 in GBM cells in this experiment, leads to the accumulation of ROS [34,35]. Excess ROS accelerate lysosomal permeabilization, and leaked lysosomal proteases deteriorate MMP, resulting in increased cytoplasmic ROS leakage, creating a vicious cycle [48]. Thus, ROS could be a factor for the SB365-induced cytotoxicity. Substantial to this assumption, 2.5 mM NAC recovered the cytotoxicity by over 50% when added 1 h after SB365 treatment (Figure 5C). Furthermore, when NAC was added 48 h after SB365 treatment, the recovery rate was over 70% (Figure 5D). These results imply that ROS was the main factor leading to cell death by SB365, and ROS presumably began to accumulate to cause cell death 24 h after SB365 treatment in parallel with MMP deterioration.

Meanwhile, 5 mM NAC failed to recover cell proliferation. This may be because of excessive eradication of ROS by the antioxidant, which performs physiological functions in cell proliferation [49]. Substantial to this assumption, 10 mM NAC augmented the cytotoxic effect of SB365 (data not shown). Additionally, even the low concentrations of NAC (0.623–2.5 mM) augmented the effect of SB365 when it was treated before SB365 (data not shown).

Attempts have been made to improve the efficacy of TMZ against GBM by combining it with other drugs. TMZ is typically combined with autophagic flux inhibitors such as chloroquine, hydroxychloroquine, or bafilomycin A1, with which it reportedly exerts synergistic effects [10]. Since SB365 inhibited autophagic flux in GBM cells, we evaluated the efficacy of the combination of SB365 and TMZ. The combination of SB365 and TMZ increased the frequency of cell death in vitro (Figure 6) and inhibited tumor growth in vivo (Figure 7). Thus, SB365 could be used in combination with TMZ in place of chloroquine, hydroxychloroquine, or bafilomycin A1, which synergistically inhibit tumor growth but have several side effects [10,50]. One concern is that SB365 exerts hemolytic activity on red blood cells of the sheep [42] and the rabbit [51], which was considered as a major drawback for its clinical development [42].

SB365 alone induced death in TMZ-resistant T98G cells (Supplementary Materials Figure S1) as effectively as in TMZ-sensitive U87-MG cells. Furthermore, in T98G cells, the combination of SB365 and TMZ additively increased cell death (Supplementary Materials Figure S7). Unfortunately, we did not determine whether SB365 downregulated the expression of MGMT genes.

In conclusion, SB365 inhibited autophagic flux, and induced CICD in GBM cells in a manner mediated by cathepsin B and mainly by ROS very likely due to autophagic flux inhibition and MMP deterioration. Moreover, SB365 and TMZ exerted an additive cytotoxic effect in vivo and in vitro. Thus, SB365 could be used in combination with TMZ for the treatment of TMZ-resistant GBM.

4. Materials and Methods

4.1. Chemicals

SB365 was supplied by SB Pharmaceutical Co. Ltd. (Gongju, Republic of Korea (ROK)). TMZ (T2577) was purchased from Sigma-Aldrich (St. Louis, MO, USA). SB365 and TMZ stock solutions (100 mM) were prepared in dimethyl sulfoxide (DMSO). The final DMSO concentration in culture media was

≤0.4%, which did not exert a toxic effect on GBM cells (data not shown). Stock solutions of cathepsin B inhibitor II (219385; Calbiochem, San Diego, CA, USA), pepstatin A (cathepsin D inhibitor, P5318; Sigma-Aldrich, and N-acetyl cysteine (NAC) (A7250; Sigma-Aldrich, Saint Louis, MO, USA) were prepared and stored at −80 °C until use.

4.2. Cell Lines and Culture Conditions

TMZ-susceptible U87-MG and TMZ-resistant T98G human GBM cells, as well as HT-29 and Huh-7 cells (Korean Cell Line Bank, Seoul, ROK) were used in this study. The cells were cultured in minimum essential Eagle's medium (EMEM) supplemented with 10% fetal bovine serum, 1% penicillin/streptomycin, and 1% non-essential amino acids (Welgene, Daegu, ROK) at 37 °C in a 5% CO_2 atmosphere in a humidified chamber.

4.3. Cell Counting Kit-8 Assay

The cytotoxicity of SB365 and TMZ was assessed using a Cell Counting Kit-8 (CCK-8; EZ-3000; Dojindo, Kumamoto, Japan) following the manufacturer's instructions. Briefly, U87-MG cells (5×10^3/well) or T98G cells (2×10^3/well) were cultured in quadruplicate in 96-well plates overnight and treated with SB365 and/or TMZ at predefined concentrations. The culture medium was discarded, and 100 µL of CCK-8 working solution (10% (v/v) CCK-8 stock solution in phosphate-buffered saline (PBS)) were added. The cells were incubated at 37 °C for 1–3 h, and the absorbance at 450 nm was measured using a SpectraMax Plus 384 spectrophotometer (Molecular Devices, Sunnyvale, CA, USA).

IC50 value was obtained, based on the CCK-8 results, by the Quest Graph™ IC50 Calculator, a four parameter logistic regression model [52], with the minimum response value fixed to zero computationally.

4.4. Apoptosis Assay

U87-MG (7.5×10^4/well) and T98G (3×10^4/well) cells were seeded in a six-well plate and cultured overnight at 37 °C in a CO_2 incubator. The cells were treated with SB365 and/or TMZ and collected in fluorescence-activated cell sorting (FACS) tubes. After washing twice with FACS buffer (0.5% BSA in PBS), the cells were resuspended in 100 µL of FACS buffer, 2 µL of annexin V were added (556419; BD Pharmingen, San Jose, CA, USA) and the plate was shaken for 15 min at room temperature. Next, 1 µL of 7-AAD was added (559925; BD Pharmingen), and the cells were subjected to FACS analysis on a FACSCalibur flow cytometer (BD Biosciences, Heidelberg, Germany).

To evaluate nuclear morphology, cells treated with SB365 for 72 h were harvested and seeded onto poly-L-lysine-coated multispot slides. The cells were washed with PBS, fixed in 4% paraformaldehyde for 20 min, and stained with 4,6-diamidino-2-phenylindole (DAPI; F6057, Sigma-Aldrich, Saint Louis, MO, USA).

4.5. Western Blotting

Cells were dissociated by pipetting in cold radioimmunoprecipitation assay (RIPA) buffer (50 mM Tris-HCl (pH 7.4), 150 mM NaCl, 1% sodium deoxychloride, 0.1% sodium dodecyl sulfate (SDS), 1% Triton X-100, 2 mM ethylenediaminetetraacetic acid (EDTA), and 1% protease inhibitors), and centrifuged at 18,000× g for 10 min at 4 °C. The supernatant was collected, and the protein concentration measured by bicinchoninic acid assay, then 20 or 100 µg (for caspase-3) of protein were mixed with RIPA buffer and 5× SDS loading dye (S2002; Biosesang, Seongnam, ROK) to a final volume of 20 µL. The mixture was boiled at 95 °C for 10 min, loaded onto a sodium dodecyl sulfate polyacrylamide gel, and electrophoresed at 50 V for stacking and 120 V for separation. The samples were subsequently transferred to a nitrocellulose membrane at 400 mA for 1 h at 4 °C and blocked in blocking buffer (5% skim milk, 0.05% Tween 20 in PBS) for 1 h at room temperature. Finally, the samples were incubated with the appropriate primary antibody in blocking buffer overnight at 4 °C, followed by the corresponding secondary antibody for 1 h at room temperature. Protein bands were visualized using an enzyme-linked chemiluminescence detection kit (DG-WF200; DoGEN, Seoul, ROK). The primary antibodies used were as follows: rabbit anti-human LC3B (NB 600-1384; Novus Biologicals, Minneapolis, MN, USA; 1:5000); rabbit anti-human beclin-1 (ab2557; Abcam, Cambridge, MA, USA; 1:5000); mouse anti-human p62 (ab56416; Abcam; 1:10,000); rabbit anti-human caspase-3 (9662), p-AKT (9271), AKT (9272), p-mTOR (2971), and mTOR (2972; Cell Signaling Technology, Inc., Danvers, MA, USA; 1:1000); and mouse anti-human β-actin (3700; Cell Signaling Technology; 1:5000). A goat anti-mouse IgG-horseradish peroxidase (HRP) (SC-2005; Santa Cruz Biotechnology, Santa Cruz, CA, USA; 1:5000) or anti-rabbit IgG-HRP (SC-2030; Santa Cruz Biotechnology; 1:5000) was used as the secondary antibody.

4.6. Lysosome Stability Assay

Lysosomal membrane stability was determined by staining SB365-treated cells with 3 µg/mL acridine orange (A8097; Sigma-Aldrich, Saint Louis, MO, USA) for 20 min at 37 °C. This metachromatic dye emits red fluorescence when it is confined in the cytosol where it is present as a monomer. When the dye penetrates into the dysfunctional lysosome, it converts into aggregates due to the acidic environment in the lysosome and emits green fluorescence. The property has been used to measure lysosomal membrane stability [53]. Flow cytometric analysis was performed to determine the red (FL3; 650 nm) and green (FL1; 510–530 nm) fluorescence of cells excited by blue (488 nm) light using a FACSCalibur instrument.

4.7. Mitochondrial Membrane Potential Assay

SB365-treated cells were stained with 2.5 µM JC-1 (T3168; Life Technologies, Carlsbad, CA, USA) for 20 min at 37 °C, and analyzed by flow cytometry. JC-1 is a lipophilic and cationic dye. It enters the mitochondria, converts from monomers to aggregates by membrane potential, and accumulates inside the mitochondrion. In FACS analysis, monomers and aggregates emit green and red fluorescence, and indicate lower and higher mitochondrial membrane potential (MMP), respectively [54].

4.8. Animal Xenograft Model

Animal experiments were approved by the Institutional Animal Care and Use Committee (SNU-150521-3-2). Seven-week-old male Balb/c-nu mice were purchased from OrientBio (Seongnam, ROK). U87-MG cells were mixed with Matrigel HC (354248; BD Biosciences) at a 50:50 volume ratio, and the mixture was inoculated into both flanks (5×10^6 cells/100 µL/flank) of the mice. When the tumor reached a volume of approximately 100–200 mm^3, the mice were assigned to control, SB365, TMZ, and SB365 + TMZ treatment groups; the mean mass of each group was similar. Next, the mice underwent intratumoral injection of SB365 (5 mg/kg) every other day and/or intraperitoneal injection of TMZ (2.5 mg/kg) or vehicle (≤3% DMSO) daily. The day of the first injection was regarded as day 0 and the injections were administered until day 21; the mice were sacrificed on day 22. The body weight and tumor volume were measured every other day. Tumor size was measured using calipers and the tumor volume was calculated as volume (V) = length (L) × width (W)^2 × 0.5.

4.9. Statistical Analysis

The Mann–Whitney U-test was used to evaluate statistical significance. Statistical analysis was performed using Statistical Package for the Social Sciences software ver. 12 (SPSS, Inc., Chicago, IL, USA). A value of $p < 0.05$ was taken to indicate statistical significance.

Supplementary Materials: The following are available online: Figure S1. In vitro cytotoxic effect of SB365 on TMZ-resistant T98G cells, Figure S2. Cell death in U87-MG cells treated with SB365. Figure S3. Cell death in T98G cells treated with SB365. Figure S4. Effect of SB365 on the cleavage of caspase-3 in T98G cells. Figure S5. Effect of SB365 on the expression of autophagy-related proteins in GBM cells. Figure S6. Recovery of SB365-induced cell death by antioxidant, NAC, in T98G cells. Figure S7. Cytotoxic effect of SB365 and TMZ on T98G cells. Figure S8. Augmentation of SB365 cytotoxicity by rapamycin pre-treatment. Table S1. Cell proliferation in U87-MG cells treated with SB365 and/or TMZ.

Author Contributions: Conceptualization, W.J.L. and Y.-i.H.; methodology, J.-M.H.; resources, J.-H.K. and H.K.; writing—original draft preparation, J.-M.H.; writing—review and editing, J.-M.H. and Y.-i.H.; supervision, Y.-i.H.

Funding: This research was supported by the Education and Research Encouragement Fund of Seoul National University Hospital (2019).

Conflicts of Interest: The authors declare no conflict of interest.

References

1. Morgan, E.R.; Norman, A.; Laing, K.; Seal, M.D. Treatment and outcomes for glioblastoma in elderly compared with non-elderly patients: A population-based study. *Curr. Oncol.* **2017**, *24*, e92–e98. [CrossRef]
2. Paolillo, M.; Boselli, C.; Schinelli, S. Glioblastoma under Siege: An Overview of Current Therapeutic Strategies. *Brain Sci.* **2018**, *8*, 15.
3. Wang, Z.; Yang, G.; Zhang, Y.Y.; Yao, Y.; Dong, L.H. A comparison between oral chemotherapy combined with radiotherapy and radiotherapy for newly diagnosed glioblastoma: A systematic review and meta-analysis. *Med. (Baltim.).* **2017**, *96*, e8444. [CrossRef]
4. Stupp, R.; Dietrich, P.Y.; Kraljevic, S.O.; Pica, A.; Maillard, I.; Maeder, P.; Meuli, R.; Janzer, R.; Pizzolato, G.; Miralbell, R.; et al. Promising survival for patients with newly diagnosed glioblastoma multiforme treated with concomitant radiation plus temozolomide followed by adjuvant temozolomide. *J. Clin. Oncol.* **2002**, *20*, 1375–1382. [CrossRef]
5. D'Atri, S.; Tentori, L.; Lacal, P.M.; Graziani, G.; Pagani, E.; Benincasa, E.; Zambruno, G.; Bonmassar, E.; Jiricny, J. Involvement of the mismatch repair system in temozolomide-induced apoptosis. *Mol. Pharm.* **1998**, *54*, 334–341. [CrossRef]
6. Ostermann, S.; Csajka, C.; Buclin, T.; Leyvraz, S.; Lejeune, F.; Decosterd, L.A.; Stupp, R. Plasma and cerebrospinal fluid population pharmacokinetics of temozolomide in malignant glioma patients. *Clin. Cancer Res.* **2004**, *10*, 3728–3736. [CrossRef]
7. Bocangel, D.B.; Finkelstein, S.; Schold, S.C.; Bhakat, K.K.; Mitra, S.; Kokkinakis, D.M. Multifaceted resistance of gliomas to temozolomide. *Clin. Cancer Res.* **2002**, *8*, 2725–2734.
8. Gilbert, M.R.; Wang, M.; Aldape, K.D.; Stupp, R.; Hegi, M.E.; Jaeckle, K.A.; Armstrong, T.S.; Wefel, J.S.; Won, M.; Blumenthal, D.T.; et al. Dose-dense temozolomide for newly diagnosed glioblastoma: A randomized phase III clinical trial. *J. Clin. Oncol.* **2013**, *31*, 4085–4091. [CrossRef]
9. Bozic, I.; Reiter, J.G.; Allen, B.; Antal, T.; Chatterjee, K.; Shah, P.; Moon, Y.S.; Yaqubie, A.; Kelly, N.; Le, D.T.; et al. Evolutionary dynamics of cancer in response to targeted combination therapy. *Elife.* **2013**, *2*, e00747. [CrossRef]
10. Rosenfeld, M.R.; Ye, X.B.; Supko, J.G.; Desideri, S.; Grossman, S.A.; Brem, S.; Mikkelson, T.; Wang, D.; Chang, Y.Y.C.; Hu, J.; et al. A phase I/II trial of hydroxychloroquine in conjunction with radiation therapy and concurrent and adjuvant temozolomide in patients with newly diagnosed glioblastoma multiforme. *Autophagy.* **2014**, *10*, 1359–1368. [CrossRef]
11. Kim, Y.; Bang, S.-C.; Lee, J.-H.; Ahn, B.-Z. Pulsatilla saponin D: The antitumor principle from Pulsatilla koreana. *Arch. Pharmacal Res.* **2004**, *27*, 915–918. [CrossRef]
12. Bang, S.C.; Lee, J.H.; Song, G.Y.; Kim, D.H.; Yoon, M.Y.; Ahn, B.Z. Antitumor activity of Pulsatilla koreana saponins and their structure-activity relationship. *Chem. Pharm Bull.* **2005**, *53*, 1451–1454. [CrossRef]
13. Hong, S.W.; Jung, K.H.; Lee, H.S.; Choi, M.J.; Son, M.K.; Zheng, H.M.; Hong, S.S. SB365 inhibits angiogenesis and induces apoptosis of hepatocellular carcinoma through modulation of PI3K/Akt/mTOR signaling pathway. *Cancer Sci.* **2012**, *103*, 1929–1937. [CrossRef]

14. Hong, S.W.; Jung, K.H.; Lee, H.S.; Son, M.K.; Yan, H.H.; Kang, N.S.; Lee, J.; Hong, S.S. SB365, Pulsatilla saponin D, targets c-Met and exerts antiangiogenic and antitumor activities. *Carcinog.* **2013**, *34*, 2156–2169. [CrossRef]
15. Son, M.K.; Jung, K.H.; Hong, S.W.; Lee, H.S.; Zheng, H.M.; Choi, M.J.; Seo, J.H.; Suh, J.K.; Hong, S.S. SB365, Pulsatilla saponin D suppresses the proliferation of human colon cancer cells and induces apoptosis by modulating the AKT/mTOR signalling pathway. *Food Chem.* **2013**, *136*, 26–33. [CrossRef]
16. Son, M.K.; Jung, K.H.; Lee, H.S.; Lee, H.; Kim, S.J.; Yan, H.H.; Ryu, Y.L.; Hong, S.S. SB365, Pulsatilla saponin D suppresses proliferation and induces apoptosis of pancreatic cancer cells. *Oncol. Rep.* **2013**, *30*, 801–808. [CrossRef]
17. Moon, K.S.; Ji, J.Y.; Cho, Y.J.; Lee, J.H.; Choi, M.S.; Kim, E.E. Therapeutic effects of SB natural anticancer drug in 50 patients with stage IV pancreatic cancer. *J. Cancer Treat. Res.* **2015**, *3*, 42–46. [CrossRef]
18. Jang, W.J.; Park, B.; Jeong, G.S.; Hong, S.S.; Jeong, C.H. SB365, Pulsatilla saponin D, suppresses the growth of gefitinib-resistant NSCLC cells with Met amplification. *Oncol. Rep.* **2014**, *32*, 2612–2618. [CrossRef]
19. Zhang, Y.L.; Bao, J.L.; Wang, K.; Jia, X.J.; Zhang, C.; Huang, B.R.; Chen, M.W.; Wan, J.B.; Su, H.X.; Wang, Y.T.; et al. Pulsatilla saponin D inhibits autophagic flux and synergistically enhances the anticancer activity of chemotherapeutic agents against hela cells. *Am. J. Chin. Med.* **2015**, *43*, 1657–1670. [CrossRef]
20. Gupta, A.; Dwivedi, T. A simplified overview of world health organization classification update of central nervous system tumors 2016. *J. Neurosci. Rural. Pr.* **2017**, *8*, 629–641. [CrossRef]
21. Xie, Y.; Bergstrom, T.; Jiang, Y.W.; Johansson, P.; Marinescu, V.D.; Lindberg, N.; Segerman, A.; Wicher, G.; Niklasson, M.; Baskaran, S.; et al. The human glioblastoma cell culture resource: validated cell models representing all molecular subtypes. *Ebiomedicine.* **2015**, *2*, 1351–1363. [CrossRef]
22. Kokkinakis, D.M.; Bocangel, D.B.; Schold, S.C.; Moschel, R.C.; Pegg, A.E. Thresholds of O-6-alkylguanine-DNA alkyltransferase which confer significant resistance of human glial tumor xenografts to treatment with 1,3-bis(2-chloroethyl)-1-nitrosourea or temozolomide. *Clin. Cancer Res.* **2001**, *7*, 421–428.
23. Hegi, M.E.; Diserens, A.; Gorlia, T.; Hamou, M.; de Tribolet, N.; Weller, M.; Kros, J.M.; Hainfellner, J.A.; Mason, W.; Mariani, L.; et al. MGMT gene silencing and benefit from temozolomide in glioblastoma. *New Engl. J. Med.* **2005**, *352*, 997–1003. [CrossRef]
24. Vermes, I.; Haanen, C.; Reutelingsperger, C. Flow cytometry of apoptotic cell death. *J. Immunol Methods.* **2000**, *243*, 167–190. [CrossRef]
25. Khan, K.H.; Blanco-Codesido, M.; Molife, L.R. Cancer therapeutics: Targeting the apoptotic pathway. *Crit. Rev. Oncol. Hemat.* **2014**, *90*, 200–219. [CrossRef]
26. Carloni, S.; Buonocore, G.; Balduini, W. Protective role of autophagy in neonatal hypoxia-ischemia induced brain injury. *Neurobiol. Dis.* **2008**, *32*, 329–339. [CrossRef]
27. Pankiv, S.; Clausen, T.H.; Lamark, T.; Brech, A.; Bruun, J.A.; Outzen, H.; Overvatn, A.; Bjorkoy, G.; Johansen, T. p62/SQSTM1 binds directly to Atg8/LC3 to facilitate degradation of ubiquitinated protein aggregates by autophagy. *J. Biol. Chem.* **2007**, *282*, 24131–24145. [CrossRef]
28. Jiang, P.; Mizushima, N. LC3- and p62-based biochemical methods for the analysis of autophagy progression in mammalian cells. *Methods.* **2015**, *75*, 13–18. [CrossRef]
29. Vallecillo-Hernández, J.; Barrachina, M.D.; Ortiz-Masiá, D.; Coll, S.; Esplugues, J.V.; Calatayud, S.; Hernández, C. Indomethacin disrupts autophagic flux by inducing lysosomal dysfunction in gastric cancer cells and increases their sensitivity to cytotoxic drugs. *Sci. Rep.-Uk.* **2018**, *8*, 3593. [CrossRef]
30. Boya, P.; Kroemer, G. Lysosomal membrane permeabilization in cell death. *Oncogene.* **2008**, *27*, 6434–6451. [CrossRef]
31. Racoma, I.O.; Meisen, W.H.; Wang, Q.E.; Kaur, B.; Wani, A.A. Thymoquinone inhibits autophagy and induces cathepsin-mediated, caspase-independent cell death in glioblastoma cells. *PloS ONE* **2013**, *8*, e72882. [CrossRef]
32. Noguchi, S.; Shibutani, S.; Fukushima, K.; Mori, T.; Igase, M.; Mizuno, T. Bosutinib, an SRC inhibitor, induces caspase-independent cell death associated with permeabilization of lysosomal membranes in melanoma cells. *Vet. Comp. Oncol.* **2018**, *16*, 69–76. [CrossRef]
33. Geng, Y.; Kohli, L.; Klocke, B.J.; Roth, K.A. Chloroquine-induced autophagic vacuole accumulation and cell death in glioma cells is p53 independent. *Neuro-Oncol.* **2010**, *12*, 473–481.

34. Bray, K.; Mathew, R.; Lau, A.; Kamphorst, J.J.; Fan, J.; Chen, J.; Chen, H.Y.; Ghavami, A.; Stein, M.; DiPaola, R.S.; et al. Autophagy suppresses rip kinase-dependent necrosis enabling survival to mTOR inhibition. *PLoS ONE* **2012**, *7*, e41831. [CrossRef]
35. Klose, J.; Stankov, M.V.; Kleine, M.; Ramackers, W.; Panayotova-Dimitrova, D.; Jager, M.D.; Klempnauer, J.; Winkler, M.; Bektas, H.; Behrens, G.M.N.; et al. Ihibition of autophagic flux by salinomycin results in anti-cancer effect in hepatocellular carcinoma cells. *PLoS ONE* **2014**, *9*, e95970. [CrossRef]
36. Li, N.Y.; Oquendo, E.; Capaldi, R.A.; Robinson, J.P.; He, Y.D.D.; Hamadeh, H.K.; Afshari, C.A.; Lightfoot-Dunn, R.; Narayanan, P.K. A systematic assessment of mitochondrial function identified novel signatures for drug-induced mitochondrial disruption in cells. *Toxicol. Sci.* **2014**, *142*, 261–273. [CrossRef]
37. Song, J.H.; Song, D.K.; Pyrzynska, B.; Petruk, K.C.; Van Meir, E.G.; Hao, C.H. TRAIL triggers apoptosis in human malignant glioma cells through extrinsic and intrinsic pathways. *Brain Pathol.* **2003**, *13*, 539–553. [CrossRef]
38. Kang, K.A.; Piao, M.J.; Hyun, Y.J.; Zhen, A.X.; Cho, S.J.; Ahn, M.J.; Yi, J.M.; Hyun, J.W. Luteolin promotes apoptotic cell death via upregulation of Nrf2 expression by DNA demethylase and the interaction of Nrf2 with p53 in human colon cancer cells. *Exp. Mol. Med.* **2019**, *51*, 40. [CrossRef]
39. Jung, C.H.; Jun, C.B.; Ro, S.-H.; Kim, Y.-M.; Otto, N.M.; Cao, J.; Kundu, M.; Kim, D.-H. ULK-Atg13-FIP200 complexes mediate mTOR signaling to the autophagy machinery. *Mol Biol Cell.* **2009**, *20*, 1992–2003. [CrossRef]
40. Gu, J.; Hu, W.; Song, Z.P.; Chen, Y.G.; Zhang, D.D.; Wang, C.Q. Rapamycin inhibits cardiac hypertrophy by promoting autophagy via the MEK/ERK/Beclin-1 pathway. *Front. Physiol.* **2016**, *7*, 104. [CrossRef]
41. Wang, C.R.; Wang, X.J.; Su, Z.J.; Fei, H.J.; Liu, X.Y.; Pan, Q.X. The novel mTOR inhibitor Torin-2 induces autophagy and downregulates the expression of UHRF1 to suppress hepatocarcinoma cell growth. *Oncol. Rep.* **2015**, *34*, 1708–1716. [CrossRef]
42. Gauthier, C.; Legault, J.; Girard-Lalancette, K.; Mshvildadze, V.; Pichette, A. Haemolytic activity, cytotoxicity and membrane cell permeabilization of semi-synthetic and natural lupane- and oleanane-type saponins. *Bioorgan Med. Chem.* **2009**, *17*, 2002–2008. [CrossRef]
43. Gilabert-Oriol, R.; Mergel, K.; Thakur, M.; von Mallinckrodt, B.; Melzig, M.F.; Fuchs, H.; Weng, A. Real-time analysis of membrane permeabilizing effects of oleanane saponins. *Bioorgan Med. Chem.* **2013**, *21*, 2387–2395. [CrossRef]
44. Zhang, Y.; Yang, N.D.; Zhou, F.; Shen, T.; Duan, T.; Zhou, J.; Shi, Y.; Zhu, X.Q.; Shen, H.M. (-)-Epigallocatechin-3-gallate induces non-apoptotic cell death in human cancer cells via ros-mediated lysosomal membrane permeabilization. *PLoS ONE* **2012**, *7*, e46749. [CrossRef]
45. Broker, L.E.; Huisman, C.; Span, S.W.; Rodriguez, J.A.; Kruyt, F.A.E.; Giaccone, G. Cathepsin B mediates caspase-independent cell death induced by microtubule stabilizing agents in non-small cell lung cancer cells. *Cancer Res.* **2004**, *64*, 27–30. [CrossRef]
46. Michallet, M.C.; Saltel, F.; Flacher, M.; Revillard, J.P.; Genestier, L. Cathepsin-dependent apoptosis triggered by supraoptimal activation of T lymphocytes: A possible mechanism of high dose tolerance. *J. Immunol.* **2004**, *172*, 5405–5414. [CrossRef]
47. Tait, S.W.G.; Ichim, G.; Green, D.R. Die another way—non-apoptotic mechanisms of cell death. *J. Cell Sci.* **2014**, *127*, 2135–2144. [CrossRef]
48. Zhao, M.; Antunes, F.; Eaton, J.W.; Brunk, U.T. Lysosomal enzymes promote mitochondrial oxidant production, cytochrome c release and apoptosis. *Eur. J. Biochem.* **2003**, *270*, 3778–3786. [CrossRef]
49. Diebold, L.; Chandel, N.S. Mitochondrial ROS regulation of proliferating cells. *Free Radic. Biol. Med.* **2016**, *100*, 86–93. [CrossRef]
50. Kanzawa, T.; Germano, I.M.; Komata, T.; Ito, H.; Kondo, Y.; Kondo, S. Role of autophagy in temozolomide-induced cytotoxicity for malignant glioma cells. *Cell Death Differ.* **2004**, *11*, 448–457. [CrossRef]
51. Chen, Z.; Duan, H.; Tong, X.; Hsu, P.; Han, L.; Morris-Natschke, S.L.; Yang, S.; Liu, W.; Lee, K.H. Cytotoxicity, hemolytic toxicity, and mechanism of action of pulsatilla saponin d and its synthetic derivatives. *J. Nat. Prod.* **2018**, *81*, 465–474. [CrossRef]
52. Khinkis, L.A.; Levasseur, L.; Faessel, H.; Greco, W.R. Optimal design for estimating parameters of the 4-parameter hill model. *Nonlinearity Biol. Toxicol Med.* **2003**, *1*, 363–377. [CrossRef]

53. Pierzynska-Mach, A.; Janowski, P.A.; Dobrucki, J.W. Evaluation of acridine orange, lysotracker red, and quinacrine as fluorescent probes for long-term tracking of acidic vesicles. *Cytom. Part A* **2014**, *85a*, 729–737. [CrossRef]
54. Salvioli, S.; Ardizzoni, A.; Franceschi, C.; Cossarizza, A. JC-1, but not DiOC6(3) or rhodamine 123, is a reliable fluorescent probe to assess $\Delta\Psi$ changes in intact cells: Implications for studies on mitochondrial functionality during apoptosis. *Febs. Lett.* **1997**, *411*, 77–82. [CrossRef]

Sample Availability: Samples of the compounds are not available from the authors.

© 2019 by the authors. Licensee MDPI, Basel, Switzerland. This article is an open access article distributed under the terms and conditions of the Creative Commons Attribution (CC BY) license (http://creativecommons.org/licenses/by/4.0/).

Article

Diallyl Disulfide Induces Apoptosis and Autophagy in Human Osteosarcoma MG-63 Cells through the PI3K/Akt/mTOR Pathway

Ziqi Yue [1], Xin Guan [1], Rui Chao [1], Cancan Huang [1], Dongfang Li [1], Panpan Yang [1], Shanshan Liu [1], Tomoka Hasegawa [2], Jie Guo [1] and Minqi Li [1,*]

[1] Department of Bone Metabolism, School of Stomatology Shandong University, Shandong Provincial Key Laboratory of Oral Tissue Regeneration, Jinan 250012, China
[2] Department of Developmental Biology of Hard Tissue, Graduate School of Dental Medicine, Hokkaido University, Sapporo 060-8586, Japan
* Correspondence: liminqi@sdu.edu.cn; Tel.: +86-531-88382095

Received: 24 May 2019; Accepted: 22 July 2019; Published: 23 July 2019

Abstract: Diallyl disulfide (DADs), a natural organic compound, is extracted from garlic and scallion and has anti-tumor effects against various tumors. This study investigated the anti-tumor activity of DADs in human osteosarcoma cells and the mechanisms. MG-63 cells were exposed to DADs (0, 20, 40, 60, 80, and 100 μM) for different lengths of time (24, 48, and 72 h). The CCK8 assay results showed that DADs inhibited osteosarcoma cell viability in a dose-and time-dependent manner. FITC-Annexin V/propidium iodide staining and flow cytometry demonstrated that the apoptotic ratio increased and the cell cycle was arrested at the G_2/M phase as the DADs concentration was increased. A Western blot analysis was employed to detect the levels of caspase-3, Bax, Bcl-2, LC3-II/LC3-I, and p62 as well as suppression of the mTOR pathway. High expression of LC3-II protein revealed that DADs induced formation of autophagosome. Furthermore, DADs-induced apoptosis was weakened after adding 3-methyladenine, demonstrating that the DADs treatment resulted in autophagy-mediated death of MG-63 cells. In addition, DADs depressed p-mTOR kinase activity, and the inhibited PI3K/Akt/mTOR pathway increased DADs-induced apoptosis and autophagy. In conclusion, our results reveal that DADs induced G_2/M arrest, apoptosis, and autophagic death of human osteosarcoma cells by inhibiting the PI3K/Akt/mTOR signaling pathway.

Keywords: diallyl disulfide; apoptosis; autophagy; osteosarcoma; PI3K/Akt/mTOR pathway

1. Introduction

Osteosarcoma (OS) is one of the most frequent bone malignancies, developing from the bone-forming mesenchymal cell lines. The rapid expansion of OS is due to the direct or indirect formation of osteoid and osseous tissues [1]. Mortality from OS in children and adolescents remains very high, and the incidence rate reaches a second peak after the age of 60 years [2]. However, early intervention and appropriate treatments, such as chemotherapy drugs, have greatly improved the survival rate of the disease [3]. Some studies have confirmed that adjuvant chemotherapy has a beneficial effect on the relapse-free survival rate of patients with OS of the extremities [4]. However, the reality is that the drug resistance of tumors is becoming more and more complex, so new anti-cancer drugs are urgently needed.

Diallyl disulfide (DADs) is a natural organic compound in garlic and scallion, that has demonstrated anti-tumor properties in a variety of tumor types. Garlic has a long history of being used as a food additive and in pharmaceutical products. However, it was not until modern times that its anti-cancer mechanisms were demonstrated in specific studies. The effects of garlic include suppressing tumor proliferation and invasion [5], inducing G_2/M arrest [6], and enhancing reactive oxygen species production [7]. Furthermore, DADs inhibit the growth of various tumors, such as colon cancer, bladder cancer, cervical cancer [8–10], and OS [5] by inducing apoptosis.

Uncontrolled cell proliferation is the most prominent feature of cancer. The integrity of the cell cycle is the basis for normal cell proliferation and is mainly regulated by cyclin dependent kinases (CDKs) and CDK inhibitor proteins. An unbalanced cell cycle promotes the occurrence and development of tumors [11,12]. Successful G_2/M transformation is the key to cell division [13]. Many plant natural compounds inhibit tumor growth by blocking the G_2/M phase [14,15].

Most natural organic compounds play a role by inducing cell death. Several modes of cell death have been described, such as apoptosis, autophagy, and others (necrosis and mitotic catastrophe). Apoptosis refers to the orderly death of cells controlled by genes with the purpose of maintaining homeostasis. The most notable features of apoptosis are pyknosis, DNA splitting, and the formation of apoptotic bodies [16]. Apoptosis is a strictly regulated multi-channel complex process that is mainly coordinated by activation of an aspartic acid-specific cysteine protease (caspase) cascade, including two main pathways: One relies on mitochondria (independent of the receptor) and the other involves the interaction between the death receptor and its ligand [17].

Autophagy is a process in which hydrolytic enzymes in lysosomes degrade proteins and organelles, including formation of phagophores and autophagosomes and fusion with lysosomes. The main function of autophagy is to promote cellular homeostasis. However, the role of autophagy in the tumor process is complex. Several studies have suggested that apoptosis and autophagy are interrelated and affect each other [18,19]. Autophagy can have positive or negative effects on tumor growth depending on the disease environment, and the survival function of autophagy may be harmful. In recent years, many natural organic compounds have exerted their anti-cancer activities by inducing autophagy of cells, which is of great importance to the further exploration and development of chemical anti-cancer therapy [20].

A great many anti-tumor drugs induce apoptosis and autophagy by inhibiting the AKT/mTOR pathway [21–23]. The PI3K/Akt/mTOR pathway is a common vulnerability in OS [24]. This signaling pathway affects most major cellular functions, so it plays a huge role in regulating basic cellular behaviors, such as growth and proliferation. The PI3K/Akt/mTOR pathway is associated with a variety of diseases, including cancer, obesity, and neurodegeneration. Early studies reported that the mTOR pathway has negative regulatory effects on apoptosis and autophagy [25–27]. A great deal of effort is currently being made to pharmacologically target this pathway [28].

In the present study, we explored the anti-cancer effect of DADs in OS MG-63 in vitro. In addition, we expounded on the potential mechanisms of apoptosis and autophagy through the mTOR signaling pathway.

2. Results

2.1. DADs Inhibit Osteosarcoma Cell Viability and Induces Cell Cycle Arrest at the G2/M Phase

The chemical structure of DADs is shown in Figure 1A. MG-63 cells were treated with different concentrations of DADs (0, 20, 40, 60, 80, and 100 µM) for 12, 48, and 72 h. Cell viability was measured with the Cell Counting Kit-8 (CCK-8) assay. As shown in Figure 1B, viability of OS cells treated with DADs was inhibited in a dose-and-time-dependent manner compared with the control group. The 20, 60, and 100 µM treatments were selected as representative doses for the in vitro and subsequent studies. The clone formation assay showed that the DADs treatment inhibited cloning of MG-63 cells (Figure 1C,D). DADs inhibited the colony counts of OS cells in a dose-dependent manner.

Cell cycle arrest may lead to inhibited proliferation, so we determined the effect of DADs on the cell cycle by flow cytometry. The G_0/G_1 phase cell population decreased, while the sub G_1 phase and the G_2/M phase increased significantly after treatment with 0, 20, 60, or 100 µM DADs for 24 h (Figure 1E,F).

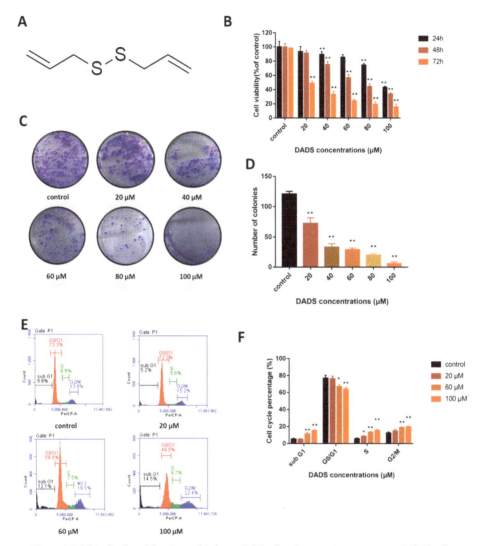

Figure 1. Inhibited cell proliferation and induces G_2/M cell cycle arrest in osteosarcoma MG-63 cells. (**A**) Chemical structure of diallyl disulfide (DADs). (**B**) Cell viability. MG-63 cells were treated with the indicated dose of DADs (0, 20, 40, 60, 80, and 100 µM) for different times (24, 48, and 72 h). Cell viability was detected by CCK8 assay (n = 3). (**C,D**) Clone formation. MG-63 cells were treated with 0, 20, 60, and 100 µM DADs, and the number of cell colonies was measured by clone formation assay 9 days later. (**E,F**) Cells were treated with DADs for 24 h, and the cell cycle was detected by flow cytometry. G_2/M cell cycle arrest was observed in MG-63 cells. The percentage of the sub G_1, G_0/G_1, S, and G_2/M phase cell populations were represented by the mean ± SD of at least three independent experiments. Statistical differences were analyzed by student's t-test (* $p < 0.05$, ** $p < 0.01$ compared with control group).

2.2. DADs Induce Apoptosis of Osteosarcoma Cells

DADs may inhibit the growth of OS cells through apoptosis. Therefore, we determined whether DADs induce OS cell apoptosis through Annexin V/propidium iodide (PI) double staining. As shown in Figure 2A,B, the flow cytometry results showed that OS cells caused a dose-dependent increase in early and late apoptotic cells after the DADs treatment. We investigated the expression of important signaling proteins during apoptosis by Western blot. After a certain period of treatment, the protein expression of caspase-3, cleaved-caspase 3, and Bax increased significantly while that of Bcl-2 decreased (Figure 2C,D).

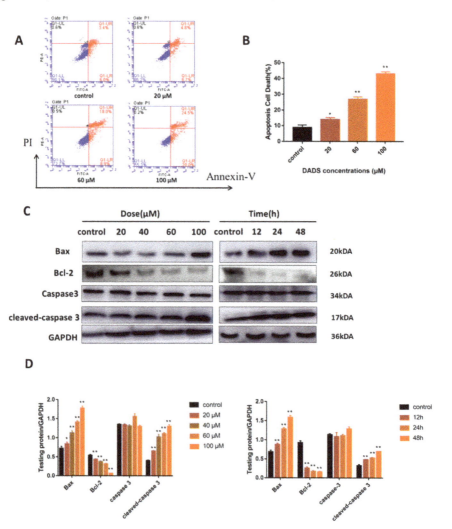

Figure 2. Induces caspase-dependent apoptosis in MG-63 cells. (**A,B**) After cells were stained by Annexin V-FITC/PI and left in dark at room temperature for 15 min, the apoptosis rate was measured by flow cytometry. Data were presented as means ± SD ($n = 3$). (**C,D**) Cells were treated with different doses of DADs for 24 h or incubated with DADs (60 μM) for various hours. The apoptosis-related proteins caspase-3, cleaved-caspase 3, Bax, and Bcl-2 were measured by Western blot. GAPDH was used as a loading control. (* $p < 0.05$, ** $p < 0.01$ compared with control group).

2.3. DADs Induce Autophagy of Osteosarcoma Cells

We continued to explore whether DADs induced autophagy in OS cells. We examined the expression of autophagy-related proteins in the DADs and control groups by Western blot analysis. The results showed that DADs increased the levels of LC3B-II, an indicator of autophagosome formation, in MG-63 cells in a dose-dependent manner relative to the controls (Figure 3A,B). Moreover, we observed an increase in p62 protein expression in the DADs-treated groups compared with that in the untreated group.

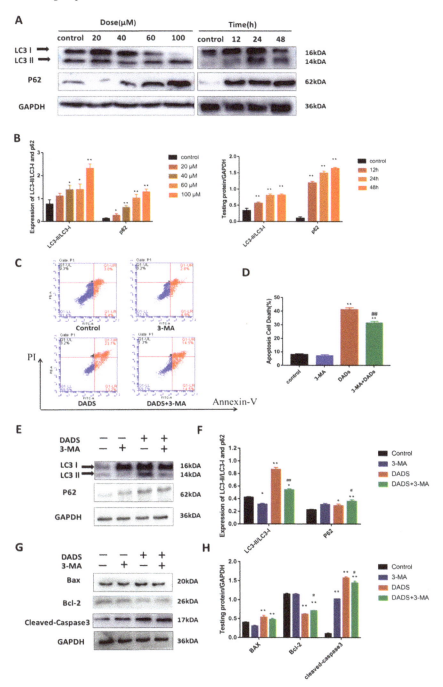

Figure 3. Triggered autophagy flux of MG-63 cells, and inhibition of autophagy reduces DADs-induced apoptosis. (**A,B**) Cells were treated with different dose of DADs for about 24 h or incubated with DADs

(60 µM) for various hours. Western blotting was used to analyze protein expression, and antibodies against LC-3-I, LC3-II, and p62 were tested. (* $p < 0.05$, ** $p < 0.01$ compared with control group). (**C,D**) Cells were pretreated with 3-MA (2.5 mM) for 2 hours and then treated with 100 µM DADs for 24 h. Apoptosis was measured by flow cytometry. The proportion of apoptotic cells from three independent experiments was shown by histograms. (**E,F**) Western blot showed the expression of apoptosis-related proteins LC3 and p62 with DADs or 3-MA treatment. (**G,H**) The Western blot results showed that caspase-3, Bax, and Bcl-2 protein expression levels after 3-MA treatment compared with that in the DADs only treatment groups. GAPDH was used as load control. (* $p < 0.05$, ** $p < 0.01$ compared with the control group, # $p < 0.05$, ## $p < 0.01$ compared with only DADs treated group).

We used the autophagy inhibitor 3-methyladenine (3-MA) to perform an experiment. The 3-MA inhibits autophagosome formation during the early stage by blocking class III phosphatidylinositol 3-kinases [29]. The level of LC3-II induced by DADs in association with 3-MA (2.5 mM) was clearly less than that observed with DADs alone (Figure 3E,F). The results of Annexin V and PI double staining showed that the percentage of apoptotic cells was less in the DADs + 3-MA group than in the DADs group. However, the apoptosis rate continued to increase regardless of 3-MA compared with the control group (Figure 3C,D). The Western blot results showed that apoptosis-related proteins decreased after 3-MA treatment compared with that in the DADs only treatment groups (Figure 3G,H).

2.4. DADs Induces Apoptosis and Autophagy by Inhibiting the PI3K/Akt/mTOR Signaling Pathway

Previous studies have confirmed that PI3K/Akt/mTOR is a signaling pathway that has a negative regulatory effect on apoptosis and autophagy. Inhibiting the mTOR pathway promotes autophagosome formation during the early stage. Therefore, we examined whether DADs stimulates autophagy by detecting activation of mTOR. As shown in Figure 4A,B, the Western blot results indicate that after treatment with DADs for 24 h, the expression of PI3K was decreased and, at the same time, the decrease in the phosphorylation of AKT protein was observed. We further observed that the exposure of MG-63 cells to DADs decreased the phosphorylated (activated) form of mTOR as well as its downstream effectors p70S6K and p-p70S6K proteins compared with that in the control group. The CCK8 assay results showed that rapamycin (mTOR inhibitor; 100 nM) significantly increased the inhibitory effect of DADs on cell viability (Figure 4C). A Western blot analysis revealed that the level of LC3-II increased compared to that in cells treated with DADs alone, whereas there was almost no difference in the level of the p62 protein. Apoptosis-related proteins increased after rapamycin treatment compared with that in the DADs group (Figure 4D,E).

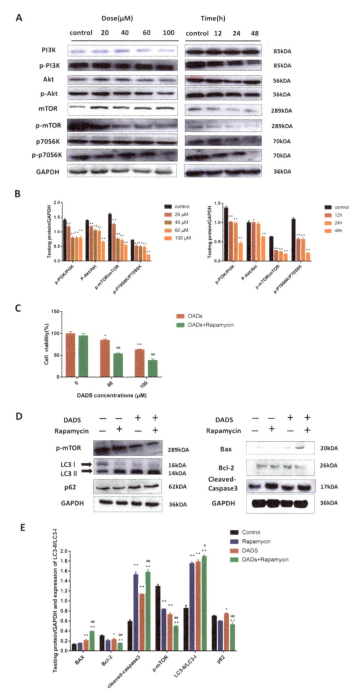

Figure 4. Induced apoptosis and autophagy of OS cells through mTOR pathway. (**A**,**B**) Expression of PI3K/Akt/mTOR pathway proteins were analyzed by Western blot. Cells were treated with DADs for

24 h. (C) Cell viability was detected by CCK8 24 h after DADs treatment (0, 60, and 100 µM). (n = 3). (D,E) Cells were pretreated with rapamycin (mTOR inhibitor) and then incubated with 100 µM DADs for 24 h. Western blot analyses were used to determine the levels of autophagy-related proteins (LC3-II/LC-3-I and p62), apoptosis-related proteins (caspase-3, Bax, and Bcl-2) and p-mTOR. GAPDH was used as load control. (* $p < 0.05$, ** $p < 0.01$ compared with the control group, # $p < 0.05$, ## $p < 0.01$ compared with only DADs treated group).

3. Discussion

Garlic is not only an important food seasoning, but also a traditional medicine. Epidemiological studies have shown that the consumption of garlic is inversely proportional to the incidence of cancer [30]. The medicinal value of garlic is dependent on its active ingredient garlicin. DADs, diallyl trisulfide, and diallyl tetrasulfide are the main components of garlicin. Among them, DADs has a wide range of anti-cancer effects [31]. Studies have shown that DADs inhibits the growth of various tumors by inducing apoptosis and cell cycle arrest [32,33].

OS has an aggressive malignant neoplasm with poor outcomes, whose cells proliferate intensively and represent very dynamic biological structure, create numerous mutations resulting in new tumor cell lines with different genotypes and phenotypes. In such malignancies, a highly variable sensitivity to therapeutics can be observed, and some cell lines develop resistance to the treatment (plant molecules including). Therefore, the biological effects of combining various plant molecules (phytochemicals) with proven cytotoxic effects administered with conventional therapy to target a substantially wider range of signaling pathways in cancer cells should be superior compared to single compounds in cancer treatment and may delay the development of resistance [34,35]. Therefore, further urgent research is needed for the identification of new molecules (including plant-derived) with excellent anticancer properties within combinational therapies against OS. In our study, we confirmed that DADs inhibited proliferation of OS cells in dose-and-time-dependent manners, caused G_2/M phase arrest, induced apoptosis, and restricted autophagic flux in OS in vitro. We also studied the mTOR pathway mechanism during apoptosis and autophagic flux.

Tumor development is initially manifested by uncontrolled cell division [36]. The entire process from the completion of one division to the end of the next division is called a cell cycle, which consists of interphase and mitotic periods. Interphase consists of the G_1, S, and G_2 phases. G_1/S and G_2/M transformation are two very important phases in the cell cycle. Cells in these two stages experience a complex and active period of change and are particularly susceptible to environmental conditions. In our study, flow cytometry revealed that the OS cell cycle was blocked after 24 h of DADs (0, 20, 60, 100 µM) treatment and the effect was concentration dependent. Li et al. demonstrated that the OS cell cycle could be blocked by diallyl disulfide in the G_2/M phase [5]. This result corresponds to the CCK8 and clone forming assay results, demonstrating that DADs inhibited cell viability and proliferative activity of OS cells.

Cell cycle arrest can induce apoptosis, and the tumor inhibitory effect of DADs also includes abnormal cell death, such as apoptosis [37] and autophagy [38]. Apoptosis is composed of a complex multi-gene regulated mechanism. The caspase activation pathway plays a key role in apoptosis. HJ, K. et al. demonstrated that DADS promoted trail-mediated apoptosis by inhibiting Bcl-2 [37]. There are two pathways that have been studied most fully: The cell surface death receptor (caspase-8) pathway and the mitochondria-initiated (caspase-9) pathway. However, caspase-3 participates in both pathways [39]. Furthermore, the anti-apoptotic Bcl-2 protein prevents mitochondria from releasing cytochrome c, thus, keeping cells alive. Bax and Bcl-2 both belong to the Bcl-2 gene family. Bcl-2 is an inhibitor of the apoptosis gene. Bax antagonizes the inhibitory effect of Bcl-2 on apoptosis, and promotes apoptosis [40]. In our study, the caspase-3 and Bax proteins both increased, whereas the Bcl-2 protein decreased with DADs treatment. Flow cytometry showed that the proportion of apoptotic cells in the DADs treatment group increased significantly in a dose-dependent manner during the same treatment time. These results reveal that DADs stimulated caspase-dependent apoptosis.

Autophagy is a non-caspase-dependent form of cell death that degrades proteins and organelles in cells through the lysosomal pathway. Autophagy mainly plays an adaptive role in the body, protecting organisms from various pathological damage, including infection, cancer, nerve degeneration, aging and heart disease [41]. LC3 is a marker of autophagy. During autophagy, LC3-I is hydrolyzed and converted to LC3-II. Enhanced expression of autophagy leads to aggregation of P62 and complete protein degradation by fusion with a lysosome [42]. Therefore, the ratio of LC3-II/LC3-I and the p62 expression level could reflect the level of autophagy. Our experimental results show that the LC3-II/LC3-I ratio was upregulated, which confirmed that autophagy was promoted. In contrast, we observed an increase in the expression of p62 protein in our study, which is usually considered a sign of inhibited autophagy activity [29]. This observation suggested that DADs blocked the autophagic flux after early stimulation of autophagy, resulting in accumulation of the protein.

However, autophagy is a multi-stage process, and the damage and preservation of cells are unclear. We used autophagy inhibitors to further explore the mechanism of autophagy. 3-MA inhibits autophagy by inhibiting formation of the autophagosome. After adding 3-MA, LC3-II expression was downregulated and p62 expression was upregulated compared with the DADs treatment alone. Notably, the apoptosis rate of the DADs + 3-MA group decreased synchronously compared with that of the DADs alone group. We suspected that the reason is that autophagy enhances caspase-dependent cell death and independently causes cell death. However, 3-MA inhibits autophagy from the initial stage, thus reducing part of the cell death [19,43].

It is commonly known that the mTOR pathway is related to autophagy and apoptosis [26,28,44]. Our study discovered that DADs inhibits the PI3K/Akt/mTOR/p70S6K signal pathway. Inhibition of the mTOR pathway can partially activate autophagy and apoptosis. PI3K/Akt/mTOR signaling pathway is a classical pathway, which not only promotes angiogenesis and cell progress, but also plays an important role in various human malignant tumors [45]. To further test whether apoptosis and autophagy induced by DADs are related to the mTOR pathway, we added rapamycin (mTOR inhibitor). The results showed that cell proliferation in the rapamycin + DADs group was lower than that of the DADs group alone, and the levels of apoptosis-related proteins and autophagy-related proteins both increased. These finding indicate that DADs may play a role in apoptosis and autophagy through the mTOR signaling pathway. After rapamycin was used to promote autophagy, the expression of LC3-II increased, whereas content of the p62 protein did not change significantly. Thus, we conclude that DADs triggered autophagy flux; however, the process was not completed as indicated by the p62 level. This probably occurred because this process turned into apoptosis during the late stage of treatment.

However, there is no doubt that further research is needed to study the specific mechanism of DADs in OS treatment. For example, p53 is an important tumor suppressor gene, which is related to cell cycle arrest and apoptosis. The mutation and deletion of P53 gene account for about 50% of all human tumors, which leads to tumor formation, metastasis and drug resistance of tumors. MG-63 expresses P53, and p53-mediated cell death is partly related to the interaction between Bcl-2 and Bax. What's more, the correlation between apoptosis and autophagy under DADs treatment has not been completely explained. Therefore, further exploration is needed to reveal the potential mechanism of DADs in the treatment of OS.

In conclusion, our study clarified the possible effects of DADs on OS cells and showed that DADs inhibited OS by causing G_2/M phase arrest and inducing apoptosis and autophagy. Moreover, DADs induced apoptosis and autophagy by inhibiting the PI3K/Akt/mTOR signaling pathway. In addition, 3-MA inhibited autophagy weakened DADs-induced cell death, indicating that DADs induced autophagy-mediated cell death. The possible mechanism of action of DADs is shown in Figure 5. This study provides insight into the clinical application of this compound and a new method to treat OS.

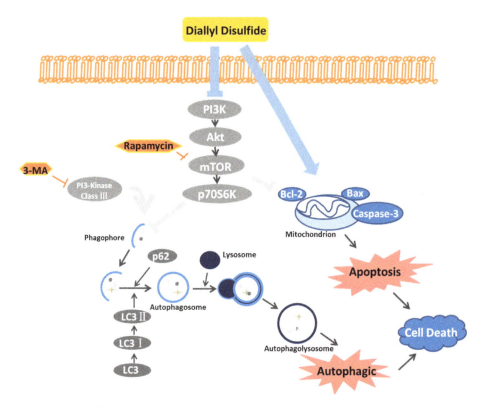

Figure 5. DADs inducing autophagy and apoptosis of human OS cells.

4. Materials and Methods

4.1. Reagents and Antibodies

Diallyl Disulfide (DADs, 30648) was purchased from Sigma-Aldrich (Shanghai, China). 3-methyladenine (3-MA, HY-19312) and rapamycin (HY-10219) were purchased from MedChemExpress (Shanghai, China). Antibodies against Bax (5023T), Bcl-2 (4223T), cleaved-caspase 3 (9664T), mTOR (2983T), p-mTOR (5536T), and p-p70S6K (108D2) were purchased from Cell Signaling Technology (CST, Shanghai, China). Anti-SQSTM1/p62 (ab109012), LC3B (ab48394), caspase-3 (ab13847), PI3K (ab180967), Akt (ab32505), p-Akt (ab192623), and p70S6K (ab32529) was purchased from Abcam (Shanghai, China). p-PI3K (AF3241) was purchased from Affinity Biosciences (Jiangsu, China). Anti-GAPDH was purchased from Proteintech (Wuhan, Hubei, China).

4.2. Cell Culture

The human osteosarcoma cell line MG-63 was obtained from the Cell Bank of the Chinese Academy of Sciences (Shanghai, China). Cells were raised in Dulbecco's Modified Eagle's Medium (DMEM/F12) supplemented with 10% fetal bovine serum (Gibco, Grand Island, NY, USA), 1% penicillin-streptomycin and 1% non-essential amino acid (NEAA) under standard culture condition (37 °C, 95% humidified air and 5% CO_2). The medium was changed every 2 days.

4.3. Cell Viability Assay

For the sake of confirming the effect of DADs on OS cell proliferation, cells were incubated in 96-well plates for 24 h at a density of 5×10^3 cells/well with 100 µL culture medium. Then, cells were treated with different dose of DADs (0, 20, 40, 60, 80, and 100 µM). After 24, 48, and 72 h, cell viability was detected by Cell Counting Kit-8 assay (MedChem Express, China). Adding 10 µL CCK8 working solution per well for about 2 h, the absorbance value of each well was measured at 450 nm with enzyme-labeled instrument. Then, the cell viability of the experimental group was calculated.

4.4. Clone Formation Assay

Cells were cultured in 6-well plates at a density of 500 cells/well and incubated under standard culture condition for 24 h. They were then treated them with different doses of DADs (0, 20, 40, 60, 80, and 100 µM), for about 9 days. Next, the medium was removed and the cell clones was washed with PBS. Shortly afterwards, they were fixed with 4 % paraformaldehyde and dyed with 0.1% crystal violet. Finally, colonies including more than 50 cells were calculated.

4.5. Cell Cycle Analysis

The cells were incubated in 6-well plates with a density of 2×10^5 cells/well. After 24 h, cells were treated with DADs (0, 20, 60, and 100 µM) for another 24 h. Then we collected cells from 6-well plates, added 70% ethanol and fixed at 4 °C overnight. After centrifugation to remove ethanol, they were washed again with PBS and the cells were stained with propidium iodide (PI) and RNase A (KeyGEN Biotech, China). The mixture was kept at 37 °C for 15 min in the dark. Finally, the cell cycle was measured by flow cytometry, and data were analyzed with BD Accuri C6 plus (Becton Dickinson, Franklin Lakes, NJ, USA) software.

4.6. Apoptosis Flow-Cytometry Assay

The OS cells were seeded into 6-well plates with a density of 2×10^5 cells/well, then treated with different concentrations of DADs (0, 20, 60, and 100 µM) for 24 h. Cells were collected with trypsin, washed twice with pre-chilled PBS, and then suspended with 500 µL binding buffer. Then, Annexin V-FITC and PI (Hanbio, Shanghai, China) were added respectively, and the cells were mixed. After being incubated at room temperature in the dark for 15 min, the samples were analyzed by flow cytometry using BD Accuri C6 plus (Becton Dickinson, Franklin Lakes, NJ, USA). According to the principle that phosphatidylserine can bind to Annexin V-FITC, we measured the proportion of early apoptotic cells. PI is a DNA-binding dye. It could emit red fluorescence but cannot pass through living cell membranes. The proportion of late apoptosis/dead cells can be determined by it.

4.7. Western Blot Analysis

Cells were scraped into EP tubes with cell scraping tools and mixed with RIPA Lysis Buffer (Betotium Institute of Biotechnology, Beijing, China). Protease and phosphatase inhibitors were added in a ratio of 100:1. All operations were carried out on ice. After measuring the protein concentration with BCA protein assays (Beyotime, Beijing, China), we added 1/4 volume of $5 \times$ SDS loading buffer in each EP tubes and heated it at 95 °C for 5 min. Proteins were separated by 6%–15% SDS-PAGE, and transferred to the polyvinylidene fluoride (PVDF) membrane. The membranes were incubated by 1:1000~5000 primary antibodies at 4 °C overnight, followed by incubation with secondary antibodies at room temperature for 1 h. Finally, the ECL detection system (SmartChemi 420, Beijing, China) was used to measure the immune reaction zone. Each experiment was repeated at least three times.

4.8. Statistical Analysis

All data were represented by the average of three independent experiments. Statistical analysis was conducted with Graghpad prism 7 software (San Diego, CA, USA). Differences between experimental groups and control groups were calculated by Student's *t-test*. $p < 0.05$ was considered to have statistical significance.

Author Contributions: Conceptualization, Z.Y.; data curation, X.G., R.C., C.H., S.L., T.H. and J.G.; formal analysis, Z.Y., X.G., R.C., C.H., D.L., P.Y., S.L. and T.H.; funding acquisition, J.G. and M.L.; investigation, X.G., R.C., C.H., P.Y., S.L., T.H. and M.L.; methodology, Z.Y., X.G., R.C., C.H., S.L., T.H., J.G. and M.L.; project administration, J.G. and M.L.; resources, X.G., C.H. and D.L.; Software, X.G., R.C., D.L. and P.Y.; supervision, J.G. and M.L.; validation, Z.Y. and M.L.; visualization, Z.Y. and X.G.; writing – original draft, Z.Y.; writing – review & editing, Z.Y.

Funding: This research received no external funding.

Acknowledgments: This study was partially supported by the Shandong Key Research and Development Project (Grant No. 2018GSF118134) to Li M. The National Nature Science Foundation of China (Grant No.81771108) and the Shandong Key Research and Development Project (Grant No. 2017GSF218017) to Guo J.

Conflicts of Interest: The authors declare no conflict of interest.

References

1. Raymond, A.K.; Jaffe, N. Osteosarcoma multidisciplinary approach to the management from the pathologist's perspective.%a raymond ak. *Cancer Treat. Res.* **2009**, *152*, 63–84. [PubMed]
2. Mirabello, L.; Troisi, R.J.; Savage, S.A. Osteosarcoma incidence and survival rates from 1973 to 2004: Data from the surveillance, epidemiology, and end results program. *Cancer* **2009**, *115*, 1531–1543. [CrossRef] [PubMed]
3. Durfee, R.A.; Mohammed, M.; Luu, H.H. Review of osteosarcoma and current management. *Rheumatol. Ther.* **2016**, *3*, 221–243. [CrossRef] [PubMed]
4. Link, M.P.; Goorin, A.M.; Miser, A.W.; Green, A.A.; Pratt, C.B.; Belasco, J.B.; Pritchard, J.; Malpas, J.S.; Baker, A.R.; Kirkpatrick, J.A.; et al. The effect of adjuvant chemotherapy on relapse-free survival in patients with osteosarcoma of the extremity. *N. Engl. J. Med.* **1986**, *314*, 1600–1606. [CrossRef] [PubMed]
5. Y, L.; Z, W.; J, L.; X, S. Diallyl disulfide suppresses foxm1-mediated proliferation and invasion in osteosarcoma by upregulating mir-134. *J. Cell. Biochem.* **2018**, *undefined*, undefined.
6. Ji, X.X.; Liu, F.; Xia, H.; He, J.; Tan, H.; Yi, L.; Su, Q. [downregulation of mcl-1 by diallyl disulfide induces g2/m arrest in human leukemia k562 cells and its mechanism]. *Zhongguo Shi Yan Xue Ye Xue Za Zhi* **2018**, *26*, 750–755. [PubMed]
7. Wu, X.J.; Kassie, F.; Mersch-Sundermann, V. The role of reactive oxygen species (ros) production on diallyl disulfide (dads) induced apoptosis and cell cycle arrest in human a549 lung carcinoma cells. *Mutat. Res.* **2005**, *579*, 115–124. [CrossRef]
8. Yang, J.S.; Chen, G.W.; Hsia, T.C.; Ho, H.C.; Ho, C.C.; Lin, M.W.; Lin, S.S.; Yeh, R.D.; Ip, S.W.; Lu, H.F.; et al. Diallyl disulfide induces apoptosis in human colon cancer cell line (colo 205) through the induction of reactive oxygen species, endoplasmic reticulum stress, caspases casade and mitochondrial-dependent pathways. *Food Chem. Toxicol.* **2009**, *47*, 171–179. [CrossRef]
9. Lu, H.F.; Sue, C.C.; Yu, C.S.; Chen, S.C.; Chen, G.W.; Chung, J.G. Diallyl disulfide (dads) induced apoptosis undergo caspase-3 activity in human bladder cancer t24 cells. *Food Chem. Toxicol.* **2004**, *42*, 1543–1552. [CrossRef]
10. Lin, Y.T.; Yang, J.S.; Lin, S.Y.; Tan, T.W.; Ho, C.C.; Hsia, T.C.; Chiu, T.H.; Yu, C.S.; Lu, H.F.; Weng, Y.S.; et al. Diallyl disulfide (dads) induces apoptosis in human cervical cancer ca ski cells via reactive oxygen species and ca2+-dependent mitochondria-dependent pathway. *Anticancer Res.* **2008**, *28*, 2791–2799.
11. Sherr, C.J. Cancer cell cycles. *Science* **1996**, *274*, 1672–1677. [CrossRef] [PubMed]
12. Hartwell, L.H.; Kastan, M.B. Cell-cycle control and cancer. *Science* **1994**, *266*, 1821–1828. [CrossRef] [PubMed]
13. Yin, X.R.; Zhang, R.; Feng, C.; Zhang, J.; Liu, D.; Xu, K.; Wang, X.; Zhang, S.Q.; Li, Z.F.; Liu, X.L.; et al. Diallyl disulfide induces g2/m arrest and promotes apoptosis through the p53/p21 and mek-erk pathways in human esophageal squamous cell carcinoma. *Oncol. Rep.* **2014**, *32*, 1748–1756. [CrossRef] [PubMed]
14. Xu, J.C.; Zhou, X.P.; Wang, X.A.; Xu, M.D.; Chen, T.; Chen, T.Y.; Zhou, P.H.; Zhang, Y.Q. Cordycepin induces apoptosis and g2/m phase arrest through the erk pathways in esophageal cancer cells. *J. Cancer* **2019**, *10*, 2415–2424.
15. Cheng, A.C.; Hsu, Y.C.; Tsai, C.C. The effects of cucurbitacin e on gadd45β-trigger g2/m arrest and jnk-independent pathway in brain cancer cells. *J. Cell. Mol. Med.* **2019**, *23*, 3512–3519.

16. Burgess, D.J. Apoptosis: Refined and lethal. *Nat. Rev. Cancer* **2013**, *13*, 79. [CrossRef] [PubMed]
17. Zimmermann, K.C.; Bonzon, C.; Green, D.R. The machinery of programmed cell death. *Pharmacol. Ther.* **2001**, *92*, 57–70. [CrossRef]
18. Nikoletopoulou, V.; Markaki, M.; Palikaras, K.; Tavernarakis, N. Crosstalk between apoptosis, necrosis and autophagy. *Bba-Mol. Cell Res.* **2013**, *1833*, 3448–3459. [CrossRef] [PubMed]
19. Eisenberg-Lerner, A.; Bialik, S.; Simon, H.U.; Kimchi, A. Life and death partners: Apoptosis, autophagy and the cross-talk between them. *Cell Death Differ.* **2009**, *16*, 966–975. [CrossRef] [PubMed]
20. Lee, H.; Venkatarame Gowda Saralamma, V.; Kim, S.; Ha, S.; Raha, S.; Lee, W.; Kim, E.; Lee, S.; Heo, J.; Kim, G. Pectolinarigenin induced cell cycle arrest, autophagy, and apoptosis in gastric cancer cell via pi3k/akt/mtor signaling pathway. *Nutrients* **2018**, *10*, 1043.
21. Wang, G.; Zhang, T.; Sun, W.; Wang, H.; Yin, F.; Wang, Z.; Zuo, D.; Sun, M.; Zhou, Z.; Lin, B.; et al. Arsenic sulfide induces apoptosis and autophagy through the activation of ros/jnk and suppression of akt/mtor signaling pathways in osteosarcoma. *Free Radic Biol. Med.* **2017**, *106*, 24–37. [CrossRef] [PubMed]
22. Ma, K.; Zhang, C.; Huang, M.Y.; Li, W.Y.; Hu, G.Q. Cinobufagin induces autophagy-mediated cell death in human osteosarcoma u2os cells through the ros/jnk/p38 signaling pathway. *Oncol. Rep.* **2016**, *36*, 90–98. [CrossRef] [PubMed]
23. Kim, S.H.; Son, K.M.; Kim, K.Y.; Yu, S.N.; Park, S.G.; Kim, Y.W.; Nam, H.W.; Suh, J.T.; Ji, J.H.; Ahn, S.C. Deoxypodophyllotoxin induces cytoprotective autophagy against apoptosis via inhibition of pi3k/akt/mtor pathway in osteosarcoma u2os cells. *Pharm. Rep.* **2017**, *69*, 878–884. [CrossRef] [PubMed]
24. Perry, J.A.; Kiezun, A.; Tonzi, P.; Van Allen, E.M.; Carter, S.L.; Baca, S.C.; Cowley, G.S.; Bhatt, A.S.; Rheinbay, E.; Pedamallu, C.S.; et al. Complementary genomic approaches highlight the pi3k/mtor pathway as a common vulnerability in osteosarcoma. *Proc. Natl. Acad. Sci. USA* **2014**, *111*, E5564–E5573. [CrossRef] [PubMed]
25. Kim, Y.C.; Guan, K.L. Mtor: A pharmacologic target for autophagy regulation. *J. Clin. Investig.* **2015**, *125*, 25–32. [CrossRef] [PubMed]
26. Saiki, S.; Sasazawa, Y.; Imamichi, Y.; Kawajiri, S.; Fujimaki, T.; Tanida, I.; Kobayashi, H.; Sato, F.; Sato, S.; Ishikawa, K.; et al. Caffeine induces apoptosis by enhancement of autophagy via pi3k/akt/mtor/p70s6k inhibition. *Autophagy* **2011**, *7*, 176–187. [CrossRef] [PubMed]
27. Chen, L.; Xu, B.S.; Liu, L.; Luo, Y.; Yin, J.; Zhou, H.Y.; Chen, W.X.; Shen, T.; Han, X.Z.; Huang, S.L. Hydrogen peroxide inhibits mtor signaling by activation of ampk alpha leading to apoptosis of neuronal cells. *Lab. Investig.* **2010**, *90*, 762–773. [CrossRef] [PubMed]
28. Laplante, M.; Sabatini, D.M. Mtor signaling in growth control and disease. *Cell* **2012**, *149*, 274–293. [CrossRef]
29. Kliosnky, D. Guidelines for the use and interpretation of assays for monitoring autophagy (3rd edition) (vol 12, pg 1, 2015). *Autophagy* **2016**, *12*, 1–222.
30. Buiatti, E.; Palli, D.; Decarli, A.; Amadori, D.; Avellini, C.; Bianchi, S.; Biserni, R.; Cipriani, F.; Cocco, P.; Giacosa, A.; et al. A case-control study of gastric cancer and diet in italy. *Int. J. Cancer* **1989**, *44*, 611–616. [CrossRef]
31. Sundaram, S.G.; Milner, J.A. Diallyl disulfide inhibits the proliferation of human tumor cells in culture. *Biochim. Biophys. Acta* **1996**, *1315*, 15–20. [CrossRef]
32. Kwon, K.B.; Yoo, S.J.; Ryu, D.G.; Yang, J.Y.; Rho, H.W.; Kim, J.S.; Park, J.W.; Kim, H.R.; Park, B.H. Induction of apoptosis by diallyl disulfide through activation of caspase-3 in human leukemia hl-60 cells. *Biochem. Pharm.* **2002**, *63*, 41–47. [CrossRef]
33. Knowles, L.M.; Milner, J.A. Depressed p34(cdc2) kinase activity and g(2)/m phase arrest induced by diallyl disulfide in hct-15 cells. *Nutr. Cancer* **1998**, *30*, 169–174. [CrossRef] [PubMed]
34. Kapinova, A.; Stefanicka, P.; Kubatka, P.; Zubor, P.; Uramova, S.; Kello, M.; Mojzis, J.; Blahutova, D.; Qaradakhi, T.; Zulli, A.; et al. Are plant-based functional foods better choice against cancer than single phytochemicals? A critical review of current breast cancer research. *Biomed. Pharmacother. = Biomed. Pharmacother.* **2017**, *96*, 1465–1477. [CrossRef] [PubMed]
35. Abotaleb, M.; Kubatka, P.; Caprnda, M.; Varghese, E.; Zolakova, B.; Zubor, P.; Opatrilova, R.; Kruzliak, P.; Stefanicka, P.; Büsselberg, D. Chemotherapeutic agents for the treatment of metastatic breast cancer: An update. *Biomed. Pharmacother. = Biomed. Pharmacother.* **2018**, *101*, 458–477. [CrossRef] [PubMed]
36. Hanahan, D.; Weinberg, R.A. Hallmarks of cancer: The next generation. *Cell* **2011**, *144*, 646–674. [CrossRef] [PubMed]

37. Kim, H.J.; Kang, S.; Kim, D.Y.; You, S.; Park, D.; Oh, S.C.; Lee, D.H. Diallyl disulfide (dads) boosts trail-mediated apoptosis in colorectal cancer cells by inhibiting bcl-2. *Food Chem. Toxicol.: Int. J. Publ. Br. Ind. Biol. Res. Assoc.* **2019**, *125*, 354–360. [CrossRef] [PubMed]
38. Wu, Y.; Hu, Y.; Zhou, H.; Zhu, J.; Tong, Z.; Qin, S.; Liu, D. Organosulfur compounds induce cytoprotective autophagy against apoptosis by inhibiting mtor phosphorylation activity in macrophages. *Acta Biochim. Et Biophys. Sin.* **2018**, *50*, 1085–1093. [CrossRef]
39. Budihardjo, I.; Oliver, H.; Lutter, M.; Luo, X.; Wang, X. Biochemical pathways of caspase activation during apoptosis. *Annu. Rev. Cell Dev. Biol.* **1999**, *15*, 269–290. [CrossRef]
40. Kroemer, G. The proto-oncogene bcl-2 and its role in regulating apoptosis. *Nat. Med.* **1997**, *3*, 614–620. [CrossRef]
41. Levine, B.; Kroemer, G. Autophagy in the pathogenesis of disease. *Cell* **2008**, *132*, 27–42. [CrossRef] [PubMed]
42. Chen, Y.; Song, F. Research advances in selective adaptor protein autophagy of p62/sequestosome-1. *Chin. J. Pharmacol. Toxicol.* **2016**, *30*, 258–265.
43. Yu, L.; Alva, A.; Su, H.; Dutt, P.; Freundt, E.; Welsh, S.; Baehrecke, E.H.; Lenardo, M.J. Regulation of an atg7-beclin 1 program of autophagic cell death by caspase-8. *Science* **2004**, *304*, 1500–1502. [CrossRef] [PubMed]
44. Kumar, D.; Shankar, S.; Srivastava, R.K. Rottlerin induces autophagy and apoptosis in prostate cancer stem cells via pi3k/akt/mtor signaling pathway. *Cancer Lett.* **2014**, *343*, 179–189. [CrossRef] [PubMed]
45. Vanhaesebroeck, B.; Guillermet-Guibert, J.; Graupera, M.; Bilanges, B. The emerging mechanisms of isoform-specific pi3k signalling. *Nat. Rev. Mol. Cell Biol.* **2010**, *11*, 329–341. [CrossRef]

Sample Availability: Samples of the compounds are available from the authors.

© 2019 by the authors. Licensee MDPI, Basel, Switzerland. This article is an open access article distributed under the terms and conditions of the Creative Commons Attribution (CC BY) license (http://creativecommons.org/licenses/by/4.0/).

Article

13-Ethylberberine Induces Apoptosis through the Mitochondria-Related Apoptotic Pathway in Radiotherapy-Resistant Breast Cancer Cells

Hana Jin, Young Shin Ko, Sang Won Park, Ki Churl Chang and Hye Jung Kim *

Department of Pharmacology, College of Medicine, Institute of Health Sciences,
Gyeongsang National University, Jinju 52727, Korea
* Correspondence: hyejungkim@gnu.ac.kr; Tel.: +82-55-772-8074; Fax: +82-55-772-8079

Academic Editor: Roberto Fabiani
Received: 5 June 2019; Accepted: 2 July 2019; Published: 4 July 2019

Abstract: Berberine is reported to have multiple biological effects, including antimicrobial, anti-inflammatory, and antitumor activities, and 13-alkyl-substituted berberines show higher activity than berberine against certain bacterial species and human cancer cell lines. In particular, 13-ethylberberine (13-EBR) was reported to have anti-inflammatory effects in endotoxin-activated macrophage and septic mouse models. Thus, in this study, we aimed to examine the anticancer effects of 13-EBR and its mechanisms in radiotherapy-resistant (RT-R) MDA-MB-231 cells derived from the highly metastatic MDA-MB-231 cells. When we compared the gene expression between MDA-MB-231 and RT-R MDA-MB-231 cells with an RNA microarray, RT-R MDA-MB-231 showed higher levels of anti-apoptotic genes and lower levels of pro-apoptotic genes compared to MDA-MB-231 cells. Accordingly, we examined the effect of 13-EBR on the induction of apoptosis in RT-R MDA-MB-231 and MDA-MB-231 cells. The results showed that 13-EBR reduced the proliferation and colony-forming ability of both MDA-MB-231 and RT-R MDA-MB-231 cells. Moreover, 13-EBR induced apoptosis by promoting both intracellular and mitochondrial reactive oxygen species (ROS) and by regulating the apoptosis-related proteins involved in the intrinsic pathway, not in the extrinsic pathway. These results suggest that 13-EBR has pro-apoptotic effects in RT-R MDA-MB-231 and MDA-MB-231 cells by inducing mitochondrial ROS production and activating the mitochondrial apoptotic pathway, providing useful insights into new potential therapeutic strategies for RT-R breast cancer treatment.

Keywords: apoptosis; 13-ethylberberine; mitochondrial ROS; RT-R breast cancer cells

1. Introduction

Breast cancer is one of the most common causes of death in women around the world [1]. Most breast cancer patients respond to conventional treatments such as surgery, chemotherapy, and radiotherapy. However, there are inherent limitations to each therapy, which cause therapeutic resistance and the relapse of cancer, eventually leading to the failure of therapy. In particular, radiotherapy has several benefits, such as improving overall survival and synergizing with surgical resection [2,3]; the relapse of cancer after radiotherapy is common and, in particular, ductal carcinoma and early/advanced invasive breast cancers show radiotherapy resistance. Moreover, because triple-negative breast cancer (TNBC), which is characterized by the absence of estrogen receptor (ER), progesterone receptor (PR), and human epidermal growth factor receptor 2 (HER2) expression, is an aggressive type of cancer that is difficult to treat, breast cancer patients suffer more if TNBC acquires radiotherapy resistance, and the patients do not survive after radiation therapy. Therefore, in a previous study, we repeatedly irradiated the MDA-MB-231 breast cancer cell line, which is a common cell model system to represent highly metastatic TNBC, to establish radiotherapy-resistant MDA-MB-231 (RT-R MDA-MB-231) cells, and we aimed to find effective therapeutics to treat RT-R MDA-MB-231 cells in the present study.

Berberine (BBR) is an isoquinoline alkaloid that is isolated from *Cotridis rhizoma* and has multiple biological activities, such as antimicrobial, anti-inflammatory, and antitumor effects [4–7]. In particular, the anticancer effects of BBR on breast cancer cells were reported; BBR induces breast cancer cell apoptosis via the activation of the apoptotic signaling pathway [8,9], the inhibition of proliferation and migration [10], the suppression of cell motility through the downregulation of related molecules [11,12], and the enhancement of chemosensitivity, which induces apoptosis [13]. Recently, it was reported that 13-alkyl-substituted berberines showed better antimicrobial activity against certain bacterial species and cytotoxic activity against human cancer cell lines than BBR [14,15]. Furthermore, among these 13-alkyl-substituted berberines, 13-ethylberberine (13-EBR) was reported to have anti-inflammatory effects in endotoxin-activated macrophage and septic mouse models [16,17]. However, the effects of 13-EBR on cancer cell growth and signaling pathways were not reported. Therefore, we tried to identify the differences between MDA-MB-231 cells and RT-R MDA-MB-231 cells in gene expression levels, and determined the anticancer effects of 13-EBR on RT-R MDA-MB-231 breast cancer cells, as well as MDA-MB-231. Moreover, we explored the associated mechanisms of 13-EBR using MDA-MB-231 and RT-R MDA-MB-231 breast cancer cells in this study.

2. Results

2.1. 13-EBR Had Anticancer Effects on RT-R MDA-MB-231 Cells and MDA-MB-231 Cells, as Demonstrated by Suppressing the Proliferation and Colony-Forming Ability

In our previous study, we showed that RT-R MDA-MB-231 cells had increased cell viability and colony-forming ability after irradiation, and exhibited higher chemoresistance compared to the MDA-MB-231 parental cells [18]. In this study, we analyzed the gene expression levels between MDA-MB-231 cells and RT-R MDA-MB-231 cells and found that RT-R MDA-MB-231 cells showed lower expression of pro-apoptotic genes and higher expression of anti-apoptotic genes than MDA-MB-231 cells (Table 1). Thus, we were interested in identifying effective anticancer drugs to treat RT-R breast cancer cells because numerous cancer patients suffer from aggressive disease and the relapse of radiotherapy-resistant cancer. Figure 1 shows that 13-EBR effectively reduced proliferation (Figure 1B) and colony formation (Figure 1C) in RT-R MDA-MB-231 cells and MDA-MB-231 cells in a dose-dependent manner compared to the controls. These results suggested that 13-EBR has anticancer effects as a result of the suppression of cell growth and colony-forming ability in both MDA-MB-231 and RT-R MDA-MB-231 cells.

Table 1. Analysis of gene expression levels between MDA-MB-231 and radiotherapy-resistant (RT-R) MDA-MB-231 cells. Total RNA was extracted from MDA-MB-231 and RT-R MDA-MB-231 cells, and the genes involved in apoptotic cell death were analyzed.

	Apoptosis-Related Genes	Fold Change (RT-R MDA-MB-231/MDA-MB-231)
Pro-apoptotic genes	Bax	0.622
	Bad	0.620
	Cytochrome c	0.576
	Cleaved caspase-3 (p17)	0.363
	Cleaved caspase-7 (p11)	0.846
Anti-apoptotic genes	Bcl-2	0.927
	Bcl-2A1	8.036
	Mcl-1	1.263

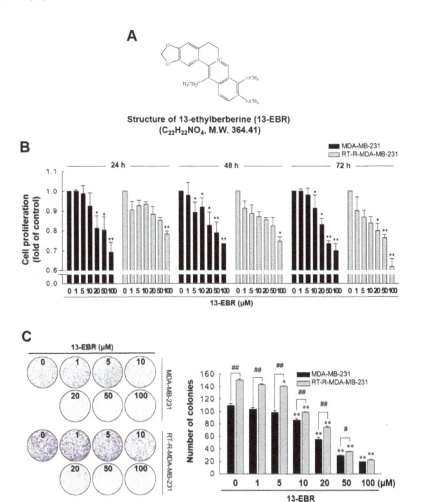

Figure 1. Chemical structure of 13-ethylberberine (13-EBR), and the effects of 13-EBR on cell proliferation, colony formation, and apoptosis in breast cancer cells. (**A**) The chemical structure of 13-EBR. (**B**) MDA-MB 231 and radiotherapy-resistant (RT-R) MDA-MB 231 cells were treated with 13-EBR at the indicated doses (1, 5, 10, 20, 50, and 100 µM) for 24–72 h, and cell proliferation was measured using the Cell Counting Kit-8 (CCK-8) reagent, as described in Section 4. The values represent the mean ± standard error of the mean (SEM) of three independent experiments; ** $p < 0.01$, * $p < 0.05$ compared to the controls (vehicle-treated cells) at each time point. (**C**) Both breast cancer cell lines (1000 cells/well) were seeded in six-well plates. The cells were stimulated with 13-EBR for 24 h at the indicated doses. Following treatment, a colony-formation assay was performed, as described in Section 4, and was quantified by counting the colonies. The values represent the mean ± SEM of three independent experiments; ** $p < 0.01$, * $p < 0.05$ compared to the control for each cell line; ## $p < 0.01$, # $p < 0.05$ compared to the MDA-MB-231 cells.

2.2. 13-EBR Upregulated Intracellular Total and Mitochondrial ROS Production in Both MDA-MB-231 and RT-R MDA-MB-231 Cells

It was reported that excessive ROS can induce apoptosis through both the extrinsic and intrinsic pathways [19]. Furthermore, excessive mitochondrial oxidant stress can induce cell death in tumor cells [20]. Thus, we examined the effects of 13-EBR on the production of intracellular total ROS, including mitochondrial ROS, in RT-R MDA-MB-231 and MDA-MB-231 cells. When both of the breast cancer cell lines were treated with 50 µM 13-EBR, the treatment significantly enhanced the intracellular total ROS and mitochondrial ROS production from early time points compared to the controls (Figure 2).

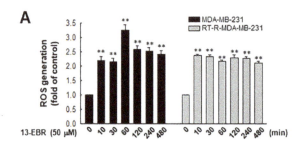

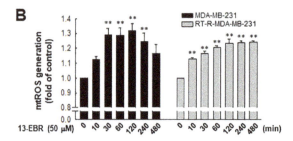

Figure 2. Effects of 13-EBR on the production of intracellular and mitochondrial reactive oxygen species (ROS) production in MDA-MB-231 and RT-R MDA-MB-231 cells. (**A**,**B**) Both cell lines were stimulated with 50 µM 13-EBR for the indicated times, and then the intracellular ROS (**A**) and mitochondrial ROS (**B**) were measured by staining with 2′,7′-dichlorodihydrofluorescein diacetate (H_2DCF-DA) and MitoSOX Red, respectively. The values represent the mean ± SEM of three independent experiments; ** $p < 0.01$ compared to the control of each cell line.

2.3. 13-EBR Induced MDA-MB-231 and RT-R MDA-MB-231 Apoptosis through a Mitochondria-Related Apoptotic Pathway, Not an Extrinsic Pathway

Next, we further examined whether 13-EBR induces apoptosis in both MDA-MB-231 and RT-R MDA-MB-231 cells by observing DNA shrinkage or nuclear fragmentation that occurs in cells undergoing apoptosis. As expected, 13-EBR stimulation induced DNA shrinkage at 10 µM and DNA fragmentation at 50 µM in both MDA-MB-232 and RT-R MDA-MB-231 cells (Figure 3A). Moreover, 13-EBR induced apoptosis, as shown by the increased sub-gap 1 (subG_1) population in both MDA-MB-231 and RT-R MDA-MB-231 cells in a dose-dependent manner compared to that in the controls (Figure 3B). In addition to the subG_1 population, the synthesis (S) and gap 2/mitosis (G_2/M) populations were remarkably changed in response to 50 µM 13-EBR treatment in MDA-MB-231 and RT-R MDA-MB-231 cells compared to the controls (Figure 3C). As mentioned above, an increase in ROS can induce apoptosis through both the extrinsic and intrinsic pathways. Thus, we investigated which apoptotic signaling pathway is involved in 13-EBR-induced apoptosis in MDA-MB-231 and RT-R MDA-MB-231 cells. The protein level of Bax, a pro-apoptotic gene, in the control group of RT-R MDA-MB-231 was little lower than that in the control of MDA-MB-231, as presented in Table 1, and 13-EBR stimulation significantly increased the protein level of Bax in both cell lines. However, 13-EBR decreased the level of *Bcl-2*, an anti-apoptotic gene, in a time-dependent manner compared to that in the controls in both cell lines. Moreover, the protein levels of cleaved caspase-9, -3, and poly(ADP ribose) polymerase (PARP) were also increased in response to 13-EBR treatment compared to the controls. However, 13-EBR did not affect the cleaved caspase-8 protein levels (Figure 4A,B). These results suggested that 13-EBR induces apoptotic cell death via the regulation of the mitochondria-related intrinsic pathway rather than an extrinsic pathway in MDA-MB-231 and RT-R MDA-MB-231 cells.

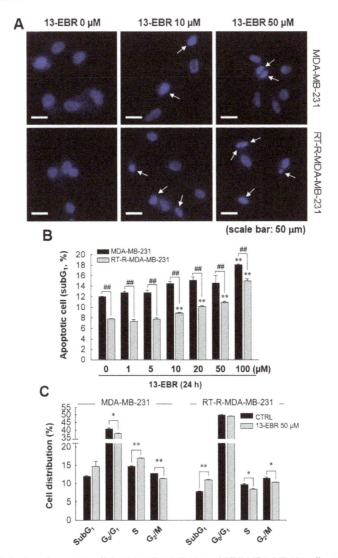

Figure 3. Induction of apoptotic cell death in MDA-MB-231 and RT-R MDA-MB-231 cells with 13-EBR through the intrinsic pathway. (**A**) Both cell lines were seeded on a cover slip that was mounted onto a self-designed perfusion chamber and then stimulated with 13-EBR at the indicated doses for 24 h. The fragmented DNA was observed, as described in Section 4. White arrows represent the fragmented DNA. (**B**) Both breast cancer cell lines were treated with 13-EBR at the indicated doses for 24 h, and then apoptotic cells were identified by analyzing the sub-gap 1 (subG$_1$) phase using a fluorescence-activated cell sorting (FACS) system, as described in Section 4. (**C**) Both cell lines were treated or were not treated with 13-EBR at 50 µM for 24 h, and then the cell distribution in the cell cycle was determined using the FACS system. The values represent the mean ± SEM of three independent experiments; ** $p < 0.01$, * $p < 0.05$ compared to the control; ## $P < 0.01$ compared to the MDA-MB-231 cells.

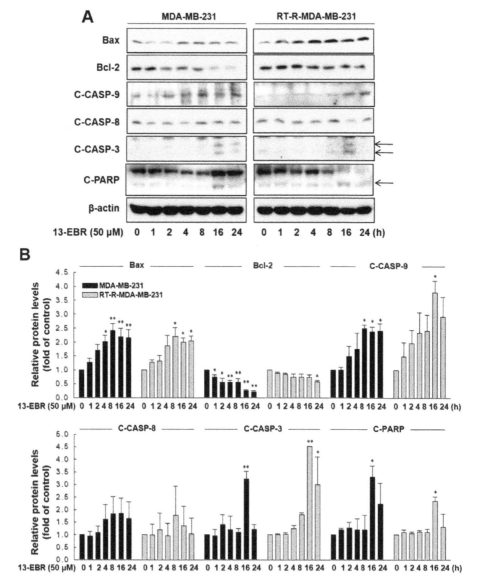

Figure 4. 13-EBR-mediated induction of apoptosis through the intrinsic pathway. (**A,B**) Both cell lines were stimulated with 50 μM 13-EBR for the indicated times, and total proteins were extracted. The Bax, Bcl-2, cleaved caspase-9 (C-CASP-9), -8 (C-CASP-8), -3 (C-CASP-3), cleaved poly(ADP ribose) polymerase (C-PARP), and β-actin protein levels were analyzed in cell lysates by Western blotting (**A**), as described in Section 4, and were quantified (**B**). The values represent the mean ± SEM of three independent experiments; ** $p < 0.01$, * $p < 0.05$ compared to the control of each cell line.

3. Discussion

TNBC refers to breast cancer that does not express ER, PR, and HER2, which is known to be more aggressive, with worse prognosis than that of other types of breast cancers that express hormone receptors [21,22]. Due to the lack of specific molecular targets, general cancer treatments are limited and not available for the treatment of TNBC patients. Thus, combinatorial therapy, which consists of surgery, chemotherapy, and radiation, is used for TNBC patients. According to clinically safe and effective therapeutic methods, a small amount of X-ray is periodically used to irradiate breast cancer, and even with combinatorial treatment, the surviving residual cancer cells eventually exhibit radiotherapy resistance [23]. Previously, we reported that not only the RT-R TNBC cell line (RT-R MDA-MB-231) but also the ER- and PR-positive breast cancer cell lines (RT-R MCF-7 and RT-R T47D) showed increased proliferation and colony formation, and were even more resistant to cancer chemotherapy than their parental breast cancer cells. Among these cells, RT-R MDA-MB-231 cells exhibited notable aggressiveness during tumor growth and invasion and showed characteristics of breast cancer stem cells [18]. Therefore, we tried to find a possible anticancer drug candidate, and, in this study, we investigated the anticancer effect of 13-EBR and the underlying mechanisms in radiotherapy-resistant TNBC.

BBR is known to be a low-toxicity and safe agent and was reported to have multiple biological functions, including antimicrobial, anti-inflammatory, and antitumor effects [4–13]. Treatment with 13-EBR, a BBR analog that is substituted at C-13 by alkyl groups, was reported to have anti-inflammatory effects [16,17]; however, the role of this molecule in tumor suppression was not reported. Thus, we aimed to determine whether 13-EBR could be a potential chemotherapeutic agent by examining the effects of 13-EBR on MDA-MB-231 cells and the radiotherapy-resistant TNBC cell line, RT-R MDA-MB-231. In this study, our results revealed that 13-EBR suppressed RT-R MDA-MB-231 and MDA-MB-231 cell proliferation and colony-forming ability (Figure 1) compared to the controls. Further studies showed that 13-EBR stimulation induced the production of ROS, including mitochondrial ROS, in both breast cancer cell lines (Figure 2). In addition, 13-EBR caused cell-cycle arrest and upregulated Bcl-2 family protein levels, such as Bax (pro-apoptotic), and reduced Bcl-2 (anti-apoptotic) levels and activated the caspase-9 and -3 pathways but not the caspase-8 pathway, suggesting that 13-EBR evokes apoptosis through the mitochondria-mediated signaling pathway in RT-R MDA-MB-231 cells (Figure 3). Although it was reported that 13-alkyl-substituted berberines showed better anti-microbe and anticancer activities than BBR [14,15], as described in Section 1, there is no evidence about which one of the two compounds has better efficacy on suppressing breast cancer cells, especially RT-R TNBC. Thus, we further compared the effects of BBR and 13-EBR on proliferation, colony formation, and cellular apoptosis in RT-R MDA-MB-231 cells. Interestingly, 13-EBR was significantly more effective in suppressing the cell proliferation and colony-forming ability and in inducing cellular apoptosis than BBR in RT-R MDA-MB-231 cells, which showed a significantly enhanced ability of colony formation and lower levels of apoptotic cell population compared to MDA-MB-231 cells (Supplementary Materials). When we determined the toxicity of 13-EBR on normal epithelial cells (MCF-10A), 13-EBR showed cytotoxicity in a dose-dependent manner; however, 20 µM 13-EBR, which showed apoptotic cell death and anti-colony forming ability in MDA-MB-231 and RT-R MDA-MB-231, showed less toxicity (cell viability over than 80%) (data not shown).

Many chemotherapeutic and radiotherapeutic agents eliminate cancer cells through the augmentation of ROS production [24,25]. A high level of mitochondrial ROS can also initiate intrinsic apoptosis, leading to the release of cytochrome c, a mitochondrial apoptogenic factor, into the cytosol [26]. In this study, we determined that 13-EBR increases the level of intracellular total ROS, including mitochondrial ROS, compared to that in the controls (Figure 2). Furthermore, cell-cycle progression is linked to cell proliferation and apoptosis. The cell cycle is divided into four phases, and the cellular decision to initiate mitosis or to be quiescent (G_0 state) occurs during the G_1 phase [27]. Thus, we determined whether 13-EBR acts as a regulator of the cell cycle in breast cancer cells, and elicited a cell-cycle arrest in the $subG_1$ phase, which was increased in response to 13-EBR

(Figure 3B,C) compared to that in the controls, suggesting that 13-EBR functions as an apoptosis inducer in RT-R TNBC and TNBC.

Apoptosis is essential for normal development and tissue homeostasis, and perturbations in the regulation of apoptosis contribute to numerous pathological conditions, including cancer, autoimmune diseases, and degenerative diseases [28]. Furthermore, apoptosis is an important target for anticancer drugs because accumulated evidence on cancer development indicated that cancerous cells are able to survive due to acquired mechanisms of apoptosis resistance in addition to uncontrolled proliferation [28]. Apoptosis is a form of programmed cell death that can be initiated through one of two of the best-understood activation mechanisms: extrinsic and intrinsic pathways [29]. The extrinsic signaling pathways that initiate apoptosis involve transmembrane receptor-mediated interactions. The extrinsic pathway-related death receptors are members of the tumor necrosis factor (TNF) receptor gene superfamily, such as cluster of differentiation 95 (CD95) (apoptosis antigen-1; APO-1/Fas) or TNF-related apoptosis-inducing ligand (TRAIL) receptors [30]. Upon ligand binding, the death receptors trigger the activation of the initiator caspase-8, and, once caspase-8 is activated, the execution phase of apoptosis is triggered through the direct cleavage of downstream effector caspases, such as caspase-3 [30]. On the other hand, the intrinsic signaling pathways that initiate apoptosis related to diverse nonreceptor-mediated stimuli produce intracellular signals that act directly on targets within the cell, which are mitochondrial-initiated events. The stress stimuli that can trigger the intrinsic pathway can occur in the absence of certain growth factors, cytokines, toxins, hypoxia, hyperthermia, viral infections, and free radicals [31]. The members of the Bcl-2 family of proteins mediate these apoptotic mitochondrial events. The Bcl-2 family proteins control mitochondrial membrane permeability and can be either pro-apoptotic or anti-apoptotic. Some of the anti-apoptotic proteins include Bcl-2, Bcl-XL, Bcl-XS, and Bcl-2 associated athanogene (BAG); some of the pro-apoptotic proteins include Bax, Bak, Bid, Bad, and Bim. The main mechanism of action of Bcl-2 family proteins is the regulation of cytochrome c release from the mitochondria via the alteration of mitochondrial membrane permeability [32]. Because stimulation with 13-EBR caused apoptotic cell death in MDA-MB-231 and RT-R MDA-MB-231 cells, we examined which apoptotic signaling pathway is involved in the 13-EBR-mediated induction of apoptosis and tried to investigate how 13-EBR affects RT-R MDA-MB-231 cells, which had lower expression of pro-apoptotic genes and higher expression of anti-apoptotic genes than MDA-MB-231 cells (Table 1). The results showed that 13-EBR treatment induced the expression of Bcl-2 family proteins and the activation of caspase-9, -3, and PARP in both MDA-MB-231 and RT-R MDA-MB-231 cells, which suggests that 13-EBR promotes RT-R TNBC apoptosis through the mitochondria-related intrinsic pathway. However, we did not observe the activation of caspase-8, which suggested that the extrinsic pathway is not involved in 13-EBR-evoked TNBC and RT-R TNBC apoptosis (Figure 4A,B). Although the induction of apoptosis in RT-R MDA-MB-231 cells is complicated due to the lower expression levels of pro-apoptotic genes and the higher expression levels of anti-apoptotic genes than in the parental cells, 13-EBR showed effective anticancer effects in both MDA-MB-231 and RT-R MDA-MB-231 cells.

Taken together, we demonstrated that 13-EBR has antiproliferative and pro-apoptotic effects in both TNBC and RT-R TNBC cells through the induction of intracellular ROS production and the mitochondria-mediated apoptosis pathway (Figure 5). Although the effects of 13-EBR were not specific to RT-R breast cancer cells, it is meaningful that 13-EBR could suppress cell proliferation and colony formation and could induce cellular apoptosis in RT-R MDA-MB-231 cells, which had higher colony-forming abilities and showed lower expression of pro-apoptotic genes and higher expression of anti-apoptotic genes than MDA-MB-231 cells. These findings provide useful insights for the exploration of new potential therapeutic strategies for both radiotherapy-resistant breast cancer treatment and also TNBC treatment.

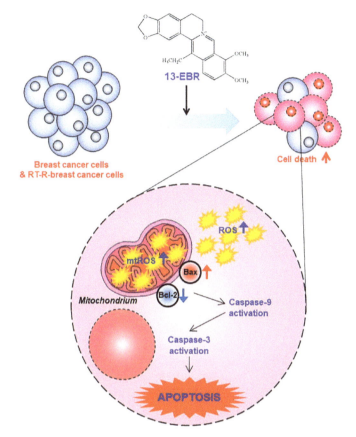

Figure 5. A proposed mechanism via which 13-EBR exhibits anticancer effects in triple-negative breast cancer (TNBC) and RT-R TNBC. Treatment with 13-EBR exhibits anticancer effects through the suppression of cell proliferation and colony-forming ability and by inducing apoptosis through the mitochondria-mediated signaling pathway in RT-R MDA-MB-231 cells, as well as MDA-MB-231 cells.

4. Materials and Methods

4.1. Materials

The 13-EBR (Figure 1A) was kindly provided by Prof. Dong-Ung Lee (Dongguk University, Gyeongju, Korea). Antibodies against Bax (ab32503), Bcl-2 (ab692), and caspase-8 (ab25901) were purchased from Abcam (Cambridge, UK), and antibodies against cleaved caspase-9 (#52873) and -3 (#9661) were obtained from Cell Signaling Technology (Beverly, MA, USA). The Cell Counting Kit-8 (CCK-8) reagent was obtained from Dongin Biotech (Seoul, Korea). MitoSOX Red was obtained from Thermo Fisher Scientific (Rockford, IL, USA). Hybond-P$^+$ polyvinylidene difluoride membranes and enhanced chemiluminescence (ECL) Western blotting detection reagents were purchased from Amersham Biosciences (Little Chalfont, UK) and Bio-Rad (Hercules, CA, USA), respectively. All other reagents, including an anti-β-actin antibody (A2066), propidium iodide (PI), 4′,6-diamidino-2-phenylindole dihydrochloride (DAPI), and 2′,7′-dichlorodihydrofluorescein diacetate (H$_2$DCF-DA), were purchased from Sigma-Aldrich (St. Louis, MO, USA).

4.2. Establishment of RT-R MDA-MB-231 Cells and Cell Culture

The human breast cancer cell line MDA-MB-231 was obtained from the Korea Cell Line Bank (Seoul, Korea). RT-R MDA-MB-231 cells were established as described by Ko et al. [22]. Briefly, cells were irradiated with 2 Gy using a 6-MV photon beam that was produced by a linear accelerator (Clinac 21EX, Varian Medical Systems, Inc., Palo Alto, CA, USA) until a final dose of 50 Gy was achieved, which is a commonly used clinical regimen for radiotherapy in patients with breast cancer. RT-R MDA-MB-231 cells were cultured in Roswell Park Memorial Institute (RPMI) 1640 medium supplemented with 10% fetal bovine serum (FBS) (HyClone Laboratories, Logan, UT, USA), 100 IU/mL penicillin, and 10 µg/mL streptomycin (HyClone Laboratories), and then incubated at 37 °C in a humidified atmosphere containing 5% CO_2 and 95% air. RT-R MDA-MB-231 cells were used within five passages.

4.3. Gene Expression Array Analysis

Total RNA was extracted from MDA-MB-231 and RT-R MDA-MB-231 cells using TRIzol reagent (Invitrogen, Carlsbad, CA, USA) according to the manufacturer's protocol. Gene expression profiling (*Bax, Bad, cytochrome c, Bcl-2, Bcl-2A1*, and *Mcl-1*) was performed with QuantiSeq 3' messenger RNA sequencing (mRNA-Seq) Service (Ebiogen, Seoul, Korea). Total proteins were also extracted from MDA-MB-231 and RT-R MDA-MB-231 cells using radioimmunoprecipitation assay (RIPA) buffer (0.1% NP-40, and 0.1% sodium dodecyl sulfate in phosphate-buffered saline (PBS)) containing a protease inhibitor cocktail (Sigma-Aldrich). Protein expression profiling (cleaved caspase-3 and -7) was analyzed by a Signaling Explorer Antibody Array (Ebiogen).

4.4. Cell Proliferation Assay

Cell proliferation was analyzed using a CCK-8 assay. The cells were seeded in 96-well flat-bottom plates (Thermo Fisher Scientific) and then treated with the indicated doses of 13-EBR and incubated at 37 °C. Then, 10 µL/well of CCK-8 reagent was added at 24, 48, and 72 h and incubated for 30 min at 37 °C. The optical density of each well was measured at a wavelength of 450 nm using a microplate reader (Tecan, Männedorf, Switzerland).

4.5. Colony-Formation Assay

MDA-MB-231 or RT-R MDA-MB-231 cells (1×10^3) were seeded in six-well flat-bottom plates (Thermo Fisher Scientific) and treated with the indicated doses of 13-EBR and then incubated at 37 °C. The culture medium was discarded following 24 h and replaced with fresh complete medium every 2–3 days. After 10 days, the medium was discarded, and each well was carefully washed with PBS. The colonies were fixed for 10 min in absolute methanol and then stained with 0.1% Giemsa staining solution at room temperature, and the number of colonies was counted.

4.6. Detection of DNA Fragmentation

Cells were seeded on a cover slip that was mounted onto a self-designed perfusion chamber (SPL, Gyeonggi-do, Korea) and then stimulated with 13-EBR at the indicated doses for 24 h at 37 °C. The cells were fixed with 4% formaldehyde (Sigma-Aldrich) for 5 min at 4 °C and then incubated with 0.1% Triton X-100 (Sigma-Aldrich) for permeabilization at 4 °C. After 10 min, the cells were washed with PBS, stained with DAPI, and the fragmented DNA was detected under a fluorescence microscope.

4.7. Flow Cytometric Analysis

For the analysis of the cell-cycle profiles, cells stimulated with 13-EBR for 24 h were fixed with 70% ethanol overnight at −80 °C and then washed with ice-cold PBS. Whole cells were then stained with PI solution (10 mM Tris (pH 8.0), 1 mM NaCl, 0.1% NP40, 0.7 µg/mL RNase A, and 0.05 mg/mL

PI) in the dark for 30 min at room temperature and then analyzed using a fluorescence-activated cell sorting (FACS) Calibur™ system (Becton Dickinson Bioscience, San Jose, CA, USA).

4.8. Measurement of Intracellular ROS and Mitochondrial ROS

Cells were seeded in 96-well plates and then treated with 50 µM 13-EBR for the indicated times. Following treatments, the cells were incubated with 5 µM H_2DCF-DA (for 30 min) or 5 µM MitoSOX Red (10 min) to determine the intracellular total ROS or mitochondrial ROS, respectively, in the dark. After incubation, the cells were washed three times with PBS. The fluorescence intensity was measured at an emission wavelength of 485 nm and an excitation wavelength of 535 nm for intracellular ROS, or at an emission wavelength of 510 nm and an excitation wavelength of 580 nm for mitochondrial ROS using a microplate fluorescence reader (Tecan).

4.9. Western Blot Analysis

For the isolation of total cell extracts, we washed cells with ice-cold PBS and lysed them in RIPA buffer containing a protease inhibitor cocktail. The suspension was centrifuged at 13,000 rpm for 15 min, and nuclear proteins were obtained by further centrifugation at 13,000 rpm for 5 min. For the isolation of total cell extracts, cells were lysed in RIPA buffer (0.1% NP-40 and 0.1% sodium dodecyl sulfate in PBS) containing a protease inhibitor cocktail. Approximately 50–70 µg aliquots of protein were subjected to 10% sodium dodecyl sulfate polyacrylamide gel electrophoresis and transferred onto Hybond-P$^+$ polyvinylidene difluoride membranes. The membranes were incubated with the indicated antibodies, and the bound antibodies were detected with horseradish peroxidase (HRP)-conjugated secondary antibodies and an ECL western blotting detection reagent (Bionote, Gyeonggi-do, Korea).

4.10. Statistical Evaluations

The treatment groups were compared using one-way analysis of variance (ANOVA) and a Newman–Keuls post hoc test. A p-value < 0.05 was considered statistically significant. All data were evaluated for normality and the homogeneity of variance, and are expressed as the mean ± standard error of the mean (SEM).

Supplementary Materials: The following are available online, Figure S1: Effects of BBR and 13-EBR on cell proliferation, colony formation, and apoptosis in breast cancer cells.

Author Contributions: H.J. performed the experiments and wrote the manuscript. Y.S.K. performed data analysis. S.W.P. and K.C.C. reviewed the manuscript and gave comments on the results. H.J.K. conceived the hypothesis, directed the project, and wrote the manuscript. All authors read and approved the final manuscript.

Funding: This work was supported by the Basic Science Research Program through the National Research Foundation of Korea (NRF) funded by the Ministry of Education, Science, and Technology (2018R1A2B6001786) and (2018R1D1A1B07049963), and by the Ministry of Science, Information and Communication Technology (ICT), and Future Planning (NRF-2015R1A5A2008833).

Conflicts of Interest: There are no conflicts of interest to declare.

References

1. Torre, L.A.; Bray, F.; Siegel, R.L.; Ferlay, J.; Lortet-Tieulent, J.; Jemal, A. Global cancer statistics, 2012. *CA Cancer J. Clin.* **2015**, *65*, 87–108. [CrossRef] [PubMed]
2. EBCTCG (Early Breast Cancer Trialists' Collaborative Group); McGale, P.; Taylor, C.; Correa, C.; Cutter, D.; Duane, F.; Ewertz, M.; Gray, R.; Mannu, G.; Peto, R.; et al. Effect of radiotherapy after mastectomy and axillary surgery on 10-year recurrence and 20-year breast cancer mortality: Meta-analysis of individual patient data for 8135 women in 22 randomised trials. *Lancet* **2014**, *383*, 2127–2135. [PubMed]
3. Deloch, L.; Derer, A.; Hartmann, J.; Frey, B.; Fietkau, R.; Gaipl, U.S. Modern Radiotherapy Concepts and the Impact of Radiation on Immune Activation. *Front. Oncol.* **2016**, *6*, 141. [CrossRef] [PubMed]
4. Kuo, C.L.; Chi, C.W.; Liu, T.Y. The anti-inflammatory potential of berberine in vitro and in vivo. *Cancer Lett.* **2004**, *203*, 127–137. [CrossRef]

5. Yu, H.H.; Kim, K.J.; Cha, J.D.; Kim, H.K.; Lee, Y.E.; Choi, N.Y.; You, Y.O. Antimicrobial activity of berberine alone and in combination with ampicillin or oxacillin against methicillin-resistant staphylococcus aureus. *J. Med. Food* **2005**, *8*, 454–461. [CrossRef] [PubMed]
6. Wu, K.; Yang, Q.; Mu, Y.; Zhou, L.; Liu, Y.; Zhou, Q.; He, B. Berberine inhibits the proliferation of colon cancer cells by inactivating Wnt/beta-catenin signaling. *Int. J. Oncol.* **2012**, *41*, 292–298.
7. Huang, Z.H.; Zheng, H.F.; Wang, W.L.; Wang, Y.; Zhong, L.F.; Wu, J.L.; Li, Q.X. Berberine targets epidermal growth factor receptor signaling to suppress prostate cancer proliferation in vitro. *Mol. Med. Rep.* **2015**, *11*, 2125–2128. [CrossRef]
8. Patil, J.B.; Kim, J.; Jayaprakasha, G.K. Berberine induces apoptosis in breast cancer cells (MCF-7) through mitochondrial-dependent pathway. *Eur. J. Pharmacol.* **2010**, *645*, 70–78. [CrossRef]
9. Zhao, Y.; Jing, Z.; Lv, J.; Zhang, Z.; Lin, J.; Cao, X.; Zhao, Z.; Liu, P.; Mao, W. Berberine activates caspase-9/cytochrome c-mediated apoptosis to suppress triple-negative breast cancer cells in vitro and in vivo. *Biomed. Pharmacother.* **2017**, *95*, 18–24. [CrossRef]
10. Ma, W.; Zhu, M.; Zhang, D.; Yang, L.; Yang, T.; Li, X.; Zhang, Y. Berberine inhibits the proliferation and migration of breast cancer ZR-75-30 cells by targeting Ephrin-B2. *Phytomedicine* **2017**, *25*, 45–51. [CrossRef]
11. Jeong, Y.; You, D.; Kang, H.G.; Yu, J.; Kim, S.W.; Nam, S.J.; Lee, J.E.; Kim, S. Berberine Suppresses Fibronectin Expression through Inhibition of c-Jun Phosphorylation in Breast Cancer Cells. *J. Breast Cancer* **2018**, *21*, 21–27. [CrossRef] [PubMed]
12. Kim, S.; Lee, J.; You, D.; Jeong, Y.; Jeon, M.; Yu, J.; Kim, S.W.; Nam, S.J.; Lee, J.E. Berberine Suppresses Cell Motility Through Downregulation of TGF-β1 in Triple Negative Breast Cancer Cells. *Cell. Physiol. Biochem.* **2018**, *45*, 795–807. [CrossRef] [PubMed]
13. Pan, Y.; Zhang, F.; Zhao, Y.; Shao, D.; Zheng, X.; Chen, Y.; He, K.; Li, J.; Chen, L. Berberine Enhances Chemosensitivity and Induces Apoptosis Through Dose-orchestrated AMPK Signaling in Breast Cancer. *J. Cancer* **2017**, *8*, 1679–1689. [CrossRef] [PubMed]
14. Iwasa, K.; Nanba, H.; Lee, D.U.; Kang, S.I. Structure-activity relationships of protoberberines having antimicrobial activity. *Planta Med.* **1998**, *64*, 748–751. [CrossRef] [PubMed]
15. Iwasa, K.; Moriyasu, M.; Yamori, T.; Turuo, T.; Lee, D.U.; Wiegrebe, W. In Vitro cytotoxicity of the protoberberine-type alkaloids. *J. Nat. Prod.* **2001**, *64*, 896–898. [CrossRef] [PubMed]
16. Lee, D.U.; Kang, Y.J.; Park, M.K.; Lee, Y.S.; Seo, H.G.; Kim, T.S.; Kim, C.H.; Chang, K.C. Effects of 13-alkyl-substituted berberine alkaloids on the expression of COX-II, TNF-alpha, iNOS, and IL-12 production in LPS-stimulated macrophages. *Life Sci.* **2003**, *73*, 1401–1412. [CrossRef]
17. Lee, D.U.; Ko, Y.S.; Kim, H.J.; Chang, K.C. 13-Ethylberberine reduces HMGB1 release through AMPK activation in LPS-activated RAW264.7 cells and protects endotoxemic mice from organ damage. *Biomed. Pharmacother.* **2017**, *86*, 48–56. [CrossRef] [PubMed]
18. Ko, Y.S.; Jin, H.; Lee, J.S.; Park, S.W.; Chang, K.C.; Kang, K.M.; Jeong, B.K.; Kim, H.J. Radioresistant breast cancer cells exhibit increased resistance to chemotherapy and enhanced invasive properties due to cancer stem cells. *Oncol. Rep.* **2018**, *40*, 3752–3762. [CrossRef]
19. Ozben, T. Oxidative stress and apoptosis: Impact on cancer therapy. *J. Pharm. Sci.* **2007**, *96*, 2181–2196. [CrossRef] [PubMed]
20. Marullo, R.; Werner, E.; Degtyareva, N.; Moore, B.; Altavilla, G.; Ramalingam, S.S.; Doetsch, P.W. Cisplatin induces a mitochondrial-ROS response that contributes to cytotoxicity depending on mitochondrial redox status and bioenergetic functions. *PLoS ONE* **2013**, *8*, e81162. [CrossRef] [PubMed]
21. Dent, R.; Trudeau, M.; Pritchard, K.I.; Hanna, W.M.; Kahn, H.K.; Sawka, C.A.; Lickley, L.A.; Rawlinson, E.; Sun, P.; Narod, S.A. Triple-negative breast cancer: Clinical features and patterns of recurrence. *Clin. Cancer Res.* **2007**, *13*, 4429–4434. [CrossRef] [PubMed]
22. Dawson, S.J.; Provenzano, E.; Caldas, C. Triple negative breast cancers: Clinical and prognostic implications. *Eur. J. Cancer* **2009**, *45*, 27–40. [CrossRef]
23. Begg, A.C.; Stewart, F.A.; Vens, C. Strategies to improve radiotherapy with targeted drugs. *Nat. Rev. Cancer* **2011**, *11*, 239–253. [CrossRef] [PubMed]
24. Renschler, M.F. The emerging role of reactive oxygen species in cancer therapy. *Eur. J. Cancer* **2004**, *40*, 1934–1940. [CrossRef] [PubMed]
25. Toler, S.M.; Noe, D.; Sharma, A. Selective enhancement of cellular oxidative stress by chloroquine: Implications for the treatment of glioblastoma multiform. *Neurosurg. Focus* **2006**, *21*, E10. [CrossRef]

26. Jacquemin, G.; Margiotta, D.; Kasahara, A.; Bassoy, E.Y.; Walch, M.; Thiery, J.; Lieberman, J.; Martinvalet, D. Granzyme B-induced mitochondrial ROS are required for apoptosis. *Cell Death Differ.* **2015**, *22*, 862–874. [CrossRef]
27. Blagosklonny, M.V.; Pardee, A.B. The restriction point of the cell cycle. *Cell Cycle* **2002**, *1*, 103–110. [CrossRef]
28. Adams, J.M.; Cory, S. The Bcl-2-regulated apoptosis: Mechanism and therapeutic potential. *Curr. Opin. Immunol.* **2007**, *19*, 488–496. [CrossRef]
29. Elmore, S. Apoptosis: A review of programmed cell death. *Toxicol. Pathol.* **2007**, *35*, 495–516. [CrossRef]
30. Tummers, B.; Green, D.R. Caspase-8: Regulating life and death. *Immunol. Rev.* **2017**, *277*, 76–89. [CrossRef]
31. Green, D.R.; Llambi, F. Cell Death Signaling. *Cold Spring Harb. Perspect. Biol.* **2015**, *7*, a006080. [CrossRef] [PubMed]
32. Edlich, F. BCL-2 proteins and apoptosis: Recent insights and unknowns. *Biochem. Biophys. Res. Commun.* **2018**, *500*, 26–34. [CrossRef] [PubMed]

Sample Availability: 13-EBR was kindly provided by Prof. Dong-Ung Lee (as mentioned in the Section 4.1. Materials). Sample of this compound is available from Prof. Lee.

 © 2019 by the authors. Licensee MDPI, Basel, Switzerland. This article is an open access article distributed under the terms and conditions of the Creative Commons Attribution (CC BY) license (http://creativecommons.org/licenses/by/4.0/).

Article

Licochalcone A Suppresses the Proliferation of Osteosarcoma Cells through Autophagy and ATM-Chk2 Activation

Tai-Shan Shen [1,†], Yung-Ken Hsu [1,†], Yi-Fu Huang [2], Hsuan-Ying Chen [2], Cheng-Pu Hsieh [1,2] and Chiu-Liang Chen [1,3,*]

1. Department of Orthopedic Surgery, Changhua Christian Hospital, Changhua 50006, Taiwan
2. Orthopedics & Sports Medicine Laboratory, Changhua Christian Hospital, Changhua 50006, Taiwan
3. Department of Nursing, Da Yeh University, Changhua 51591, Taiwan
* Correspondence: 111111@cch.org.tw; Tel.: +886-4-7238595; Fax: +886-4-7228289
† These authors have contributed equally to this work.

Academic Editor: Roberto Fabiani
Received: 27 May 2019; Accepted: 28 June 2019; Published: 2 July 2019

Abstract: Licochalcone A, a flavonoid extracted from licorice root, has been shown to exhibit broad anti-inflammatory, anti-bacterial, anticancer, and antioxidative bioactivity. In this study, we investigated the antitumor activity of Licochalcone A against human osteosarcoma cell lines. The data showed that Licochalcone A significantly suppressed cell viability in MTT assay and colony formation assay in osteosarcoma cell lines. Exposure to Licochalcone A blocked cell cycle progression at the G2/M transition and induced extrinsic apoptotic pathway in osteosarcoma cell lines. Furthermore, we found the Licochalcone A exposure resulted in rapid ATM and Chk2 activation, and high levels of nuclear foci of phosphorylated Chk2 at Thr 68 site in osteosarcoma cell lines. In addition, Licochalcone A exposure significantly induced autophagy in osteosarcoma cell lines. When Licochalcone A-induced autophagy was blocked by the autophagy inhibitor chloroquine, the expression of activated caspase-3 and Annexin V positive cells were reduced, and cell viability was rescued in Licochalcone A-treated osteosarcoma cell lines. These data indicate that the activation of ATM-Chk2 checkpoint pathway and autophagy may contribute to Licochalcone A-induced anti-proliferating effect in osteosarcoma cell lines. Our findings display the possibility that Licochalcone A may serve as a potential therapeutic agent against osteosarcoma.

Keywords: Licochalcone A; ATM-Chk2; autophagy; osteosarcoma

---

1. Introduction

Osteosarcoma is the most common primary tumor of bone. It is a highly malignant form of bone cancer characterized by osteoid production. Osteosarcoma arises predominantly in adolescents and children, with a second incidence peak in the elderly [1,2]. Osteosarcoma often originates from long bones including the distal femur, proximal tibia, and proximal humerus. It is characterized by high malignancy, frequent recurrence, and distant metastasis. By next-generation sequencing, several groups have revealed huge somatically mutated genes in osteosarcoma samples from patients. There are several cancer-causing genes showing high frequency of mutation in osteosarcoma samples including TP53, RB1, BRCA2, and DLG2 [3].

The major treatment for osteosarcoma is surgery. However, the survival rate of patients with osteosarcoma treated with surgery alone is about 15–17% [4]. In the early 1970s, chemotherapy was introduced as adjuvant treatment to facilitate surgical resection. The common chemotherapy protocols comprise of drugs namely, cisplatin, doxorubicin, and high-dose methotrexate. This incorporation has results in an overall 5-years survival rates that approach 70%. Unfortunately, 30% of patients diagnosed with osteosarcoma will not survive for more than 5 years. Treatment often fails due to the development of metastasis, chemo-resistance, and relapse of disease. A total of 30–40% of patients with localized osteosarcoma will develop a local or distant recurrence [5], resulting in only 23–29% overall 5 year survival rates in these patients [6].This outcome has remained virtually unchanged over the past 30 years. Therefore, novel strategies and effective drugs are urgently required, especially for patients suffering from advanced osteosarcoma.

Recent progress has focused on the chemotherapy by natural compounds for their anti-growth activity against cancer cells. These compounds may exhibit less adverse effects compared to synthetic chemicals [7,8]. Licorice (Glycyrrhiza glabra) is a well-known herb named for its unique sweet flavor. It is utilized to add flavor to foods, beverages, and tobacco, and is widely used as an herbal medicine. Licorice is used for gastritis, ulcers, cough, bronchitis, and inflammation [9]. Licochalcone A is an oxygenated chalcone (a type of natural phenol) (Figure 1A) that can be isolated from the roots of Glycyrrhiza species (such as *G. glabra, G. inflata,* and *G. eurycarpa*) which belong to the plant family of Fabaceae [10]. It has been demonstrated to possess antiviral [11] and antimicrobial activities [12]. In addition, literatures have shown that Licochalcone A has antioxidant [13], anti-angiogenesis [14], anti-inflammation [15], and anti-tumor effects [16]. Licochalcone A induces cell cycle arrest at S and G2/M phase and triggers intrinsic and extrinsic apoptosis in oral squamous cell carcinoma cells [17]. Licochalcone A suppresses the proliferation of lung cancer cell via G2/M cell cycle arrest and ER stress [18]. Licochalcone A inhibits PI3K/Akt/mTOR activation, and promotes autophagy and apoptosis in breast cancer cells [19] and cervical cancer cells [20]. Licochalcone A induces apoptotic cell death via p38 activation in human nasopharyngeal carcinoma cells [21] and head and neck squamous carcinoma cells [22]. In this study, we evaluated the potential anti-tumor effect of Licochalcone A against osteosarcoma.

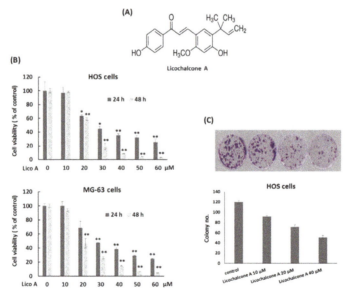

Figure 1. Licochalcone A inhibits cell viability of osteosarcoma. (**A**) The chemical structure of Licochalcone A. (**B**) Licochalcone A inhibits osteosarcoma cell growth in a dose-dependent manner. MTT assays were performed with osteosarcoma HOS and MG-63 cells exposed to Licochalcone A (Lico A) in the indicated concentrations. Experiments were conducted with three biological replicates per treatment, and the values represent the mean ± SD. (*) $p < 0.01$ and (**) $p < 0.001$ as compared with the untreated cells. (**C**) Licochalcone A suppresses colony formation of osteosarcoma cell lines. HOS cells were plated in colony formation assays after treatment with Licochalcone A for 7 h. Five hundred cells were plated per dish. All experiments were performed in triplicate, and the figure above shows a representative example.

2. Results

2.1. Licochalcone A Inhibits Osteosarcoma Cell Viability and Proliferation

Mutations in TP53 have been observed in 50–90% of osteosarcoma. It is most frequently mutated gene in osteosarcoma [3]. To mimic this genetic background in in vitro study, osteosarcoma HOS cells (R156P p53 mutation) [23] and MG-63 (mutant-p53, harboring a rearrangement in intron 1) [24,25] were used. Cell viability of osteosarcoma cell lines after exposure to various concentrations of Licochalcone A (0–60 µM) was detected by the MTT assay. The data showed that Licochalcone A clearly inhibited cell viability of osteosarcoma HOS cells and MG-63 cells at the concentrations of 20–60 µM following exposure for 24 h and 48 h compared with the control group (Figure 1B). The half maximal inhibitory concentration (IC_{50}) calculated based on data of the MTT assays for HOS cells were 29.43 µM at 24 h and 22.48 µM at 48 h, and those for MG-63 cells were 31.16 µM at 24 h and 22.39 µM at 48 h. Next, the colony formation assay was performed to examine the effect of Licochalcone A on cell proliferating capacity. The results showed that the treatment with Licochalcone A reduced colony number at the concentrations of 10–40 µM compared with the control group in osteosarcomas HOS cells (Figure 1C). These data indicate that Licochalcone A significantly inhibits the cell viability of osteosarcoma cell lines in a dose-dependent manner.

2.2. Licochalcone A Induces Apoptosis and Cell Arrest

To determine whether programmed cell death was involved in the anti-proliferative effect of Licochalcone A, we analyzed the rate of apoptosis cells in Licochalcone A-treated HOS cells and MG-63 cells by Annexin V and PI staining observed by flow cytometry. The data showed that the rate of Annexin V positive cells was significantly increased after exposure to Licochalcone A (30 µM or 40 µM) for 24 h in both lines of osteosarcoma cells (Figure 2A), indicating Licochalcone A has the potential to induce apoptosis in osteosarcoma cell lines. To determine whether caspase activation was involved in Licochalcone A–induced apoptosis, we measured the protein levels of the activated forms of caspase-3, -8, and -9 and PARP by Western blot analysis in treated HOS cells and MG-63 cells. The data showed that treatment with Licochalcone A (20–40 µM) for 24 h resulted in up-regulated activated forms of caspase 8, caspase 3, and PARP, but decreased activated forms of caspase 9 and Bax (Figure 2B). Besides, we also observed that treatment with Licochalcone A both resulted in down-regulation of pro-survival protein Bcl-2 and inhibitors of the apoptosis protein (IAP) family such as XIAP and survivin (Figure 2B). These findings suggest that Licochalcone A induces apoptosis by caspase 8 and caspase 3 signaling pathway.

Cell cycle distribution of HOS cells or MG-63 cells treated with Licochalcone A (30 µM, a concentration close to IC_{50} at 24 h for both cell lines) for different time points were analyzed by flow cytometry. The results showed that a significant accumulation of 4N cells (G2/M phase cells) was induced in Licochalcone A-treated HOS cells (Figure 3A) and MG-63 cells (Figure 3B). Furthermore, we evaluated the expression of proteins that regulate the G2/M phase transition by Western blot assay. The data showed that the protein level of phospho-cdc2, cdc2, and Cdc25C were decreased in treated both cell lines, but Cyclin B1 protein levels had no apparent change in HOS cells and was decreased in MG-63 cells (Figure 3C), indicating Licochalcone A induces G2/M phase arrest of osteosarcoma cell lines.

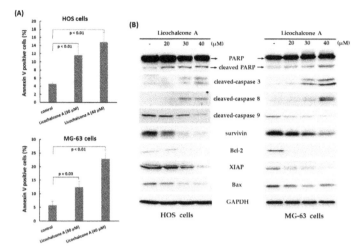

Figure 2. Licochalcone A induces apoptosis in osteosarcoma cells. Osteosarcoma HOS cells or MG-63 cells were treated with Licochalcone A (30 µM) for 24 h. To detect apoptosis, the HOS cells or MG-63 cells were stained with Annexin V and propidium iodide (PI), and analyzed using flow cytometry. Quantitative results of Annexin V positive cells are shown (**A**). Expression of apoptosis-related proteins was measured by Western blotting (**B**). Experiments were conducted with three biological replicates per condition, and the values represent the mean ± SD.

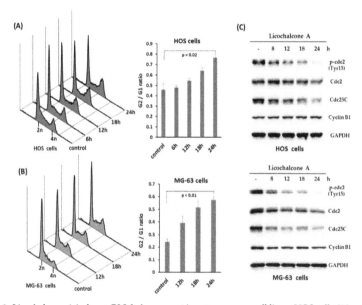

Figure 3. Licochalcone A induces G2/M phase arrest in osteosarcoma cell lines. HOS cells (**A**) or MG-63 cells (**B**) were treated with Licochalcone A (30 µM), and harvested in the indicated time points. The cells were stained with propidium iodide and analyzed by flow cytometer. 2n corresponds to G1 phase cells and 4n corresponds to the G2/M phase cells. The ration of G2/M to G1 phase cells were showed in right panel. Experiments were conducted with three biological replicates per condition, and the values represent the mean ± SD. (**C**) Licochalcone A decreases the expression of p-cdc2, cdc2, and cdc25c in osteosarcoma cell lines. HOS cells or MG-63 cells were treated with Licochalcone A (30 µM), and harvested in the indicated time points. The treated cells were analyzed by Western blotting using the indicated antibodies.

2.3. Activation of Chk2 and ATM in Response to Licochalcone A

To identify the mechanism involving in Licochalcone A-induced G2/M phase arrest in osteosarcoma cells, we tested the activation of ATM-Chk2 and ATR-Chk1, the primary pathway for activating G2/M checkpoint. The data showed that the treatment with Licochalcone A induced rapid phosphorylation of Chk2 (at threonine 68) and ATM (at serine 1981) (Figure 4A), but did not induce Chk1 phosphorylation (at serine 345) (data not shown). Interestingly, we observed the threonine 68-phosphorylated form of Chk2 formed distinct nuclear foci in response to Licochalcone A treatment in HOS cells (Figure 4B). These data suggest that ATM-Chk2 pathway may contribute to Licochalcone A-induced G2/M phase arrest in osteosarcoma cells. Recently, it has been reported that oxidative stress can activate ATM [26]. We examined the cellular redox status following Licochalcone A treatment in osteosarcoma MG-63 cells using 2′,7′-dichlorofluorescin diacetate (DCFDA), a fluorogenic dye that measures reactive oxygen species (ROS) within the cell. The data showed that ROS levels were significantly elevated in Licochalcone A-treated cells. Thus, we propose oxidative stress to be involved in Licochalcone A-mediated activation of ATM.

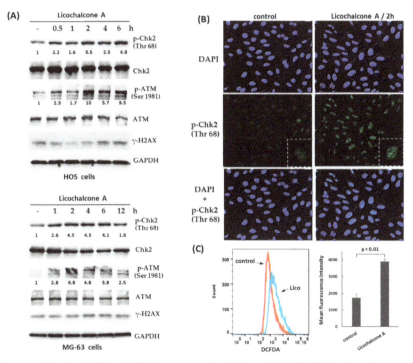

Figure 4. Activation of Chk2 and ATM in response to Licochalcone A. (**A**) ATM-Chk2 pathway is activated in Licochalcone A -treated HOS cells. HOS cells or MG-63 cells were treated with Licochalcone A (30 µM), and harvested in the indicated time points. The treated cells were analyzed by Western blotting using the indicated antibodies. The levels of p-Chk2 (Thr 68) and p-ATM (Ser 1981) were quantified and are shown below each blot. (**B**) Licochalcone A induces phospho-Chk2 T68 foci. HOS cells were treated with Licochalcone A (30 µM), and 2 h later were fixed with formaldehyde, permealized with Triton X-100, and then immunostained with antibody to phospho-Chk2 T68 (green color) and 4′,6-diamidino-2-phenylindole (DAPI) for labeling nucleus (blue color). (**C**) Licochalcone A enhances reactive oxygen species (ROS) generation. MG-63 cells were treated with Licochalcone A (Lico) for 2 h. Intracellular ROS was analyzed by flow cytometry using 2′,7′-dichlorofluorescin diacetate (DCFDA) staining and is shown as median fluorescence intensity. Mean ± SD. is plotted for 3 replicates from each condition.

2.4. Autophagy is Involved in Licochalcone A—Induced Apoptosis

Autophagy is a cellular process used to recycle or degrade proteins and cytoplasmic organelles in response to stress. Accumulating evidence has revealed that a large number of natural compounds are involved in autophagy modulation, either inducing or inhibiting autophagy [27]. Therefore, we decided to demonstrate the interaction between Licochalcone A and autophagy. First, we analyzed the formation of LC3A/B-II, the marker protein for autophagosomes by Western blotting assay. The data showed that the protein level of LC3A/B-II was significantly induced by Licochalcone A exposure in osteosarcoma HOS cells and MG-63 cells (Figure 5A), suggesting Licochalcone A has potential to induced autophagy. This result was further confirmed by LC3 puncta formation assay using confocal microscopy (Figure 5B). In addition, the marked reduction of actin filaments was observed in Licochalcone A-treated cells (Figure 5B). Next, we examined the role of autophagy in Licochalcone A-treated osteosarcoma cells. Once the autophagy was blocked by autophagy inhibitors, chloroquine, the cleaved caspase 3 and Annexin V positive cells were reduced (Figure 6A,B), and cell viability was rescued (Figure 6C) in Licochalcone A-treated HOS cells, indicating that the autophagy is associated with Licochalcone A-induced apoptosis in osteosarcoma HOS cells.

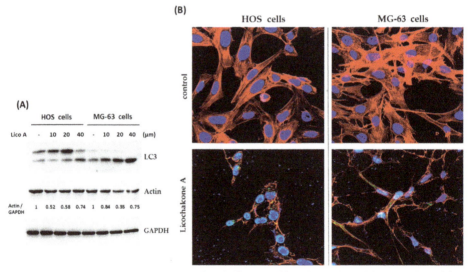

Figure 5. Autophagy is induced in Licochalcone A -treated osteosarcoma cells. HOS cells and MG-63 cells were treated with indicated concentrations of Licochalcone A (Lico A) for 24 h. The treated cells were analyzed by Western blotting using the indicated antibodies (**A**), or were immunostained with LC3 antibody for autophagy formation (green color), phalloidin-iFluor 594 reagent for labeling actin filaments (red color), and 4′,6-diamidino-2-phenylindole (DAPI) for labeling nucleus (blue color) (**B**). The levels of actin were quantified and are shown below the blot in Western blotting.

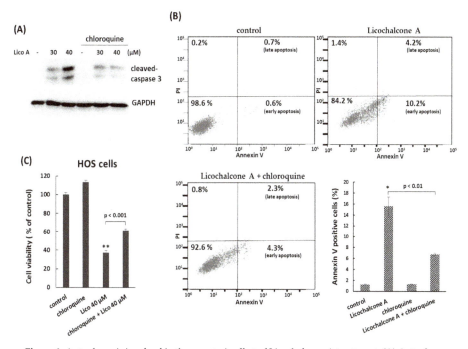

Figure 6. Autophagy is involved in the apoptosis effect of Licochalcone A treatment. (**A**) Autophagy inhibitor, chloroquine, suppresses Licochalcone-induced caspase 3 activation. HOS cells were treated with indicated concentrations of Licochalcone A (Lico A) with or without chloroquine for 24 h. The treated cells were analyzed by Western blotting using the indicated antibodies. (**B**) Autophagy inhibitor chloroquine suppresses Licochalcone-induced apoptosis. HOS cells were treated with Licochalcone A (Lico A) (40 µM) with or without chloroquine for 24 h. To detect apoptosis, the HOS cells were stained with Annexin V and propidium iodide (PI), and analyzed using flow cytometry. Quantitative results of Annexin V positive cells are shown in the lower panels. (**C**) Autophagy inhibitor chloroquine rescues the anti-proliferative effect of Licochalcone A treatment. HOS cells were treated as described in (**B**). MTT assay was performed to determine cell viability in treated cells. Experiments were conducted with three biological replicates per treatment, and the values represent the mean ± SD. (*) $p < 0.01$ and (**) $p < 0.001$ as compared with the control group.

3. Discussion

In this study, we demonstrated that Licochalcone A suppressed cell viability through the induction of cell cycle arrest at G2/M phase and caspase 8/3-dependent apoptosis in osteosarcoma cell lines. Our data further indicated that the activation of ATM/Chk2 and autophagy may be involved in Licochalcone A-induced anti-proliferating effect in osteosarcoma cell lines.

Apoptosis can be conducted in two major signaling pathways: The extrinsic pathways and the intrinsic pathways. The extrinsic pathways involve transmembrane receptor-mediated interactions, and are modulated by caspase 8 and caspase 3. The intrinsic pathways involve mitochondrial, and are controlled by several proteins including Bax, Bcl-2 protein family, cytochrome c, caspase 9, and caspase 3 [28,29]. In this report, we showed that Licochalcone A treatment induced cleaved-caspase 8 and caspase 3, but decreased the levels of Bax and cleaved-caspase 9, indicating that Licochalcone A triggers apoptosis that mediated by the extrinsic pathways in osteosarcoma cell lines. However, several literatures report that caspase 9-mediated intrinsic pathways could be induced by Licochalcone A in other cell types such as glioma stem cells [16], and nasopharyngeal carcinoma cells [21].

In response to DNA damage, ATM/Chk2 and ATR/Chk1 pathways have the central roles in maintaining genome stability by inducing cell cycle arrest, apoptosis, and DNA repair [30]. As shown in Figure 4A, the activation of ATM and Chk2 could be observed as early as 0.5–1 h after Licochalcone A treatment, whereas the level of γH2AX, the DNA double-strand breaks biomarker, didn't be significantly elevated. Therefore, we propose that Licochalcone A-mediated activation of ATM/Chk2 is unlikely due to direct damage of DNA. However, it was still uncertain how Licochalcone A activated ATM/Chk2 pathways. Recently, it has been reported oxidative stress such as H_2O_2 treatment can activate ATM-Chk2 in the absence of DNA double-strand breaks [31]. Our data showed that reactive oxygen species (ROS) levels were significantly elevated in Licochalcone A-treated cells. However, the role of oxidative stress in Licochalcone A-mediated activation of ATM-Chk2 should be further confirmed in the future.

Autophagy is an important catabolic process used to degrade or recycle proteins and cytoplasmic organelles in response to stress. Nevertheless, the cellular outcome for inducing autophagy is different and depends on the context of cells [32–34]. The literature shows that Licochalcone A induces autophagy in cervical cancer cells, and treatment with autophagy inhibitors enhances Licochalcone A-induced apoptosis in these cells [20]. However, in osteosarcoma cell lines, we showed Licochalcone A-induced apoptosis was suppressed, and cell viability was rescued as the induced autophagy was blocked by autophagy inhibitors. We suggest Licochalcone A-induced autophagy has the positive effect in promoting apoptosis in osteosarcoma cell lines. In addition, reactive oxygen species (ROS) have been shown to be a general inducer of autophagy [35]. It is possible that the elevated ROS by Licochalcone A treatment may contribute to induce autophagy formation in osteosarcoma cell lines.

In conclusion, the present study demonstrated that Licochalcone A exhibits antitumor activity in vitro by inhibiting cell viability, arresting cell cycle progression and inducing apoptosis in osteosarcoma cells. It has been reported that Licochalcone A has less cytotoxic effect on normal cells (HK-2 and WI-38) [20]. Therefore, Licochalcone A may serve as a potential therapeutic agent against osteosarcoma

4. Materials and Methods

4.1. Cell Culture and Chemicals

Human HOS and MG-63 osteosarcoma cells were purchased from the Bioresource Collection and Research Center (Hsinchu, Taiwan). HOS and MG-63 cells were maintained in Minimum Essential Medium (#11095-080; Gibco, Carlsbad, CA, USA) supplemented with 10% fetal bovine serum (Gibco, Carlsbad, CA, USA), 1% penicillin/streptomycin (Gibco, Carlsbad, CA, USA). PCR mycoplasma detection kit (BSMP-101, Bio-Smart, Hsinchu, Taiwan) was used to detect mycoplasma infection every three months. The used cells are mycoplasma-negative. Licochalcone A were obtained as a power from Santa Cruz Biotechnology (sc-319884). The purity of Licochalcone A was more than 96%.

4.2. Colony Formation Assay

HOS cells were briefly treated with Licochalcone A at various concentrations (0, 10, 20, and 40 μM). Once the expression of proteins that regulate the G2/M phase transition such as Cdc25C were decreased, starting at 6–8 h after Licochalcone A treatment (Figure 3C), the treated cells were washed by PBS three times, trypsinized, plated and maintained onto 35 mm dishes (500 cells/dish) with drug-free complete medium and cultured for another 10–14 days to allow colony formation. Colonies were fixed in 70% ethanol and stained by 1% crystal violet solution before counting.

4.3. Cell Viability Assay

The cytotoxic activity of Licochalcone A was tested using the MTT assay. The human osteosarcoma HOS and MG-63 cells were seeded in 24-well plates for overnight. The cells were treated with different concentrations of Licochalcone A for 24 h or 48 h. At the end of the assay time, the cells was incubated with 15 µL of MTT solution (5 mg/mL) (Invitrogen, Carlsbad, CA, USA) for 2 h at 37 °C. After removing the cultured medium, 200 µL of dimethyl sulfoxide (DMSO) was added to each well. Absorbance at 590 nm of the dissolved formazan product was read on an automated microplate spectrophotometer (Thermo Multiskan SPECTRUM, Thermo Fisher Scientific, Waltham, MA, USA). The half maximal inhibitory concentration (IC_{50}) for HOS and MG-63 cells were calculated by the "Forecast" function in Microsoft Excel.

4.4. Cell Cycle Analysis

The cells were washed by PBS two times, trypsinized, fixed with ice-cold 100% ethanol and kept on −20 °C for overnight. Cells were rehydrated with cold PBS, and then resuspended in PBS with propium iodine (40 µg/mL) (#P4170; Sigma-Aldrich, St. Louis, MO, USA) and Ribonuclease A (0.2 µg/mL) at room temperature for 30 min in the dark. The content of DNA in each sample were analyzed by a Cytomics™ FC500 flow cytometer (Beckman Coulter; Brea, CA, USA).

4.5. Apoptosis Assay

Licochalcone A-induced apoptosis in HOS cells was determined by flow cytometry using the Annexin-V-FITC staining kit (Becton Dickinson, San Jose, CA, USA) according to the manufacturer's protocol. Briefly, the treated cells were trypsinized, and washed twice by cold PBS. The cells were incubated with 100 µL of 1× binding buffer with 5 µL of FITC Annexin V and 5 µL of propidium iodide for 15 min at room temperature (RT) in the dark. After incubation, 400 µL of 1× binding buffer was added to each tube, and the fluorescence was measured by a Cytomics™ FC500 flow cytometer (Beckman Coulter, Miami, FL, USA).

4.6. Immunofluorescence Assays

For immunofluorescence analysis, cells were grown on glass coverslips and fixed in a 4% formaldehyde solution for 15 min at room temperature. After rinsing three times with PBS for 5 min each, the cells were blocked in blocking buffer (1XPBS, 5% normal serum, 0.3% Triton X-100) for 60 min at room temperature. After removing blocking solution, the primary antibodies were diluted in antibody dilution buffer (1XPBS, 1% BSA, 0.3% Triton X-100), and incubated on cells overnight at 4 °C. The coverslips were washed with PBS three times and incubated in fluorescent-dye conjugated secondary antibody diluted in antibody dilution buffer for 2 h at room temperature. The cells were washed with PBS three times and counterstained with DAPI. The cells were examined and photographed by immunofluorescence microscopy. The primary antibodies was used in this assay included Phospho-Chk2 (Thr68) (#2197; Cell Signaling, Danvers, MA, USA), LC3A/B (#12741; Cell Signaling, Danvers, MA, USA). Phalloidin-iFluor 594 reagent (ab176757; Abcam, Cambridge, UK) were used for labeling actin filaments

4.7. Western Blot Analysis

Whole cell lysates were produced using TEGN buffer (10 mM Tris, pH 7.5, 1 mM EDTA, 420 mM NaCl, 10% glycerol, and 0.5% Nonidet P-40) containing proteases inhibitor cocktail (Roche, Mannheim, Germany), phosphatase inhibitors (Roche, Mannheim, Germany), and 1 mM dithiothreitol (DTT). For Western blotting, the cell lysates were boiled in protein sample buffer (2 Mβ-mercaptoethanol, 12% sodium dodecyl sulfate (SDS), 0.5 M Tris, pH 6.8, 0.5 mg/mL bromophenol blue, and 30% glycerol). The samples were analyzed by 8–12% SDS-polyacrylamide gel electrophoresis (PAGE). The primary antibodies were listed as following: cleaved caspase-3 (#9661; Cell Signaling, Danvers, MA, USA), cleaved caspase-8 (#9496; Cell Signaling, Danvers, MA, USA), cleaved caspase-9 (#9508; Cell Signaling, Danvers, MA, USA), Bcl-2 (#9258; Cell Signaling, Danvers, MA, USA), XIAP (#2042; Cell Signaling, Danvers, MA, USA), Survivin (#2808; Cell Signaling, Danvers, MA, USA), GAPDH (#2118; Cell Signaling, Danvers, MA, USA), actin (A2066; Sigma-Aldrich, St. Louis, MO, USA), PARP (#9542; Cell Signaling, Danvers, MA, USA), Bax (#2772; Cell Signaling, Danvers, MA, USA), Phospho-cdc2 (Tyr15)

(#4539; Cell Signaling, Danvers, MA, USA), Phospho-Chk2 (Thr68) (#2197; Cell Signaling, Danvers, MA, USA), LC3A/B (#12741; Cell Signaling, Danvers, MA, USA), cdc2 (06-923SP; Millipore, Temecula, CA, USA), p-ATM (Ser1981) (sc-47739; Santa Cruz, CA, USA), ATM (GTX70103; GeneTex, Irvine, CA, USA), Chk2 (#05-649; EMD Millipore, Temecula, CA, USA).

4.8. Detection of Intracellular ROS

Cellular ROS levels were measured in live cells by 2′,7′-dichlorofluorescin diacetate (DCFDA) (#ab113851, Abcam, Cambridge, UK). After treatment with Licochalcone A, the cells were washed with serum free media, and incubate at 37 °C incubator for 30min with 10 µM DCFDA in serum free media. Finally, the cells were trypsinized and suspended in phosphate-buffered saline (PBS). The fluorescence was detected by a Cytomics™ FC500 flow cytometer (Beckman Coulter, Miami, FL, USA).

4.9. Statistical Analysis

The experimental data are expressed as mean ± standard deviation. Statistical differences between groups were conducted using the Student's t-test. A *p*-value less than 0.05 was considered to indicate a statistically significant difference. All statistical analyses were performed using the software package GraphPad Prism (Version 4.0, GraphPad Software; San Diego, CA, USA).

Author Contributions: Conceived and designed the experiments: T.-S.S., Y.-K.H., Y.-F.H., H.-Y.C., C.-P.H., C.-L.C.; Performed the experiments: T.-S.S., Y.-K.H., Y.-F.H., H.-Y.C.; Analyzed the data: T.-S.S., Y.-F.H., H.-Y.C., C.-P.H., C.-L.C.; Contributed reagents/materials/analysis tools: T.-S.S., Y.-F.H; Wrote the paper: Y.-F.H. and C.-L.C.

Funding: This study was supported by grants 106-CCH-IRP-020 from the Changhua Christian Hospital Research Foundation, Changhua City, Taiwan.

Conflicts of Interest: The authors declare no conflict of interest.

References

1. Mirabello, L.; Troisi, R.J.; Savage, S.A. Osteosarcoma Incidence and Survival Rates from 1973 to 2004: Data from the Surveillance, Epidemiology, and End Results Program. *Cancer* **2009**, *115*, 1531–1543. [CrossRef] [PubMed]
2. Otoukesh, B.; Boddouhi, B.; Moghtadaei, M.; Kaghazian, P.; Kaghazian, M. Novel Molecular Insights and New Therapeutic Strategies in Osteosarcoma. *Cancer Cell Int.* **2018**, *18*, 158. [CrossRef] [PubMed]
3. Rickel, K.; Fang, F.; Tao, J. Molecular Genetics of Osteosarcoma. *Bone* **2017**, *102*, 69–79. [CrossRef] [PubMed]
4. Bernthal, N.M.; Federman, N.; Eilber, F.R.; Nelson, S.D.; Eckardt, J.J.; Eilber, F.C.; Tap, W.D. Long-Term Results (>25 Years) of a Randomized, Prospective Clinical Trial Evaluating Chemotherapy in Patients with High-Grade, Operable Osteosarcoma. *Cancer* **2012**, *118*, 5888–5893. [CrossRef] [PubMed]
5. Kempf-Bielack, B.; Bielack, S.S.; Jurgens, H.; Branscheid, D.; Berdel, W.E.; Exner, G.U.; Gobel, U.; Helmke, K.; Jundt, G.; Kabisch, H.; et al. Osteosarcoma Relapse after Combined Modality Therapy: An Analysis of Unselected Patients in the Cooperative Osteosarcoma Study Group (Coss). *J. Clin. Oncol.* **2005**, *23*, 559–568. [CrossRef] [PubMed]
6. Carrle, D.; Bielack, S. Osteosarcoma Lung Metastases Detection and Principles of Multimodal Therapy. *Cancer Treat. Res.* **2009**, *152*, 165–184. [PubMed]
7. Siveen, K.S.; Uddin, S.; Mohammad, R.M. Targeting Acute Myeloid Leukemia Stem Cell Signaling by Natural Products. *Mol. Cancer* **2017**, *16*, 13. [CrossRef]
8. Chinembiri, T.N.; du Plessis, L.H.; Gerber, M.; Hamman, J.H.; du Plessis, J. Review of Natural Compounds for Potential Skin Cancer Treatment. *Molecules* **2014**, *19*, 11679–11721. [CrossRef]
9. Wang, L.; Yang, R.; Yuan, B.; Liu, Y.; Liu, C. The Antiviral and Antimicrobial Activities of Licorice, a Widely-Used Chinese Herb. *Acta Pharm. Sin. B* **2015**, *5*, 310–315. [CrossRef]
10. Sidhu, P.; Shankargouda, S.; Rath, A.; Ramamurthy, P.H.; Fernandes, B.; Singh, A.K. Therapeutic Benefits of Liquorice in Dentistry. *J. Ayurveda Integr. Med.* **2018**. [CrossRef]

11. Adianti, M.; Aoki, C.; Komoto, M.; Deng, L.; Shoji, I.; Wahyuni, T.S.; Lusida, M.I.; Soetjipto; Fuchino, H.; Kawahara, N.; et al. Anti-Hepatitis C Virus Compounds Obtained from Glycyrrhiza Uralensis and Other Glycyrrhiza Species. *Microbiol. Immunol.* **2014**, *58*, 180–187. [CrossRef] [PubMed]
12. Si, H.; Xu, C.; Zhang, J.; Zhang, X.; Li, B.; Zhou, X.; Zhang, J. Licochalcone A: An Effective and Low-Toxicity Compound against Toxoplasma Gondii in Vitro and in Vivo. *Int. J. Parasitol. Drugs Drug Resist.* **2018**, *8*, 238–245. [CrossRef] [PubMed]
13. Liang, M.; Li, X.; Ouyang, X.; Xie, H.; Chen, D. Antioxidant Mechanisms of Echinatin and Licochalcone A. *Molecules* **2019**, *24*, 3. [CrossRef] [PubMed]
14. Kim, Y.H.; Shin, E.K.; Kim, D.H.; Lee, H.H.; Park, J.H.; Kim, J.K. Antiangiogenic Effect of Licochalcone A. *Biochem. Pharmacol.* **2010**, *80*, 1152–1159. [CrossRef] [PubMed]
15. Jia, T.; Qiao, J.; Guan, D.; Chen, T. Anti-Inflammatory Effects of Licochalcone a on Il-1beta-Stimulated Human Osteoarthritis Chondrocytes. *Inflammation* **2017**, *40*, 1894–1902. [CrossRef] [PubMed]
16. Kuramoto, K.; Suzuki, S.; Sakaki, H.; Takeda, H.; Sanomachi, T.; Seino, S.; Narita, Y.; Kayama, T.; Kitanaka, C.; Okada, M. Licochalcone a Specifically Induces Cell Death in Glioma Stem Cells Via Mitochondrial Dysfunction. *FEBS Open Bio* **2017**, *7*, 835–844. [CrossRef]
17. Zeng, G.; Shen, H.; Yang, Y.; Cai, X.; Xun, W. Licochalcone a as a Potent Antitumor Agent Suppresses Growth of Human Oral Cancer Scc-25 Cells in Vitro Via Caspase-3 Dependent Pathways. *Tumour Biol.* **2014**, *35*, 6549–6555. [CrossRef]
18. Qiu, C.; Zhang, T.; Zhang, W.; Zhou, L.; Yu, B.; Wang, W.; Yang, Z.; Liu, Z.; Zou, P.; Liang, G. Licochalcone a Inhibits the Proliferation of Human Lung Cancer Cell Lines A549 and H460 by Inducing G2/M Cell Cycle Arrest and Er Stress. *Int. J. Mol. Sci.* **2017**, *18*, 1761. [CrossRef]
19. Xue, L.; Zhang, W.J.; Fan, Q.X.; Wang, L.X. Licochalcone a Inhibits Pi3k/Akt/Mtor Signaling Pathway Activation and Promotes Autophagy in Breast Cancer Cells. *Oncol. Lett.* **2018**, *15*, 1869–1873. [CrossRef]
20. Tsai, J.P.; Lee, C.H.; Ying, T.H.; Lin, C.L.; Lin, C.L.; Hsueh, J.T.; Hsieh, Y.H. Licochalcone a Induces Autophagy through Pi3k/Akt/Mtor Inactivation and Autophagy Suppression Enhances Licochalcone a-Induced Apoptosis of Human Cervical Cancer Cells. *Oncotarget* **2015**, *6*, 28851–28866. [CrossRef]
21. Chuang, C.Y.; Tang, C.M.; Ho, H.Y.; Hsin, C.H.; Weng, C.J.; Yang, S.F.; Chen, P.N.; Lin, C.W. Licochalcone a Induces Apoptotic Cell Death Via Jnk/P38 Activation in Human Nasopharyngeal Carcinoma Cells. *Environ. Toxicol.* **2019**, *34*, 853–860. [CrossRef] [PubMed]
22. Park, M.R.; Kim, S.G.; Cho, I.A.; Oh, D.; Kang, K.R.; Lee, S.Y.; Moon, S.M.; Cho, S.S.; Yoon, G.; Kim, C.S.; et al. Licochalcone-a Induces Intrinsic and Extrinsic Apoptosis Via Erk1/2 and P38 Phosphorylation-Mediated Trail Expression in Head and Neck Squamous Carcinoma Fadu Cells. *Food Chem. Toxicol.* **2015**, *77*, 34–43. [CrossRef] [PubMed]
23. Ottaviano, L.; Schaefer, K.L.; Gajewski, M.; Huckenbeck, W.; Baldus, S.; Rogel, U.; Mackintosh, C.; de Alava, E.; Myklebost, O.; Kresse, S.H. Molecular Characterization of Commonly Used Cell Lines for Bone Tumor Research: A Trans-European Eurobonet Effort. *Genes Chromosomes Cancer* **2010**, *49*, 40–51. [CrossRef] [PubMed]
24. Chandar, N.; Billig, B.; McMaster, J.; Novak, J. Inactivation of P53 Gene in Human and Murine Osteosarcoma Cells. *Br. J. Cancer* **1992**, *65*, 208–214. [CrossRef] [PubMed]
25. Novello, C.; Pazzaglia, L.; Conti, A.; Quattrini, I.; Pollino, S.; Perego, P.; Picci, P.; Benassi, M.S. P53-Dependent Activation of Microrna-34a in Response to Etoposide-Induced DNA Damage in Osteosarcoma Cell Lines Not Impaired by Dominant Negative P53 Expression. *PLoS One* **2014**, *9*, e114757. [CrossRef]
26. Guo, Z.; Deshpande, R.; Paull, T.T. Atm Activation in the Presence of Oxidative Stress. *Cell Cycle* **2010**, *9*, 4805–4811. [CrossRef]
27. Wang, P.; Zhu, L.; Sun, D.; Gan, F.; Gao, S.; Yin, Y.; Chen, L. Natural Products as Modulator of Autophagy with Potential Clinical Prospects. *Apoptosis* **2017**, *22*, 325–356. [CrossRef]
28. Elmore, S. Apoptosis: A Review of Programmed Cell Death. *Toxicol. Pathol.* **2007**, *35*, 495–516. [CrossRef]
29. Ouyang, L.; Shi, Z.; Zhao, S.; Wang, F.T.; Zhou, T.T.; Liu, B.; Bao, J.K. Programmed Cell Death Pathways in Cancer: A Review of Apoptosis, Autophagy and Programmed Necrosis. *Cell Prolif.* **2012**, *45*, 487–498. [CrossRef]
30. Awasthi, P.; Foiani, M.; Kumar, A. Atm and Atr Signaling at a Glance. *J. Cell Sci.* **2015**, *128*, 4255–4262. [CrossRef]

31. Guo, Z.; Kozlov, S.; Lavin, M.F.; Person, M.D.; Paull, T.T. Atm Activation by Oxidative Stress. *Science* **2010**, *330*, 517–521. [CrossRef] [PubMed]
32. Lin, L.; Baehrecke, E.H. Autophagy, Cell Death, and Cancer. *Mol. Cell Oncol.* **2015**, *2*, e985913. [CrossRef] [PubMed]
33. Lo, Y.C.; Lin, Y.C.; Huang, Y.F.; Hsieh, C.P.; Wu, C.C.; Chang, I.L.; Chen, C.L.; Cheng, C.H.; Chen, H.Y. Carnosol-Induced Ros Inhibits Cell Viability of Human Osteosarcoma by Apoptosis and Autophagy. *Am. J. Chin. Med.* **2017**, *45*, 1761–1772. [CrossRef] [PubMed]
34. Ge, X.Y.; Yang, L.Q.; Jiang, Y.; Yang, W.W.; Fu, J.; Li, S.L. Reactive Oxygen Species and Autophagy Associated Apoptosis and Limitation of Clonogenic Survival Induced by Zoledronic Acid in Salivary Adenoid Cystic Carcinoma Cell Line Sacc-83. *PLoS ONE* **2014**, *9*, e101207. [CrossRef] [PubMed]
35. Mrakovcic, M.; Bohner, L.; Hanisch, M.; Frohlich, L.F. Epigenetic Targeting of Autophagy via Hdac Inhibition in Tumor Cells: Role of P53. *Int. J. Mol. Sci.* **2018**, *19*, 3952. [CrossRef] [PubMed]

Sample Availability: Not Available.

© 2019 by the authors. Licensee MDPI, Basel, Switzerland. This article is an open access article distributed under the terms and conditions of the Creative Commons Attribution (CC BY) license (http://creativecommons.org/licenses/by/4.0/).

Article

A Bifunctional Molecule with Lectin and Protease Inhibitor Activities Isolated from *Crataeva tapia* Bark Significantly Affects Cocultures of Mesenchymal Stem Cells and Glioblastoma Cells

Camila Ramalho Bonturi [1], Mariana Cristina Cabral Silva [1], Helena Motaln [2], Bruno Ramos Salu [1], Rodrigo da Silva Ferreira [1], Fabricio Pereira Batista [1], Maria Tereza dos Santos Correia [3], Patrícia Maria Guedes Paiva [3], Tamara Lah Turnšek [4,*] and Maria Luiza Vilela Oliva [1,*]

[1] Department of Biochemistry, Federal University of São Paulo, 04044-020 São Paulo, SP, Brazil; camilabntr@gmail.com (C.R.B.); mariana.cabral@kroton.com.br (M.C.C.S.); bruno_salu@hotmail.com (B.R.S.); rodrigobioq@gmail.com (R.d.S.F.); fabriciopbat@gmail.com (F.P.B.)
[2] Department of Biotechnology, Jozef Stefan Institute, 1000 Ljubljana, Slovenia; helena.motaln@ijs.si
[3] Department of Biochemistry, Federal University of Pernambuco, 50670-910 Recife, PE, Brazil; mtscorreia@gmail.com (M.T.d.S.C.); ppaiva63@yahoo.com.br (P.M.G.P.)
[4] Department of Genetic Toxicology and Cancer Biology, National Institute of Biology, 1000 Ljubljana, Slovenia
* Correspondence: tamara.lah@nib.si (T.L.T.); olivaml.bioq@epm.br (M.L.V.O.); Tel.: +55-00386-05-9232-713 or +55-00386-41-651-629 (T.L.T.); +55-11-55764445 (M.L.V.O.)

Academic Editor: Roberto Fabiani
Received: 19 April 2019; Accepted: 27 May 2019; Published: 4 June 2019

Abstract: Currently available drugs for treatment of glioblastoma, the most aggressive brain tumor, remain inefficient, thus a plethora of natural compounds have already been shown to have antimalignant effects. However, these have not been tested for their impact on tumor cells in their microenvironment-simulated cell models, e.g., mesenchymal stem cells in coculture with glioblastoma cell U87 (GB). Mesenchymal stem cells (MSC) chemotactically infiltrate the glioblastoma microenvironment. Our previous studies have shown that bone-marrow derived MSCs impair U87 growth and invasion via paracrine and cell–cell contact-mediated cross-talk. Here, we report on a plant-derived protein, obtained from *Crataeva tapia* tree Bark Lectin (CrataBL), having protease inhibitory/lectin activities, and demonstrate its effects on glioblastoma cells U87 alone and their cocultures with MSCs. CrataBL inhibited U87 cell invasion and adhesion. Using a simplified model of the stromal microenvironment, i.e., GB/MSC direct cocultures, we demonstrated that CrataBL, when added in increased concentrations, caused cell cycle arrest and decreased cocultured cells' viability and proliferation, but not invasion. The cocultured cells' phenotypes were affected by CrataBL via a variety of secreted immunomodulatory cytokines, i.e., G-CSF, GM-CSF, IL-6, IL-8, and VEGF. We hypothesize that CrataBL plays a role by boosting the modulatory effects of MSCs on these glioblastoma cell lines and thus the effects of this and other natural lectins and/or inhibitors would certainly be different in the tumor microenvironment compared to tumor cells alone. We have provided clear evidence that it makes much more sense testing these potential therapeutic adjuvants in cocultures, mimicking heterogeneous tumor–stroma interactions with cancer cells in vivo. As such, CrataBL is suggested as a new candidate to approach adjuvant treatment of this deadly tumor.

Keywords: CrataBL; glioblastoma; mesenchymal stem cells; microenvironment; plant lectin; protease inhibitor

1. Introduction

In nature, plant lectins are essential for their viral, bacterial, and fungal defense properties. Legume lectins emerge in human diet, being present in foods like tomato, peanut, banana, lentil, soybean, rice, potato, and others. When consumed, lectins' biological activity may be preserved, as they are resistant to gut digestion, enabling their possible function, even in the circulation system [1]. Plants lectins are used as potential biomarkers for several diseases as they are associated with antimicrobial, insecticidal, and antitumor activity [1–3]. As lectins bind to carbohydrates moieties, they detect large and heterogeneous group of proteins in tumor cancer tissues, since glycosylated proteins are commonly upregulated in cancer, where their various functions tend to increase tumor progression and other aberrant conditions [4–6]. CrataBL is a natural compound isolated from *Crataeva tapia* tree. The "BL" in CrataBL stands for bark lectin origin, as the protein was extracted from *Capparaceae* bark grown in northeastern Brazil [7]. It has a dual function, as besides its lectin activity, it acts as a 20 kDa Kunitz-type serine protease inhibitor, inhibiting trypsin (43 µM) and human factor Xa (8.6 µM) [8]. CrataBL lectin capacity was demonstrated by its specificity to bind sulfated oligosaccharides [7–9]. This protein affects the development of larvae of *Callosobruchus maculatus* [2], prolongation of blood coagulation, reduction of occlusion time of arterial flow [9], and glycemia in diabetic mice [10]. In a cancer model, it has been reported that CrataBL induced apoptosis of prostate cancer cell lines DU-145 and PC-3 [7].

Glioblastoma (GB) are classified as rare cancers, yet they the most common and aggressive among brain cancers with still no efficient treatment, reflected in their early recurrence. Tumor cell-autonomous heterogeneity [11] and treatment-dependent plasticity of recurrent vs. primary glioma subtypes hamper efficient radiation and chemotherapy [11,12]. Although several synthetics and natural compounds have been proposed as adjuvant therapeutics to standard radiotherapy, no effective breakthrough in treatment of GB has yet been achieved. On the other hand, tumor cell-nonautonomous heterogeneity of GB, comprising a plethora of stromal cells infiltrating the tumor, presents the obstacle to successful treatment. Thus, novel strategies, including ones involving mimetics of the GB microenvironment, should be considered [12,13].

Human mesenchymal stem cells (MSCs) are adult, nonhematopoietic, multipotent progenitor cells, originally isolated from the bone marrow, which are traditionally characterized in vitro by their plastic adherence, trimesenchymal differentiation, and expression of a panel of distinguishing surface markers [14]. The interaction of human mesenchymal stem cells (hMSCs) and tumor cells has been investigated in various contexts. MSCs are considered as cellular treatment vectors based on their capacity to migrate towards a malignant lesion. However, concerns about the unpredictable behavior of transplanted MSCs are accumulating markers [13]. Mesenchymal stem cells are part of GB stromal components and have also been investigated for cellular GB treatment due to their ability to modulate glioblastoma cell phenotype [15,16]. However, the mechanisms by which MSCs affect various types of cancers remain controversial and include mediation by cytokines, their receptors, and growth factors [15–20]. Of these, priming of toll-like receptors TLR3 and TLR4 seems to significantly affect MSC interactions with tumor cells and we propose this to be the key role in MSC and GB cross-talk via CCL2/MCP1 cytokines [21,22]. On the other hand, MSCs may be capable of delivering therapeutics to the brain, as they can transverse the blood–brain barrier under pathological conditions [17,20–22]. Considering that, we aimed to characterize CrataBL's effects on glioblastoma and its microenvironment, mimicked here by direct coculturing of GB cells with MSCs.

2. Results

2.1. CrataBL Impaired Cell Viability and Induced Cell Death

CrataBL had stronger effects on the viability of MSC than on the U87 cells. Whereas the viability of MSC was affected after 24 h treatment only at the 100 µM dose, the viability of U87 cells remained unchanged; however, CrataBL significantly impaired the viability of both cell lines after 48 h treatment. On the contrary, treatment with CrataBL reduced the viability of cells in cocultures after 24 h in a dose-dependent manner (Figure 1A). This indicates that MSCs enhance the effect of CrataBL in decreasing the viability of U87 cells in coculture. To verify whether the reduction of cell viability resulted from the induction of cell death, flow cytometry coupled to annexin V (AN) and propidium iodide (PI) staining was used to distinguish the live (AN^-/PI^-), early apoptotic (AN^+/PI^-), late apoptotic (AN^+/PI^+), and necrotic (AN^-/PI^+) cells in cultures exposed to CrataBL for 24 h. Although the viability of MSCs was affected by CrataBL, no significance in the number of early, late, or necrotic cells was detected after 24 h treatment. Consistent with the viability data, no decrease in the apoptotic or necrotic cells counts was detected in treated U87 cells, suggesting that CrataBL alone does not induce cell death pathways in these cells. In contrast, in cocultures treated with 50 µM and 100 µM of CrataBL, an increase in late apoptotic and necrotic cells was detected (Figure 1B). The fact that the mixed cocultures were more sensitive to the inhibitor than each of the individual cell types may suggest that their interactions would render the heterogeneous tumor more refractory to CrataBL.

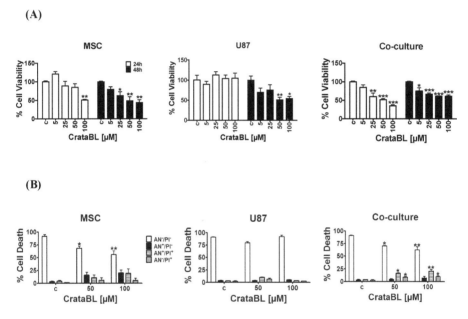

Figure 1. Effects of CrataBL on the cell viability and cells death of MSCs, U87 cells, and cocultured cells. Cell viability was measured by MTT assay at increasing concentrations of the inhibitor CrataBL at 5 µM, 25 µM, 50 µM, and 100 µM concentrations after 24 and 48 h of treatment. The absorbance values were normalized to nontreated control cells (c), as described in the Methods section. Panels represent (**A**) MSC, U87 glioblastoma cells, and direct cocultures. Cell death (**B**) was evaluated by flow cytometry as described in the Methods section, in MSC, U87 cells and cocultures after 24 h of CrataBL treatment. AN^-/PI^- represent viable cells; AN^+/PI^- represent early apoptotic; AN^+/PI^+ and AN^-/PI^+ represent late apoptotic and necrotic cells. Significance among experimental groups was considered as * $p < 0.05$, ** $p < 0.005$, and *** $p < 0.0005$.

2.2. CrataBL Affected Proliferation and Cell Cycle

The proliferation, measured by BrdU incorporation, was affected already after 24 h for nearly 40% and only in the high dose after 48 h. In contrast, CrataBL inhibited proliferation of U87 cells already at lower doses after 48 h and in the high dose after 24 h. In the cocultures, CrataBL treatment strongly decreased their proliferation upon 48 h (Figure 2A) in a similar pattern as observed with U87 cells. Flow cytometry analysis was used to monitor the percentage of cells in each cell cycle phases (G_1, S, G_2, or M phase) in response to CrataBL treatment. The percentage of treated MSCs residing in the S phase of the cell cycle was increased, and in U87, the percentage of the cells in G_0/G_1 phase was decreased. No alterations in the cell cycle in the cocultures (Figure 2B) were observed. Still, the percentage of cells residing in the resting, sub-G_0 phase showed tend to increase in U87 cell monoculture and in coculture conditions upon CrataBL treatment, possibly meaning that the cells were stopped at G_0/G_1 checkpoints, but CrataBL had a minor, non significant effect on cocultured cells.

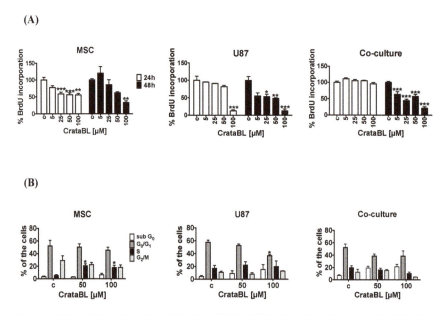

Figure 2. Effects of CrataBL on the proliferation of MSCs, U87 cells, and cocultured cells. Proliferation was determined by cell cycle assay using flow cytometry. Panels represent (**A**) MSC, U87 glioblastoma cells, and direct cocultured cells, in the presence of increasing concentrations of CrataBL inhibitor as measured by BrdU assay, relative (%) to nontreated control cells (c). Cell cycle phase distribution was determined by propidium iodide staining followed by flow cytometry analyzes (**B**). Panels represent MSC, U87 GB cells, and cocultured cells, treated with CrataBL (50 μM and 100 μM) for 24 h and then permeabilized. Percentages of cells in sub G_0, G_0/G_1, S, and G_2/M phase were determined using Accuri C6 cytometer, with at least 10,000 events collected. Significance was considered at * $p < 0.05$, ** $p < 0.005$, and *** $p < 0.0005$.

2.3. CrataBL Reduced Invasion and Metalloprotease Activity

Cancer cell invasion is characterized by degradation of the surrounding extracellular matrix and their spread into the adjacent tissues. While MSC appeared not to be affected by the 50 μM and 100 μM inhibitor treatment after 24 h, U87 cells at the same CrataBL concentrations were significantly less invasive. In cocultures, the inhibitory activity followed the U87 cells' pattern (Figure 3A), pointing to the fact that CrataBL selectively inhibited invasion of cancer cells in the cocultures.

Next, we evaluated the CrataBL treatment on MMP-9 and MMP-2 since these gelatinolytic MMPs play important a role in invasion by degrading native extracellular matrix protein components or otherwise affecting protease signaling [23]. As shown in Figure 3B, the gelatinolytic activity of MMPs appears as a translucent band in the images, and the band intensity was analyzed by densitometry, as described in the Methods section. The molecular weight of MMP-9 was predicted as ±84/92 kDa and MMP-2 ±64/72 kDa. Using a metalloprotease activity blocking buffer, no bands were observed, suggesting that the observed gelatinolytic activities corresponded to MMPs (data not shown). As demonstrated in Figure 3B, CrataBL treatment only inhibited MMP-9 and MMP-2 activity in coculture media, which parallels the decreased invasion in cocultured cells. No CrataBL effect was observed on MMP-9 and MMP-2 activity of MSCs and only a trend of invasion decrease was observed in U87 cells. These data support the notion that MMPs' activity is only inhibited by CrataBL upon MSC interaction with U87 cells in coculture.

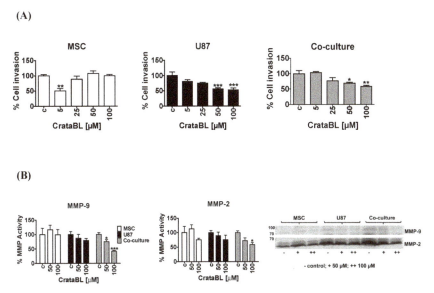

Figure 3. The effects of CrataBL on the invasion and metalloproteases activity secretion of MSCs, U87 cells, and cocultured cells. The invasion through the matrigel layer was evaluated as the percentage of invasion normalized to nontreated cells as control (c) in MSC, U87 glioblastoma cells, and cocultured cells (**A**). CrataBL influenced metalloprotease activity as observed by gelatin zymography gel assay and quantified for metalloproteases MMP-9 ±84/92 kDa and MMP-2 ±64/72 kDa (**B**). Densitometry analyses showed a diminished activity of MMP-2 and MMP-9 in the cocultures. (−) indicates control, (+) indicates 50 µM of CrataBL and (++) 100 µM of CrataBL in MSC, U87 and coculture. Significance was considered at * $p < 0.05$, ** $p < 0.005$, and *** $p < 0.0005$.

2.4. CrataBL Reduced Cell Adhesion

During tumor progression, the synthesis and composition of extracellular matrix (ECM) components by cancer cells are altered and are partially degraded, e.g., an increase of collagen fibers, culminating in cancer invasion and metastatic spread. Such alterations in the extracellular matrix network, consisting of collagen I and IV, fibronectin and in brain tumors, in particular, the most abundant ECM component laminin, may contribute to altered molecular mechanisms of invasion [24]. In MSCs, no significant alterations in cell adhesion by CrataBL treatment were observed, regardless of the substrate used. As expected, U87 cells' adhesion decreased on laminin, the native brain substrate for glioblastoma cells upon treatment with CrataBL. Accordingly, in coculture conditions, impaired cell adhesion on collagen I and IV, as well as in laminin, but not fibronectin, was observed when cells were treated with higher concentrations (50 and 100 µM) of CrataBL (Figure 4).

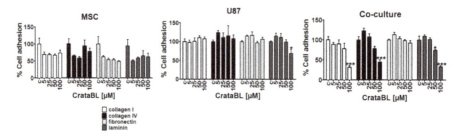

Figure 4. The effects of CrataBL on cell adhesion. Cell adhesion to collagen I, collagen IV, laminin, and fibronectin coatings was evaluated, as described in Methods on MSC, U87 glioblastoma cells, and cocultured cells at 5 µM, 25 µM, 50 µM, and 100 µM CrataBL concentrations, normalized to nontreated cells as control (c). Significance was considered at * $p < 0.05$, and *** $p < 0.0005$.

2.5. CrataBL Altered Cytokine Levels and Nitric Oxide Release

To verify whether the CrataBL inhibitory effects are possibly mediated by the inflammatory cytokines and chemokines playing a role in GB progression [25], the selected chemokines', i.e., G-CSF, GM-CSF, IL-6, IL-8, MCP-1, and VEGF, as well as nitric oxide (NO), release was analyzed in mono and cocultured media upon CrataBL treatment. This inhibitor decreased the levels of all the above cytokines/chemokines in MSC, U87 cells, and cocultures (Figure 5A). The decrease was more abrupt in cocultures when compared to monocultured cells. These results may be linked to the increase in nitric oxide (NO) production upon addition of CrataBL after 24 h and 48 h in MSC and in U87 cell monocultures. In U87 cell culture, a nearly four-fold increase in NO release was detected under 50 µM CrataBL treatments after 48 h. In coculture, the changes in NO release were less prominent, due to yet unknown interactions between MSC and U87 cells (Figure 5B).

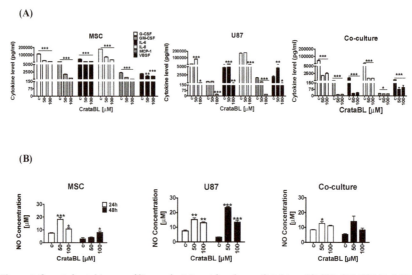

Figure 5. Secreted cytokines profiling and nitric oxide release. Cytokines (G-CSF, GM-CSF, IL-6, IL-8, MCP-1, and VEGF) release following CrataBL treatment with 50 µM and 100 µM after 24 h in the media of (**A**) MSC, U87 glioblastoma cells, and cocultured cells. Nitric oxide release was analyzed by chemiluminescence in (**B**) MSC, U87 glioblastoma cells, and cocultured cells after 24 h and 48 h of CrataBL treatment (50 µM and 100 µM), compared to nontreated controls (c). Significance was considered at * $p < 0.05$, ** $p < 0.005$, and *** $p < 0.0005$.

3. Discussion

Although much attention is given to cell autonomous mechanisms of tumor progression [11], non-cell-autonomous mechanisms, particularly interactions between tumor cells and stromal cells, are increasingly recognized as important contributors to tumor growth and resistance to therapy [12]. Compared with other cancers, the stroma of GB is not well understood, due to the uniqueness of the brain and it is thought to be composed of reactive astrocytes, endothelial cells, and immune cells. However, the contributions of other cell types, such as MSCs, have not been studied to a great extent. MSCs are also known for their ability to migrate to zones of tissue injury and tumors, being attracted to inflammatory cytokines which are produced in the cancer microenvironment. We have previously demonstrated that glioblastoma cells, as well as MSCs, differentially express connexins and that they interact via gap-junctional coupling. Besides this so-called functional syncytium formation, Schichor et al. [26] have also provided evidence of cell fusion events (structural syncytium). Direct and indirect interactions between mesenchymal stem cells (MSCs) and cancer cells in vitro and in vivo have been shown to either negatively or positively affect the malignant phenotype of cancer cells [27–29], even inducing their senescence and dormancy [27] as well as cell fusion in direct cultures, forming cancerous hybrid cells and exhibiting entosis (cell cannibalism), as observed recently by Oliveira et al. [18].

We have demonstrated that CrataBL diminished the viability of MSCs to a greater extent than in U87 cells, this being in agreement with previous reports by Ferreira et al. [8] in prostate cancer. However, the inhibition of mixed cells grown in 1:1 coculture, mimicking glioblastoma in vivo, where MSCs would infiltrate the tumor was even more affected by CrataBL than U87 cells alone. This indicates that there is cellular cross-talk between these two cell types, rendering the coculture less viable (Figure 1).

Furthermore, we investigated the processes that may affect cell viability, such as cell death being induced in the cocultures of U87/MSC. An increase in the percentage of late apoptotic and necrotic cells was observed in U87/MSC cocultures upon 24 h of CrataBL treatment (Figure 1A). We believe that at this time point, MSCs in monocultures and in the cocultures were affected first. This lowers total coculture cell viability in 24 h. However, the total cell proliferation (division, Figure 2A), due to more efficiently dividing U87 cells, in this coculture compensated for the lowered proliferation of MSCs after 48 h. This corroborates previous observations in U87/MSC [18]. It is noteworthy that BrdU incorporation does not discriminate the types of cells in coculture, and U87 stopped dividing only after a prolonged time, both alone and in the coculture (48 h), demonstrating that these cells are more resistant to CrataBL. The decrease of total cell viability, as shown in Figure 1A, in cocultured cells at 24 h is thus due to dying MSCs, as these cells were more vulnerable to the apoptotic effect of the inhibitor, as one would expect (Figure 1B). Interestingly, these results are in contrast to our previous experiments with another protease inhibitor (EcTI), isolated from *Enterolobium contortisiliquum*, also of the same Kunitz-type family, but having no lectin-binding properties [30]. EcTI did not cause U87 cell death, neither in mono, nor in coculture, suggesting that the proteolysis inhibition per se does not play a role in apoptosis. However, CrataBL increased the numbers of cells in the G_0/G_1 checkpoint after 24 h treatment, in U87 monocultured cells, and in mixed cocultured cells, indicating that senescence is linked to cell cycle checkpoint G_0 phase arrest.

As CrataBL impaired metabolic activity, cancer cell cycle progression, and proliferation of the coculture, we were interested whether the remaining viable U87 cells' invasiveness was affected by CrataBL. In the invasion experiments, we did not discriminate between U87 and MSCs that were collectively determined before and after treatment with CrataBL for 15 min. As the inhibitor did not significantly affect MSCs alone, but did affect U87 cells above 50 μM (Figure 3A), we may speculate that even if the invaded cells contained both U87 cells and MSCs, CrataBL would first bind to and inhibit the more invasive U87 cells, as they also express higher levels of metalloproteases (Figure 3B). These cells also express other proteolytic enzymes, as reported by Breznik et al. [16]. In that study, we used a differential dye and gene labeling of the MSCs and U87 cells, respectively, in 2D and in 3D cocultures. We have shown that MSCs in coculture did impair U87 invasion, but enhanced the invasion of U373 under the same coculture conditions (1:1 ratio of cells), which was associated with an increase in other proteolytic enzymes, cathepsin B and urokinase, besides MMPs. In another study, Oliveira et al. [18] used the same coculture system to expose the coculture to the inflammatory peptide bradykinin. After that, the invasion of U87 cells (but not MSCs) out of the cocultures was increased and, furthermore, after separation of U87 cells from the three-day cocultures, we found that this was due to increased expression of EMT-related genes' induction in the glioblastoma cells.

Here, we also found that the activities of MMP-2 and MMP-9 are involved in glioblastoma cells' invasion [31], as the activities of MMP-2 and MMP-9 were impaired in CrataBL-treated cocultures, but remained unchanged in CrataBL-treated MSC and U87 cell monocultures. This confirms that MMPs may enhance MSCs and U87 cells, as observed in our previous studies [16].

The adherence of cancer cells to extracellular matrix proteins, which in the brain is comprised mostly of laminin, and to a lesser extent of fibronectin and collagen type IV, which is needed for their subsequent degradation by GB cells, enables them to invade adjacent tissues. Here, we found that in U87/MSC cocultures, CrataBL treatment significantly inhibited U87 and U87/MSC cocultured cells' invasion, but did not affect monocultured MSCs' migration. Impaired invasion of U87 cells in indirect cocultures with bone marrow MSCs has previously been reported, however, in the experiments by Breznik et al. [16] and based on transcriptomes alteration of these two cell lines upon paracrine interactions. We conclude that not only MMP induction, but also enhanced binding to some of the induced adhesion molecules, such as ephrin, may play a role in cell adhesion and repulsion processes, adding to the generally reduced invasion in indirect cocultured U87-MG cells [32]. This is in contrast to the observed effects of CrataBL potentiating the MMP-induced inhibition of U87 cell adhesion to matrix proteins. Several studies have already highlighted the importance of these matrix proteins in brain tumor malignancy, angiogenesis, proliferation, and patient survival [23]. In this study, the adhesion of MSC and U87 cells alone to three ECM-related substrates remained unaltered upon CrataBL treatment, but was inhibited in U87 coculture pretreatment, especially with laminin, where the adhesion was significantly decreased at a higher (100 µM) CrataBL concentrations. CrataBL more effectively decreased U87 cell adhesion compared to the other Kunitz-type inhibitor, EcTI, which only slightly decreased the adhesion of U87 cells to fibronectin and MSC cocultured U87 cells to collagen IV [30]. CrataBL seems to interact in a more complex manner with cells to impair their adhesion to extracellular matrix components than EcTI. In parallel to adhesion, both matrix proteases were more inhibited by CrataBL compared to EcTI-treated cocultures.

Inflammatory cytokines play important roles in cancer progression, and selected cytokine therapies have already been used [32,33]. Due to accumulating evidence on the immunomodulatory effects of activated MSCs, the enhanced secretion of chemokine and cytokines when in contact with cancer cells has been observed. In indirect cocultures, we found CCL2/MCP to be the most significantly regulated chemokine in U87/MSC paracrine signaling, in addition to several other chemokines [32,34,35] that may account for changed cocultured cells' phenotypes by affecting several genes associated with proliferation (Pmepa-1, NF-κB, IL-6, IL-1b), invasion (EphB2, Sod2, Pcdh18, Col7A1, Gja1, Mmp1/2), and senescence (Kiaa1199, SerpinB2). These have generally been found as either inhibitory or promoting cancer progression, due to several reasons, as reviewed in Lee and Hong [36]. Here, we found that in response to CrataBL treatment, the levels of selected chemokines and selected inflammatory cytokines', including IL-6, IL-8, GM-CSF, VEGF, MCP-1, and G-CSF, secretion were effectively reduced in the mono- and cocultures, demonstrating the antitumorigenic effect of CrataBL. The cytokines IL-6 and IL-8 are both known to enhance the invasiveness of GB cells and increase angiogenesis, whereas GM-CSF is involved in proliferation and growth, and VEGF promotes angiogenesis by HIF-1alpha induction in GB. Likewise, MCP-1 (CCL-2, CC-chemokine ligand 2) facilitates GB progression, as a monocyte-attracting chemokine. GM-CSF enables glioma cells to proliferate and invade adjacent tissue [33].

Similarly, nitric oxide (NO), a multifaceted small molecule associated with the anti or procancer activity, is known to mediate the attraction of immune cells to the wound site and to act as an antiproliferative agent in cancer cells. NO may be produced constitutively or is induced after inflammatory conditions, vascular injury and/or cell proliferation [37,38]. In GB, NO release was related to the sensitivity of cancer cells to temozolomide therapy [38]. Nitric oxide release was enhanced in U87 cells as well as in MSCs treated with CrataBL. In cocultures, however, a significant increase in NO release after CrataBL treatment was diminished compared to MSCs of U86 alone. Altogether, these results confirm that the antiproliferative effect of CrataBL on U87 cells and GB/MSC cocultures may also be mediated via NO release. In GB, the MSC treatment was effective in animal experiments [39,40]. The fact that the mixed cocultures were more sensitive to the inhibitor than each of the individual cell types with respect to cell processes suggests that their interactions would render the heterogeneous tumor mass more resistant to CrataBL. It has also been demonstrated in vivo (animal and clinical studies) that heterogeneous tumors are less sensitive to treatment, due to stromal cells supporting tumor resistance.

Taken together, our research was focused whether CrataBL acts on cocultures that resemble the in vivo GB tumor microenvironment more (or less) efficiently than on monocultures of MSCs and U87 cells. We have shown that, indeed, CrataBL decreased cell proliferation, enhanced the transition from early to late apoptosis, and slightly enhanced cell proliferation in the cocultures vs. monocultures. Further, it inhibited the adhesion of mixed cells, but did not significantly inhibit the invasion out of the cocultures. CrataBL significantly affected NO release from cocultures after longer exposure, which would have an effect on cytokine release and immune response, which is undoubtedly different in a heterogeneous tumor environment compared to isolated tumor cells. Thus, we emphasize that it makes much more sense to test inhibitors or other natural compounds in cocultures that mimic heterogonous tumors, than in isolated cancer cells, as is usually done. Finally, we present the potential beneficial effects of the natural bifunctional plant protein CrataBL. These should be proven in preclinical testing using patient-derived GB cells and in animal studies prior to its clinical application.

4. Materials and Methods

4.1. Chemicals and Reagents

Sodium chloride, ammonium sulfate, bovine serum albumin ≥96% purity, Dulbecco's Modified Eagle's Medium (DMEM 5921), L-glutamine, Na-pyruvate, nonessential amino acids, streptomycin, penicillin, 3-(4,5-dimethylthiazol-2-yl)-2,5-diphenyltetrazolium bromide (MTT), trypsin 0.025%, triton X-100, 8.0 µm pore diameter inserts, toluidine blue, heat-inactivated fetal bovine serum (FBS), collagen I, collagen IV, fibronectin, laminin, and phosphate buffered saline 10 mM, pH 7.4 (PBS 1×) were all purchased from Sigma Aldrich (St. Louis, MO, USA). Gelatin was purchased from Calbiochem®. 5-Bromo-2′-Deoxyuridine (BrdU), RNase, propidium iodide, annexin-AlexaFluor488, microBCA, C18 column (Vydac) and coomassie brilliant blue were purchased from ThermoFisher Scientific (Grand Island, NY, USA). CM-cellulose resin and Superdex 75 column were purchased from GE Healthcare.

4.2. CrataBL Purification

Plant material was collected in Northeast of Brazil in Recife, and identified by the specialist at *Instituto Agronômico de Pernambuco*. A voucher specimen was deposited at the herbarium of the same *Instituto* (n° 61.415). The inhibitor was purified according to Araujo et al., [7] with some modifications. Briefly, the protein was extracted from *Crataeva tapia* bark powder with 0.15 M NaCl (1:20 w/v) during 12 h agitation. After filtration, the solution was centrifuged at 4000× g, 15 min, 4 °C. The soluble crude extract was submitted to ammonium sulfate fractionation, first 0–30% and last step involving 30–60% (w/v). Lectin content was resuspended and dialyzed in 10 mM citrate phosphate buffer, pH 5.5, and applied to CM-cellulose chromatography column, previously equilibrated with the same dialysis buffer. Absorbed proteins were evaluated by spectrophotometry analyses (A_{280}), being CrataBL eluted in dialysis buffer with 0.5 M NaCl. The eluted fraction was injected into a second chromatography (molecular exclusion) Superdex 75 column coupled to an Äkta Avant System (GE Healthcare). Total protein was determined by Lowry protocol following a standard curve with bovine serum albumin. The homogeneity of CrataBL was evaluated by reverse phase chromatography (HPLC) using a C18 column (Vydac) in a linear gradient of acetonitrile in trifluoroacetic acid and by SDS-polyacrylamide gel electrophoresis. The CrataBL activity was analyzed by phenol-sulfuric acid carbohydrates assay in a microplate, as described by Masuko et al. [41].

4.3. Cell Lines and Cocultures

Glioblastoma U87-MG cells and bone marrow mesenchymal stem cells (MSCs) were purchased from ATTC® HTB-14™ and Lonza BioScience Walkersville Inc., respectively. MSCs and U87-MG were cultured in DMEM low glucose medium supplemented with 200 mM L-glutamine, 100 mM Na-pyruvate, 1× nonessential amino acids, 100 µg/mL streptomycin, 100 IU/mL penicillin, and 20% (v/v) heat-inactivated fetal bovine serum (FBS), whereas U87-MG cells were cultured in 10% (v/v) FBS. The medium was changed every three days. MSC up to passage ten and U87-MG cells up to passage 70 were used. Experiments were performed in triplicates using the medium with 10% FBS. Monolayer cocultures of MSCs and GB cells (U87) were prepared by mixing the cells in 1:1 ratio (MSC/GB cells) that were seeded into monolayer culture plates with formats, corresponding to each particular experiment. The cells were analyzed after 24 and 48 h of direct coculturing.

4.4. Cell Viability Assay

The inhibitory effect of CrataBL on the metabolic activity of mesenchymal stem cells, U87 cells, and direct cocultures was measured by reducing NAD(P)H-dependent cellular oxidoreductase enzymes that reflect the number of viable cells present. These enzymes are capable of reducing the tetrazolium dye MTT 3-(4,5-dimethylthiazol-2-yl)-2,5-diphenyltetrazolium bromide to its insoluble formazan form and was performed in triplicate to confirm. Cells were seeded into a 96-well plate: 2000 MSCs, 5000 U87 cells, and in coculture (2000 MSC together with 5000 U87 cells) in 100 µL of medium/well and incubated at 37 °C and 5% (v/v) CO_2 overnight. Different concentrations of CrataBL (5–100 µM) were used to treat cells for 24 h and 48 h. MTT substrate (5 mg/mL), 10 µL, was added to the cells in culture upon treatment, followed by 2 h of incubation at 37 °C. The media was then discarded and addition of 100 µL of DMSO enhanced dissolution of crystal formazan, which allowed absorbance (OD) recording at 540 nm using a microplate spectrophotometer (Spectra max Plus 384, Molecular Devices, CA, USA). Percentage of cell viability was calculated in relation to control: Viability (%) = samples OD/control OD × 100%.

4.5. Cell Proliferation

BrdU labeling (5-bromo 2′-deoxyuridine) was used to detect new DNA strand synthesis. MSC (2000 cells/well), U87 (5000 cells/well) or coculture cells (mix of MSC and U87 cells used in monoculture) were seeded into a 96-black well plate and incubated for 24 h at 37 °C and 5% (v/v) CO_2. CrataBL was added in concentrations of 5, 25, 50, and 100 µM for 24 h and 48 h. Then, BrdU label solution (10 µM) was added per well and incubated for another 4 h at 37 °C. Solution was removed and cells were fixed and denatured for 30 min at room temperature, by a FixDenat solution provided by ThermoFisher Scientific Company—Cell Proliferation ELISA Chemiluminescence Kit. The level of incorporated BrdU was measured by chemiluminescence using anti-BrdU-peroxidase conjugated antibody, diluted 1:100, followed by 120 min incubation at room temperature. After washing, substrate was added and the plate was stirring for approximately 5 min on a shaker. Spectrophotometer FlexStation Multi-Mode Microplate Reader (Molecular Devices) at 405 nm was used to read the light emission of the samples.

4.6. Cell Cycle and Cell Death Assay

Cell cycle was analyzed by seeding 100,000 U87 cells, 40,000 MSCs and the mixed cocultures, into 6-well plates containing medium plus 10% (v/v) FBS, followed by incubation time (24 h at 37 °C and 5% (v/v) CO_2). For cell cycle, cells were maintained in 0.2% (v/v) of FBS for 24 h prior CrataBL treatment (50 and 100 µM) and changed after treatment to medium with 10% FBS for additional 24 h. For analysis, cells were trypsinized and centrifuged at 3000× g for 5 min and fixed in ice cold 70% ethanol. For cell cycle cells were stained with a solution composed of 0.1% (v/v) of triton X-100, 100 µg/mL of RNase and 10 µg/mL of propidium iodide in PBS 1×, after 2 h of cell fixation in cold ethanol 70%. For cell death, 50 µL binding buffer was added to the pelleted cells, containing 2.5 µg/mL of FITC annexin and 5 µg/mL of PI. Dot plots and histograms were analyzed in BD Accuri C6, collecting at least 30,000 events for cell cycle and 10,000 events for cell death, per each experimental condition (BD, San Jose, CA, USA).

4.7. Invasion Assay

Cell invasion assay was evaluated by the Boyden Chamber method [42] using inserts with 8.0 µm pore diameter. Inserts were placed into 24-well plates and coated with Matrigel (1:6 in DMEM free FBS) for 30 min at 37 °C to allow polymerization. Cells (MSC 20,000, U87 50,000 and coculture mix) were plated in 250 µL of FBS free medium, after pre-incubation with CrataBL (5–100 µM) for 15 min. Then 400 µL of DMEM complete medium was added to the lower chamber and the plate was kept at 37 °C, 5% (v/v) CO_2 for 24 h. Then inserts were washed with PBS 1× and non-invaded cells were gently removed with a cotton swab from the upper surface of the chamber. Membranes were fixed in cold methanol and stained using 1% (w/v) of toluidine blue for 30 min or overnight, followed by PBS 1× wash. Invaded cells were counted in at least 10 visual fields, under an inverted microscope (Leica, Camera 3000 G and software Leica Application Suite).

4.8. Zymography Activity of Metalloproteases

Zymography was performed by electrophoresis with the gel containing 0.2% (w/v) of gelatin. Cells were seeded into the 96-well plate, 5000 U87 cells/well, 2000 MSC/well and mixed cells for coculture/well and incubated at 37 °C and 5% (v/v) CO_2. Upon 24 h, the medium was collected and quantified by MicroBCA assay. 100 µg of total protein was loaded per each lane onto a 7.5% SDS-polyacrylamide separating gels containing gelatin and 5% stacking gel. Gels were run using a BioRad PowerPac apparatus at 100 V, for about 90 min. Then they were washed for 20 min in 2.5% Triton X-100 and incubated in metalloproteases activation buffer (50 mM Tris/HCl, pH 8. 0, 5 mM $CaCl_2$, 2 µM $ZnCl_2$) or alternatively, in metalloprotease blocking activity buffer (0.5 mM EDTA plus 0.5 mM phenanthroline monohydrate for 16 h at 37 °C. After incubation, the gels were stained with Coomassie solution (40% methanol, 10% acetic acid and 0.1% (w/v) Coomassie brilliant blue), distained with 10% acetic acid and 40% methanol, and scanned. Gelatinolytic activity of metalloproteases appears as a translucent band with a blue background. In the images, the color scale was inverted to white/black and the band intensity was analyzed by densitometry method with ImageJ software [43].

4.9. Cell Adhesion Assay

To measure the adhesion of the cells to the extracellular matrix, different substrates were used for well coating: collagen I (8 µg/well), collagen IV (4 µg/well), fibronectin (4 µg/well), and laminin (4 µg/well). Cells were plated (in mixed cocultured of 20,000 MSCs and 50,000 U87 cells) into 96-well plate in the presence of 100 µL of CrataBL inhibitor (5–100 µM). They were incubated for 4 h at 37 °C and 5% (v/v) CO_2. After incubation 1% (w/v) BSA was added to the cells and they were left for 1 h at 37 °C, followed by washing in PBS 1×. Adhered cells were fixed with 70% cold methanol (v/v) for 40 min, washed with PBS 1× and stained with 1% of toluidine blue (w/v) for 30 min. Upon three additional washes with PBS 1×, 1 µL of 1% SDS (w/v) was added to each well containing 100 µL of PBS 1× for 30 min at 37 °C to solubilize the cells. Absorbance was recorded at 540 nm by microplate reader spectrophotometer (Spectra Max Plus 384, Molecular Devices, Atascadero, CA, USA) [44].

4.10. Cytokine Measurement

Cytokine profile of the media was determined in Luminex MAP, using Milliplex MAP Human Cytokine/Chemokine Magnetic Bead Panel (Merck, Kenilworth, NJ, USA) containing G-CSF, GM-CSF, IL-6, IL-8, MCP-1, and VEGF [45]. Cells were seeded into 6-well plate (20,000 MSCs, 50,000 U87 cells, and a mix of cells for coculture) and incubated at 37 °C and 5% (v/v) CO_2. At 24 h, CrataBL (50 and 100 µM) inhibitor was added to the cells and the medium was collected after 24 h of treatment. MicroBCA assay was used to determine protein concentrations in media samples against a standard FBS curve, and the samples with 40 µg of the total protein were utilized to perform cytokine analyzes. Medium containing FBS 10% was used as a blank in the measurements.

4.11. Nitric Oxide Cell Release

Nitric oxide (NO) release was measured using indirect conversion of nitric oxide to nitrite by a chemiluminescence reaction, in Nitric Oxide Analyzer (NOA^{TM} 208i–Sievers). The NO analyzer was calibrated using a standard sodium nitrite curve ranging from 0.5 µM to 100 µM. Cells were plated into 96-well plate, 5000 cells for U87, 2000 cells for MSCs, and mixed cells for coculture, and incubated for 24 h at 37 °C and 5% (v/v) CO_2. Cells were treated with CrataBL (50 and 100 µM) for an additional 24 h and the medium was collected for NO measurement. The protein content of the medium was quantified by MicroBCA assay. The samples containing 100 µg of total protein was evaluated. The medium containing 10% FBS, lacking exposure to the cells, was used as a control.

4.12. Statistical Analyses

All experiments were performed in triplicate and independently repeated at least three times. The statistical analyses were expressed as the means ± standard deviation (SD) and analyzed using GraphPad Prisma Software. Comparisons among the variables, measured in defined experimental groups were conducted using one-way ANOVA, followed by Tukey´s test. Statistical significance was defined as * $p < 0.05$, ** $p < 0.005$, and *** $p < 0.0005$.

Author Contributions: C.R.B. and M.C.C.S.: conceptualization, project administration and investigation; M.T.d.S.C., P.M.G.P., F.P.B. and R.d.S.F.: resources; B.R.S.: investigation and formal analysis; C.R.B., H.M., M.L.V.O. and T.L.T: writing-original draft; M.L.V.O., H.M. and T.L.T.: supervision and writing-review & editing.

Funding: This work was supported by Fundação de Amparo à Pesquisa do Estado de São Paulo (FAPESP) [2017/07972-9 and 2017/06630-7]; Coordenação de Aperfeiçoamento de Pessoal de Nível Superior-Brasil (CAPES)-Finance Code 001; Conselho Nacional de Desenvolvimento Científico e Tecnológico (CNPq) [401452/2016-6] and the Slovenian Research Agency to support part this work by the Programme P1-0245.

Conflicts of Interest: The authors reported no conflict of interest.

References

1. Mejía, E.G.; Prisecaru, V.I. Lectins as Bioactive Plant Proteins: A Potential in Cancer Treatment. *Crit. Rev. Food Sci. Nutr.* **2005**, *45*, 425–445. [CrossRef] [PubMed]
2. Nunes, N.N.; Ferreira, R.S.; Silva-Lucca, R.A.; de Sá, L.F.; de Oliveira, A.E.; Correia, M.T.; Paiva, P.M.; Wlodawer, A.; Oliva, M.L. Potential of the Lectin/Inhibitor Isolated from *Crataeva tapia* Bark (CrataBL) for Controlling *Callosobruchus maculatus* Larva Development. *J. Agric. Food Chem.* **2015**, *63*, 10431–10436. [CrossRef] [PubMed]
3. Lagarda-Diaz, I.; Guzman-Partida, A.M.; Vazquez-Moreno, L. Legume Lectins: Proteins with Diverse Applications. *Int. J. Mol. Sci.* **2017**, *18*, 1242. [CrossRef] [PubMed]
4. Zhang, F.; Walcott, B.; Zhou, D.; Gustchina, A.; Lasanajak, Y.; Smith, D.F.; Ferreira, R.S.; Correia, M.T.S.; Paiva, P.M.G.; Bovin, N.V.; et al. Structural Studies on the Interaction of *Crataeva tapia* Bark Protein with Heparin and other Glycosaminoglycans. *Biochemistry* **2013**, *52*, 2148–2156. [CrossRef]
5. Macedo, M.L.; Oliveira, C.F.; Oliveira, C.T. Insecticidal Activity of Plant Lectins and Potential Application in Crop Protection. *Molecules* **2015**, *20*, 2014–2033. [CrossRef]

6. Hashim, O.H.; Jayapalan, J.J.; Lee, C.S. Lectins: An effective tool for screening of potential cancer biomarkers. *Peer J.* **2017**, *5*, e3784. [CrossRef]
7. Araújo, R.M.; Ferreira, R.S.; Napoleão, T.H.; Carneiro-da-Cunha, M.; Coelho, L.C.; Correia, M.T.; Oliva, M.L.; Paiva, P.M. *Crataeva tapia* bark lectin is an affinity adsorbent and insecticidal agent. *Plant. Sci.* **2012**, *183*, 20–26. [CrossRef]
8. Ferreira, R.S.; Zhou, D.; Ferreira, J.G.; Silva, M.C.; Silva-Lucca, R.A.; Mentele, R.; Paredes-Gamero, E.J.; Bertolin, T.C.; dos Santos Correia, M.T.; Paiva, P.M.; et al. Crystal Structure of *Crataeva tapia* Bark Protein (CrataBL) and Its Effect in Human Prostate Cancer Cell Lines. *PLoS ONE* **2013**, *8*, e64426. [CrossRef]
9. Salu, B.R.; Ferreira, R.S.; Brito, M.V.; Ottaiano, T.F.; Cruz, J.W.; Silva, M.C.; Correia, M.T.; Paiva, P.M.; Maffei, F.H.; Oliva, M.L. CrataBL, a lectin and Factor Xa inhibitor, plays a role in blood coagulation and impairs thrombus formation. *Biol. Chem.* **2014**, *395*, 1027–1035. [CrossRef]
10. Da Rocha, A.A.; Araújo, T.F.; da Fonseca, C.S.; da Mota, D.L.; de Medeiros, P.L.; Paiva, P.M.; Coelho, L.C.; Correia, M.T.; Lima, V.L. Lectin from *Crataeva tapia* Bark Improves Tissue Damages and Plasma Hyperglycemia in Alloxan-Induced Diabetic Mice. *Evid. Based Complement. Alternat. Med.* **2013**, *2013*, 869305. [CrossRef]
11. Sottoriva, A.; Verhoeff, J.J.; Borovski, T.; McWeeney, S.K.; Naumov, L.; Medema, J.P.; Sloot, P.M.; Vermeulen, L. Cancer stem cell tumor model reveals invasive morphology and increased phenotypical heterogeneity. *Cancer Res.* **2010**, *70*, 46–56. [CrossRef] [PubMed]
12. Bao, S.; Wu, Q.; McLendon, R.E.; Hao, Y.; Shi, Q.; Hjelmeland, A.B.; Dewhirst, M.W.; Bigner, D.D.; Rich, J.N. Glioma stem cells promote radioresistance by preferential activation of the DNA damage response. *Nature* **2006**, *444*, 756–760. [CrossRef] [PubMed]
13. Kucerova, L.; Matuskova, M.; Hlubinova, K.; Altanerova, V.; Altaner, C. Tumor cell behaviour modulation by mesenchymal stromal cells. *Mol. Cancer* **2010**, *9*, 129. [CrossRef] [PubMed]
14. Ullah, I.; Subbarao, R.B.; Rho, G.J. Human mesenchymal stem cells—Current trends and future prospective. *Biosci. Rep.* **2015**, *35*, e00191. [CrossRef] [PubMed]
15. Motaln, H.; Turnšek, T.L. Cytokines play a key role in communication between mesenchymal stem cells and brain cancer cells. *Protein Peptide Lett.* **2015**, *22*, 322–331. [CrossRef]
16. Breznik, B.; Motaln, H.; Vittori, M.; Rotter, A.; Turnšek, T.L. Mesenchymal stem cells differentially affect the invasion of distinct glioblastoma cell lines. *Oncotarget* **2017**, *8*, 25482–25499. [CrossRef]
17. Kang, S.-G.; Figueroa, J.; Gao, F.; Marini, F.C.; Hossain, A.; Sulman, E.; Gumin, J. Mesenchymal Stem Cells Isolated From Human Gliomas Increase Proliferation and Maintain Stemness of Glioma Stem Cells Through the IL 6/gp130/STAT3 Pathway. *Stem Cells* **2015**, *33*, 2400–2415. [CrossRef]
18. Oliveira, M.N.; Pillat, M.M.; Motaln, H.; Ulrich, H.; Lah, T.T. Kinin-B1 Receptor Stimulation Promotes Invasion and is Involved in Cell-Cell Interaction of Co-Cultured Glioblastoma and Mesenchymal Stem Cells. *Sci. Rep.* **2018**, *8*, 1299. [CrossRef]
19. Rasulov, M.F.; Vasilchenkov, A.V.; Onishchenko, N.A.; Krasheninnikov, M.E.; Kravchenko, V.I.; Gorshenin, T.L.; Pidtsan, R.E.; Potapov, I.V. First experience of the use bone marrow mesenchymal stem cells for the treatment of a patient with deep skin burns. *Bull. Exp. Biol. Med.* **2005**, *139*, 141–144. [CrossRef]
20. Chamberlain, G.; Fox, J.; Ashton, B.; Middleton, J. Concise review: Mesenchymal Stem Cells: Their Phenotype, Differentiation Capacity Immunological Features, and Potential for Homing. *Stem Cells* **2007**, *25*, 2739–2749. [CrossRef]
21. Tajnšek, U.; Motaln, H.; Levic, N.; Rotter, A.; Lah, T.T. *Trends in Stem Cell Proliferation and Cancer Research*; Resende, R.R., Ulrich, H., Eds.; Springer: Dordrecht, The Netherlands, 2013. [CrossRef]
22. Kim, S.M.; Woo, J.S.; Jeong, C.H.; Ryu, C.H.; Lim, J.Y.; Jeun, S.S. Effective combination therapy for malignant glioma with TRAIL-secreting mesenchymal stem cells and lipoxygenase inhibitor MK886. *Cancer Res.* **2012**, *72*, 4807–4817. [CrossRef]
23. Gialeli, C.; Theocharis, A.D.; Karamanos, N.K. Roles of matrix metalloproteinases in cancer progression and their pharmacological targeting. *FEBS J.* **2011**, *278*, 16–27. [CrossRef]
24. Mammoto, T.; Jiang, A.; Jiang, E.; Panigrahy, D.; Kieran, M.W.; Mammoto, A. Role of Collagen Matrix in Tumor Angiogenesis and Glioblastoma Multiforme Progression. *Am. J. Pathol.* **2013**, *183*, 1293–1305. [CrossRef]
25. Anestakis, D.; Petanidis, S.; Kalyvas, S.; Nday, C.M.; Tsave, O.; Kioseoglou, E.; Salifoglou, A. Mechanisms and Applications of Interleukins in Cancer Immunotherapy. *Int. J. Mol. Sci.* **2015**, *16*, 1691–1710. [CrossRef]

26. Schichor, C.; Albrecht, V.; Korte, B.; Buchner, A.; Riesenberg, R.; Mysliwietz, J.; Paron, I.; Motaln, H.; Turnsek, T.L.; Jürchott, K.; et al. Mesenchymal stem cells and glioma cells form a structural as well as a functional syncytium in vitro. *Exp. Neurol.* **2012**, *234*, 208–219. [CrossRef]
27. Bartosh, T.J.; Ullah, M.; Zeitouni, S.; Beaver, J.; Prockop, D.J. Cancer cells enter dormancy after cannibalizing mesenchymal stem/stromal cells (MSCs). *Proc. Natl. Acad. Sci. USA* **2018**, *113*, E6447–E6456. [CrossRef]
28. Melzer, C.; Yang, Y.; Hass, R. Interaction of MSC with tumor cells. *J. Cell Commun. Signal.* **2016**, *14*, 20. [CrossRef]
29. Barcellos-de-Souza, P.; Gori, V.; Bambi, F.; Chiarugi, P. Tumor microenvironment: Bone marrow-mesenchymal stem cells as key players. *Biochim. Biophys. Acta* **2013**, *1836*, 321–335. [CrossRef]
30. Bonturi, C.R.; Motaln, H.; Silva, M.C.C.; Salu, B.R.; Brito, M.V.; Costa, L.A.L.; Torquato, H.F.V.; Nunes, N.N.S.; Paredes-Gamero, E.J.; Turnsek, T.L.; et al. Could a plant derived protein potentiate the anticancer effects of a stem cell in brain cancer? *Oncotarget* **2018**, *30*, 21296–21312. [CrossRef]
31. Kessenbrock, K.; Plaks, V.; Werb, Z. Matrix Metalloproteinases: Regulators of the Tumor Microenvironment. *Cell* **2010**, *141*, 52–67. [CrossRef]
32. Motaln, H.; Gruden, K.; Hren, M.; Schichor, C.; Primon, M.; Rotter, A.; Lah, T.T. Human mesenchymal stem cells exploit the immune response mediating chemokines to impact the phenotype of glioblastoma. *Cell Transplant.* **2012**, *21*, 1529–1545. [CrossRef]
33. Albulescu, R.; Codrici, E.; Popescu, I.D.; Mihai, S.; Necula, L.G.; Petrescu, D.; Teodoru, M.; Tanase, C.P. Cytokine Patterns in Brain Tumour Progression. *Mediators Inflamm.* **2013**, *2013*, 979748. [CrossRef]
34. Dinarello, C.A. Proinflammatory cytokines. *Chest* **2000**, *118*, 503–508. [CrossRef]
35. Wcisło-Dziadecka, D.; Zbiciak, M.; Brzezińska-Wcisło, L.; Mazurek, U. Anti-cytokine therapy for psoriasis-not only TNF-α blockers. Overview of reports on the effectiveness of therapy with IL-12/IL-23 and T and B lymphocyte inhibitors. *Postepy. Hig. Med. Dosw.* **2016**, *70*, 1198–1205.
36. Lee, H.; Hong, I. Double-edged sword of mesenchymal stem cells: Cancer-promoting versus therapeutic potential. *Cancer Sci.* **2017**, *108*, 1939–1946. [CrossRef]
37. Schwentker, A.; Vodovotz, Y.; Weller, R.; Billiar, T.R. Nitric oxide and wound repair: Role of cytokines? *Nitric Oxide* **2002**, *7*, 1–10. [CrossRef]
38. Altieri, R.; Fontanella, M.; Agnoletti, A.; Altieri, R.; Fontanella, M.; Agnoletti, A.; Panciani, P.P.; Spena, G.; Crobeddu, E.; Pilloni, G.; et al. Role of Nitric Oxide in Glioblastoma Therapy: Another Step to Resolve the Terrible Puzzle? *J. Trans. Med.* **2015**, *12*, 54–59.
39. Koç, O.N.; Gerson, S.L.; Cooper, B.W.; Dyhouse, S.M.; Haynesworth, S.E.; Caplan, A.I.; Lazarus, H.M. Rapid hematopoietic recovery after coinfusion of autologous-blood stem cells and culture-expanded marrow mesenchymal stem cells in advanced breast cancer patients receiving high-dose chemotherapy. *J. Clin. Oncol.* **2000**, *18*, 307–316. [CrossRef]
40. Shah, K. Mesenchymal stem cells engineered for cancer therapy. *Adv. Drug Deliv. Rev.* **2012**, *64*, 739–748. [CrossRef]
41. Masuko, T.; Minami, A.; Iwasaki, N.; Majima, T.; Nishimura, S.; Lee, Y.C. Carbohydrate analysis by a phenol-sulfuric acid method in microplate format. *Anal. Biochem.* **2005**, *339*, 69–72. [CrossRef]
42. Falasca, M.; Raimondi, C.; Maffucci, T. Boyden chamber. *Methods Mol. Biol.* **2011**, *769*, 87–95. [CrossRef]
43. Lisboa, R.A.; Andrade, M.V.; Cunha-Melo, J.R. Zimography is an effective method for detection of matrix metalloproteinase 2 (MMP-2) activity in cultured human fibroblasts. *Acta Cir. Bras.* **2013**, *28*, 216–220. [CrossRef]
44. Weitz-Schmidt, G.; Chreng, S. Cell adhesion assays. *Methods Mol. Biol.* **2012**, *757*, 15–30. [CrossRef]
45. Leng, S.X.; McElhaney, J.E.; Walston, J.D.; Xie, D.; Fedarko, N.S.; Kuchel, G.A. Elisa and Multiplex Technologies for Cytokine Measurement in Inflammation and Aging Research. *J. Gerontol. A Biol. Sci. Med. Sci.* **2008**, *63*, 879–884. [CrossRef]

Sample Availability: Samples of the compounds are not available from the authors.

© 2019 by the authors. Licensee MDPI, Basel, Switzerland. This article is an open access article distributed under the terms and conditions of the Creative Commons Attribution (CC BY) license (http://creativecommons.org/licenses/by/4.0/).

Article

Paris Polyphylla Inhibits Colorectal Cancer Cells via Inducing Autophagy and Enhancing the Efficacy of Chemotherapeutic Drug Doxorubicin

Liang-Tzung Lin [1,2], Wu-Ching Uen [3,4], Chen-Yen Choong [5], Yeu-Ching Shi [5], Bao-Hong Lee [5], Cheng-Jeng Tai [5,6] and Chen-Jei Tai [7,8,9,*]

1. Department of Microbiology and Immunology, School of Medicine, College of Medicine, Taipei Medical University, Taipei 11042, Taiwan; ltlin@tmu.edu.tw
2. Graduate Institute of Medical Sciences, College of Medicine, Taipei Medical University, Taipei 11042, Taiwan
3. School of Medicine, Fujen Catholic University, New Taipei City 24205, Taiwan; m002047@ms.skh.org.tw
4. Department of Hematology and Oncology, Shin Kong Wu Ho-Su Memorial Hospital, Taipei 11042, Taiwan
5. Division of Hematology and Oncology, Department of Internal Medicine, Taipei Medicine University Hospital, Taipei 11042, Taiwan; chenyen@tmu.edu.tw (C.-Y.C.); jasmineycs@yahoo.com.tw (Y.-C.S.); f96b47117@ntu.edu.tw (B.-H.L.); cjtai@tmu.edu.tw (C.-J.T.)
6. Division of Hematology and Oncology, Department of Internal Medicine, School of Medicine, College of Medicine, Taipei Medical University, Taipei 11042, Taiwan
7. Department of Chinese Medicine, Taipei University Hospital, Taipei 11042, Taiwan
8. Traditional Herbal Medicine Research Center, Taipei Medical University Hospital, Taipei 11042, Taiwan
9. Department of Obstetrics and Gynecology, School of Medicine, College of Medicine, Taipei Medical University, Taipei 11042, Taiwan
* Correspondence: chenjtai@tmu.edu.tw; Tel.: +886-02-27372181-3903; Fax: +886-02-2736-3051

Academic Editor: Roberto Fabiani
Received: 1 May 2019; Accepted: 31 May 2019; Published: 3 June 2019

Abstract: Colorectal cancer is one of the most common cancers worldwide and chemotherapy is the main approach for the treatment of advanced and recurrent cases. Developing an effective complementary therapy could help to improve tumor suppression efficiency and control adverse effects from chemotherapy. *Paris polyphylla* is a folk medicine for treating various forms of cancer, but its effect on colorectal cancer is largely unexplored. The aim of the present study is to investigate the tumor suppression efficacy and the mechanism of action of the ethanolic extract from *P. polyphylla* (EEPP) in DLD-1 human colorectal carcinoma cells and to evaluate its combined effect with chemotherapeutic drug doxorubicin. The data indicated that EEPP induced DLD-1 cell death via the upregulation of the autophagy markers, without triggering p53- and caspase-3-dependent apoptosis. Moreover, EEPP treatment in combination with doxorubicin enhanced cytotoxicity in these tumor cells. Pennogenin 3-*O*-beta-chacotrioside and polyphyllin VI were isolated from EEPP and identified as the main candidate active components. Our results suggest that EEPP deserves further evaluation for development as complementary chemotherapy for colorectal cancer.

Keywords: folk medicine; DLD-1 cells; doxorubicin; chemotherapy; drug resistance

1. Introduction

Paris polyphylla is a well-known herbal medicine used in China and Taiwan, primarily to treat fevers, headaches, burns, and wounds, and for neutralizing snake poison [1]. The plant extract was documented to exert anti-cancer activity both in vivo and in vitro [2]. Numerous natural steroidal saponins isolated from herbs show potential apoptosis-promoting activity against several cancer cells types [3–5]. In addition, *P. polyphylla* treatment can inhibit epithelial–mesenchymal transition (EMT) and invasion in breast cancer [6] and lung cancer cells [3–5]. Recently, *P. polyphylla* extract was also found to inhibit ovarian carcinoma cell growth [7].

The use of complementary and alternative medicine is now a very popular option to support conventional therapy in many countries [8–10]. For example, many herbal formulas and remedies based on traditional Chinese medicine are well accepted among cancer patients with Chinese background [11–13]. Traditional Chinese medicine (TCM) is based on the use of natural products and well-established theoretical approaches. TCM provides many potential candidates as effective drugs for integrated cancer chemotherapy, such as TJ-41 (Bu-Zhong-Yi-Qi-Tang) and PHY906 (Huang-Qin-Tang) [11,12]. In TCM practice, a therapeutic formula is normally prepared as an aqueous extract mixed with various medical herbs. One major herb in this formula is responsible for relieving the target symptom, whereas other medicinal herbs are added to enhance the therapeutic effects or reduce the side effects of the major herb [13].

Colorectal cancer is one of the most common cancer types worldwide with particularly high incidences in developed countries [14]. In Taiwan, colorectal cancer is the most common type of cancer and the third most common cause of cancer-related deaths [15]. Currently, surgery is still the only curative treatment for colorectal cancer. Although 75–80% of newly diagnosed cases are localized or regional tumors, around 50% of patients suffer recurrence after surgery [16,17]. Adjuvant therapy such as postoperative chemotherapy is used to eliminate remaining lesions and help control the risk of recurrence. Chemotherapy is also one of the main treatment approaches in advanced and recurrent cases while often associated with adverse side effects in patients, particularly in the elderly population [12,13]. Various drug resistance problems in colorectal cancer cases also reduce the response rates. These clinical features limit the use of chemotherapy in patients. Any effective drug which promotes the tumor suppression efficacy of chemotherapeutic regimens or eases the associated adverse effects may serve as an appropriate candidate to establish integrated chemotherapy and improve clinical outcomes in cancer patients. Combining standard chemotherapeutics with antitumor drugs to induce tumor cell death via other molecular pathways would not only improve tumor suppression efficiency but also reduce the doses of chemotherapeutic drugs, which could help control adverse effects and may slow the development of drug resistance. Due to the use of chemotherapy as the main approach for advanced and recurrent cancers, developing effective complementary drugs could help improve tumor suppression efficiency and control adverse effects from chemotherapy. DLD-1 is a colorectal adenocarcinoma cell line similar to HT-29 and Caco-2 cell lines [16], which are established from tumorigenic epithelial tissue. In this study, we investigated the effect of the ethanolic extracts of *P. polyphylla* (EEPP) on the suppression of DLD-1 colorectal carcinoma cells with or without chemotherapeutic drug (doxorubicin) treatment.

2. Results and Discussion

2.1. Treatment Effect of P. polyphylla on Colorectal Cancer Cell Growth

As shown in Figure 1A, compared to the untreated group, cell viability of DLD-1 colorectal carcinoma cells were decreased after treatment with 3.13–50 µg/mL EEPP for 24 or 48 h in a dose-dependent manner. On the other hand, the aqueous extract of *P. polyphylla* (AEPP) required higher doses to inhibit the growth of colorectal cancer cells. In addition, EEPP treatment, particularly at 6.25 µg/mL, induced apparent morphological alterations in the DLD-1 cells compared to the untreated group (Figure 1B). These results indicate that EEPP treatment induced cytotoxicity in colorectal carcinoma cells, suggesting that EEPP treatment causes DLD-1 colorectal cancer cell death.

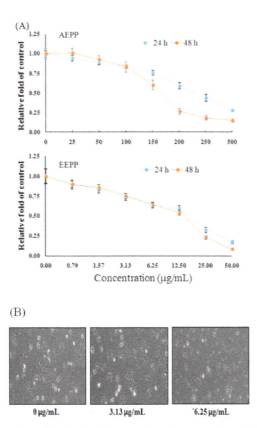

Figure 1. Inhibitory effect of *Paris polyphylla* on colorectal cancer cells. (**A**) Inhibitory effect of aqueous extract of *P. polyphylla* (AEPP) or ethanolic extract of *P. polyphylla* (EEPP) on DLD-1 colorectal carcinoma cells after treatment for 24 and 48 h, respectively. Data are shown as means ± SD (n = 3). (**B**) The morphological appearance of DLD-1 colorectal carcinoma cells after 24 h of EEPP treatment.

One approach in developing integrated chemotherapy is to choose a drug which enhances tumor cell suppression efficiency by increasing cytotoxicity using a different cell death mechanism from the other drugs used in the regimen. In general, the tumor suppression mechanisms of current chemotherapeutic drugs are mainly based on disruption of cell-cycle processes, resulting in cell apoptosis. Next, we sought to examine the possible mechanism through which EEPP causes DLD-1 colorectal cancer cell death. To this end, we examined the effect of EEPP treatment on cell-cycle regulation in the DLD-1 cells. As indicated in Figure 2A, treatment of the DLD-1 cells with 3.13–13.5 μg/mL EEPP for 12 h demonstrated a similar cell-cycle distribution pattern to the control group, suggesting that EEPP does not disrupt the cell-cycle progression in the DLD-1 colorectal cancer cells. To further determine the cell death pathway involved in EEPP-induced colorectal carcinoma cytotoxicity, we tested for DNA fragmentation associated with apoptosis. As indicated in Figure 2B, EEPP treatment did not induce DNA fragmentation in the DLD-1 cells. Together, these results suggest that the EEPP-mediated inhibition of the DLD-1 colorectal cancer cell growth does not involve apoptosis.

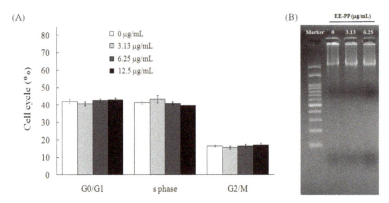

Figure 2. No effects of EEPP on (**A**) cell-cycle distribution and (**B**) DNA ladder in DLD-1 colorectal carcinoma cells. Data are shown as means ± SD ($n = 3$).

2.2. EEPP Treatment Causes Autophagic Cell Death in Colorectal Carcinoma Cells

Apart from apoptosis, autophagy also plays crucial roles in cancer cell survival and death, and is gaining increasing interest in cancer research. Autophagy, also termed type II programmed cell death (PCD), is a physiologic process that allows sequestration and degradation of the cytoplasmic contents through the lysosomal machinery [18]. Autophagy allows recycling of cellular components and ensures cellular energy supplement during nutrition starvation, infection, and other stress conditions [19]. Several lines of studies suggest cytotoxic agents including chemotherapeutic agents induce cancer cell autophagy [20–22]. To investigate whether autophagy is implicated in the EEPP-induced DLD-1 colorectal carcinoma cell death, cells were treated with EEPP for 24 h for evaluating the expression levels of the autophagy-related proteins including Beclin-1, microtubule-associated protein-1 light chain-3 (LC3), and p62 (a marker for autophagic degradation) [23,24], as well as the apoptosis-associated proteins such as Bax (Bcl2-associated X protein), p53 (tumor protein p53), Akt (Protein Kinase B), and Bcl-2 (B-cell lymphoma 2). As shown in Figure 3, in contrast to the untreated control groups, autophagy markers such as LC3 and Beclin-1 proteins were increased after treating with EEPP for 24 h in a dose-dependent manner (Figure 3A, E and F). On the other hand, Akt level was downregulated in EEPP-treated DLD-1 cells after 24 h of treatment (Figure 3A, D), whereas the expression of p62 (Figure 3A, H), and the apoptosis markers such as p53 (Figure 3A, G), Bax (Figure 3A, B), and Bcl-2 (Figure 3A, C) proteins were not affected by EEPP treatment. These results indicated that EEPP treatment induced autophagic cell death in the DLD-1 cells.

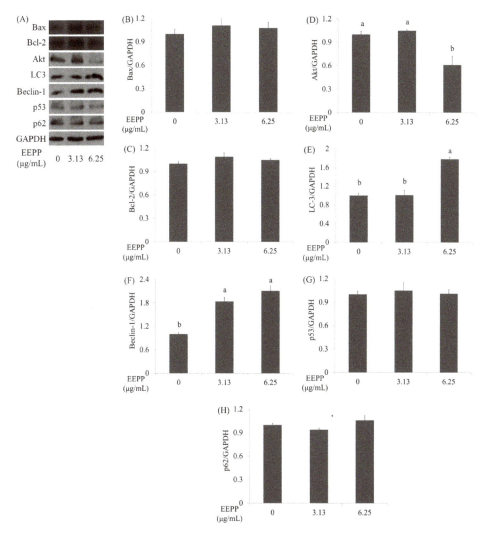

Figure 3. The effects of EEPP on apoptosis and autophagy markers. (**A**) The effects of EEPP on Bax, Bcl-2, Akt, LC-3, Beclin-1, p53, and p62 levels in DLD-1 colorectal carcinoma cells after 24 h of treatment. (**B–H**) Quantitative analysis for each protein levels. Data are shown as means ± SD ($n = 3$). The significant differences are denoted by different letters ($p < 0.05$).

2.3. Effect of EEPP–Doxorubicin Combination Treatment on Autophagy Induction in Colorectal Carcinoma Cells

Since EEPP induces autophagic cell death in DLD-1 cells, the present study further examined the potential effect of EEPP in combination with the chemotherapeutic drug doxorubicin (Dox) on DLD-1 cells. Dox functions as a topoisomerase II inhibitor and interferes with DNA/RNA synthesis in tumor cells [25]. Colorectal carcinoma cells were treated with various doses of Dox alone or in combination with EEPP for 24 h. Figure 4A illustrates that Dox treatment dose-dependently decreased cell viability in DLD-1 cells. When compared with Dox treatment alone, EEPP (3.13 μg/mL) combined with Dox treatment displayed stronger inhibitory activity against the DLD-1 cells, indicating that EEPP synergizes with Dox to inhibit the DLD-1 colorectal cancer cell growth.

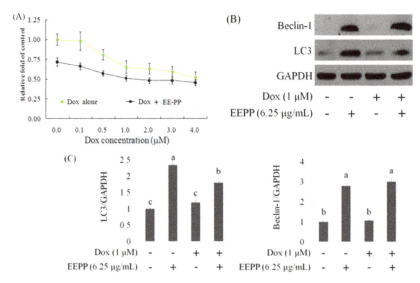

Figure 4. The combined effect of EEPP with doxorubicin (**A**). The suppressive effect of EEPP (3.13 µg/mL) combined with doxorubicin (Dox) for 24 h in DLD-1 colorectal carcinoma cells (**B** and **C**). The upregulation of Beclin-1 and LC3 expressions in EEPP-treated DLD-1 carcinoma cells. Data are shown as means ± SD (n = 3). The significant differences are denoted by diferent letters (p < 0.05).

Given that Dox is a well-known chemotherapeutic drug that induces apoptosis via the activation of p53 and caspase-3 signaling pathways in many tumor cells, we speculate that the increased cancer cell death resulting from Dox–EEPP combination treatment could be due to the potentiation of the triggered cell death pathways. To examine this hypothesis, DLD-1 cells were treated with 6.25 µg/mL EEPP alone, 1 µM Dox alone, or Dox in combination with 6.25 µg/mL EEPP for 24 h for Western blot analysis against the apoptosis- and autophagy-related proteins. In contrast to p53 and caspase-3, whose expressions were unaltered by the combination therapy (data not shown), Dox in combination with EEPP increased both LC3 and Beclin-1 protein expressions compared to Dox alone (Figure 4B, C). This concomitant increase in autophagy markers is likely due to the presence of EEPP, which alone also upregulated the autophagy markers. Together, these results suggested that EEPP may potentially enhance the anti-tumor effect in human colorectal carcinoma cells when combined with Dox.

2.4. Isolation and Identification of Active Compounds from EEPP

After showing that EEPP induces autophagic cell death in DLD-1 cells, we next sought to identify the active components of the extract responsible for its cytotoxicity. We used an octadecylsilyl column to separate EEPP into five fractions by different percentages of methanol elution (Figure 5A), after which the cytotoxicity of the fractions against the DLD-1 cells was tested. Cell viability of DLD-1 colorectal carcinoma cells was decreased after treating with the 80% methanolic fraction for 24 h (0% group: survival at 78.6%; 20% group: survival at 73.7%; 60% group: survival at 58.1%; 80% group: survival at 21.7%; 100% group: survival at 38.7%). We further isolated the active components from the 80% methanolic fraction by LC–MS and confirmed the active compounds by NMR. Pennogenin 3-O-beta-chacotrioside and polyphyllin VI were the two main compounds isolated from the 80% methanolic fraction (Figure 5B), with purity up to 95% (Figure 6).

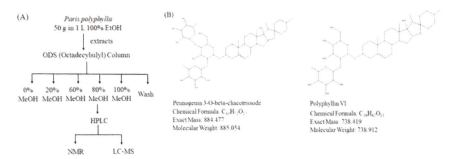

Figure 5. Isolation of active ingredients from EEPP. (**A**) The flowchart for identification of active compounds obtained from EEPP. (**B**) Pennogenin 3-O-beta-chacotrioside and polyphyllin VI were isolated and confirmed by NMR and LC–MS.

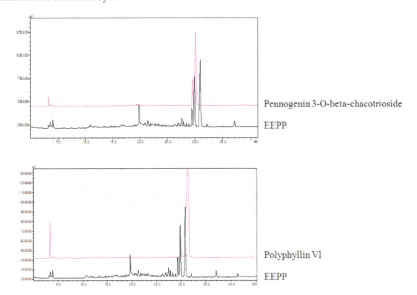

Figure 6. The purity of pennogenin 3-O-beta-chacotrioside and polyphyllin VI isolated from EEPP.

Next, the concentrations of pennogenin 3-O-beta-chacotrioside and polyphyllin VI were calculated according to EEPP, and the cell viability of DLD-1 colorectal carcinoma cells was determined. The data showed that, when compared to the untreated group, treatment of cells for 24 h with EEPP (6.25 μg/mL), pennogenin 3-O-beta-chacotrioside (1.8 μM), or polyphyllin VI (1.4 μM) decreased DLD-1 cell viability, indicating that pennogenin 3-O-beta-chacotrioside and polyphyllin VI are the two main active compounds from EEPP involved in colorectal cancer cell inhibition (Figure 7).

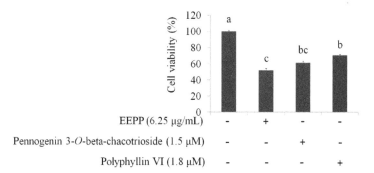

Figure 7. The suppression of DLD-1 colorectal carcinoma cells treated with EEPP (6.25 µg/mL), pennogenin 3-O-beta-chacotrioside (1.8 µM), or polyphyllin VI (1.4 µM) for 24 h. The concentrations of pennogenin 3-O-beta-chacotrioside and polyphyllin VI were calculated according to EEPP. Data are shown as means ± SD (n = 3). The significant differences are denoted by different letters ($p < 0.05$).

Finally, we asked whether the active components (pennogenin 3-O-beta-chacotrioside and polyphyllin VI) could also modulate the expression of the autophagy-related proteins in the DLD-1 cells. As shown in Figure 8A,B, both pennogenin 3-O-beta-chacotrioside and polyphyllin VI treatments for 24 h markedly increased the expressions of LC3 and Beclin-1, suggesting that these compounds, similar to EEPP, also inhibit colorectal cancer cell death by modulating autophagy. In conclusion, our results suggest that EEPP deserves further evaluation for development as complementary chemotherapy for colorectal cancer, and pennogenin 3-O-β-chacotrioside and polyphyllin VI identified as the main candidate active components in EEPP. Schematics of the mode of action of *Paris polyphylla* ethanol extract on DLD-1 colorectal cancer cells is shown in Figure 9.

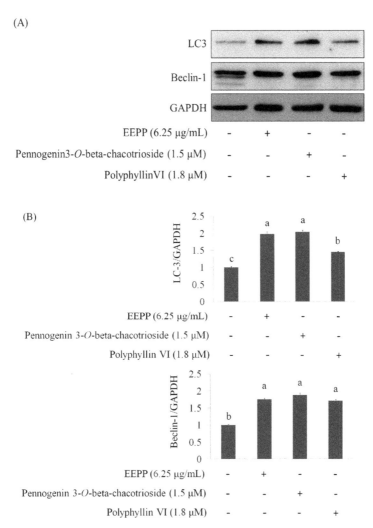

Figure 8. Elevation of autophagy markers in DLD-1 colorectal carcinoma cells treated with pennogenin 3-O-β-chacotrioside or polyphyllin VI for 24 h. The concentrations of pennogenin 3-O-β-chacotrioside and polyphyllin VI were calculated according to EEPP. (**A**) Pennogenin 3-O-β-chacotrioside or polyphyllin VI markedly increased the expressions of LC3 and Beclin-1. (**B**) Quantitative analysis for each protein levels. Data are shown as means ± SD ($n = 3$). The significant differences are denoted by different letters ($p < 0.05$).

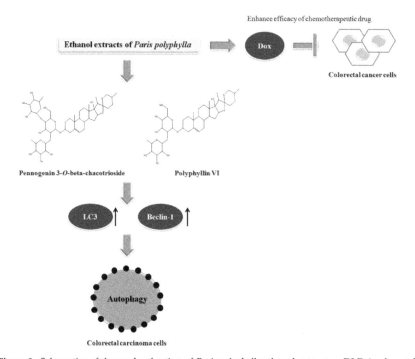

Figure 9. Schematics of the mode of action of *Paris polyphylla* ethanol extract on DLD-1 colorectal cancer cells.

3. Materials and Methods

3.1. Chemicals

P. polyphylla was purchased from Taiwan Indigena Botanica Co., Ltd (Taipei, Taiwan), and 10 g of the herb was extracted with ethanol (100 mL) three times at room temperature for 24 h. After evaporating the solvents through freeze-drying, a residue was obtained and stored at −20 °C. Crystal violet, doxorubicin, Propidium iodide (PI), sodium dodecyl sulfate (SDS), Triton X-100, trypsin, and trypan blue were purchased from Sigma Chemical Co. (St. Louis, MO, USA). Fetal bovine serum (FBS) was purchased from Life Technologies (Auckland, New Zealand). Dimethyl sulfoxide was purchased from Wako Pure Chemical Industries (Saitama, Japan). Anti-caspase-3, anti-Bax, anti-Bcl2, anti-p62, anti-p53, anti-LC-3, and anti-GAPDH (Glyceraldehyde 3-phosphate dehydrogenase) antibodies were purchased from Santa Cruz (Santa Cruz, CA, USA). Pennogenin 3-O-beta-chacotrioside was purchased from BioCrickBioTech (Chengdu, Sichuan, China). Polyphyllin VI was purchased from Chem Faces (Wuhan, Hubei, China).

3.2. Cell Culture

The human colorectal carcinoma cell line DLD-1 (Bioresource Collection and Research Center, HsinChu, Taiwan) was grown in Dulbecco's modified Eagle's medium (Gibco BRL, Grand Island, NY, USA) containing 2 mM L-glutamine and 1.5 g/L sodium bicarbonate, supplemented with 10% FBS and 2% penicillin–streptomycin (10,000 U/mL penicillin and 10 mg/mL streptomycin). The cells were cultured in a humidified incubator at 37 °C under 5% CO_2.

3.3. Cell Viability

The cytotoxic effect of EEPP against DLD-1 cells was measured using a crystal violet staining assay. Cells were seeded on 24-well plates (3×10^4 cells per well) and treated with various EEPP concentrations for 24 h. The medium was then removed, washed with phosphate-buffered saline (PBS), stained with 2 g/L crystal violet in phosphate-buffered formaldehyde for 20 min, and washed with water. The crystal violet bound to the cells was dissolved in 20 g/L SDS solution and its absorbance was measured at 600 nm.

3.4. Cell Cycle

After 12 h of exposure to 3.13–12.5 µg/mL EEPP, the medium was aspirated and adherent cells were harvested and centrifuged at 300× g for 5 min. Cells were washed with PBS, fixed with 700 mL/L ice-cold ethanol at −20 °C overnight, and then stained with PI at room temperature for 30 min. The cell-cycle distribution was analyzed by flow cytometry using an FACScan-LSR flow cytometer equipped with CellQuest software (BD Biosciences, San Jose, CA, USA) [26].

3.5. DNA Ladder

DLD-1 cells were treated with EEPP for 24 h; the cells were then harvested by scraping with a disposable cell lifter, suspended in PBS, and centrifuged for 10 min (250× g) at 4 °C, and the pellet was suspended in 0.1 mL of hypotonic lysing buffer (10 mM Tris, pH 7.4; 10 mM EDTA, pH 8.0; 0.5% Triton X-100). The cells were incubated for 10 min at 4 °C, and the resultant lysate was centrifuged for 30 min (13,000× g) at 4 °C. The supernatant, which contained fragmented DNA, was digested and incubated for 1 h at 37 °C with 5 mg/mL RNase A and then incubated for 1.5 h at 50 °C with 2.5 mg/mL proteinase K. DNA was precipitated with 0.5 volume equivalent of 10 M ammonium acetate and 2.5-fold volume equivalent of ethanol at −20 °C overnight. The precipitate was centrifuged at 13,000× g for 30 min at 4 °C. The resultant pellet was air-dried and resuspended in 10 mM Tris buffer (pH 7.4) containing 1 mM EDTA. An aliquot equivalent to 1×10^6 cells was electrophoresed at 50 V for 1 h in 1.5% agarose gel in 90 mM Tris-borate buffer containing 2 mM EDTA (pH 8.0). After electrophoresis, the gel was stained with ethidium bromide (0.5 µg/mL), and the nucleic acids were visualized with an ultraviolet transilluminator [27].

3.6. Western Blot

Cells were rinsed with ice-cold PBS and lysed by RIPA lysis buffer with protease and phosphatase inhibitors for 20 min on ice. Then, the cells were centrifuged at 12,000× g for 10 min at 4 °C. Protein extracts (20 µg) were resolved using SDS polyacrylamide gel electrophoresis (SDS-PAGE; 200 V, 45 min). The protein bands were electrotransferred to nitrocellulose membranes (80 V, 120 min). Membranes were then treated with a 5% enhanced chemiluminescence (ECL) blocking agent (GE Healthcare Bio-Sciences) in saline buffer (TBS-T) containing 0.1% Tween-20, 10 mM Tris-HCl, 150 mM NaCl, 1 mM $CaCl_2$, and 1 mM $MgCl_2$ at a pH of 7.4 for 1 h, and then incubated with the primary antibody overnight at 4 °C. Subsequently, membranes were washed three times in TBS-T and bound antibodies were detected using appropriate horseradish peroxidase-conjugated secondary antibodies, followed by analysis in an ECL plus Western blotting detection system (GE Healthcare Bio-Science) [28].

3.7. Method of Isolation and Identification of Active Compounds

Firstly, 50 g of *Paris polyphylla* was dissolved in 1 L of 100% ethanol and extracted. The extracts were then separated using an ODS (octadecylsilyl) column into different parts. After eluting with different concentrations of methanol, 80% methanol-treated parts were isolated and detected by HPLC. Pennogenin 3-O-beta-chacotrioside and polyphyllin VI were the two major active compounds in the extracts, identified by LC–MS and NMR.

3.8. Statistical Analysis

Results were expressed as means ± SD. Comparisons among groups were made using one-way ANOVA. The differences between mean values in all groups were tested through Duncan's multiple-range test (SPSS statistical software package, version 17.0, SPSS, Chicago, IL, USA). A *p*-value less than 0.05 was considered as a significant difference between means.

4. Conclusions

The present study demonstrated that EEPP induced autophagic cell death in colorectal cancer cells and that EEPP combined with Dox might exert a more potent anti-cancer effect against these tumor cells. We suggest that EEPP and its active ingredients pennogenin 3-*O*-beta-chacotrioside and polyphyllin VI could be further explored as potential candidates for the development of complementary chemotherapy against colorectal cancer.

Author Contributions: L.-T.L. and C.-J.T. performed the design for the overall study and analyzed the data. W.-C.U. performed most of the biochemical assays, and L.-T.L. revised the manuscript. C.-Y.C., B.-H.L., and C.-J.T. were involved in the experimental design and provided significant scientific suggestions and draft corrections before submission. The corresponding author C.-J.T. was responsible for financial resources and funds for the project, supervision of the research activities, and submission of the manuscript. The corresponding author C.-J.T. led the research group and drafted corrections.

Funding: This research work and subsidiary spending were supported by the Taipei Medical University and Taipei Medical University Hospital (108-TMU-TMUH-05), and the Ministry of Science and Technology (MOST-106-2320-B-038-017) (Taiwan, R.O.C.).

Conflicts of Interest: The authors declare no conflicts of interest.

Abbreviations

EEPP	ethanolic extract of *Paris polyphylla*
LC-3	light chain-3
TCM	traditional Chinese medicine
PCD	programmed cell death
Dox	doxorubicin

References

1. Man, S.; Gao, W.; Wei, C.; Liu, C. Anticancer drugs from traditional toxic Chinese medicines. *Phytother. Res.* **2012**, *26*, 1449–1465. [CrossRef]
2. Zhang, C.; Jia, X.; Bao, J.; Chen, S.; Wang, K.; Zhang, Y.; Li, P.; Wan, J.B.; Su, H.; Wang, Y.; et al. Polyphyllin VII induces apoptosis in Hep G2 cells through ROS-mediated mitochondrial dysfunction and MAPK pathways. *BMC Compl. Alt. Med.* **2016**, *16*, 58. [CrossRef]
3. Li, Y.; Gu, J.F.; Zou, X.; Wu, J.; Zhang, M.H.; Jiang, J. The anti-lung cancer activities of steroidal saponins of *P. polyphylla* Smith var. chinensis (Franch.) Hara through enhanced immunostimulation in experimental Lewis tumor-bearing C57BL/6 mice and induction of apoptosis in the A549 cell line. *Molecules* **2013**, *18*, 12916–12936. [CrossRef]
4. He, H.; Zheng, L.; Sun, Y.P.; Zhang, G.W.; Yue, Z.G. Steroidal saponins from *Paris polyphylla* suppress adhesion migration and invasion of human lung cancer A549 cells via down-regulating MMP-2 and MMP-9. *Asian Pac. J. Cancer Prev.* **2014**, *15*, 10911–10916. [CrossRef] [PubMed]
5. Li, Y.H.; Sun, Y.; Fan, L.; Zhang, F.; Meng, J.; Han, J. Paris saponin VII inhibits growth of colorectal cancer cells through Ras signaling pathway. *Biochem. Pharmacol.* **2014**, *88*, 150–157. [CrossRef] [PubMed]
6. Li, F.R.; Jiao, P.; Yao, S.T.; Sang, H.; Qin, S.C.; Zhang, W.; Zhang, Y.B.; Gao, L.L. *Paris polyphylla* Smith extract induces apoptosis and activates cancer suppressor gene connexin26 expression. *Asian Pac. J. Cancer Prev.* **2012**, *13*, 205–209. [CrossRef] [PubMed]
7. Wang, C.W.; Tai, C.J.; Choong, C.Y.; Lin, Y.C.; Lee, B.H.; Shi, Y.C.; Tai, C.J. Aqueous extract of *Paris polyphylla* (AEPP) inhibits ovarian cancer via suppression of peroxisome proliferator-activated receptor-gamma coactivator (PGC)-1alpha. *Molecules* **2016**, *21*, 727. [CrossRef]

8. Ernst, E.; Cassileth, B.C. The prevalence of complementary/alternative medicine in cancer: A systematic review. *Cancer* **1998**, *83*, 777–782. [CrossRef]
9. Patterson, R.E.; Neuhouser, M.L.; Hedderson, M.M. Types of alternative medicine used by patients with breast, colorectal, or prostate cancer: Predictors, motives, and costs. *J. Alternat. Complement. Med.* **2002**, *8*, 477–485. [CrossRef] [PubMed]
10. Richardson, M.A.; Sanders, T.; Palmer, J.L.; Greisinger, A.; Singletary, S.E. Complementary/alternative medicine use in a comprehensive cancer center and the implications for oncology. *J. Clin. Oncol.* **2000**, *18*, 2505–2514. [CrossRef]
11. Xu, W.; Towers, A.D.; Li, P.; Collet, J.P. Traditional Chinese medicine in cancer care: Perspectives and experiences of patients and professionals in China. *Eur. J. Cancer Care* **2006**, *15*, 397–403. [CrossRef] [PubMed]
12. Qi, F.; Li, A.; Inagaki, Y. Chinese herbal medicines as adjuvant treatment during chemo- or radio-therapy for cancer. *Biosci. Trends* **2010**, *4*, 297–307. [PubMed]
13. Youns, M.; Hoheisel, J.D.; Efferth, T. Traditional Chinese Medicines (TCMs) for molecular targeted therapies of tumours. *Curr. Drug Discov. Technol.* **2010**, *7*, 37–45. [CrossRef]
14. Coleman, M.P.; Quaresma, M.; Berrino, F. Cancer survival in five continents: A worldwide population-based study (CONCORD). *Lancet Oncol.* **2008**, *9*, 730–756. [CrossRef]
15. *Cancer Registry Annual Report*; Department of Health, Executive Yuan: Taiwan, 2009. Available online: https://www.mohw.gov.tw/lp-137-2.html (accessed on 23 May 2019).
16. De Dosso, S.; Sessa, C.; Saletti, P. Adjuvant therapy for colorectal cancer: Present and perspectives. *Cancer Treatment Rev.* **2009**, *35*, 160–166. [CrossRef]
17. Kopetz, S.; Freitas, D.; Calabrich, A.F.C.; Hoff, P.M. Adjuvant chemotherapy for stage II colorectal cancer. *Oncology* **2008**, *22*, 260–270.
18. Ye, M.X.; Zhao, Y.L.; Li, Y.; Miao, Q.; Li, Z.K.; Ren, X.L.; Song, L.Q.; Yin, H.; Zhang, J. Curcumin reverses cis-platin resistance and promotes human lung adenocarcinoma A549/DDP cell apoptosis through HIF-1α and caspase-3 mechanisms. *Phytomedicine* **2012**, *19*, 779–787. [CrossRef]
19. Klionsky, D.J.; Emr, S.D. Autophagy as a regulated pathway of cellular degradation. *Science* **2000**, *290*, 1717–1721. [CrossRef]
20. Mizushima, N.; Ohsumi, Y.; Yoshimori, T. Autophagosome formation in mammalian cells. *Cell Struct. Funct.* **2002**, *27*, 421–429. [CrossRef]
21. Liu, Y.L.; Yang, P.M.; Shun, C.T.; Wu, M.S.; Weng, J.R.; Chen, C.C. Autophagy potentiates the anti-cancer effects of the histone deacetylase inhibitors in hepatocellular carcinoma. *Autophagy* **2010**, *6*, 1057–1065. [CrossRef]
22. Nahimana, A.; Attinger, A.; Aubry, D.; Greaney, P.; Ireson, C.; Thougaard, A.V.; Tjørnelund, J.; Dawson, K.M.; Dupuis, M.; Duchosal, M.A. The NAD biosynthesis inhibitor APO866 has potent antitumor activity against hematologic malignancies. *Blood* **2009**, *113*, 3276–3286. [CrossRef]
23. Kim, J.Y.; Cho, T.J.; Woo, B.H.; Choi, K.U.; Lee, C.H.; Ryu, M.H.; Park, H.R. Curcumin-induced autophagy contributes to the decreased survival of oral cancer cells. *Arch. Oral Biol.* **2012**, *57*, 1018–1025. [CrossRef]
24. Tai, C.J.; Wang, C.K.; Tai, C.J.; Lin, Y.F.; Lin, C.S.; Jian, J.Y.; Chang, Y.J.; Chang, C.C. Aqueous extract of *Solanum nigrum* leaves induces autophagy and enhances cytotoxicity of cisplati, doxorubicin, docetaxel, and 5-fluorouracil in human colorectal carcinoma cells. *Evid. Based Complement. Alternat. Med.* **2013**, *2013*, 514719. [PubMed]
25. Cummings, J.; Smyth, J.F. DNA topoisomerase I and II as targets for rational design of new anticancer drugs. *Ann. Oncol.* **1993**, *4*, 533–543. [CrossRef] [PubMed]
26. Hsu, W.H.; Lee, B.H.; Pan, T.M. Red mold dioscorea-induced G2/M arrest and apoptosis in human oral cancer cells. *J. Sci. Food Agric.* **2010**, *90*, 2709–2715. [CrossRef] [PubMed]

27. Bushell, M.; Poncet, D.; Marissen, W.E.; Flotow, H.; Lloyd, R.E.; Clemens, M.J.; Morley, S.J. Cleavage of polypeptide chain initiation fctor eIF4GI during apoptosis in lymphoma cells: Characterization of an internal fragment generated by caspase-3-mediated cleavage. *Cell. Death Differ.* **2000**, *7*, 628–636. [CrossRef] [PubMed]
28. Hsu, W.H.; Lee, B.H.; Huang, Y.C.; Hsu, Y.W.; Pan, T.M. Ankaflavin, a novel Nrf-2 activator for attenuating allergic airway inflammation. *Free Radic. Biol. Med.* **2012**, *53*, 1643–1651. [CrossRef]

Sample Availability: Not available.

© 2019 by the authors. Licensee MDPI, Basel, Switzerland. This article is an open access article distributed under the terms and conditions of the Creative Commons Attribution (CC BY) license (http://creativecommons.org/licenses/by/4.0/).

Article

Cytostatic and Cytotoxic Natural Products against Cancer Cell Models

Taotao Ling [1], Walter H. Lang [1], Julie Maier [1], Marizza Quintana Centurion [2] and Fatima Rivas [1,*]

[1] Department of Chemical Biology and Therapeutics, St. Jude Children's Research Hospital. 262 Danny Thomas Place. Memphis, TN 38105-3678, USA; Taotao.Ling@STJUDE.ORG (T.L.); Walter.Lang@STJUDE.ORG (W.H.L.); Julie.Maier@STJUDE.ORG (J.M.)

[2] Dirección de Investigación Biológica/Museo Nacional de Historia Natural del Paraguay, Casilla de Correo 19.004. Sucursal 1, Campus UNA. 169 CDP San Lorenzo, Central XI, Paraguay; bmquintana2008@hotmail.com

* Correspondence: Fatima.rivas@stjude.org; Tel.: +1-901-595-6504; Fax: +1-901-595-5715

Academic Editor: Roberto Fabiani
Received: 29 April 2019; Accepted: 24 May 2019; Published: 26 May 2019

Abstract: The increasing prevalence of drug resistant and/or high-risk cancers indicate further drug discovery research is required to improve patient outcome. This study outlines a simplified approach to identify lead compounds from natural products against several cancer cell lines, and provides the basis to better understand structure activity relationship of the natural product cephalotaxine. Using high-throughput screening, a natural product library containing fractions and pure compounds was interrogated for proliferation inhibition in acute lymphoblastic leukemia cellular models (SUP-B15 and KOPN-8). Initial hits were verified in control and counter screens, and those with EC_{50} values ranging from nanomolar to low micromolar were further characterized via mass spectrometry, NMR, and cytotoxicity measurements. Most of the active compounds were alkaloid natural products including cephalotaxine and homoharringtonine, which were validated as protein synthesis inhibitors with significant potency against several cancer cell lines. A generated BODIPY-cephalotaxine probe provides insight into the mode of action of cephalotaxine and further rationale for its weaker potency when compared to homoharringtonine. The steroidal natural products (ecdysone and muristerone A) also showed modest biological activity and protein synthesis inhibition. Altogether, these findings demonstrate that natural products continue to provide insight into structure and function of molecules with therapeutic potential against drug resistant cancer cell models.

Keywords: natural product alkaloids; cephalotaxine; protein synthesis inhibition; antiproliferation agents; cancer

---

1. Introduction

Cancer is a complex chronic disease characterized by abnormal signaling processes that leads to aberrant cellular growth, causing premature death worldwide [1,2]. Natural products have a strong track record in the development of anti-cancer agents, thus many drug discovery programs continue to exploit this rich source of molecular scaffolds [3]. The re-emergence of natural products for drug discovery in the genomics era enables the use of advanced cellular models that recapitulate the disease of interest. Natural products, particularly alkaloids, are commonly used in ethnopharmacology and several are in clinical use (Figure 1).

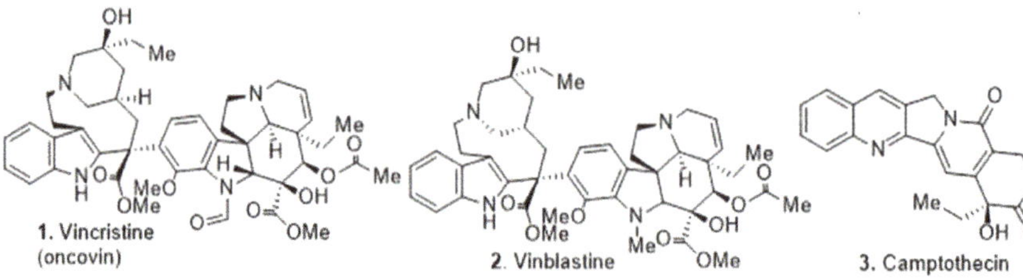

Figure 1. Natural product-derived alkaloids in clinical use.

Acute lymphoblastic leukemia (ALL) is among the most common pediatric cancers, occurring in approximately 1:1500 children [4]. Specific genetic aberrations define B cell precursor ALL subtypes with distinct biological and clinical characteristics. A class of genetic aberrations comprises tyrosine kinase-activating lesions, including translocations and rearrangements of tyrosine kinase and cytokine receptor genes such as the Philadelphia chromosome (Ph/BCR/ABL+) lesion among other genetic abnormalities leading to drug resistance. While response to high-risk drug treatment varies, drug resistance or disease recurrence are responsible for frequent causes of treatment failure [4–6].

The application of cellular screening systems and technological advancements in cell and molecular biology enable the chain of translatability by using disease-relevant cell models that display a more truthful phenotype, therefore aiding in the identification of new molecular scaffolds with clinical potential. Following this paradigm, the main focus of this study was to identify compound hits against high-risk cellular models of ALL (SUP-B15 and KOPN-8) while displaying therapeutic index (TI) when compared to normal cells (BJ and PBMCs), using an enriched library of natural product fractions [7–10] from a subgroup of alkaloid-producing plants (Annonaceae, Aristolochiaceae, Berberidaceae, Eupomatiaceae Fabaceae, Fumariaceae, Lauraceeae, Magnoliaceae, Monimiaceae, Nelumbonaceae, Papaveraceae, Ranunculaceae, Combretaceae, Rutaceae, Araliaceae, Apiaceae, Rubiaceae, and Araceae families), which were collected in collaboration with Museo Nacional de Historia Natural del Paraguay where several of these specimens were collected and have been deposited.

2. Results

2.1. Cytostatic/Cytotoxic Evaluation via Cell Proliferation Assay

A natural product library containing fractions (3K) and pure compounds (2K) were evaluated in a single point cell proliferation assay (CTG), as established in our group [11–13], and the Z' values were consistently higher than 0.5, revealing a large separation between positive and negative controls (Supplementary Material, Figures S1 and S2). From the single point primary screen, 22 compounds made the cutoff of 50% inhibition at 100 µM. Then, the most potent fractions were validated by dose-response CTG assay and the corresponding compounds elucidated by NMR and mass spectrometry (Figure 2). The biological activities of the pure natural products were confirmed and EC_{50} values are shown in Table 1 (Supplementary Material, Figures S3 and S4), providing several compounds with promising anti-proliferative effects against some of the high-risk cancer cell models disclosed in this work.

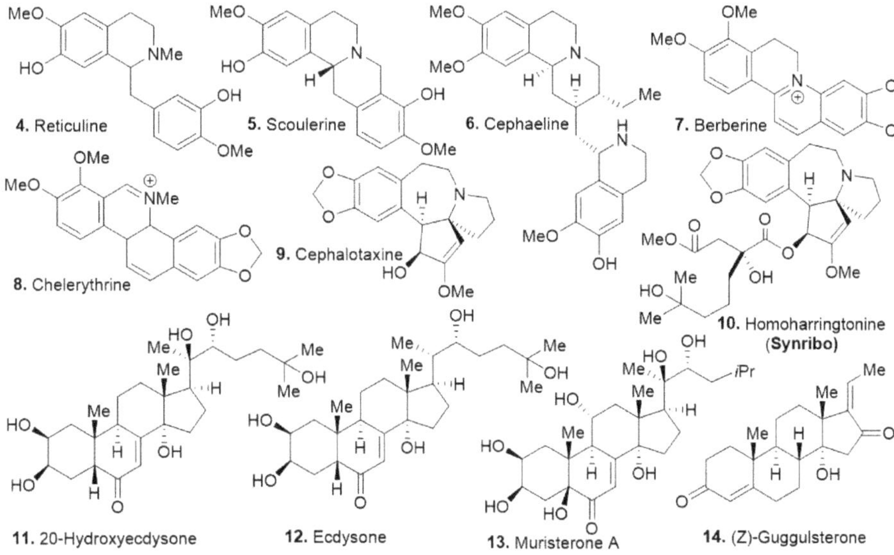

Figure 2. Identified natural products with cytostatic and cytotoxicity activity against ALL cell lines.

Table 1. CTG viability assay (CellTiter-Glo, 72 h) [1], which quantitates the amount of ATP present to determine cell viability. Hit compounds (**1–13**) were evaluated against several ALL cell lines and non-cancerous cell lines (BJ/PBMC cells) to determine therapeutic index.

	SUP-B15 (EC_{50} μM)	KOPN-8 (EC_{50} μM)	NALM-06 (EC_{50} μM)	UoC-B1 (EC_{50} μM)	BJ (EC_{50} μM)	PBMC (EC_{50} μM)	TI (PBMC/SUP-B15)
1	0.1201	0.41 ± 0.05	0.41 ± 0.15	0.0401 ± 0.01	>38.8788	>38.8788	>316
4	38 ± 6.1	32.1 ± 3.5	18.3 ± 2.5	>23.3766	>46.7532	>86.7532	>2.25
5	29 ± 3.7	32 ± 7.6	8.3832 ± 1.8	17.28 ± 3.8	>43.2900	>43.2900	>1.48
6	9.83 ± 0.5	0.1602 ± 0.05	18.45 ± 3.1	>21.6450	>43.2900	>43.2900	>4.3
7	6.05 ± 0.8	4.0732 ± 0.45	3.7229 ± 0.5	2.88 ± 1.1	>51.9481	28.0668	4.6
8	2.77 ± 0.36	0.8208 ± 0.22	1.9 ± 0.2	1.96 ± 0.28	29.8382	7.8914	2.8
9	10.43 ± 0.75	23.12 ± 5.2	9.5 ± 2.7	17.6 ± 5.8	>43.2900	>83	>7.7
10	0.045 ± 0.005	0.0799 ± 0.01	0.0322 ± 0.01	0.0129 ± 0.002	>43.2900	6.3499	140
11	29 ± 6	>43.2900	>43.2900	>21.6450	>43.2900	>43.2900	>1.48
12	20.1 ± 3.2	>43.2900	>43.2900	>21.6450	>43.2900	>43.2900	>2.15
13	9.5 ± 2.1	>43.2900	0.518 ± 0.15	>21.6450	>43.2900	>43.2900	>4.5
14	17.2 ± 4.5	>43.2900	>43.2900	>21.6450	>43.2900	>43.2900	>2.5

[1] Table represents mean ± SEM of triplicates. Analysis via Pipeline pilot or GraphPad Prism program.

The identified compounds exhibit half maximal effective concentration (EC_{50}) in the low micromolar range (<10 μM) against cancer cell lines, while displaying low or no cytotoxicity against non-cancerous cells (therapeutic index, TI > 5). Assessment in a broad range of non-cancerous tissue provides insightful information, thus BJ and PBMC cells were selected to determine TI (a higher TI is desired). Several leukemia cellular models were utilized for the viability assay to evaluate the scope of activity, but the study was particularly focused on a subset of ALL, namely KOPN-8 and SUP-B15 models, which carry specific genomic lesions. KOPN-8 carries the MLL-ENL fusion gene and SUP-B15 was established from a pediatric ALL relapsed patient (with the m-BCR ALL variant of the BCR-ABL1 fusion gene) [14].

The gene for the histone methyltransferase (MLL) participates in chromosomal translocations that eventually create MLL-fusion proteins associated with very aggressive forms of childhood acute leukemia, which serve as an independent dismal prognostic factor for this patient cohort [4–6]. Therefore, the identification of compounds against models with MLL gene fusions can provide insight into the discovery of new therapies. The chemical treatment response was similar for the ALL cell models, as shown in Table 1. Further evaluation demonstrated the most active compounds induced cell death in a concentration- dependent manner.

The identified compounds share properties with ample opportunity for improvement and further formulation to improve the compounds' physicochemical properties and suitability for specific delivery methods. While the compounds show similar antiproliferative activities, no distinctive structural similarities were recorded for the most potent compounds other than their shared alkaloid core between 4–10, and the steroidal core for compounds 11–14. Among these compounds, homoharringtonine (compound 10 [15–17]), a known potent protein synthesis inhibitor already approved for clinical use by the United States Food and Drug Administration against chronic myeloid leukemia (CML) [18–20] was identified. It shows high potency in the pre-B ALL cell models (SUP-B15 and KOPN-8) in agreement with previous reports [21,22]. However, it was not clear whether cephalotaxine (compound 9) worked via the same mechanism as compound 10. While the steroidal compounds (11–14) displayed much weaker potency, it was important to further evaluate their potential against these cell lines. Under these experimental conditions, compound 10 showed the best potency against SUP-B15, NALM-06, and UoC-B1 (a glucocorticoid resistant cell line) with an EC_{50} of 0.0452 µM, 0.0322 µM, and 0.0129 µM, respectively, with no observable activity against BJ at the highest tested concentration (43.3 µM). Overall, the natural product fractions library provided molecular scaffolds with specific cell death mechanisms that renders acceptable therapeutic index as observed herein.

2.2. Cell Cycle Arrest and Apoptosis

To determine if cell cycle progression was affected by the most active compounds 8–13, KOPN-8 and SUP-B15 cancer cells were investigated. Silvestrol, a protein synthesis inhibitor currently under clinical trials which exhibits significant cytotoxic activity against several human cancer cell lines such as oral carcinoma, melanoma, acute myelogenous leukemia, and cervical cancer with IC_{50} values in the low micromolar range [23–26], was used as a positive control.

Proliferating cells proceed through various phases of the cell cycle (G0, G1, S, G2, and M phase), and protein synthesis inhibitors can regulate cell cycle and cell proliferation. No serum starvation or phase-synchronization methods were used for these cells, in order to avoid secondary effects due to such manipulation [27,28]. The DMSO control indicates that both cell lines were primarily at the G0/G1 and S phase at the time of the experiment and only a small number (10%) were entering the G2/M phase (Supplementary Material, Figures S7 and S8). Silvestrol showed cell arrest in the G1/G0 phase of both cell lines. While compound 8 displayed a significant cell arrest in G1/G0 in KOPN-8, the cell arrest was more significantly increased in the percentage of cells (~20%) in S phase, with a significant reduction in the percentage of cells (~10%) in G2/M phase of SUP-B15. Interestingly, compound 10's profile was almost identical to silvestrol at the same concentration (5 µM) in both cell lines. However, cephalotaxine, which had shown antiproliferation effects by CTG at 10 µM, showed no significant cell arrest. Compounds 12 and 13 had minimal effect on the cell cycle in KOPN-8, with only a small effect in the S phase by compound 13. Conversely, compound 12 arrested G2/M with a significant reduction in the percentage of cells (~10%) in the S phase in the SUP-B15 cell line (Figure 3A, B). Similar results were obtained in colorectal and hepatocellular cancer cells treated with silvestrol, where most cells were stalled in the early stages of the cycle [26]. Since bypass of the G1 phase of the cycle due to DNA damage leads to apoptosis, to determine whether the compounds induce programmed cell death in KOPN-8 and SUP-B15, the cells were double stained with Annexin V and PI dyes to determine the percentage of cells in early vs late apoptosis and viable cells (Supplementary Material, Figures S5 and S6). As shown in Figure 3C, D, there is a significant decrease in live cells for compounds 8–10 in

KOPN-8 and SUP-B15. Most of the cells treated by compound **8** were not viable, while treatment with compound **9** only shows a small decrease in live cells (Annexin-, PI-), accompanied with an increase in apoptotic cells or dead cells during the 24 h treatment. Data presented on compounds **12** and **13** did not capture significant increase of late apoptotic state or cell death upon treatment for either cell line, but there was a small decrease in live cells by compound **12**. These results in combination with viability data (CTG) suggests that these steroidal compounds (**11–14**) are likely cytostatic agents and should be more effective in combination therapy. Data are shown from at least three independent experiments and statistical analysis of data was performed using Graph Pad Prism 7. The differences between the groups and negative control were analyzed by Tukey's test, with standard error bars representing the standard deviation of the mean (± SD).

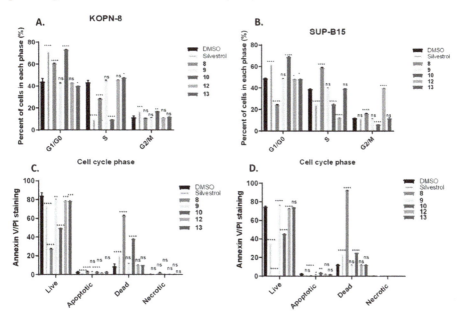

Figure 3. Analysis of apoptosis and cell cycle of compounds **8–10**, **12**, and **13** by Annexin V/Propidium Iodide (PI) flow cytometry assay after 24 h treatment using KOPN-8 and SUP-B15 cellular models. Negative control (DMSO), positive control (silvestrol, 5 µM), compounds **8** (5 µM), **9** (10 µM), **10** (5 µM), **12** (10 µM), **13** (10 µM). **A.** KOPN-8. **B.** SUP-B15. **C.** Annexin V/PI of KOPN-8. **D.** Annexin V/PI of SUP-B15. Bars depict mean and SD of at least three independent experiments. **** $p < 0.0001$, *** $p < 0.0004$, ** $p < 0.0086$, * $p < 0.025$ and ns (no statistical significant) according to Tukey's test when compared to DMSO control.

In addition to Annexin V staining for apoptosis induction, an alternate method of apoptosis detection method such as caspase 3/7 activity assay was performed. To validate these results, ApoTox-Glo triplex assay (Promega) was performed to determine viable cells (GF-AFC), apoptotic cells (caspase 3/7), or cytotoxicity by membrane integrity (bis-AAF-RF110) as an alternative cell death modality induced by compounds **9** and **10**. Both compounds increased caspase 3/7 activity, and decreased viability of SUP-B15 cells, further validating the cell death inducing effects of these compounds (Figure 4). The data indicate that compounds **9** and **10** might undergo similar cell death mechanisms, albeit at different concentrations, and suggests apoptosis is dependent on caspase activity in this cell line [29].

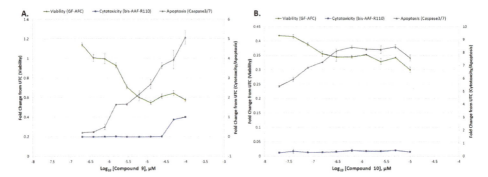

Figure 4. The ApoTox-Glo triplex assay against SUP-B15 to determine viable cells (using GF-AFC as fluorescent readout), apoptotic cells (caspase 3/7Glo as luminescent readout) or cytotoxicity by membrane integrity (bis-AAF-RF110 as fluorescent readout) upon compound treatment for 36 h. **A.** compound **9** (0.2–100 µM). **B.** compound **10** (0.02–10 µM). Graphs show decrease in cell viability, while increasing apoptotic activity with little cytotoxic effects by membrane integrity evaluation under the evaluated conditions (time and concentration).

To conduct live cell imaging studies (protein synthesis and co-localization studies), the adherent triple negative breast cellular models were selected as such studies are currently not feasible with suspension cells (leukemia cells). The first step was to evaluate if the compounds display cytotoxicity against the adherent cells. Propidium iodide (PI) assay was performed for the drug resistant solid tumor cellular models (triple negative breast cancer models: SUM149 and MDA-MB-231) for 48 h (Figure 5) [30,31]. All the tested compounds had better efficacy against SUM149 with promising activity at 12 µM, except for compound **14**, which displayed the weakest activity against the tested cancer cell lines. Compound **9** demonstrated 50% reduction of cell viability even at 3 µM concentration in SUM149, while only compounds **9** and **10** showed substantial activity in MDA-MB-231 cell line (drastically reducing cell viability at the highest concentration).

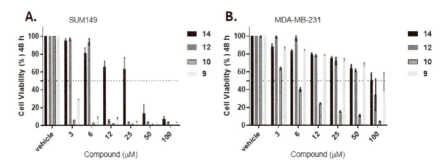

Figure 5. Propidium iodide (PI) assay of compounds **9–10**, **12** and **14** after 48 h treatment against breast cancer cell models to determine apoptotic effects. **A.** SUM-149, and **B.** MDA-MB-231. Bars represent mean ± SEM of at least three biological replicates.

2.3. Protein Synthesis Evaluation

Protein synthesis is essential in cell growth, proliferation, signaling, differentiation, and death [32]. To interrogate whether the compounds inhibited protein synthesis as nascent proteins are generated, protein synthesis levels were monitored using a commercially available kit (EZClick™ Global Protein Synthesis Assay Kit). The assay includes a robust chemical method based on an alkyne containing o-propargyl-puromycin probe, which stops translation by forming covalent conjugates with nascent polypeptide chains. Truncated polypeptides are rapidly turned over by the proteasome and can be detected based on the subsequent click reaction with a fluorescent azide [33]. To test whether these compounds (**4–14**) induced de novo protein synthesis inhibition, compounds were tested for a 2 h exposure time, under the same conditions as the known protein synthesis inhibitor cycloheximide (CHX) in MDA-MB-231 cell model following an in-cell-click de novo protein synthesis assay [33–35]. Partial protein synthesis inhibition was observed for compound **4** (Supplementary Material, Figure S9), and compounds **7–8** showed no inhibition (data not shown). The results show that compound **9** inhibits protein synthesis at 10 µM while compound **10** inhibits de novo protein synthesis at lower concentrations (1–5 µM, higher concentrations induce immediate cell detachment) under these experimental conditions (Figure 6A, B). Compounds **12** and **13** showed little to no protein synthesis inhibition, as shown in Figure 6E–G along with their relative quantification. The remaining compounds (**11** and **14**) showed no protein synthesis inhibition.

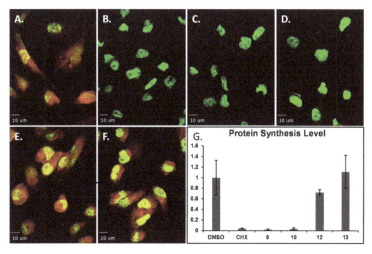

Figure 6. Representative images of protein synthesis inhibition EZClick™ assay using MDA-MB-231 cellular model. Cells were treated with compounds for 1.5 h prior to click reaction followed by staining. **A.** Vehicle (DMSO). **B.** Positive control **CHX** (1 µM). **C.** Compound **9** (5 µM). **D.** Compound **10** (10 µM). **E.** Compound **12** (10 µM). **F.** Compound **13** (10 µM). **G.** Relative quantification. Scale bar: 10 µm.

2.4. Probe Synthesis and Evaluation

Fluorescent labels are generally used in bio-orthogonal labelling for co-localization studies in order to better understand the compounds' mode of action [36]. Small-molecule fluorophores are the dominant method of choice due to their relative ease of use and excellent sensitivity, together with good spatial and temporal resolution. Washout studies of the corresponding BODIPY-FL ester have reliably shown that the reagent does not accumulate in the cell. Thus, 3-BODIPY-FL was treated with 2,4,6-trichlorobenzoyl chloride and Et$_3$N for 1 h in DCM, followed by addition of cephalotaxine along with DMAP in DCM for 16 h RT to provide probe **9a** in 87% overall yield as shown in Scheme 1 (Supplementary Material, Figures S11–S14).

It is important to determine the intracellular accumulation of compound **9** as this may aid in identifying its intracellular interactions. To facilitate these experiments, a series of orthogonal fluorescent organelle markers were used for co-localization studies. The organelle marker set included indicators for the lysosome (Lysotracker Red), the mitochondria (MitoTracker Deep Red), and the endoplasmic reticulum (ER tracker Blue/White). Where applicable, nuclei were stained with Hoechst (Blue). All these markers were compatible with BODIPY-FL (green) to allow for flexible combinations. The MDA-MB-231 adherent cellular model was used for the live cell co-localization studies as this cellular model displays well-defined organelle morphologies that can be consistently and positively identified [37]. First, to evaluate the accumulation of compound **9a** after 30 min treatment at 1 µM, washout experiments were conducted as shown in Figure 7A in the presence of Hoechst nuclear stain. Distinct green specks were observed near the nucleus of the cell. A similar observation was made using lysosome tracker (red, Figure 7B) and merging of both images clearly shows co-localization, observed as yellow (Figure 7C). Next MitoTracker-Deep Red (purple) was evaluated (Figure 7D) but no co-localization was detected with probe **9a** as seen by the absence of white staining in the merged image. This demonstrates that probe **9a** co-localizes with lysosomes, but not with mitochondria inside the cell (Figure 7E).

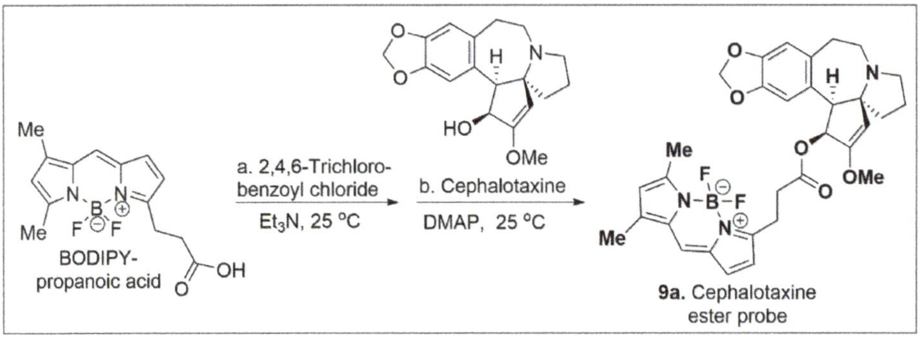

Scheme 1. Synthesis of compound **9a**. (a) 2.0 equiv Et$_3$N, 25 °C, 1 h; (b) 1. 0 equiv cephalotaxine 0.1 equiv DMAP, DCM, 25 °C, 16 h.

To gain further information regarding intracellular compound localization, studies were extended to include a marker for the ER (Figure 8). The staining pattern for the individual probes are shown in Figure 8A (ER), Figure 8B (Lysosome) and Figure 8D (probe **9a**). Co-localization of ER with probe **9a** (Figure 8C, yellow specks) and lysosome tracker was observed as seen in Figure 7. Co-localization of probe **9a** with ER-Tracker Blue/White (turquoise color in Figure 8E) was also observed. Moreover, when all three channels were merged, white specks were observed, indicating co-localization of probe **9a** with cellular structures that are stained by both ER and lysosomal marker dyes (Figure 8F). The data show either partial compound accumulation in the ER, while lysosomal structures are removing probe **9a** bound to the ER or active probe **9a** removal by lysosomal action before the compound can become active in the ER. These mechanisms provide a plausible explanation for the lower potency of compound **9** compared to compound **10**. Finally, in vitro ADME (absorption, distribution, metabolism, and elimination) profiles [11] were evaluated for compounds **9** and **10** to compare with the obtained cellular data (Supplementary Material, Figure S10). The aqueous solubility properties at physiological pH (7.40) of compounds **9** and **10** were good compared to controls. Simulated gastric fluid (SGF) stability was robust for compounds **9** and **10** with $t_{1/2} > 33$ h. Metabolic stability studies in mouse and human liver microsomes indicate that compound **9** is rapidly degraded, while compound **10** displays t1/2 of at least 1 h in both models. However, compound **9** showed remarkable stability in both mouse and human plasma stability assay ($t_{1/2} > 25$ h), while compound **10** showed poor stability in mouse plasma, but moderate stability in human plasma ($t_{1/2} < 1$ h, $t_{1/2} > 12$ h respectively). PAMPA assay indicated favorable permeability properties for both compounds, but caco-2 permeability assay suggests that compound **9** undergoes efflux (as the efflux ratio, B2A/A2B is closer to 2) while no efflux is suspected for compound **10** (B2A/A2B < 1) (Supplementary Material, Figure S10). The combined findings indicate that the steric-effects of the hydroxysuccinate appendage of compound **10** renders it less prone to degradation by lysosomal activity or other competing cellular removal mechanisms.

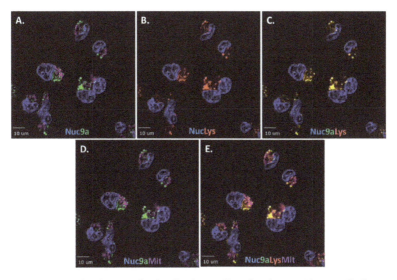

Figure 7. Representative images of co-localization studies of probe **9a** with organelle fluorescent trackers in live MDA-MB-231 cells. **A.** Hoechst nuclear stain (blue) and probe **9a** (1 µM, green). **B.** Lysosome tracker (red) and nuclear stain (blue). **C.** Merge of probe **9a**, lysosome and nuclear stains. **D.** MitoTracker Deep Red (purple), nuclear stain (blue) and compound **9a** (green). **E.** Nuclear stain, probe **9a**, Lysosome (red) and MitoTracker Deep Red. Scale bar: 10 µm.

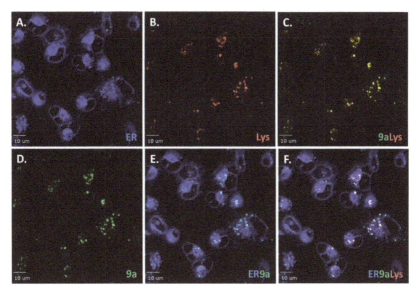

Figure 8. Representative images of co-localization studies of probe **9a** with organelle fluorescent stains in live cells. **A.** ER tracker Blue/White (blue). **B.** Lysosome tracker (red). **C.** Merged image of lysosomal and probe **9a** staining. **D.** Probe **9a** (2 µM, green). **E.** Merged image of ER tracker and probe **9a** staining. **F.** Merged image of ER tracker, Lysosome tracker and probe **9a** staining. Scale bar: 10 µm.

3. Discussion

The combined studies indicate that natural products continue to play an important role in drug discovery. Herein, several natural products were identified to have significant cytotoxicity against aggressive ALL cellular models by affecting cell cycle and inducing cell death via caspase activation, cell cycle arrest, and apoptosis.

Furthermore, this study highlights potential mechanistic processes responsible for the bioactive properties of cephalotaxine, compound **9**. Protein synthesis assays indicate that both compounds **9** and **10** inhibit protein synthesis, however compound **9** caused minimal cell death when compared to compound **10**. Compound **10** acts only on the initial step of protein translation and does not inhibit protein synthesis from mRNAs that have already commenced translation, unlike peptidyl transferase inhibitors such as cycloheximide (CHX), which inhibits peptide formation on mRNAs that are actively translated [34,35]. Previous medicinal chemistry campaigns [38–42] evaluating compound **10** had made observations that compound **9** was less potent than compound **10**. However, it was speculated that the elaborated ester side chain of compound **10** was required for interacting with the target. Co-crystallization studies of compound **10** with the large ribosomal subunit from H. marismortui, a member of the Halobacteriaceae family, have shown that protein translation is halted by preventing the initial elongation step of protein synthesis via interaction with the ribosomal A-site [42]. While both core and side chain contribute to the interactions, an important interaction of the side chain is its hydrophobic interaction with the base of U2506, which appears to lock the drug in its binding site in this model. The amine, which is protonated under physiological conditions forms a hydrogen bond with the carbonyl of C2452 [42]. This interaction should be feasible for compound **9**. However, a synthesized BODIPY-cephalotaxine probe **9a** demonstrates that while cephalotaxine (compound **9**) localized to the ER to inhibit protein synthesis, the probe **9a** also co-localized with the lysosome. Thus, this provides a potential mechanism for rapid removal of compound **9** via lysosome, which would render it less effective at inducing cell death.

4. Materials and Methods

4.1. Cell Culture

All human cell lines were incubated at 37 °C in a 5% CO_2 atmosphere and maintained under sterile conditions [43]. Cells were tested for Mycoplasma (Lonza, Alpharetta, GA, USA) using the manufacturer's conditions prior to experiments. Cell lines were purchased from American Type Culture Collection (ATCC, Manassas, VA, USA) or Leibniz-Institute Deutsche Sammlung von Mikroorganismen und Zellkulturen GmbH (DSMZ, Braunschweig, Germany) and cultured without antibiotics unless stated. Leukemia cells were cultured in RPMI and supplemented with 10% fetal bovine serum (FBS, Hyclone, Logan, UT, USA). SUP-B15, and KOPN-8 lines were cultured in RPMI supplemented with 10% FBS (Hyclone), 1% GlutaMAX™, 1% Pen/Step, and 0.1% β-mercaptoethanol. BJ cells were cultured in EMEM media supplemented with 10% FBS (Hyclone). Breast cancer cells (MDA-MB-231 & SUM149) were cultured in DMEM and Ham's F12 respectively, supplemented with 10% FBS, 1% GlutaMAX™, 1% Pen/Step, and 2 µM cortisol/1 µg/mL insulin for SUM149. Both cells were grown to 80% confluence densities as recommended by ATCC. PBMCs were supplemented with concanavalin-A (5 µg/mL) and IL-2 (50 U/ml). To test general cytotoxicity, the following cell numbers were used for 384 well plates: KOPN-8 (ACC 552, infant human B cell precursor acute lymphoblastic leukemia with MLL-MLLt1/ENL fusion, 1000 cells/well), SUP-B15 (ACC389, human B cell precursor acute lymphoblastic leukemia of pediatric second relapse carrying the ALL-variant (m-bcr) of BCR-ABL1 fusion gene (e1-a2), 1600 cells/well), NALM-06 (DSMZ ACC128, non-T/non-B ALL at relapse with P15INK4B and P16INK4A deletions, 1200 cells/well), UoC-B1 (pediatric BCP-ALL at second relapse with TCF3/E2A-HFL fusion, 1000 cells/well), BJ (CRL-2522, normal human foreskin fibroblast cells, 400 cells/well) and PBMCs (periphery blood monocyte cells from healthy donors, IBC#BDC046, 10,000 cells/well).

4.2. Natural Product Compound Library

A 5000-fraction library was assembled from 48,000 natural product fractions and 2000 pure natural products in the collection of St. Jude Children's Research Hospital which includes bioactive natural products purchased from Sigma-Aldrich (St. Louis, MO, USA), ChromaDex (Longmont, CO, USA), Med Chem Express (Monmouth Junction, NJ, USA), and ChemBridge (San Diego, CA, USA). All fractions were dissolved in DMSO at 2 mM based on mass spectrometry, arrayed in 384 polypropylene plates and stored at $-20\ °C$.

4.3. CellTiter-Glo Viability Assay (CTG)

Cytotoxicity evaluation was performed using the CellTiter-Glo Luminescent Cell Viability Assay kit (G7570, Promega, Madison, WI, USA), according to the manufacturer's instructions. Briefly, the cell concentrations used were experimentally determined to ensure logarithmic growth during the 72-h duration of the experiment, and avoid adverse effects on cell growth by DMSO exposure. 1×10^3–4.8×10^3 or 4×10^2–1.2×10^3 cells/well were seeded in 96 or 384-well white flat-bottomed plates (3610 or 8804BC, Corning, Corning, NY, USA) in 100 µL or 30 µL/well, respectively. The plates were incubated at 37 °C in 5% CO_2 for 24 h before drugging. Test compounds (10 mM in DMSO) in nine 3-fold serial dilutions were dispensed via pintool (Biomek liquid handler, Beckman, Indianapolis, USA) to assay plates. The final concentration of DMSO was 0.3% (v/v) in each well. The positive controls included staurosporine (10 µM), and gambogic acid (10 µM). The plates were incubated for 72 h at 37 °C in 5% CO_2, then quenched with CellTiter-Glo® (Promega, Madison, WI, USA, 50 µL/96 or 30 µL/384), centrifuged at 1000 rpm for 1 min and incubated at RT for 20 min. Luminescence was recorded with a plate reader (Envision, Perkin Elmer, Waltham, MA, USA). The mean luminescence of each experimental treatment group was normalized as a percentage of the mean intensity of untreated controls. EC_{50} values were calculated by Pipeline Pilot software (Accelrys, Enterprise Platform, San Diego, CA, USA).

4.4. Annexin V-FITC Apoptosis and Cell Cycle

The samples were probed with AnnexinV-FITC (Roche/Boehringer Mannheim, Indianapolis, IN, USA) according to the manufacturer's instructions. KOPN-8 or SUB-P15 cells were plated (1.00×10^6 cells/plate) and incubated for 12 h and incubated at 37 °C. Then, cells were treated with compounds or controls for 24 h. Cells were stained with AnnexinV-FITC, PI and the staining profiles were determined with FACScan and Cell-Quest software. For cellular DNA content, the same cell treatment as above was performed. Then, cells were fixed in cold 75% ethanol, treated with RNase and then stained with PI solution (50 µg/mL). Cell cycle distribution was analyzed with the FACSCalibur analyzer (BD Biosciences, Franklin Lakes, NJ, USA) and Cell-Quest software. The percentage of DNA content at different phases of the cell cycle was analyzed with ModFit-software (version 5.0, Verity Software House, Topsham, ME, USA).

4.5. ApoTox-Glo^{TM} Triplex Assay

Cells (9.5×10^4 cells/well in 75 µL) were dispensed in 96-well black flat bottom (8807BC, Corning) plates. The cells were incubated for 12 h at 37 °C and treated with compounds (25 µL) for 36 h. DMSO was used as a negative control and camptothecan was used as the positive control. Then, the experiment was stopped by adding the Viability/Cytotoxicity Reagent, and briefly mixed by orbital shaking (300–500 rpm for ~30 s). Plates were incubated for 30 min at 37 °C and fluorescence was measured at the wavelength 400Ex/505Em for viability and 485Ex/520Em for cytotoxicity in an Envision plate reader (Perkin Elmer). Finally, the Caspase-Glo®3/7 Reagent was added, and the plates were mixed by orbital shaking (300–500 rpm for ~30 s), followed by an additional 30 min incubation at RT. Luminescence was recorded with a plate reader (Envision, Perkin Elmer) to capture caspase 3/7 activation and determine apoptosis induction.

4.6. Protein Synthesis in Cell-Click Assay

Biovision's EZClick™ protein synthesis monitoring assay kit (EZClick™ Global Protein Synthesis Assay Kit, Catalog # K715-100, Milpitas, CA, USA) was used according to manufacturer's protocol. The assay was conducted in 8-well chambered coverslips (ibidi GmbH µ-slide # 80826, Martinsried, Germany). Cells were plated at 2×10^4 per well and incubated at 37 °C for 12 h. Then, the cells were treated with compounds for 1.5 h and processed and stained according to protocol (representative images shown in Figure 4). Protein synthesis activity is shown in red, DNA staining is shown in green. Cycloheximide (CHX) was used as a positive control and DMSO as the negative control. Images shown were taken with a Marianas CSU-X spinning disk confocal imaging system (3i, Denver, CO, USA) configured with a Zeiss Axio Observer microscope (Carl Zeiss Inc., Thornwood, NY, USA) with diode lasers, at 63× magnification and resolution of 512 × 512 pixels (3i). For each field of view, an image stack consisting of 20 optical sections were taken at 63× magnification. The total intensity of red and green staining was quantified for the entire image stack. Six fields of view were analyzed per condition. Images shown represent a single optical section for each field of view. For each stack, the red intensity (protein synthesis) was adjusted to the green intensity (DNA content) in each image. The numbers from six images were used to calculate the average and standard deviation for each condition. Numbers were normalized to the negative control value arbitrarily set to one. The Slide viewer 6 software package was used for rendering and analysis of the images.

*4.7. Probe **9a** Evaluation*

Co-localization studies were conducted in 8-well chambered coverslips (ibidi GmbH µ-slide # 80826) for MDA-MB-231. Coverslips were first coated with 0.1% Gelatin for 30 min for better cell adherence. Cells were then plated in phenol red free medium at a density of 4×10^4 cells per well and incubated at 37 °C overnight. Then the cells were treated with 1–10 µM of **9a** probe and/or organelle tracker for 1 h at the following final concentrations: 1 µM ER-Tracker™ Blue-White DPX (E12353, Invitrogen); 100 nM LysoTracker™ Red (L7528, Invitrogen, Carlsbad, CA, USA); 250 nM MitoTracker™ Deep Red FM (M22426, Invitrogen), Invitrogen). Thirty minutes after addition of the probe and tracking dyes, nuclear stain Hoechst33342 (H3570, Invitrogen) was added to samples not stained with ER-Tracker Blue/White at a final concentration of 500 nM. After incubation for a total of 1 h, cells were washed twice with fresh medium, followed by live imaging on a Marianas CSU-X spinning disk confocal imaging system configured with a Zeiss Axio Observer microscope with diode lasers, at 63× magnification and resolution of 512 × 512 pixels (3i).

4.8. Statistical Analysis

Statistical analysis of data was performed using GraphPad Prism (Version 7.0 San Diego, CA, USA) and Microsoft Excel software (Office 2010, Microsoft Corp., Redmond, WA, USA). The statistical methods used were repeated-measures analysis of variance and Tukey's test for paired data when appropriate; a p value less than 0.025 was considered statistically significant and no statistical significant was depicted by ns. Standard error bars represented the standard deviation of the mean ($\pm$ SD).

5. Conclusions

In summary our study demonstrates that natural products continue to provide potential hits against aggressive high risk ALL. A screen of a focused natural product fractions library identified several active alkaloids and steroidal compounds against high-risk ALL cellular models. The alkaloids **8**, **9**, and **10** (chelerythrine, cephalotaxine, homoharringtonine) and steroidal compounds **12** and **13** (ecdysone, and muristerone A) were validated by viability and apoptotic assays as cytotoxic and cytostatic agents, respectively. Our studies indicate these natural products (compound **8–13**) have promising therapeutic index in normal cells (BJ and PBMCs), therefore further development is warranted to improve their efficacy against these cancer subtype cell lines. Solid tumor cell models (breast cancer) were utilized for microscopy studies (not feasible in suspension cell lines) to better understand the properties of compounds **9** and **10**. Our study shows that both compounds **9** and **10** inhibit protein synthesis, but compound **10** is more efficacious at inducing cell death. By synthesizing a tool compound, probe **9a**, we demonstrate for the first time that compound **9** is rapidly removed via lysosome activity, limiting its cytotoxic effects in the cell. Thus, providing insight into the mechanism of these alkaloid compounds, and highlighting potential sites for future derivatization to improve the activity/subcellular stability of this family of natural products against high risk ALL models.

Supplementary Materials: The supplementary materials are available online.

Author Contributions: Conceptualization, F.R. and T.L.; methodology, T.L. and W.H.L.; Data analysis, J.M.; and plant data curation, F.R and M.Q.C. All authors contributed to the writing—review and editing of the manuscript.

Funding: This research was funded by ALSAC St Jude Children's Research Hospital and the St Jude imaging center is supported in part by the Cancer Center Support Grant (P30CA021765) from the National Cancer Institute.

Acknowledgments: The authors thank the Analytical Technologies Center Core Facility, the Flow and Cell Cycle Facility and the Cell and Tissue Imaging Center for technical assistance at St. Jude Children's Research Hospital.

Conflicts of Interest: The authors declare no conflict of interest.

References

1. Newman, D.J.; Cragg, G.M. Natural Products as Sources of New Drugs from 1981 to 2014. *J. Nat. Prod.* **2016**, *79*, 629–661. [CrossRef] [PubMed]
2. Gordon, M.; Cragg, D.; Kingston, G.I.; David, M.N. *Anticancer Agents from Natural Products*, 2nd ed.; CRC Press/Taylor & Francis Group: Boca Raton, FL, USA, 2012.
3. Harvey, A.L.; Edrada-Ebel, R.; Quinn, R.J. The re-emergence of natural products for drug discovery in the genomics era. *Nat. Rev. Drug. Discov.* **2015**, *14*, 111–129. [CrossRef] [PubMed]
4. Andersson, A.K.; Ma, J.; Wang, J.; Chen, X.; Gedman, A.L.; Dang, J.; Nakitandwe, J.; Holmfeldt, L.; Parker, M.; Easton, J.; et al. Jude Children's Research Hospital–Washington University Pediatric Cancer Genome Project. *Nat. Genet.* **2015**, *47*, 330–337. [CrossRef] [PubMed]
5. Mullighan, C.G.; Miller, C.B.; Radtke, I.; Phillips, L.A.; Dalton, J.; Ma, J.; White, D.; Hughes, T.P.; Le Beau, M.M.; Pui, C.-H.; et al. BCR-ABL1 Lymphoblastic Leukaemia is Characterized by the Deletion of Ikaros. *Nature.* **2008**, *453*, 110–114. [CrossRef] [PubMed]
6. Webersinke, H.R.G. Molecular Pathogenesis of Philadelphia-positive Chronic Myeloid Leukemia–Is It all BCR-ABL? *Curr. Cancer Drug Tar.* **2011**, *11*, 3–19.
7. Wassermann, A.M.; Lounkine, E.; Hoepfner, D.; Le Goff, G.; King, F.J.; Studer, C.; Peltier, J.M.; Grippo, M.L.; Prindle, V.; Tao, J.; et al. Dark Chemical Matter as a Promising Starting point for Drug Lead Discovery. *Nat. Chem. Biol.* **2015**, *11*, 958–966. [CrossRef]
8. Jones, L.H.; Bunnage, M.E. Applications of Chemogenomic Library Screening in Drug Discovery. *Nat. Rev. Drug Discov.* **2017**, *16*, 285–296. [CrossRef]
9. Gezici, S.; Şekeroğlu, N. Current Perspectives in the Application of Medicinal Plants Against Cancer: Novel Therapeutic Agents. *Anticancer Agents Med Chem.* **2018**. [CrossRef] [PubMed]
10. He, C.Y.; Fu, J.; Shou, J.W.; Zhao, Z.X.; Ren, L.; Wang, Y.; Jiang, J.D. In Vitro Study of the Metabolic Characteristics of Eight Isoquinoline Alkaloids from Natural Plants in Rat Gut Microbiota. *Molecules* **2017**, *22*, 932. [CrossRef] [PubMed]
11. Hadi, V.; Hotard, M.; Ling, T.; Salinas, Y.G.; Palacios, G.; Connelly, M.; Rivas, F. Evaluation of Jatropha Isabelli Natural Products and their Synthetic Analogs as Potential Antimalarial Therapeutic Agents. *Eur. J. Med. Chem.* **2013**, *65*, 376–380. [CrossRef]
12. Mitachi, K.; Salinas, Y.G.; Connelly, M.; Jensen, N.; Ling, T.; Rivas, F. Synthesis and Structure-Activity Relationship of Disubstituted Benzamides as a Novel Class of Antimalarial Agents. *Bioorganic Med. Chem. Lett.* **2012**, *22*, 4536–4539. [CrossRef] [PubMed]
13. Ling, T.; Lang, W.; Feng, X.; Das, S.; Maier, J.; Jeffries, C.; Shelat, A.; Rivas, F. Novel Vitexin-Inspired Scaffold Against Leukemia. *Eur. J. Med. Chem.* **2018**, *146*, 501–510. [CrossRef]

14. Barretina, J.G.; Caponigro, G.; Stransky, N.; Venkatesan, K.; Margolin, A.A.; Kim, S.; Wilson, C.J.; Lehar, J.; Kryukov, G.V.; Sonkin, D.; et al. The Cancer Cell Line Encyclopedia Enables Predictive Modelling of Anticancer Drug Sensitivity. *Nature.* **2012**, *483*, 603–607. [CrossRef]
15. Powell, R.G.; Weisleder, D.; Smith, C.R. Jr.; Rohwedder, W.K. Structures of Harringtonine, Isoharringtonine, and Homoharringtonine. *Tetrahedron Lett.* **1970**, *11*, 815–818. [CrossRef]
16. Takeda, S.; Yajima, N.; Kitazato, K.; Unemi, N. Antitumor Activities of Harringtonine and Homoharringtonine, Cephalotaxus Alkaloids which are Active Principles from Plant by Intraperitoneal and Oral Administration. *J. Pharmacobiodyn.* **1982**, *5*, 841–847. [CrossRef]
17. Abdelkafi, H.; Nay, B. Natural Products from Cephalotaxus sp.: Chemical Diversity and Synthetic Aspects. *Nat. Prod. Rep.* **2012**, *29*, 845–869. [CrossRef]
18. Berman, E. Omacetaxine: The FDA Decision. *Clin Adv Hematol Oncol.* **2011**, *9*, 57–58. [PubMed]
19. Chen, X.; Tang, Y.; Chen, J.; Chen, R.; Gu, L.; Xue, H.; Pan, C.; Tang, J.; Shen, S. Homoharringtonine is a Safe and Effective Substitute for Anthracyclines in Children Younger than 2 Years Old with Acute Myeloid Leukemia. *Front Med.* **2019**. [CrossRef] [PubMed]
20. Ju, X.; Beaudry, C.M. Total Synthesis of (-)-Cephalotaxine and (-)-Homoharringtonine via Furan Oxidation-Transannular Mannich Cyclization. *Angew. Chem. Int. Ed. Engl.* **2019**. [CrossRef]
21. Min, X.; Na, Z.; Yanan, L.; Chunrui, L. *De Novo* Acute Megakaryoblastic Leukemia with p210 BCR/ABL and t(1;16) Translocation but not t(9;22) Ph Chromosome. *J. Hematol. Oncol.* **2011**, *4*, 45. [CrossRef]
22. Jiang, T.L.; Liu, R.H.; Salmon, S.E. Comparative in vitro Antitumor Activity of Homoharringtonine and Harringtonine against Clonogenic Human Tumor Cells. *Invest. New. Drugs.* **1983**, *1*, 21–25. [CrossRef] [PubMed]
23. Hwang, B.Y.; Su, B.N.; Chai, H.; Mi, Q.; Kardono, L.B.; Afriastini, J.J.; Riswan, S.; Santarsiero, B.D.; Mesecar, A.D.; Wild, R.; et al. Silvestrol and Episilvestrol, Potential Anticancer Rocaglate Derivatives from Aglaia Silvestris. *J. Org. Chem.* **2004**, *69*, 3350–3358. [CrossRef] [PubMed]
24. Rodrigo, C.M.; Cencic, R.; Roche, S.P.; Pelletier, J.; Porco, J.A. Synthesis of Rocaglamide Hydroxamates and Related Compounds as Eukaryotic Translation Inhibitors: Synthetic and Biological Studies. *J. Med. Chem.* **2012**, *55*, 558–562. [CrossRef]
25. Kogure, T.; Kinghorn, A.D.; Yan, I.; Bolon, B.; Lucas, D.M.; Grever, M.R.; Patel, T. Therapeutic Potential of the Translation Inhibitor Silvestrol in Hepatocellular Cancer. *PLoS ONE.* **2013**, *8*, e76136. [CrossRef] [PubMed]
26. Lucas, D.M.; Edwards, R.B.; Lozanski, G.; West, D.A.; Shin, J.D.; Vargo, M.A.; Davis, M.E.; Rozewski, D.M.; Johnson, A.J.; Su, B.N.; et al. The Novel Plant-derived agent Silvestrol has B-Cell Selective Activity in Chronic Lymphocytic Leukemia and Acute Lymphoblastic Leukemia in Vitro and *in Vivo*. *Blood.* **2009**, *113*, 4656–4666. [CrossRef]
27. Kim, U.; Shu, C.W.; Dane, K.Y.; Daugherty, P.S.; Wang, J.Y.; Soh, H.T. Selection of Mammalian Cells Based on their Cell-cycle Phase using Dielectrophoresis. *Proc. Natl. Acad. Sci. USA* **2007**, *104*, 20708–20712. [CrossRef]
28. Crowley, L.C.; Marfell, B.J.; Scott, A.P.; Waterhouse, N.J. Quantitation of Apoptosis and Necrosis by Annexin V Binding, Propidium Iodide Uptake, and Flow Cytometry. *Cold Spring Harb. Protoc.* **2016**. [CrossRef]
29. Shalini, S.; Dorstyn, L.; Dawar, S.; Kumar, S. Old, New and Emerging Functions of Caspases. *Cell Death Differ.* **2015**, *22*, 526–539. [CrossRef]
30. Kastan, M.B.; Bartek, J. Cell-cycle Checkpoints and Cancer. *Nature.* **2004**, *432*, 316–323. [CrossRef] [PubMed]
31. Riccardi, C.; Nicoletti, I. Analysis of Apoptosis by Propidium Iodide Staining and Flow Cytometry. *Nat. Protoc.* **2006**, *1*, 1458–1461. [CrossRef]
32. Joazeiro, C.A.P. Mechanisms and Functions of Ribosome-Associated Protein Quality Control. *Nat. Rev. Mol. Cell. Biol.* **2019**. [CrossRef] [PubMed]
33. Liu, J.; Xu, Y.; Stoleru, D.; Salic, A. Imaging Protein Synthesis in Cells and Tissues with an Alkyne Analog of Puromycin. *Proc. Natl. Acad. Sci. USA* **2012**, *109*, 413–418. [CrossRef]
34. Cuyàs, E.; Martin-Castillo, B.; Corominas-Faja, B.; Massaguer, A.; Bosch-Barrera, J.; Menendez, J.A. Anti-Protozoal and Anti-Bacterial Antibiotics that Inhibit Protein Synthesis Kill Cancer Subtypes Enriched for Stem Cell-like Properties. *Cell Cycle.* **2015**, *14*, 3527–3532. [CrossRef] [PubMed]
35. Burger, K.; Mühl, B.; Harasim, T.; Rohrmoser, M.; Malamoussi, A.; Orban, M.; Kellner, M.; Gruber-Eber, A.; Kremmer, E.; Hölzel, M.; et al. Chemotherapeutic Drugs Inhibit Ribosome Biogenesis at Various Levels. *J. Biol. Chem.* **2010**, *285*, 12416–12425. [CrossRef]

36. Zhu, H.; Fan, J.; Du, J.; Peng, X. Fluorescent Probes for Sensing and Imaging within Specific Cellular Organelles. *Acc. Chem. Res.* **2016**, *49*, 2115–2126. [CrossRef]
37. Bhute, V.J.; Ma, Y.; Bao, X.; Palecek, S.P. The Poly (ADP-Ribose) Polymerase Inhibitor Veliparib and Radiation Cause Significant Cell Line Dependent Metabolic Changes in Breast Cancer Cells. *Sci. Rep.* **2016**, *6*, 36061. [CrossRef]
38. Smith, C.R.; Powell, R.G.; Suffness, M. Development of Homoharringtonine. *J. Clin. Oncol.* **1986**, *4*, 1283. [CrossRef]
39. Zhong, S.B.; Liu, W.C.; Li, R.L.; Ling, Y.Z.; Li, C.H.; Tu, G.Z.; Ma, L.B.; Hong, S.L. Studies on Semi-synthesis of Cephalotaxine Esters and Correlation of their Structures with Antitumor Activity. *Yao Xue Xue Bao.* **1994**, *29*, 33–38. [PubMed]
40. Fresno, M.; Jiménez, A.; Vázquez, D. Inhibition of Translation in Eukaryotic Systems by Harringtonine. *Eur. J. Biochem.* **1977**, *72*, 323–330. [CrossRef] [PubMed]
41. Kim, J.E.; Song, Y.J. Anti-varicella-zoster Virus Activity of Cephalotaxine Esters *in Vitro*. *J. Microbiol.* **2019**, *57*, 74–79. [CrossRef]
42. Gürel, G.; Blaha, G.; Moore, P.B.; Steitz, T.A. U2504 Determines the Species Specificity of the A-site Cleft Antibiotics: The Structures of Tiamulin, Homoharringtonine, and Bruceantin Bound to the Ribosome. *J. Mol. Biol.* **2009**, *389*, 46–56. [CrossRef] [PubMed]
43. Hay, R.J.; Caputo, J.L.; Macy, M.L. *ATCC Quality Control Methods for Cell Lines*, 2nd ed.; ATCC: Manassas, VA, USA, 1992.

Sample Availability: Samples of the compounds are available from the authors.

© 2019 by the authors. Licensee MDPI, Basel, Switzerland. This article is an open access article distributed under the terms and conditions of the Creative Commons Attribution (CC BY) license (http://creativecommons.org/licenses/by/4.0/).

Article

Effect of Cu/Mn-Fortification on In Vitro Activities of the Peptic Hydrolysate of Bovine Lactoferrin against Human Gastric Cancer BGC-823 Cells

Li-Ying Bo [1], Tie-Jing Li [1,*] and Xin-Huai Zhao [2,*]

[1] Key Laboratory of Dairy Science, Ministry of Education, Northeast Agricultural University, Harbin 150030, China; Boliying1746@126.com
[2] Department of Food Science, Northeast Agricultural University, Harbin 150030, China
* Correspondence: tiejingli@163.com (T.-J.L.); zhaoxh@neau.edu.cn or xhzhao63@sina.com.cn (X.-H.Z.); Tel.: +86-451-5519-1813 (X.-H.Z.)

Academic Editor: Roberto Fabiani
Received: 24 February 2019; Accepted: 23 March 2019; Published: 27 March 2019

Abstract: Bovine lactoferrin hydrolysate (BLH) was prepared with pepsin, fortified with Cu^{2+} (Mn^{2+}) 0.64 and 1.28 (0.28 and 0.56) mg/g protein, and then assessed for their activity against human gastric cancer BGC-823 cells. BLH and the four fortified BLH products dose- and time-dependently had growth inhibition on the cells in both short- and long-time experiments. These samples at dose level of 25 mg/mL could stop cell-cycle progression at the G0/G1-phase, damage mitochondrial membrane, and induce cell apoptosis. In total, the fortified BLH products had higher activities in the cells than BLH alone. Moreover, higher Cu/Mn fortification level brought higher effects, and Mn was more effective than Cu to increase these effects. In the treated cells, the apoptosis-related proteins such as Bad, Bax, p53, cytochrome c, caspase-3, and caspase-9 were up-regulated, while Bcl-2 was down-regulated. Caspase-3 activation was also evidenced using a caspase-3 inhibitor, z-VAD-fmk. Thus, Cu- and especially Mn-fortification of BLH brought health benefits such as increased anti-cancer activity in the BGC-823 cells via activating the apoptosis-related proteins to induce cell apoptosis.

Keywords: lactoferrin hydrolysate; copper; manganese; gastric cancer cells; anti-cancer activity; molecular mechanism

1. Introduction

Dietary proteins provide both essential amino acids and energy for the body, and also have several health benefits by the release of so-called bioactive peptides [1], because these peptides have various physiological functions such as anti-cancer, anti-hypertensive, anti-oxidant, mineral-binding, and other effects [2,3]. The solid fraction from yogurt exerts growth inhibition on initial tumor cells, while the peptide fraction from algae protein has anti-cancer activity against the gastric cancer AGS cells through arresting the cells in the post-G1-phase [4,5]. An important Fe-binding protein lactoferrin (LF) and its derivatives have also been assessed for their bio-activities. LF and a LF derivative lactoferricin B have anti-cancer activities in the gastric cancer SGC-7901, AGS cells and oral squamous cell carcinoma [6–8]. Lactoferricin B is also well-known for its anti-bacterial effect against a wide variety of Gram-positive and Gram-negative bacteria [9,10]. From a chemical point of view, proteins have various functional groups such as −OH, −SH, −NH, etc., and thus can interact with some macro-elements and trace elements, resulting in changed nutritive values and bio-activities. For LF, Cu supplementation increases immuno-modulation in both murine splenocytes and RAW264.7 macrophages, while Fe addition can enhance growth inhibition and apoptosis induction in the HepG2 cells infected with HBV [11,12]. LF in the stomach is digested by a proteolytic enzyme pepsin; after that, the yielded LF hydrolysate might also have opportunity to interact with other dietary components including those multivalent trace metal ions. To the best of our knowledge, very few data are available on the effect of interaction between LF hydrolysate and trace metals on anti-cancer activity of LF hydrolysate in some cancer cells, and such study clearly deserves consideration in the scientific community.

Both Cu and Mn are commonly regarded as essential elements to the body. Cu plays crucial roles in the functions of proteins and many enzymes involved in energy metabolism, DNA synthesis, and respiration [13]. For example, Cu is a critical cofactor of the well-known superoxide dismutase and cytochrome oxidase [14]. Mn is also necessary for a series of physiological processes such as the metabolism of carbohydrates, lipids, and amino acids, and has important role as the cofactor of several enzymes in metabolism in the brain [15]. Both Cu^{2+} and Mn^{2+} can complex with some organic materials, which have been studied for their anti-cancer, immune, and anti-oxidant effects [16,17]. Two Cu complexes can inhibi tumor cell growth, while the Mn complex of N-substituted di(picolyl)amine can inhibit the growth of both U251 and HeLa cells via interfering with mitochondrial functions [18,19]. When LF hydrolysate in the stomach interacts with Cu^{2+} or Mn^{2+}, potential changes in its activity against gastric cancer BGC-823 cells are promising. However, to the best of our knowledge, these changes are still not assessed.

In our previous study [20], we used the well-differentiated gastric cancer AGS cells as model cells to evaluate the effects of Cu^{2+} and Mn^{2+} fortification on anti-cancer activity of a peptic bovine lactoferrin hydrolysate (namely BLH). Both Cu- and Mn-fortification were evident to increase BLH's anti-cancer activity against the AGS cells, through two events: enhanced apoptosis induction and autophagy inhibition. However, it is regarded that higher degree of cancer cell differentiation generally accompanies lower degree of malignancy. The low-differentiated gastric cancer cells thus deserve another investigation. In the present study, the low-differentiated gastric cancer BGC-823 cells were used as model cells. BLH was also fortified with $CuCl_2$ and $MnSO_4$ of two levels, and then they were assessed and compared for their anti-cancer activity changes using growth inhibition, cell-cycle arrest, mitochondrial membrane disruption, and apoptosis induction as evaluation indices. Furthermore, expression changes of several apoptosis-related proteins were assayed to disclose possible molecular mechanism responsible for the anti-cancer activity changes of the Cu/Mn-fortified BLH.

2. Results

2.1. Chemical Features of LF, BLH, and Mixtures I–IV

In this study, the used bovine LF and BLH had protein contents of about 957.3 and 923.4 g/kg, and Fe contents of about 140.6 and 130.3 mg/kg (Table 1), respectively. Compared with bovine LF, BLH had higher $-NH_2$ content (0.93 versus 0.49 mmol/g protein), due to the conducted peptic digestion. BLH was also measured with a DH value of 5.1 ± 0.1%. Due to Cu/Mn fortification, the prepared BLF-Cu mixtures (i.e., Mixtures I–II) or BLF-Mn mixtures (i.e., Mixtures III–IV) in this study contained more Cu or Mn than BLH. Thus, activity changes of these mixtures in the assessed BGC-823 cells mainly arose from the fortified Cu or Mn ions.

Table 1. Chemical features of the bovine LF and prepared BLH (dry basis).

Samples	Protein (g/kg)	$-NH_2$ (mmol/g Protein)	Degree of Hydrolysis (%)	Fe (mg/kg)
LF	957.3 ± 3.1	0.49 ± 0.01	0	140.6 ± 2.8
BLH	923.4 ± 2.7	0.93 ± 0.03	5.1 ± 0.1	130.3 ± 3.3

2.2. Growth Inhibition of BLH and Mixtures I–IV on the BGC-823 Cells

In this study, 5-FU as positive control could obviously inhibit the growth of BGC-823 cells: at 200 μmol/L, it resulted in growth inhibition values of 43.5 (24 h) and 58.7% (48 h) (Figure 1). BLH and its mixtures also exerted growth inhibition on the cells (Figure 1). BLH time- and dose-dependently showed growth inhibition values of 5.3–44.7%. Mixtures I–IV also time- and dose-dependently inhibited cell growth, and were more effective than BLH, bringing increased growth inhibition values ranging from 6.3% to 84.5%. Mixtures III–IV showed higher inhibition on the cells than Mixtures I–II (growth inhibition values 11.3–84.5% versus 6.3–62.3%). It was also seen that Mixture I (or Mixture III) had weaker growth inhibition than Mixture II (or Mixture IV), based on these measured growth inhibition values. These results indicated that it was the fortified Cu and especially Mn conferred BLH with higher growth inhibition on the cells, while higher Cu/Mn fortification levels led to higher inhibitory effect. All assessed samples at dose levels other than 25 mg/mL gave too weak or too strong growth inhibition on the cells; thus, they were only used at 25 mg/mL with treatment time of 24 h in later assays.

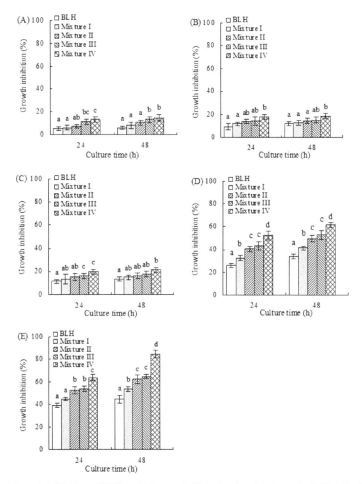

Figure 1. Growth inhibition of BLH and Mixtures I–IV at five dose levels on the BGC-823 cells with treatment times of 24 and 48 h. Mixtures I–II represent bovine lactoferrin hydrolysate (BLH) fortified with Cu^{2+} at 0.64 and 1.28 mg/g protein, while Mixtures III–IV represent BLH fortified with Mn^{2+} at 0.28 and 0.56 mg/g protein, respectively. (**A**–**E**) The mixtures were used at concentrations of 10, 15, 20, 25, and 30 mg/mL, respectively. Different letters like a, b, c, and d above the columns in the same culture time show that the means of different groups were significantly different ($p < 0.05$) by one-way analysis of variance.

When BLH and Mixtures I–IV were used at dose level of 25 mg/mL to assay their long-term growth inhibition on the cells (10 and 20 days), the results showed that Mixtures I–IV also had higher anti-proliferative effects on the cells than BLH (Figure 2). Based on the observed sizes and numbers of cell colonies, it was evident that Mixtures III–IV possessed higher activity than Mixtures I–II, while Mixture IV (or Mixture II) had higher effect than Mixture III (or Mixture I). That is, Mn was more effective than Cu to enhance long-term growth inhibition of BLH, and higher Cu/Mn fortification levels also resulted in higher long-term anti-proliferation.

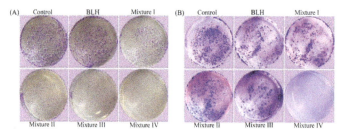

Figure 2. Long-term anti-proliferation of BLH and Mixtures I–IV on the BGC-823 cells with culture times of: 10 days (**A**); and 20 days (**B**).

2.3. Effects of BLH and Mixtures I–IV on Cell-Cycle Progression of the BGC-823 Cells

To further investigate whether BLH and Mixtures I–IV might cause cell growth inhibition via disturbing cell-cycle progression, flow cytometry analysis was done to detect cell-cycle distribution. Mixtures I–IV with treatment time of 24 h resulted in higher cell proportions at the G0/G1-phase than BLH did (63.1–69.3% versus 61.2%) (Figure 3). Of note, the cells treated by Mixtures I–II or Mixtures III–IV had different G0/G1-phase proportions (63.1–65.6% versus 67.5–69.3%). Mixtures I–IV were thus more efficient than BLH to arrest cell-cycle progression at the G0/G1-phase. Overall, Mn fortification led to greater cell-cycle arrest than Cu fortification, and higher Cu/Mn fortification level caused greater cell-cycle arrest at the G0/G1-phase. It is thus concluded that Cu and especially Mn endowed BLH with higher ability to stop cell-cycle progression at the G0/G1-phase, and thereby caused cell growth inhibition.

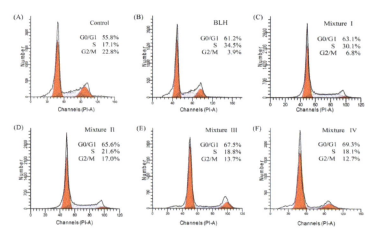

Figure 3. Cell-cycle distribution of the BGC-823 cells: without any treatment (**A**); or treated with BLH (**B**) and Mixtures I–IV (**C–F**) at dose level of 25 mg/mL.

2.4. Apoptosis Induction of BLH and Mixtures I–IV to the BGC-823 Cells

The classic Hoechst 33258 staining was used to observe the morphologic features of the BGC-823 cells exposed to BLH and Mixtures I–IV with treatment time of 24 h (Figure 4), to further disclose briefly if these samples had potential apoptosis induction to the cells. The control cells without any sample treatment had many cells in the observation vision; moreover, most of the control cells were observed to be dimly blue but only a few cells were apoptotic cells (Figure 4A). The cells exposed to BLH and especially Mixtures I–IV had decreased cell numbers in the observation vision, and increased numbers of apoptotic cells (brilliant blue together with chromatin condensation and nuclear fragmentation) were also observed (Figure 4B–F). These results suggest that BLH and Mixtures I–IV could cause cell apoptosis.

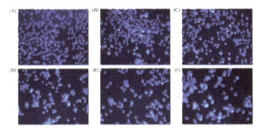

Figure 4. Observed morphology of the BGC-823 cells: without any treatment (**A**); or treated with BLH (**B**) and Mixtures I–IV (**C–F**) at dose level of 25 mg/mL by a fluorescence microscope at 200× magnification.

Apoptosis induction of BLH and Mixtures I–IV in the BGC-823 cells was then assayed by the classic flow cytometry technique, based on measured total apoptotic cell proportions (i.e., Q2 + Q4). The results (Figure 5) show that these samples all had apoptosis induction in the treated cells. The control cells had total apoptotic proportion of 4.3%. The cells exposed to Mixtures I–IV showed higher total apoptotic proportions (28.6%, 33.2%, 40.7%, and 42.7%, respectively) than those exposed to BLH alone (25.3%). Mixture IV (or Mixture II) more obviously caused cell apoptosis than Mixture III (or Mixture I). It was thus proposed that Mn fortification was more effective than Cu fortification to endow BLH with higher apoptosis induction, and higher Cu/Mn fortification level also brought higher activity. For these assessed samples, the order of apoptosis induction was completely consistent with the order of cell-cycle arrest (Figure 5), suggesting that both apoptosis induction and cell-cycle arrest contributed to the assayed growth inhibition.

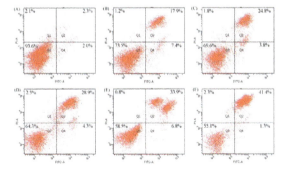

Figure 5. Cell proportions of the BGC-823 cells: without any treatment (**A**); or treated with BLH (**B**) and Mixtures I–IV (**C–F**) at dose level of 25 mg/mL. Q1–Q4 represent necrotic, late apoptotic, intact, and early apoptotic cells, respectively.

2.5. Mitochondrial Membrane Disruption of the BGC-823 Cells by BLH and Mixtures I–IV

Mitochondrial membrane potential (MMP) of the BGC-823 cells exposed to BLH and Mixtures I–IV were analyzed using flow cytometry and JC-1 dye staining, to further verify whether the treated cells had mitochondrial dysfunction. The cells treated by BLH had decreased MMP (cell proportion of red fluorescence 84.6%, Figure 6B), compared with the control cells without sample treatment (95.5%, Figure 6A). Moreover, the cells treated with Mixtures III–IV had lower cell proportions of red fluorescence (68.7% and 62.8%, Figure 6E,F) than those treated with Mixtures I–II (red fluorescence of 78.8% and 71.6%, Figure 6C,D). Mixtures I–II and especially Mixtures III–IV thereby brought greater MMP loss in the treated cells. It was thus demonstrated that these samples caused mitochondrial membrane disruption, and then led to the release of cytochrome c to trigger cell apoptosis. It was also seen from these measured data that Mn fortification was more efficient than Cu fortification to induce MMP loss, and higher Cu/Mn fortification levels brought increased MMP loss.

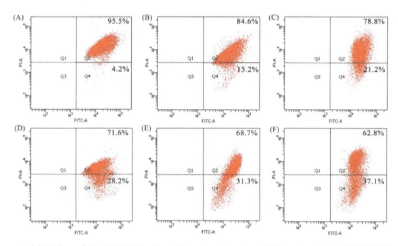

Figure 6. MMP loss of the BGC-823 cells: without any treatment (**A**); or treated with BLH (**B**) and Mixtures I–IV (**C–F**) at dose level of 25 mg/mL.

2.6. Expression Changes of Apoptosis-related Proteins in the BGC-823 Cells

Serial Western-blot assays were done to evaluate expression levels of seven proteins in the treated cells that have been classified as apoptosis-related proteins. In total, BLH and Mixtures I–IV in the cells could up-regulate Bax, Bad, p53, and cytochrome c expression and down-regulate Bcl-2 expression, together with caspase-3 and caspase-9 activation; however, these samples did not cause clear change in caspase-8 expression (Figure 7A). Mn fortification was more efficient than Cu fortification to regulate the expression of these proteins. Mixtures I–IV thus had enhanced anti-cancer activities against the BGC-823 cells than BLH alone, mainly via mediating the expression of these apoptosis-related proteins. Using the caspase-3 inhibitor z-VAD-fmk in the cells could provide further evidence (Figure 7B). When the cells were treated by the z-VAD-fmk, Mixture II and especially Mixture IV showed the ability to increase the expression of Bad (relative expression folds 1.29 and 1.30 vs. 1.15) and Bax (relative expression folds 1.23 and 1.96 vs. 1.18). These results suggest that both Mixture II and Mixture IV indeed were able to induce caspase-3 activation or cell apoptosis. BLH and Mixtures I–IV were thus suggested to induce cell apoptosis via the caspase-3-dependent pathway (Figure 8).

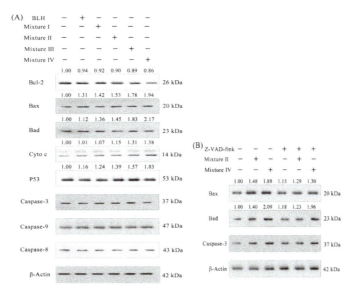

Figure 7. Expression changes of the apoptosis-related proteins in the BGC-823 cells treated with BLH and the Mixtures I–IV (**A**), respectively or treated with Mixture II or Mixture IV in the absence or presence of a caspase-3 inhibitor z-VAD-fmk (**B**).

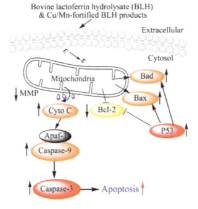

Figure 8. Proposed mechanism responsible for apoptosis induction of BLH and its Cu/Mn mixtures.

3. Discussion

Food hydrolysates possess in vitro anti-cancer activities to many cancer cells such as PC-3, DU-145, H-1299, and Hela cells [21–23]. Bovine LF as one of the most important bioactive proteins in milk has anti-cancer activity to cancer cells, but is regarded to be harmless to normal cells [24–26]. It has been demonstrated that bovine BLH has growth inhibition in gastric cancer and oral squamous cell carcinoma [7,8], can inhibit metastasis of liver and lung cancer cells in the mice [27], and displays anti-cancer effects in colon cancer cells [28]. In this study, BLH and the Cu/Mn-fortified Mixtures I–IV all had anti-cancer activities against the BGC-823 cells with clear growth inhibition, cell-cycle block, and apoptosis induction. The present results are thus consistent with the reported ones. When BLH was fortified with Cu or Mn ions, the resultant mixtures had enhanced anti-cancer effects in the cells. Similarly, the Fe-fortified bovine LF has enhanced growth inhibition on the HepG2 cells infected with HBV [12]. Two previous studies also verify that catechin, epicatechin, epigallocatechin, and particularly epigallacatechin-3-gallate in the presence of Cu can induce apoptosis of a breast cancer cell line MDA-MB-231 [29,30]. It is reasonable that the fortified Cu/Mn contributed these enhanced effects. Mn was always more efficient than Cu to increase these measured effects, which is important but was unsolved in the present study.

In general, protein hydrolysates exert anti-cancer effects via different pathways including anti-proliferation, cell-cycle arrest, apoptosis induction, and others. Rapid growth of cancer cells is achieved by cell continuous division, while cell-cycle is a programmed process of cell division. Thus, stopping cell-cycle progression at a certain cell phase is an important way to inhibit the growth of cancer cells [31]. The hydrolysates derived from donkey milk thus can arrest cell-cycle progression of human lung cancer A549 cells at the G0/G1-phase, while those from roe also can arrest cell-cycle of human oral cancer cells Ca9-22 and CAL27 at the sub-G1-phase [32,33]. Meanwhile, cell apoptosis is a critical mechanism of programmed cell death and, therefore, the induced cell apoptosis is a promising strategy for cancer treatment [34]. Protein hydrolysates derived from giant grouper (*Epinephelus Lanceolatus*) can induce apoptosis of human oral cancer cells, while those from tuna cooking juice induce apoptosis in human breast cancer MCF-7 cells [33,35]. These mentioned findings all support that BLH and Mixtures I–IV had cell-cycle arrest and apoptosis induction, and thereby led to growth inhibition in the cells.

In this study, the treated cells had changed morphologic features and especially MMP loss. This fact suggests potential disruption of mitochondrial membrane and subsequently release of cytochrome c. BLH and Mixtures I–IV thus could induce the apoptosis of the BGC-823 cells via the classic caspase-3-dependent pathway (or mitochondrial pathway). Cytochrome c released (a positive event of cell apoptosis) from the mitochondria into the cytosol activates Apaf-1 and caspase-9, leading to caspase-3 activation and thereby cell apoptosis [36]. Apoptosis of cancer cells requires effective activation of a tumor suppressor p53 [37]. P53 is able to up-regulate pro-apoptotic proteins Bax and Bad, resulting in the increased permeability of mitochondrial membrane, cytochrome c release, and the activation of apoptogenic factors apaf-1. However, another anti-apoptotic protein Bcl-2 has a function to reduce cytochrome c release, which can be suppressed by p53 [38]. The peptides from rapeseed can up-regulate p53 and Bax but down-regulate Bcl-2 expression in HepG2 cells, while rice protein hydrolysates can induce H9c2 myocardiocytes apoptosis through the Bcl-2/Bax pathway [39,40]. More importantly, a previous study demonstrating a short-term cooperation of 3,4-dihydroxy-trans-stilbene and exogenous Cu also showed preferential apoptosis induction of HepG2 cells via mitochondria apoptosis pathway [41]. In this study, these assessed samples up-regulated the pro-apoptotic proteins Bad, Bax, and p53 but down-regulated the anti-apoptotic protein Bcl-2, and then increased cytochrome c release in the cytosol, which subsequently triggered the activation of caspase-9 and caspase-3 as well as cell apoptosis. However, caspase-8 expression, which represents the activation of the extrinsic apoptosis pathway, had no significant change in the cells (Figures 6 and 7). This fact demonstrated that BLH and its fortified mixtures only activated the intrinsic but not extrinsic apoptosis pathway in the BGC-823 cells. Z-VAD-fmk as a classic caspase-3 inhibitor can suppress caspase-3 activation and inhibit the thapsigargin-induced cell death in human breast cancer cells MDA-MB-468 [42]. In this study, both Mixture II and Mixture IV decreased the suppression of z-VAD-fmk on caspase-3 activation via enhancing Bad and Bax expression (Figure 6), verifying that the disclosed apoptosis mechanism indeed was a caspase-3-dependent pathway. Mixtures I–II and especially Mixtures III–IV led to greater expression regulation on these apoptosis-related proteins than BLH did, and therefore exerted higher anti-cancer activity in the cells. However, whether BLH and the fortified mixtures could display anti-cancer effects via other pathways or mechanisms should be disclosed in the future. Moreover, whether these samples might have anti-cancer effects on other cancer cells is still unsolved.

4. Materials and Methods

4.1. Materials

Bovine LF was purchased from MILEI Gmbh (Leutkirch, Germany). The Dulbecco's modified Eagle's medium (DMEM) and porcine gastric mucosa pepsin (CAS: 9001-75-6) were purchased from Sigma-Aldrich Co. Ltd. (St. Louis, MO, USA), while the fetal bovine serum (FBS) was bought from Wisent Inc. (Montreal, QC Canada). Dextran T-70, phosphate-buffered saline (PBS), and Hoechst 33258 dye were bought from Solarbio Science and Technology Co. Ltd. (Beijing, China). 5-Fluorouracil (5-FU) was bought from Jinyao Pharmaceutical Co. Ltd. (Tianjin, China). Annexin V-FITC Apoptosis Detection Kit, Cell Cycle Analysis Kit, BCA Protein Assay Kit, RIPA Lysis Buffer, Hoechst 33258 dye, crystal violet dye, JC-1 dye, and phenylmethanesulfonyl fluoride (PMSF) were all purchased from Beyotime Institute of Biotechnology (Shanghai, China). Cell Counting Kit-8 (CCK-8) was bought from Dojindo Molecular Technologies, Inc. (Kyushu, Japan). Caspase-3 inhibitor z-VAD-fmk, primary anti-bodies (β-actin, caspase-3, caspase-9, caspase-8, Bad, Bax, p53, cytochrome c, and Bcl-2), and secondary anti-body were bought from Cell Signaling Technology, Inc. (Boston, MA, USA). Other chemicals used in this study were analytical grade. Ultrapure water was generated from Milli-Q Plus (Millipore Corporation, New York, NY, USA), and used in this study.

The BGC-823 cells were purchased from Cell Bank of the Chinese Academy of Sciences (Shanghai, China), and cultured at 37 °C in the DMEM with 10% FBS, 100 units/mL penicillin, and 100 μg/mL streptomycin, using a humidified incubator with 5% CO_2.

4.2. Sample Preparation

BLH was prepared as previously described [43]. In brief, 5.0 g bovine LF was dissolved in 100 mL water, adjusted to pH 2.5 using 1 mol/L HCl, added with pepsin of 750 units/g protein, kept at 37 °C for 4 h, heated at 80 °C for 15 min to inactive pepsin, cooled to 20 °C, neutralized to 7.0 using 1 mol/L $NaHCO_3$, and centrifuged at 12,000× g for 30 min at 4 °C. The collected supernatant (i.e., BLH) was freeze-dried with a freeze-dryer (ALPHA 1-4 LSCplus, Marin Christ, Osterode, Germany), ground into powder, and then stored at −20 °C until use.

BLH was dissolved in water, and added with $CuCl_2$ (or $MnSO_4$) solution to achieve final Cu (or Mn) levels of 0.64 and 1.28 (or 0.28 and 0.56) mg/g protein. Mixture I and Mixture II were designated as the Cu-fortified BLH with 0.64 and 1.28 Cu mg/g protein, while Mixture III and Mixture IV were designated as the Mn-fortified BLH with 0.28 and 0.56 Mn mg/g protein, respectively.

4.3. Sample Analyses

The protein contents of the samples were assayed using the Kjidahl method and a conversion factor of 6.38, while Fe content was detected using the o-phenanthroline method [44]. The content of free amino groups ($-NH_2$) was measured using the o-pthaldialdehyde method together with standard L-leucine solutions of 0–36 mg/mL [45]. Degree of hydrolysis of BLH was calculated as previously described [46]. A spectrophotometer (UV-2401PC, Shimadzu, Kyoto, Japan) was used in these spectrometric analyses.

4.4. Assay of Cytotoxic Effect

The cells (2×10^4 cells per well) were seeded in 96-well plates in 100 μL medium, and incubated for 24 h. The medium was replaced by 200 μL fresh medium containing BLH or Mixtures I–IV at dose levels of 10−30 mg/mL, followed by an incubation of 24 and 48 h and medium removal. CCK-8 solution of 100 μL (10 μL CCK-8 in 90 μL medium) was added into each well, followed by another incubation of 1.5 h. Optical density values were measured at 450 nm with a microplate reader (Bio-Rad Laboratories, Hercules, CA, USA), and used to calculate growth inhibition as previously described [20]. The cells exposed to 200 μmol/L 5-FU were designed as positive control, while those exposed to the media with 5% FBS were designed as negative control without any growth inhibition (i.e., 100% viability).

4.5. Colony Formation Assay

To evaluate long-term growth inhibition of these samples, the cells (1×10^3 cells per well) were seeded in 6-well plates, and treated with the medium containing the assessed samples at dose level of 25 mg/mL for 24 h. Then, the medium with 5% FBS was replaced every 3 days. After an incubation of 10 or 20 days, the cells were fixed with methanol, stained with crystal violet dye, dried overnight, and then photographed with an EOS 6D Canon digital camera (Canon Inc., Tokyo, Japan).

4.6. Assay of Cell-Cycle Progression

The cells (1×10^6 cells per dish) were seeded on 100-mm cell culture dish, incubated for 24 h with 10 mL medium, treated with 10 mL per dish fresh medium containing the assessed samples at dose level of 25 mg/mL for 24 h, harvested, washed twice with the cold PBS (10 mmol/L, pH 7.3), fixed with 70% cold ethanol by shaking once every 15 min overnight at 4 °C, washed with the cold PBS again, resuspended with binding buffer (500 µL), and stained with 10 µL RNase A and 25 µL propidium iodide (PI) for 30 min at 37 °C in the dark. The cells treated with the medium were designated as negative control. Cell proportions in the G0/G1-, S-, and G2/M-phases were measured using a flow cytometer (FACS Calibur, Becton Dickson, San Jose, CA, USA), and analyzed with the ModFit software (Verity Software House, Topsham, ME, USA).

4.7. Hoechst 33258 Staining

The cells (1×10^6 cells per well) were seeded in 6-well plates with 2 mL medium, incubated for 24 h, and treated with medium containing the assessed samples at dose level of 25 mg/mL for 24 h. After removal of the medium, the cells were fixed by methanol for 5 min, washed twice with PBS, stained with Hoechst 33258 dye for 5 min in the dark at 22 °C, and observed under a fluorescence microscope (Type Eclipice-Ti-S, Nikon, Japan) with a magnification of 200×.

4.8. Assay of Mitochondrial Membrane Potential

Changes of mitochondrial membrane potential (MMP) of the treated cells were detected using the flow cytometer and JC-1 dye. The cells (5×10^5 cells per well) were seeded in 6-well plates with 2 mL medium, cultured for 24 h, treated with the medium containing the samples at dose level of 25 mg/mL for 24 h, harvested, stained with JC-1 dye at 37 °C for 20 min, and then measured with the flow cytometer (FACS Calibur, Becton Dickson).

4.9. Assay of Apoptosis Induction

The cells (2×10^4 cells per well) were seeded in 6-well plates with 2 mL medium, and incubated for 24 h. After medium removal, the cells were treated with the medium containing the samples at dose level of 25 mg/mL for 24 h. The cells treated with the medium consisting of 5% FBS served as negative control. After that, an AnnexinV-FITC/PI Apoptosis Detection Kit was used according to kit instruction. The cells were harvested, resuspended in 500 µL of the Annexin V-FITC binding buffer consisting of 5 µL Annexin V-FITC and 10 µL PI at 20 °C for 30 min in the dark, and assayed by the flow cytometry (FACS Calibur, Becton Dickson) to detect the intact (Q3), early apoptotic (Q4), late apoptotic (Q2), and necrotic (Q1) cell proportions.

4.10. Western-Blot Assay

The cells (5×10^6 cells per dish) were seeded on 100-mm cell culture dishes with 10 mL medium, incubated for 24 h, treated with the medium containing the samples at dose level of 25 mg/mL for 24 h, harvested by trypsin-EDTA, washed three times with the cold PBS, and lysed on ice for 30 min with 100 µL the RIPA Lysis Buffer supplemented with 1 mmol/L PMSF. The lysate was centrifuged at 12,000× g at 4 °C for 5 min. The supernatant was collected as total cellular protein. Then, protein content was measured using the BCA Protein Assay Kit. Protein (20 µg) of total protein extracts were separated on a 10−15% SDS-PAGE gel and transferred to the PVDF membrane. The blots were blocked with 5% BSA, probed with the primary anti-body (dilution 1:3000) in blocking buffer at 4 °C overnight. The bands were incubated with the anti-rabbit secondary anti-body horseradish peroxidase conjugate. The enhanced chemiluminescence was covered on the PVDF membrane, and the signal was detected using a Chemi Scope 6300 (Clinx Science Instrument, Shanghai, China).

4.11. Statistical Analysis

All data from three independent experiments were analyzed by the SPSS 16.0 software (SPSS Inc., Chicago, IL, USA) and one-way analysis of variance (ANOVA) with Duncan's multiple range tests, and expressed as means or means ± standard deviations.

5. Conclusions

This study found that Cu^{2+} and especially Mn^{2+} fortification of a peptic bovine lactoferrin hydrolysate BLH led to desired changes for its in vitro anti-cancer effects on human gastric cancer BGC-823 cells. Compared with BLH itself, the Cu/Mn fortified BLH had increased growth inhibition, arrested more cells in the G0/G1-phase, disrupted mitochondrial membrane greatly, and promoted cell apoptosis. Furthermore, Cu/Mn fortification led to expression changes of seven apoptosis-related proteins in the cells, and thereby triggered cell apoptosis via the mitochondrial pathway. Mn^{2+} was always more efficient than Cu^{2+} to increase these assayed activities, while higher metal level consistently resulted in enhanced activities. Fortification of trace metal ions thus suggests endowing BLH with increased anti-cancer action in the BGC-823 cells.

Author Contributions: L.-Y.B. performed the experiments; T.-J.L. designed the bovine lactoferrin hydrolysis; X.-H.Z. conceived and designed the experiments, and analyzed the data; and L.-Y.B. and X.-H.Z. wrote the paper.

Funding: This research was fund by the Innovative Research Team of Higher Education of Heilongjiang Province (Project No. 2010td11).

Acknowledgments: The authors thank Li-Ling Yue from Qiqihar Medical University for her kindly help in western-blot assay as well as the anonymous referees for their valuable advice.

Conflicts of Interest: The authors declare no conflict of interest.

Abbreviations

BLH	Bovine lactoferrin hydrolysate
CCK-8	Cell counting kit-8
DMEM	Dulbecco's modified Eagle's medium
EDTA	Ethylenediamine tetra-acetic acid
FBS	Fetal bovine serum
5-FU	5-Fluorouracil
LF	Lactoferrin
MMP	Mitochondrial membrane potential
PBS	Sodium phosphate buffered solution
PI	Propidium iodide
PMSF	Phenylmethanesulfonyl fluoride

References

1. Bhat, Z.F.; Kumar, S.; Bhat, H.F. Bioactive peptides of animal origin: a review. *J. Food Sci. Tech.* **2015**, *52*, 5377–5392. [CrossRef]
2. Chalamaiah, M.; Kumar, B.D.; Hemalatha, R.; Jyothirmayi, T. Fish protein hydrolysates: proximate composition, amino acid composition, antioxidant activities and applications: a review. *Food Chem.* **2012**, *135*, 3020–3038. [CrossRef] [PubMed]
3. García, M.C.; Puchalska, P.; Esteve, C.; Marine, M.L. Vegetable foods: a cheap source of proteins and peptides with antihypertensive, antioxidant, and other less occurrence bioactivities. *Talanta* **2013**, *106*, 328–349. [CrossRef]
4. Reddy, G.V.; Friend, B.A.; Shahani, K.M.; Farmer, R.E. Antitumor activity of yogurt components. *J. Food Protect.* **1983**, *46*, 8–11. [CrossRef]
5. Sheih, I.C.; Fang, T.J.; Wu, T.K.; Lin, P.H. Anticancer and antioxidant activities of the peptides fraction from algae protein waste. *J. Agric. Food Chem.* **2010**, *58*, 1202–1207. [CrossRef] [PubMed]
6. Xu, X.X.; Jiang, H.R.; Li, H.B.; Zhang, T.N.; Zhou, Q.; Liu, N. Apoptosis of stomach cancer cell SGC-7901 and regulation of Akt signaling way induced by bovine lactoferrin. *J. Dairy Sci.* **2010**, *93*, 2344–2350. [CrossRef] [PubMed]
7. Pan, W.R.; Chen, P.W.; Chen, Y.L.; Hsu, H.C.; Lin, C.C.; Chen, W.J. Bovine lactoferricin B induces apoptosis of human gastric cancer cell line AGS by inhibition of autophagy at a late stage. *J. Dairy Sci.* **2013**, *96*, 7511–7520. [CrossRef] [PubMed]
8. Chea, C.; Miyauchi, M.; Inubushi, T.; Ayuningtyas, N.F.; Subarnbhesaj, A.; Nguyen, P.T.; Shrestha, M.; Haing, S.; Ohta, K.; Takata, T. Molecular mechanism of inhibitory effects of bovine lactoferrin on the growth of oral squamous cell carcinoma. *PLoS ONE* **2018**, *13*, 1–19. [CrossRef] [PubMed]
9. Wakabayashi, H.; Bellamy, W.; Takase, M.; Tomita, M. Inactivation of Listeria monocytogenes by lactoferricin, a potent antimicrobial peptide derived from cow's milk. *J. Food Protect.* **1992**, *55*, 238–240. [CrossRef]
10. Shin, K.; Yamauchi, K.; Teraguchi, S.; Hayasawa, H.; Tomita, M.; Otsuka, Y.; Yamazak, S. Antibacterial activity of bovine lactoferrin and its peptides against *Enterohaemorrhagic Escherichia coli* O157: H7. *Lett Appl. Microbiol.* **1998**, *26*, 407–411. [CrossRef]
11. Zhao, H.J.; Zhao, X.H. Modulatory effect of the supplemented copper ion on in vitro activity of bovine lactoferrin to murine splenocytes and RAW264.7 macrophages. *Biol. Trace Res.* **2018**. [CrossRef]
12. Li, S.T.; Zhou, H.B.; Huang, G.R.; Liu, N. Inhibition of HBV infection by bovine lactoferrin and iron-, zinc-saturated lactoferrin. *Med. Microbiol. Immun.* **2009**, *198*, 19–25. [CrossRef]
13. Bogden, J.D. *The Essential Trace Elements and Minerals*; Humana Press: Totowa, NJ, USA, 2000.
14. Mertz, W. The essential trace elements. *Science* **1981**, *1213*, 1332–1338. [CrossRef]
15. Nordberg, M.; Nordberg, G.F. Trace element research-historical and future aspects. *J. Trace Elem. Med. Biol.* **2016**, *38*, 46–52. [CrossRef] [PubMed]
16. Beard, J.L. Iron biology in immune function, muscle metabolism and neuronal functioning. *J. Nutr.* **2001**, *131*, 568–579. [CrossRef] [PubMed]
17. Tisato, F.; Marzano, C.; Porchia, M.; Pellei, M.; Santini, C. Copper in diseases and treatments, and copper-based anticancer strategies. *Med. Res. Rev.* **2010**, *30*, 708–749. [CrossRef] [PubMed]
18. Leng, J.; Shang, Z.M.; Quan, Y. High energy transition state complex increased anticancer activity: a case study on Cu^{II}-complexes. *Inorg. Chem. Commun.* **2018**, *91*, 119–123. [CrossRef]
19. Zhou, D.F.; Chen, Q.Y.; Qi, Y.; Fu, H.J.; Li, Z.; Zhao, K.D.; Gao, J. Anticancer activity, attenuation on the absorption of calcium in mitochondria, and catalase activity for manganese complexes of N-substituted di(picolyl)amine. *Inorg. Chem.* **2011**, *50*, 6929–6937. [CrossRef] [PubMed]
20. Bo, L.Y.; Li, T.J.; Zhao, X.H. Copper or magnesium supplementation endows the peptic hydrolysate from bovine lactoferrin with enhanced activity to human gastric cancer AGS cells. *Bio. Trace Elem. Res.* **2018**, 1–11.
21. Duarte, D.C.; Nicolau, A.; Teixeira, J.A.; Rodrigues, L.R. The effect of bovine milk lactoferrin on human breast cancer cell lines. *J. Dairy Sci.* **2011**, *94*, 66–76. [CrossRef]
22. Chi, C.F.; Hu, F.Y.; Wang, B.; Li, T.; Ding, G.F. Antioxidant and anticancer peptides from the protein hydrolysate of blood clam (Tegillarca granosa) muscle. *J. Funct. Foods* **2015**, *15*, 301–313.

23. Pan, X.; Zhao, Y.Q.; Hu, F.Y.; Chi, C.F.; Wang, B. Anticancer activity of hexapeptide from skate (Raja porosa) cartilage protein hydrolysate in Hela cells. *Mar. Drugs* **2016**, *14*, 153. [CrossRef] [PubMed]
24. Arias, M.; Hilchie, A.L.; Haney, E.F.; Bolscher, J.G.; Hyndman, M.E.; Hancock, R.E.; Vogel, H.J. Anticancer activities of bovine and human lactoferricin-derived peptides. *Biochem. Cell Biol.* **2017**, *95*, 91–98.
25. Chalamaiah, M.; Yu, W.; Wu, J. Immunomodulatory and anticancer protein hydrolysates (peptides) from food proteins: A review. *Food Chem.* **2018**, *245*, 205–222.
26. Guedes, J.P.; Pereira, C.S.; Rodrigues, L.R.; Cortereal, M. Bovine milk lactoferrin selectively kills highly metastatic prostate cancer PC-3 and osteosarcoma MG-63 cells *in vitro*. *Front Oncol.* **2018**, *8*, 1–12.
27. Tomita, M.; Wakabayashi, H.; Shin, K.; Yamauchi, K.; Yaeshima, T.; Iwatsuki, K. Twenty-five years of research on bovine lactoferrin applications. *Biochimie* **2009**, *91*, 52–57. [CrossRef] [PubMed]
28. Freiburghaus, C.; Janicke, B.; Lindmark-Mansson, H.; Oredsson, S.M.; Paulsson, M.A. Lactoferricin treatment decreases the rate of cell proliferation of a human colon cancer cell line. *J. Dairy Sci.* **2009**, *92*, 2477–2484. [CrossRef]
29. Farhan, M.; Khan, H.Y.; Oves, M.; Al-Harrasi, A.; Rehmani, N.; Arif, H.; Hadi, S.M.; Ahmad, A. Cancer therapy by catechins involves redox cycling of copper ions and generation of reactive oxygen species. *Toxins* **2016**, *8*, 37. [CrossRef] [PubMed]
30. Farhan, M.; Oves, M.; Chibber, S.; Hadi, S.M.; Ahmad, A. Mobilization of nuclear copper by green tea polyphenol epicatechin-3-gallate and subsequent prooxidant breakage of cellular DNA: implications for cancer chemotherapy. *Int. J. Mol. Sci.* **2017**, *18*, 34. [CrossRef]
31. Amin, A.R.; Kucuk, O.; Khuri, F.R.; Shin, D.M. Perspectives for cancer prevention with natural compounds. *J. Clin. Oncol.* **2009**, *27*, 2712–2725. [CrossRef] [PubMed]
32. Mao, X.Y.; Gu, J.N.; Sun, Y.; Xu, S.P.; Zhang, X.Y.; Yang, H.Y.; Ren, F.Z. Anti-proliferative and anti-tumour effect of active components in donkey milk on A549 human lung cancer cells. *Int. Dairy J.* **2009**, *19*, 703–708. [CrossRef]
33. Yang, J.I.; Tang, J.Y.; Liu, Y.S.; Wang, H.R.; Lee, S.Y.; Yen, C.Y.; Chang, H.W. Roe protein hydrolysates of giant grouper (Epinephelus Lanceolatus) inhibit cell proliferation of oral cancer cells involving apoptosis and oxidative stress. *Biomed. Res. Int.* **2016**, *23*, 1–12. [CrossRef]
34. Ouyang, L.; Shi, Z.; Zhao, S.; Wang, F.T.; Zhou, T.T.; Liu, B.; Bao, J.K. Programmed cell death pathways in cancer: a review of apoptosis, autophagy and programmed necrosis. *Cell Proliferat.* **2012**, *45*, 487–498. [CrossRef]
35. Hung, C.C.; Yang, Y.H.; Kuo, P.F.; Hsu, K.C. Protein hydrolysates from tuna cooking juice inhibit cell growth and induce apoptosis of human breast cancer cell line MCF-7. *J. Funct. Foods* **2014**, *11*, 563–570.
36. Kluck, R.M.; Bossy-Wetzel, E.; Green, D.R.; Newmeyer, D.D. The release of cytochrome c from mitochondria: a primary site for Bcl-2 regulation of apoptosis. *Science* **1997**, *275*, 1132–1136. [CrossRef]
37. Hollstein, M.; Sidransky, D.; Vogelstein, B.; Harris, C.C. P53 mutations in human cancers. *Science* **1991**, *253*, 9–53.
38. Donovan, M.; Cotter, T.G. Control of mitochondrial integrity by Bcl-2 family members and caspase-independent cell death. *Bioch. Bioph. Acta* **2004**, *1644*, 133–147. [CrossRef]
39. Yang, T.; Zhu, H.; Zhou, H.; Lin, Q.L.; Li, W.J.; Liu, J.W. Rice protein hydrolysate attenuates hydrogen peroxide induced apoptosis of myocardiocytes H9c2 through the Bcl-2/Bax pathway. *Food Res. Int.* **2012**, *48*, 736–741. [CrossRef]
40. Wang, L.; Zhang, J.; Yuan, Q.; Xie, H.; Shi, J.; Ju, X. Separation and purification of an anti-tumor peptide from rapeseed (Brassica campestris L.) and the effect on cell apoptosis. *Food Funct.* **2016**, *7*, 2239–2248. [CrossRef]
41. Dai, F.; Wang, Q.; Fan, G.J.; Du, Y.T.; Zhou, B. Ros-deriven and preferential killing of HepG2 over L-02 cells by a short-term cooperation of Cu(II) and a catechol-type reveratrol analog. *Food Chem.* **2018**, *250*, 213–220. [CrossRef] [PubMed]
42. Qi, X.M.; He, L.L.; Zhong, H.Y.; Distelhors, C.W. Baculovirus p35 and z-VAD-fmk inhibit thapsigargin-induced apoptosis of breast cancer cells. *Oncogene* **1997**, *15*, 1207–1212. [CrossRef] [PubMed]
43. Bellamy, W.; Takase, M.; Wakabayashi, H.; Kawase, K.; Tomita, M. Antibacterial spectrum of lactoferricin B, a potent bactericidal peptide derived from the N-terminal region of bovine lactoferrin. *J. Appl. Bacterial* **1992**, *73*, 472–479. [CrossRef]
44. AOAC. *Official Methods of Analysis of Association of Official Analytical Chemists International*, 18th ed.; AOAC International: Gaithersburg, MD, USA, 2005.

45. Church, F.C.; Swaisgood, H.E.; Porter, D.H.; Catignani, G.L. Spectrophotometric assay using o-phthaldialdehyde for determination of proteolysis in milk and isolated milk proteins. *J. Dairy Sci.* **1983**, *66*, 1219–1227. [CrossRef]
46. Nagy, A.; Marciniak-Darmochwal, K.; Krawczuk, S.; Gelencser, E. Influence of glycation and pepsin hydrolysis on immunoreactivity of albumin/globulin fraction of herbicide resistant wheat line. *Czech J. Food Sci.* **2009**, *27*, 320–329. [CrossRef]

Sample Availability: Samples of the compounds bovine lactoferrin and bovine lactoferrin hydrolysate are available from the authors.

© 2019 by the authors. Licensee MDPI, Basel, Switzerland. This article is an open access article distributed under the terms and conditions of the Creative Commons Attribution (CC BY) license (http://creativecommons.org/licenses/by/4.0/).

Article

CLE-10 from *Carpesium abrotanoides* L. Suppresses the Growth of Human Breast Cancer Cells (MDA-MB-231) In Vitro by Inducing Apoptosis and Pro-Death Autophagy Via the PI3K/Akt/mTOR Signaling Pathway

Li Tian, Fan Cheng, Lei Wang, Wen Qin, Kun Zou * and Jianfeng Chen *

Hubei Key Laboratory of Natural Products Research and Development, China Three Gorges University, Yichang 443002, China; litian0401@126.com (L.T.); fancy1351@163.com (F.C.); wang-lei1989@hotmail.com (L.W.); shmily900920@163.com (W.Q.)
* Correspondence: kzou@ctgu.edu.cn (K.Z.); chenjianfeng2003@126.com (J.C.); Tel./Fax: +86-0717-6397478 (K.Z.)

Academic Editor: Roberto Fabiani
Received: 22 February 2019; Accepted: 16 March 2019; Published: 20 March 2019

Abstract: Background: The antitumor activity of CLE-10 (4-epi-isoinuviscolide), a sesquiterpene lactone compound, isolated from *Carpesium abrotanoides* L. has rarely been reported. The aim of this study is to investigate the antitumor activity of CLE-10 and give a greater explanation of its underlying mechanisms. Methods: The cytotoxicity of CLE-10 was evaluated using MTT assay. Autophagy was detected by the formation of mRFP-GFP-LC3 fluorescence puncta and observed using transmission electron microscopy, while flow cytometry was employed to detect apoptosis. The protein expressions were detected through Western blotting. Results: CLE-10 induced pro-death autophagy and apoptosis in MDA-MB-231 cells by increasing the protein expression of LC3-II, p-ULK1, Bax, and Bad, as well as downregulating p-PI3K, p-Akt, p-mTOR, p62, LC3-I, Bcl-2, and Bcl-xl. CLE-10 that was pretreated with 3-methyladenine (3-MA) or chloroquine (CQ) weakened the upregulation of the protein expression of p-ULK1, or the downregulation of p62, p-mTOR, and decreased the level of cytotoxicity against MDA-MB-231 cells. Meanwhile, rapamycin enhanced the effect of CLE-10 on the expression of autophagy-related protein and its cytotoxicity, with the IC_{50} value of CLE-10 decreasing from 4.07 μM to 2.38 μM. Conclusion: CLE-10 induced pro-death autophagy and apoptosis in MDA-MB-231 cells by upregulating the protein expressions of LC3-II, p-ULK1, Bax, and Bad and downregulating p-PI3K, p-Akt, p-mTOR, p62, Bcl-2, and Bcl-xl.

Keywords: autophagy; apoptosis; PI3K/AKT/mTOR; CLE-10; LC3; MDA-MB-231

1. Introduction

Breast cancer is one of the most prevalent cancers diagnosed in women worldwide, causing more than 500,000 deaths every year, especially in more developed regions [1]. Breast cancer treatment has developed from single surgery to multidisciplinary treatment including radiotherapy, chemotherapy, and endocrine therapy, significantly improving the prognosis of breast cancer [2]. However, the cancer often metastasizes or recurs because of drug resistance and toxicity [3]. Therefore, it benefits public health to explore novel anti-breast cancer agents.

Aiming at apoptosis and autophagy, two different types of programmed cell death (PCD) with their own distinctive features are significant in cancer chemotherapy. Apoptosis, triggered by extrinsically or mitochondria-mediated pathways, is a crucial cytotoxic mechanism of anticancer agents [4]. Moreover, Bcl-2 family proteins are involved in the mitochondria-mediated pathways. Autophagy is a highly conserved cellular activity during which cytoplasmic components including organelles or proteins are degraded and recycled [5]. The basal level of autophagy is usually appeared under certain stimuli or stress, contributing to the maintenance of normal cellular homeostasis. In that circumstance, autophagy acts to promote cell survival [6]. Recent researches have indicated that multiple anticancer treatments lead to excessive activation of autophagy and further result in cancer cell death [7]. In the process of autophagy, the soluble cytoplasmic form of LC3-I is transformed into its membrane associated form LC3-II, involved in the formation of autophagosomes, and finally degraded by autolysosomes [8]. Therefore, it is essential to calculate the amount of LC3-II degraded by lysosomes by comparing LC3-II levels with or without lysosomal protease inhibitor. Sequestosome1 (p62/SQSTM 1), acting as a vital adaptor of target cargo in the process of autophagy, also interacts with other proteins related to autophagy such as LC3 and beclin1 [9,10].

Accumulated evidence has revealed that the activation of the PI3K/Akt/mTOR signaling pathway leads to the occurrence of malignant tumors, indicating that the targeted suppression of certain components in this pathway might be a potential therapeutic strategy for cancer treatment [11,12]. Class1 I PI3K is a heterodimer of the p85 and p110 subunits with dual activities of lipid kinases and protein kinases. Class I PI3 K activates the serine/threonine kinase Akt, and Akt directly activates mTOR via mTORC1 at S2448 or indirectly through TSC2. The inactivation of TSC2 leads to the phosphorylation of Akt, which promotes cell survival by upregulating mTORC1 activity through cascaded signaling molecules to inhibit apoptosis and autophagy [13]. mTOR, acting as an autophagy inhibitor, prevents ULk1 activation and disrupts the mutual effect between ULk1 and AMPK. Conversely, ULK1 also suppresses mTOR by phosphorylation [14].

Natural products are the major resources for new cancer therapies. CLE-10, a sesquiterpene lactone compound, was obtained from *Carpesium abrotanoides* L. (CAL), a traditional Chinese herb, which has been employed to reduce fever or insect bites [15]. Moreover, a compound from the composite plant possesses antifungal, antioxidant, and cytotoxicity properties [16,17]. Studies have shown that CLE-10 isolated from *Inula britannica* or *Carpesium faberi* exhibits cytotoxic activity against several human cancer cells, and the IC_{50} value of CLE-10 on another breast cancer cell, MCF-7, was 45.97 ± 1.21 μM [18,19]. Although there is a wide interest in and extensive use of this medicinal herb, the underlying antitumor mechanism of CLE-10 is rarely reported. In this study, we found that CLE-10 inhibited the proliferation of breast cancer cells (MDA-MB-231) by inducing apoptosis and pro-death autophagy through the PI3K/Akt/mTOR signaling pathway.

2. Results

2.1. MTT Assay

The structure of CLE-10 is presented in Figure 1 [20]. We investigated the cytotoxicity of CLE-10 on various cell lines. As described in Figure 2, MDA-MB-231 cells were the most sensitive to CLE-10 among the cell lines examined, with an IC_{50} value of 4.07 μM. In addition, CLE-10 showed lower cytotoxicity on normal cells (Figure 2b).

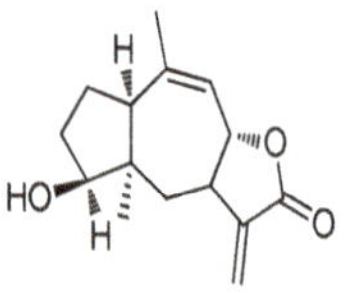

Figure 1. The chemical structure of CLE-10.

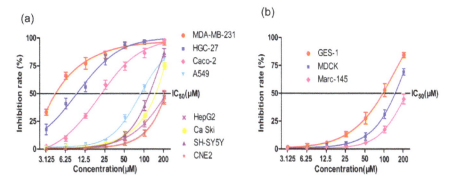

Figure 2. The cytotoxicity of CLE-10 on (**a**) MDA-MB-231, CaCo-2, A549, HepG-2, Caski, SH-SY5Y, HGC-27, CNE-2, (**b**) GES-1, MDCK, and Marc-145 by MTT assay. The data were representative results of three independent tests.

2.2. Inhibition of Autophagy Relieved CLE-10-Induced Cell Death

To corroborate the impact of autophagy on CLE-10-induced MDA-MB-231 cell death, CLE-10 was used, pretreated with an autophagy inhibitor (chloroquine (CQ), 3-methyladenine (3-MA)) and a mTOR agonist (rapamycin). The concentration of inhibitors was at a safe level with no cytotoxicity against MDA-MB-231 cells. Autophagy inhibitors 3-MA or CQ weakened the inhibition of CLE-10 on the growth of MDA-MB-231 cells with the IC_{50} levels of 6.91 µM and 6.49 µM, respectively (Figure 3). There was no distinct difference on the inhibitory effect between the CLE-10 + 3-MA group and the CLE-10 + CQ group. Furthermore, rapamycin significantly enhanced the inhibitory effect of CLE-10, especially at a low CLE-10 concentration with the IC_{50} of 2.38 µM, indicating that CLE-10 induced autophagy, leading to breast cancer cell death rather than a protective mechanism.

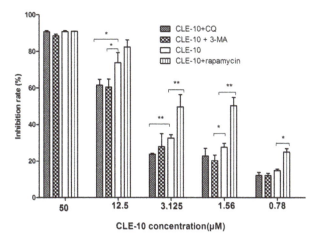

Figure 3. The inhibitory effect of CLE-10 pretreated with or without autophagy inhibitors (5 mM 3-methyladenine (3-MA), 20 µM chloroquine (CQ)) or inducer (100 nM rapamycin) on the proliferation and growth of MDA-MB-232 cells for 48 h. (* $p < 0.05$, ** $p < 0.01$, compared with the CLE-10 group).

2.3. CLE-10 Induced MDA-MB-231 Cell Apoptosis

The rates of apoptotis (accumulating both in Annexin V-Enzo Gold-positive/Necrosis Detection Reagent-negative (early apoptosis) and Annexin V-Enzo Gold-positive/Necrosis Detection

Reagent-positive (late apoptosis)) were 5.94%, 23.44%, and 50.98% (Figure 4a,b) after treating with CLE-10 (0, 10, and 15 µM) for 24 h. These data indicated that CLE-10 exerted obvious apoptosis in a dose-dependent manner. In addition, Western blot revealed that CLE-10 downregulated Bcl-2 and Bcl-xl expressions and upregulated the expression of Bax and Bad (Figure 4b).

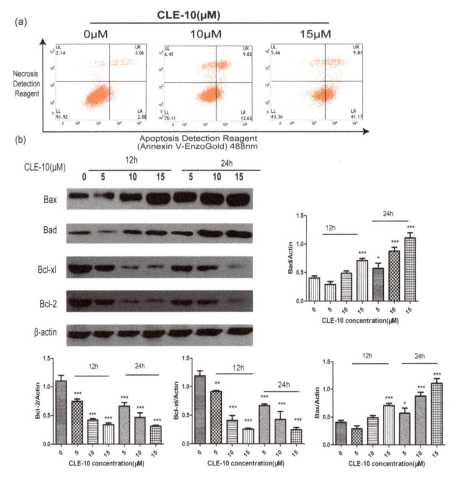

Figure 4. CLE-10 induced apoptosis in MDA-MB-231 cells. (**a**) Apoptosis induced by CLE-10 (0, 10, 15 µM) in MDA-MB-231 cells was detected by flow cytometry. (**b**) Representative Western blotting bands of Bcl-2, Bcl-xl, Bax, and Bad in MDA-MB-231 cells. Next to the bands are protein expression levels (* $p < 0.05$, ** $p < 0.01$, *** $p < 0.001$ compared with the 0 µM CLE-10 group).

2.4. CLE-10 Induced MDA-MB-231 Cell Autophagy

To confirm whether CLE-10 induced autophagy in MDA-MB-231 cells, TEM observation and the formation of mRFP-GFP-LC3 puncta were employed. Transmission electron microscopy indicated an enhanced presence of autophagosomes, autophagic vesicles, and autolysosomes in cells treated with 15 µM CLE-10 for 24 h (Figure 5a). Condensed cytoplasm, membrane invagination, and the disappearance of microvilli were observed at the same time. To further research the autophagy flux induced by CLE-10, MDA-MB-231 cells were transfected with mRFP-GFP-LC3 adenovirus before treatment with CLE-10. Autophagic flux was analyzed by the evaluation of mRFP-LC3 and

GFP-LC3 puncta locations. GFP fluorescence (green dots), quenched easily in autolysosomes, is visible only in autophagosomes. Meanwhile, mRFP (red dots) signals can be observed under the environment of autophagosomes and autolysosomes. When the two colors merged together, yellow dots representing autophagosomes and red dots representing autolysosomes can be observed. As illustrated in Figure 5b,c, obvious enhancement in the amount of yellow puncta (autophagosomes) and red-only puncta (autolysosomes) appeared and the number of red dots was greater than the number of yellow dots, indicating that CLE-10 accelerated autophagic flux without suppressing the function of lysosomes or the fusion of autophagosomes and lysosomes. Therefore CLE-10 served as an autophagy inducer.

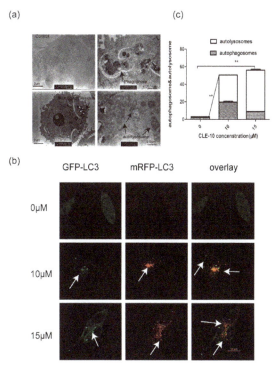

Figure 5. CLE-10-induced MDA-MB-231 cell death was mediated by autophagy. (**a**) Autophagy vesicles, autophagosomes, and autophagy lysosomes were observed in MDA-MB-231 cells after treatment with CLE-10. (**b**) MDA-MB-231 cells transfected with mRFP-GFP-LC3 adenovirus were detected with a confocal fluorescence microscopy. (**c**) The number of autophagosomes (yellow dots) and autolysosomes (red-only dots) in the control group and the CLE-10 group (** $p < 0.01$).

2.5. CLE-10 Inhibited the PI3K/Akt/mTOR Signaling Pathway in MDA-MB-231 Cells

The PI3K/Akt/mTOR signaling pathway is closely related to the enhancement of autophagy and is always activated in cancers, including breast cancer [21]. Therefore, we investigated whether the phosphorylation of PI3K, Akt, mTOR, and autophagy-related proteins (ULK1, p62, LC3) was involved in autophagy induced by CLE-10 in MDA-MB-231 cells. As shown in Figure 6, CLE-10 treatment for 12 h or 24 h in MDA-MB-231 cells downregulated PI3k, Akt, mTOR, and ULK1 phosphorylation in a dose-dependent manner but did not exert a significant effect on the expression of these total proteins. Meanwhile CLE-10 also reduced the protein expression of p62, LC3-I, and increased the protein expression of LC3-II.

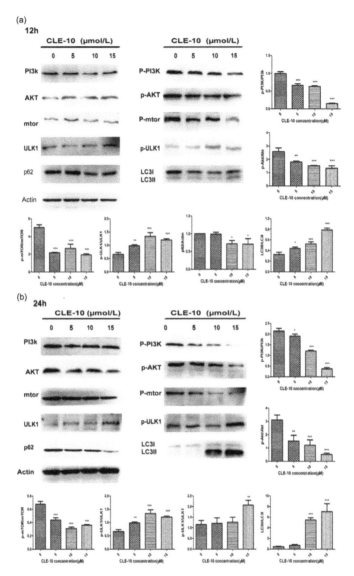

Figure 6. Influence of CLE-10 on the expression of the PI3K/Akt/mTOR signal pathway and autophagy-related proteins. (**a**) After 12 h of treatment with CLE-10, PI3K, Akt, mTOR, p-PI3K, p-Akt, p-mTOR, ULK1, p-ULK1, p62, LC3-I, and LC3-II expressions in MDA-MB-231 cells were analyzed by Western blot. (**b**) After 24 h of treatment of CLE-10, PI3K, Akt, mTOR, p-PI3K, p-Akt, p-mTOR, ULK1, p-ULK1, p62, LC3-I, amd LC3-II expressions in MDA-MB-231 cells were detected by Western blot. Data are expressed as the mean ± SD (n = 3). (* p < 0.05, ** p < 0.01, *** p < 0.001 compared with the 0 μM CLE-10 group).

2.6. The Effect of CLE-10 Combined with 3-MA, CQ, or Rapamycin on the Protein Expression of mTOR, ULK1, p62, and LC3

In addition, we detected autophagy-related proteins mTOR, p-mTOR, LC3, p-ULK1, ULK1, and p62 expression with and without autophagy inhibitors CQ, 3-MA, and mTOR inhibitor rapamycin by Western blot. As shown in Figure 7, the CLE-10, rapamycin, and CLE-10 + rapamycin treatments increased p-ULK1 and LC3-II expression as well as downregulated p62, p-mTOR, and LC3-I protein expression, with no obvious influence on the protein expression of mTOR and ULK1 as compared with the control group. The CLE-10 + 3-MA and CLE-10 + CQ treatments inhibited the upregulation of p-ULK1 expression and the downregulation of p-mTOR and p62 protein expression compared with the CLE-10 group. The CLE-10+3-MA treatment inhibited the protein expression of LC3-II while the CLE-10 + CQ treatment increased the protein expression of LC3-II compared with the CLE-10 group.

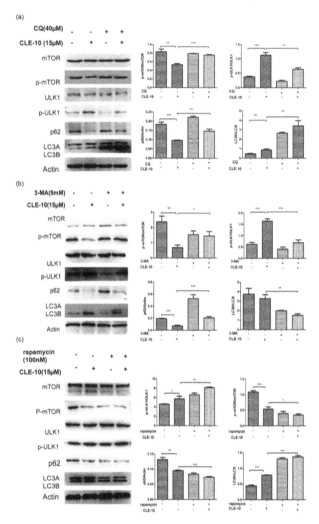

Figure 7. Effect of CLE-10 on autophagy-related proteins LC3-I/II, p-ULK1, ULk1, p62, mTOR, and p-mTOR were detected with and without autophagy inhibitor 3-MA, CQ, as well as the mTOR agonist rapamycin by Western blot analysis. (**a**) CLE-10 was used in combination with CQ. (**b**) CLE-10 was used in combination with 3-MA. (**c**) CLE-10 was used in combination with rapamycin. (* $p < 0.05$, ** $p < 0.01$, *** $p < 0.001$ compared with the control group or the CLE-10 group).

3. Discussion

The antitumor activity of CLE-10, a sesquiterpene lactone compound isolated from *Carpesium abrotanoides* L., has rarely been reported [19]. Therefore, we examined the cytotoxicity of CLE-10 and its underlying mechanisms. In this study, we demonstrated that CLE-10 had effective cytotoxicity against multiple human tumor cell lines tested, especially against MDA-MB-231 cells (IC_{50} = 4.07 µM). Meanwhile, CLE-10 showed lower cytotoxicity on normal cells, suggesting the selectivity of CLE-10 against tumor cells. Our results showed that CLE-10 induced apoptosis in MDA-MB-231 cells as evidenced by a growing number of apoptotic cells after flow cytometry (FCM) detection. This conclusion was further supported by increased expressions of Bax and Bad as well as downregulated Bcl-xl and Bcl-2 expressions. Bcl-2 family proteins are classified into three types. One type of protein inhibits apoptosis, such as Bcl-xl and Bcl-2, whereas a second type (Bak, Bax) accelerates apoptosis and another diverse type of BH3-only proteins (BIKn, Bad) can bind and control the anti-apoptotic Bcl-2 proteins to show the same function as the second type [22].

Meanwhile, CLE-10-induced MDA-MB-231 cell death and apoptosis was mediated by autophagy. TEM observation revealed significant morphological changes that were characteristic of autophagy such as membrane invagination and the existence of autophagic vacuoles and autolysosomes. Autophagy is an essential intracellular degradation process, during which lysosomes degrade and recycle cellular components, providing energy and new materials to maintain cellular environmental homeostasis [23]. The whole process of autophagy is called autophagy flux. Emerging evidence supports that impaired autophagic flux is related to a number of human diseases including tumorigenesis, cardiovascular system disease, and neurodegenerative disease [24,25]. To further evaluate the status of autophagy flux in MDA-MB-231 cells after CLE-10 treatment, a tandem mRFP-GFP-LC3 adenovirus was applied. The increasing number of green GFP-LC3 dots and red mRFP-LC3 puncta illustrated the existence of autophagy. Merged images indicated that CLE-10 increased autolysosomes more than autophagosomes, thus stimulating autophagic flux.

Generally, it is uncertain whether autophagy serves as a cell death or survival mechanism, or a bystander in dying cells [26]. In order to study the roles of autophagy induced by CLE-10 in MDA-MB-231 cells, CLE-10 was pretreated with an autophagy inhibitor (3-MA, CQ) or a mTOR inhibitor rapamycin. Autophagy inhibitors 3-MA and CQ weakened the effect of CLE-10 on the growth of MDA-MB-231 cells, while rapamycin enhanced the cytotoxicity of CLE-10, with the IC_{50} value decreasing from 4.07 µM to 2.38 µM. The results revealed that autophagy induced by CLE-10 in MDA-MB-231 cells contributed to cell death rather than promoted cell survival.

Accumulated evidence indicated that apoptosis and autophagy could be induced via the same upstream signals that affect the occurrence, proliferation, and treatment of cancer, such as PI3K/Akt/mTOR, p53, and Bcl-2 signaling pathways [27,28]. PI3K/Akt/mTOR are significant kinases activated by various stimuli. They regulate essential cellular functions including proliferation, transcription, translation, growth, and survival. PI3K, Akt, mTOR, and autophagy-related protein (ULK1, LC3, p62) expressions were analyzed in order to confirm whether this pathway was involved in CLE-10-induced cell death. CLE-10 treatment for 12 h or 24 h in MDA-MB-231 cells downregulated p-PI3K, p-Akt, p-mTOR, p62, and LC3-I and increased the protein expression of p-ULK1 and LC3-II in a dose-dependent manner, but did not exert a notable impact on the expression of these total proteins, thus suggesting that CLE-10 inhibited the PI3K/Akt/mTOR signal pathway in MDA-MB-231 cells.

Moreover, mTOR, ULK1, LC3, and p62 are involved in the process of autophagosome formation or degradation. The formation of the Atg1/ULK1/ATG13/FIP200 protein-kinase complex is essential in the initiation of phagophore formation. Meanwhile, ULK1 is activated by decreased mTORC1 signaling or increased AMPK activity, leading to the phosphorylation of ATG13 and FIP200 [29]. There are three types of LC3 present in the cell: pro-LC3, LC3-I, and LC3-II. LC3-I protein continues to exist in the cytoplasm, while LC3-II integrates into both sides of the membrane, forming autophagosomes, and LC3-II is degraded with the membrane by lysosomal enzymes [30]. The level of LC3-II is related to the amount of autophagosomes in the cell; thus, LC3-II acts as a reliable marker of autophagy [31].

p62, a traditional receptor of autophagy, is involved in ubiquitinated cargoes delivering autophagic degradation [32]. Moreover, the decreasing expression of p62 activates autophagy and the deletion of p62 leads to LC3-II formation, aggresome or autophagosome impairment, cell damage, and finally cell death [33].

It is significant that the expression of mTOR, ULK1, LC3-II, and p62 in the autophagy process was detected with or without 3-MA and CQ. When CLE-10 was used to treat MDA-MB-231 cells alone, the increasing expression of LC3-II and p-ULK1 and the decreasing p62 expression were detected compared to the 0 μm CLE-10 group, suggesting that CLE-10 induced autophagy. Meanwhile the CLE-10 + 3-MA group weakened the upregulation of protein expression of ULK1 and LC3-II or the downregulation of p62 compared to the CLE-10 group. The CLE-10 + CQ group increased the protein expression of LC3-II compared to the CLE-10 group, indicating that CLE-10 increased the formation of autophagosomes. In general, CLE-10 exhibited a similar effect on the autophagy-related proteins as rapamycin and the influence of CLE-10 on the autophagy-related proteins could be weakened by 3-MA or CQ.

4. Material and Methods

4.1. Materials

CLE-10 was isolated from CAL, as mentioned in our previous article [20]. The purified CLE-10 (over 99% pure) was dissolved in dimethylsulfoxide (DMSO) as a 25 mM solution and stored at 4 °C. Once used, the CLE-10 solution was diluted with a culture medium to a desired concentration.

4.2. Cell Lines and Cultures

The human cervical carcinoma Caski, human mammary carcinoma cell line MDA-MB-231, human lung cancer cell line A549, human colorectal carcinoma cell line CaCo-2, human nasopharyngeal carcinoma CNE-2, human SH-SY5Y neuroblastoma cell line, human hepatocellular carcinoma HepG-2 cell lines, human gastric carcinoma HGC-27 cell lines, monkey embryonic renal epithelial cell line (Marc-145), human gastric cell line (GES-1), and Madin–Darby canine kidney cell line (MDCK) were bought from China Type Culture Collection in Shanghai and preserved in our laboratory. Cells were cultured in RPMI-1640, L-15, or DMEM culture medium with 10% fetal bovine serum at 37 °C in 5% CO_2.

4.3. MTT Assays

One hundred microliters of the cell suspension diluted with culture medium (0.8–1 × 10^5 cell/mL) were added to the 96-well microplates. Twelve hours later, 100 μL of culture medium with different concentrations of CLE-10 from 3.12 μM to 100 μM were added to each well, which were incubated for another 48 h. Then, 20 μL of 5 mg/mL MTT (Sigma-Aldrich, Shanghai, China) reagent was added per well. After culturing the cells for 4 h, they were gently removed the medium and then 150 μL DMSO solution (Sigma-Aldrich) was added to each well. The absorbance was detected by a microplate reader (Tecan Shanghai, China) at 490 nm, subtracting the baseline reading.

4.4. CLE-10 Pretreated with 3-MA, CQ, Rapamycin Using MTT Assays

Cells (8–9 × 10^4/mL) were seeded into 96-wel plates in 100 μL of medium and were cultured for 12 h. The cells were divided into five groups. Cells treated with normal culture media were regarded as the control group. For the CLE-10 group, cells were treated with medium containing CLE-10 at various concentrations (3.12–100 μM). For inhibitor groups, cells were pretreated with a selective autophagy inhibitor (3-MA (5 mM), CQ (20 μM)) or the mTOR agonist rapamycin (100 nM) for 6 h, then the supernatant was gently removed and replaced by culture medium containing CLE-10 at various concentrations (3.12–100 μM) for 48 h. Next, 20 μL of 5 mg/mL MTT was added to all the wells for another 4 h. The medium was gently removed, after which 150 μL of DMSO was added to each well. The absorbance of each well was detected by a microplate reader at 490 nm by subtracting the baseline reading. Autophagy inhibitors 3-methyladenine (3-MA) and chloroquine (CQ) were obtained from MCE China, as was the mTOR inhibitor (rapamycin).

4.5. Flow Cytometry (FCM)

Apoptosis was detected with an Annexin V-Enzo Gold apoptosis detection kit (Enzo Life Sciences, Beijing, China). MDA-MB-231 cells were treated with CLE-10 (0, 10, 15 µM) for 24 h. Cells were gathered by trypsinization and washed using cold phosphate-buffered saline (PBS) twice. Then cells were resuspended in buffer or buffer containing Annexin V-EnzoGold or Necrosis Detection Reagent according to the instruction at room temperature, avoiding light for 15 min, and analyzed by flow cytometry (BD Bioscience).

4.6. Western Blot Analysis

MDA-MB-231 cells were gathered and the total proteins in each sample was obtained and quantified using the bicinchoninic acid (BCA) protein concentration assay kit (Biyuntian, Beijing, China) after treatment with different concentrations of CLE-10 (0, 5, 10, 15 µM) for 24 h. Equal amounts of protein (50 µg) was separated by 6–15% SDS-PAGE gels and transferred to polyvinylidene fluoride (PVDF) membranes (Beijing Labgic Technology Co., Ltd.). Membranes were blocked with 5% milk (BD) for 2 h. After washing, the membranes were probed overnight using the specific primary antibodies LC3, PI3K (p85), AKT, mTOR, p-PI3K (Tyr 508), p-AKT (Ser473), p-mTOR, p-ULK1(Ser 757), p62, Bax, Bad, Bcl-2, and Bcl-xl (Cell Signaling Technology) at 4 °C, followed by secondary antibody. Chemiluminescence was performed with a chemiluminescence developing solution (Biyuntian, Beijing, China) on Kodak X-ray films or using a chemiluminescence image analysis system (Tanon 5200). LC3, p-ULK1, ULk1, p62, mTOR, and p-mTOR were also detected with or without an autophagy inhibitor (CQ, 3-MA) or the mTOR agonist rapamycin. Cells were treated with 3-MA (5 mM), CQ (40 µM), or rapamycin (100 nm), CLE-10 (15 µM), 3-MA (5 mM) + CLE-10 (10 µM), CQ (40 µM) + CLE-10 (15 µM), and rapamycin (100 nm) + CLE-10 (15 µM); each for 24 h. Western blot analysis was used to detect the relative protein expression as described above.

4.7. Transmission Electron Microscopy (TEM)

For TEM analysis, CLE-10 (0, 15 µM) was applied for 24 h, after which MDA-MB-231 cells were gathered and fixed in ice-cold 2.5% glutaraldehyde containing 0.1 M cacodylate buffer with a pH of 7.4. In 1% phosphate-buffered osmium tetroxide, cells were subsequently stained with 3% aqueous uranyl acetate. Cells were then dehydrated in an increasing gradient of ethanol solution and embedded in epoxy resin. Ultrathin sections were gained and later stained. Electron micrographs were observed and imaged using a transmission electron microscope (H7650; Hitachi, Tokyo, Japan).

4.8. Tandem mRFP-GFP-LC3 Transfection

A tandem mRFP-GFP-LC3 adenovirus (Hanheng Biotechnology Co Ltd., Shanghai, China) was transfected into incubated MDA-MB-231 cells in a culture dish (Wuxi NEST Biotechnology, China) for 6.5 h at a multiplicity of infection (MOI) of 200 before receiving CLE-10 (30 µM) treatments. After treating for 24 h, cell images were obtained by laser confocal fluorescence microscopy (LCFM) (Olympus FV1200, Japan).

4.9. Statistical Analysis

The data of all the experiments were described as means ± SD. Statistical analyses were determined using the Graphpad prism 5.0 statistical software. Statistical differences were evaluated using unpaired the Student's t-test and the ANOVA method, and were considered to be significantly different at $p < 0.05$.

5. Conclusions

In this study, our results showed that CLE-10 significantly suppressed the growth of human breast cancer cells (MDA-MB-231). Furthermore, CLE-10 affected many autophagy- and apoptosis-related proteins including p-ULK1, LC3-II, p62, Bax, Bad, Bcl-2, and Bcl-xl, and suppressed the PI3K/AKT/mTOR pathway, ultimately inducing apoptosis and pro-death autophagy in MDA-MB-231 cells. In addition, autophagy inhibitors (3-MA or CQ) weakened while rapamycin enhanced the effect of CLE-10 on the autophagy-related proteins in addition to the cytotoxicity of CLE-10 against MDA-MB-231 cells. Our results revealed a potential mechanism of CLE-10-induced cell death, providing a guide for the use of CLE-10 in in vitro studies.

Author Contributions: K.Z., J.C., and L.T. designed the experiments; L.T. performed most of the experiments while W.Q. performed part of the MTT assay; F.C. and L.W. prepared the extract of the plant materials. L.T. and J.C. revised the manuscript.

Funding: This research was supported by the National Natural Science Foundation of China (NO. 81773952) and the Natural Science Foundation of Shennongjia Forestry District (No. SNJKJ2016029).

Conflicts of Interest: The authors have declared that there is no conflict of interest.

References

1. Globocan, 2012, Breast Cancer Fact Sheet 2008. Available online: http://globocan.iarc.fr/factsheets/ (accessed on 19 March 2019).
2. Jafari, S.H.; Saadatpour, Z.; Salmaninejad, A.; Momeni, F.; Mokhtari, M.; Nahand, J.S.; Rahmati, M.; Mirzaei, H.; Kianmehr, M. Breast cancer diagnosis: Imaging techniques and biochemical markers. *J. Cell. Physiol.* **2018**, *233*, 5200–5213. [CrossRef]
3. Wang, M.; Wang, Y.; Zhong, J. Side population cells and drug resistance in breast cancer. *Mol. Med. Rep.* **2015**, *11*, 4297–4302. [CrossRef] [PubMed]
4. Kiliçarslan, A.; Kahraman, A.; Akkiz, H.; Yildiz Menziletoğlu, S.; Fingas, C.D.; Gerken, G.; Canbay, A. Apoptosis in selected liver diseases. *Turk. J. Gastroenterol.* **2009**, *20*, 171–179. [CrossRef]
5. Tanida, I. Autophagosome formation and molecular mechanism of autophagy. *Antioxid. Redox Signal.* **2011**, *14*, 2201–2214. [CrossRef] [PubMed]
6. Mizushima, N.; Levine, B.; Cuervo, A.M.; Klionsky, D.J. Autophagy fights disease through cellular self-digestion. *Nature* **2011**, *451*, 1069–1075. [CrossRef]
7. Denton, D.; Nicolson, S.; Kumar, S. Cell death by autophagy: Facts and apparent artefacts. *Cell Death Differ.* **2012**, *19*, 87–95. [CrossRef]
8. Huang, R.; Liu, W. Identifying an essential role of nuclear LC3 for autophagy. *Autophagy* **2015**, *11*, 852–853. [CrossRef] [PubMed]
9. Komatsu, M. Potential role of p62 in tumor development. *Autophagy* **2011**, *7*, 1088–1090. [CrossRef]
10. Islam, M.A.; Sooro, M.A.; Zhang, P. Autophagic regulation of p62 is critical for cancer therapy. *Int. J. Mol. Sci.* **2018**, *19*, 1405. [CrossRef]
11. Bahrami, A.; Khazaei, M.; Shahidsales, S.; Hassanian, S.M.; Hasanzadeh, M.; Maftouh, M.; Ferns, G.A.; Avan, A. The therapeutic potential of PI3k/Akt/mTOR inhibitors in breast cancer: Rational and progress. *J. Cell Biochem.* **2017**, *119*, 213–222. [CrossRef]
12. Hosokawa, N.; Hara, T.; Kaizuka, T.; Kishi, C.; Takamura, A.; Miura, Y.; Lemura, S.I.; Natsume, T.; Takehana, K.; Yamada, N.; et al. Nutrient-dependent mTORC1 association with the ULK1–Atg13–FIP200 complex required for autophagy. *Mol. Biol. Cell* **2009**, *20*, 1981–1991. [CrossRef]
13. Inoki, K.; Li, Y.; Zhu, T.Q.; Wu, J.; Guan, K.L. TSC2 is phosphorylated and inhibited by Akt and suppresses mTOR signaling. *Nat. Cell Biol.* **2002**, *4*, 648–657. [CrossRef]
14. Shang, L.B.; Wang, X.D. AMPK and mTOR coordinate the regulation of ULK1 and mammalian autophagy initiation. *Autophagy* **2011**, *7*, 924–926. [CrossRef]
15. Zhang, J.P.; Wang, G.W.; Tian, X.H.; Yang, Y.X.; Liu, Q.X.; Chen, L.P.; Li, H.L.; Zhang, W.D. The genus Carpesium: A review of its ethnopharmacology phytochemistry and pharmacology. *J. Ethnopharmacol.* **2015**, *163*, 173–191. [CrossRef]

16. Feng, J.T.; Hao, W.; Ren, S.X.; He, J.; Yong, L.; Xing, Z. Synthesis and anti-fungal activities of carabrol ester derivatives. *J. Agric. Food Chem.* **2012**, *60*, 3817–3823. [CrossRef]
17. Wang, J.F.; He, W.J.; Zhang, X.X.; Zhao, B.Q.; Liu, Y.H.; Zhou, X.J. Dicarabrol, a new dimeric sesquiterpene from *Carpesium abrotanoides* L. *Bioorg. Med. Chem. Lett.* **2015**, *25*, 4082–4084. [CrossRef]
18. Fischedick, J.T.; Pesic, M.; Podolski-Renic, A.; Bankovic, J.; de Vos, R.C.H.; Peric, M.; Todorovic, S.; Tanic, N. Cytotoxic activity of sesquiterpene lactones from Inula britannica on human cancer cell lines. *Phytochem. Lett.* **2013**, *6*, 246–252. [CrossRef]
19. Li, X.W.; Weng, L.; Gao, X.; Zhao, Y.; Pang, F.; Liu, J.H.; Zhang, H.F.; Hu, J.F. Antiproliferative and apoptotic sesquiterpene lactones from carpesium faberi. *Bioorg. Med. Chem. Lett.* **2011**, *21*, 366–372. [CrossRef]
20. Wang, L.; Qin, W.; Tian, L.; Zhang, X.X.; Lin, F.; Cheng, F.; Chen, J.F.; Liu, C.X.; Guo, Z.Y.; Peter, P.; et al. Caroguaianolide A-E, five new cytotoxic sesquiterpene lactones from *Carpesium abrotanoides* L. *Fitoterapia* **2018**, *127*, 349–355. [CrossRef]
21. Park, J.H.; Kim, K.P.; Ko, J.J.; Park, K.S. PI3K/Akt/mTOR activation by suppression of ELK3 mediates chemosensitivity of MDA-MB-231 cells to doxorubicin by inhibiting autophagy. *Biochem. Biophys. Res. Commun.* **2016**, *477*, 277–282. [CrossRef]
22. Youle, R.J.; Strasser, A. The Bcl-2 protein family: Opposing activities that mediate cell death. *Nat. Rev. Mol. Cell Biol.* **2008**, *9*, 47–59. [CrossRef]
23. Shen, H.M.; Codogno, P. Autophagic cell death: Loch Ness monster or endangered species? *Autophagy* **2011**, *7*, 457–465. [CrossRef]
24. Munz, C. Enhancing immunity through autophagy. *Annu. Rev. Immunol.* **2009**, *27*, 423–449. [CrossRef]
25. Levine, B.; Kroemer, G. Autophagy in the pathogenesis of disease. *Cell* **2008**, *132*, 27–42. [CrossRef]
26. Chen, Y.; Azad, M.B.; Gibson, S.B. Methods for detecting autophagy and determining autophagy-induced cell death. *Can. J. Physiol. Pharmacol.* **2010**, *88*, 285–295. [CrossRef]
27. Fu, H.; Wang, C.; Yang, D.; Zhang, X.; Wei, Z.; Xu, J.; Hu, Z.; Zhang, Y.; Wang, W.; Yan, R.; et al. Curcumin regulates proliferation, autophagy and apoptosis in gastric cancer cells by affecting PI3k and p53 signaling. *J. Cell Physiol.* **2018**, *233*, 4634–4642. [CrossRef]
28. Kumar, D.; Das, B.; Sen, R.; Kundu, P.; Manna, A.; Sarkar, A.; Chowdhury, C.; Chatterjee, M.; Das, P. Andrographolide analogue induces apoptosis and autophagy mediated cell death in U937 cells by inhibition of PI3k/Akt/mTOR pathway. *PLoS ONE* **2015**, *10*, e0139657. [CrossRef]
29. Kenney, D.L.; Benarroch, E.E. The autophagy-lysosomal pathway: General concepts and clinical implications. *Neurology* **2015**, *85*, 634–645. [CrossRef]
30. Lee, Y.K.; Lee, J.A. Role of the mammalian Atg8/LC3 family in autophagy: Differential and compensatory roles in the spatiotemporal regulation of autophagy. *Bmb. Rep.* **2016**, *49*, 424–430. [CrossRef]
31. Tanida, I.; Minematsu-Ikeguchi, N.; Ueno, T.; Kominami, E. Lysosomal turnover, but not a cellular level of endogenous LC3 is a marker for autophagy. *Autophagy* **2005**, *1*, 84–91. [CrossRef]
32. Wei, J.L.; Lin, Y.; Wei, F.H.; Lin, J.G.; Zi, G.X.; Hong, L.W.; Chen, Y.; Liu, H.F. P62 links the autophagy pathway and the ubiqutin–proteasome system upon ubiquitinated protein degradation. *Cell Mol. Biol. Lett.* **2016**, *21*, 29. [CrossRef]
33. Su, H.; Wang, X. Autophagy and p62 in cardiac protein quality control. *Autophagy* **2011**, *7*, 1382–1383. [CrossRef]

Sample Availability: Samples of the compounds are available from the authors.

© 2019 by the authors. Licensee MDPI, Basel, Switzerland. This article is an open access article distributed under the terms and conditions of the Creative Commons Attribution (CC BY) license (http://creativecommons.org/licenses/by/4.0/).

Article

Antrodin C, an NADPH Dependent Metabolism, Encourages Crosstalk between Autophagy and Apoptosis in Lung Carcinoma Cells by Use of an AMPK Inhibition-Independent Blockade of the Akt/mTOR Pathway

Hairui Yang [1,2,3], Xu Bai [1,3], Henan Zhang [1], Jingsong Zhang [1], Yingying Wu [1], Chuanhong Tang [1], Yanfang Liu [1], Yan Yang [1], Zhendong Liu [4], Wei Jia [1,*] and Wenhan Wang [1,*]

- [1] National Engineering Research Center of Edible Fungi, Key Laboratory of Applied Mycological Resources and Utilization of Ministry of Agriculture, Shanghai Key Laboratory of Agricultural Genetics and Breeding; Institute of Edible Fungi, Shanghai Academy of Agricultural Sciences, Shanghai 201403, China; yanghairui1112@163.com (H.Y.); xubai1994@outlook.com (X.B.); henanhaoyun@126.com (H.Z.); zhangjinsong0313@126.com (J.Z.); wuyingying@saas.sh.cn (Y.W.); tangchuanhong123@163.com (C.T.); aliu-1980@163.com (Y.L.); yangyan@saas.sh.cn (Y.Y.)
- [2] WuXi App Tec Co, Ltd., Shanghai 200131, China
- [3] College of Life Sciences, Shihezi University, Shihezi 832003, China
- [4] Food Science College, Tibet Agriculture & Animal Husbandry University, Linzhi 860000, China; liu304418091@126.com
- * Correspondence: jiawei@saas.sh.cn (W.J.); wangwenhan@saas.sh.cn (W.W.); Tel.: +86-21-62200754 (W.J.); +86-21-62200754 (W.W.)

Received: 3 February 2019; Accepted: 6 March 2019; Published: 12 March 2019

Abstract: The current study aims to explore the possible anti-lung carcinoma activity of ADC as well as the underlying mechanisms by which ADC exerts its actions in NSCLC. Findings showed that ADC potently inhibited the viability of SPCA-1, induced apoptosis triggered by ROS, and arrested the cell cycle at the G2/M phase via a P53 signaling pathway. Interestingly, phenomena such as autophagosomes accumulation, conversion of the LC3-I to LC3-II, etc., indicated that autophagy could be activated by ADC. The blockage of autophagy-augmented ADC induced inhibition of cell proliferation, while autophagy activation restored cell death, indicating that autophagy had a protective effect against cell death which was induced by ADC treatment. Meanwhile, ADC treatment suppressed both the Akt/mTOR and AMPK signaling pathways. The joint action of both ADC and the autophagy inhibitor significantly increased the death of SPCA-1. An in vitro phase I metabolic stability assay showed that ADC was highly metabolized in SD rat liver microsomes and moderately metabolized in human liver microsomes, which will assist in predicting the outcomes of clinical pharmacokinetics and toxicity studies. These findings imply that blocking the Akt/mTOR signaling pathway, which was independent of AMPK inhibition, could activate ADC-induced protective autophagy in non-small-cell lung cancer cells.

Keywords: antrodin C; apoptosis; autophagy; AKT; mTOR; metabolic stability

1. Introduction

According to the report, 1.8 million people globally are diagnosed with lung cancer, and 1.6 million people died from this cancer in 2012 [1]. Amongst all cancers, the mortality of lung cancer ranks first and second among men and women, respectively [1]. The most frequent types of lung carcinoma are small-cell lung carcinoma (SCLC) and non-small-cell lung carcinoma (NSCLC), with NSCLC accounting for almost 80% of all lung carcinoma [2,3]. Chemotherapy is one of the major treatments for cancer, although it is known to have strong side effects [4]. Multi-drug resistance and dose limiting adverse reactions are the main reasons for the failure of chemotherapy in NSCLC [5]. Therefore, the development of anti-NSCLC effects and low side effects as well as the safe and long-term use of biologically active substances has become a popular topic of inquiry. Natural bioactive agents, particularly those from medicinal mushrooms, have been considered to have major potential in this field, due to their attractive anticancer effect and hypotoxicity [6–8].

Apoptosis is marked by morphological and biochemical changes including chromatin condensation, chromosome DNA fragmentation, global mRNA decay, etc., and is essentially a caspase-dependent proteolysis process [9,10]. Autophagy is a dynamic procedure which involves the bulk degradation of damaged cytoplasmic organelles and proteins via a lysosome-mediated signaling pathway. Excessive autophagy has the potential to lead to type II programmed cell death, while emerging evidence has shown that autophagy may have a cytoprotective effect on malignant cells [11–14]. It is known that interaction effects between autophagy and apoptosis are complex, sometimes being opposing, and sometimes reinforcing. However, this interaction is key for determining the survival of malignant cells. For this reason, there is a strong possibility that the effect of anti-tumor drugs has a close relationship with the relationship between autophagy and apoptosis.

Taiwanofungus camphorates (M.ZangC.H.Su) Sheng H. Wu et al. is a treasured Taiwanese mushroom which only parasitizes in the inner cavity of the endemic species *Cinnamomum kanehirae* Hayata, Lauraceae or the bull camphor tree [15,16]. *Taiwanofungus camphoratus* is known as the ruby in Taiwan's forest as a result of its excellent biological activities, which include antihepatotoxic, anticancer, anti-inflammatory, antihypertensive, neuroprotective, and antioxidant properties [17–19]. In 2016, its anticancer effect was useful for locating antroquinonol, a ubiquinone derivative isolated from the fruiting body of *T. camphoratus*, and was successfully entered into Phase II clinical trials for treating NSCLC due to its excellent anti-lung cancer effect [20]. Antrodin C (ADC), from mycelium of *T. camphoratus*, is a maleimide derivative. According to reports, more than 80% of all bioactive mushroom compounds are isolated from their fruiting bodies. However, compounds from mycelial are considered to have great future potential due to their low cost and a vast market demand [18]. Our preliminary experiments have also shown an anti-tumor effect of ADC on lung cells which was better than for other malignant cells and is similar to the anti-tumor activity of antroquinonol. Metabolic stability has a close relationship with drug clearance, and so candidate compounds for new drugs are in general analyzed in vitro [21]. In vitro stability analysis has the advantages of being relatively low cost and convenient, which can help to reduce the high cost of new drug development [22]. However, there is as yet no literature on the metabolic stability of ADC.

Therefore, our research aimed to ascertain: firstly, whether ADC could inhibit the proliferation of SPCA-1 cells; secondly, whether it is possible to define the precise mechanism of the inhibitory action; and thirdly, to evaluate phase I of the metabolic stability in vitro.

2. Results

2.1. Effects of ADC In Vitro Cell Proliferation of SPCA-1 and BEAS-2B

The effects of ADC on SPCA-1 cell proliferation were analyzed using alamarBlue®. In this study, ADC was incubated with SPCA-1 cells for 72 h, after which the cell proliferation rate was reduced in a dose-dependent manner (Figure 1A). Particularly, at a concentration of 300 µM, ADC treatment could lead to a 71.41% decrease in cell proliferation when compared with untreated cells. The IC_{50} of ADC was 120.14 µM. These results suggest that ADC could demonstrate an inhibitory effect on SPCA-1 cells.

Low cytotoxicity to normal cells is a key criterion for screening anticancer lead compounds. BEAS-2B cells were isolated from normal human bronchial epithelium as a model system for research of normal human lung epithelium. Therefore, tumor cytotoxicity without damage on normal lung cells was performed by alamarBlue® assay in this study. As shown in Figure 1B, except for 300 uM, the ADC had no inhibition effect on BEAS-2B at 72 h. In this study, the cytotoxicity of ADC to normal cells was very low in vitro. However, cytotoxicity of ADC in vivo needs to be tested in future research.

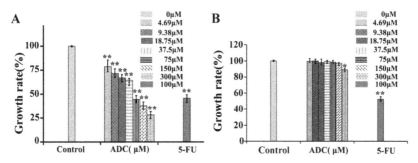

Figure 1. In vitro cell growth–inhibitory activity of ADC. SPCA-1 (**A**) and BEAS-2B (**B**) cell growth inhibition rates are shown after the cells were treated with agents at the indicated concentration for 72 h. The different agents were dissolved and applied in DMSO. 5-FU was used as a positive control * $p < 0.05$, ** $p < 0.01$ vs. control.

2.2. Effects of ADC In Vitro on the Colony Forming Ability of SPCA-1 Cells

The colony formation experiment was carried out in order to assess cancer cells' susceptibility and viability in the presence of ADC in an anchorage-independent environment. Results showed that the colony formation ability of SPCA-1 significantly decreased with ADC. As shown in Figure 2, compared with untreated cells, 240 μM of ADC induced a 76% to 50% decrease in the number of colonies, while 75 μM 5-Fu induced a 74% to 32% decrease in the number of colonies. Result indicate that ADC could significantly suppress the susceptibility and viability of SPCA-1 in vitro.

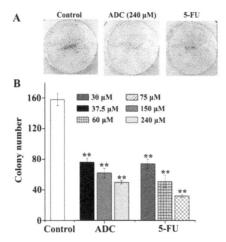

Figure 2. Colony formation assay. (**A**) ADC inhibited colony formation in SPCA-1 cells. After being treated with or without ADC, cells were seeded at 100 cells per plate and allowed to form colonies. After two weeks, the numbers of colonies were counted and recorded. (**B**) Quantification of colony formation. ADC treatment resulted in a significant decrease in colony numbers of cells when compared with untreated cells. 5-FU was used as a positive control ** $p < 0.01$ vs. control.

2.3. Effects of ADC In Vitro on Cell Migration of SPCA-1 Cells

Migration induced by ADC was analyzed by measuring wound closure in a wound healing experiment. The edge of the wounded area and the wound closure was photographed at 0, 24, and 48 h. When comparing ADC treatment, a remarkable increase in wound closure was observed in the untreated cells. As shown in Figure 3, following treatment with ADC, wound closure barely moved, whereas in the untreated cells, the wound closure distance shortened 0.17 times and 0.13 times when treated with 150 µM ADC for 24 and 48 h, respectively. This result suggested that ADC can significantly reduce the cell migration ability of SPCA-1 cells.

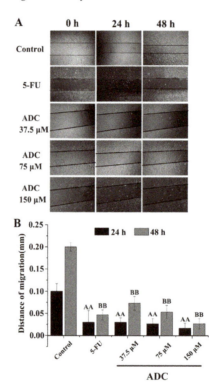

Figure 3. Cell migration assay. A scratch was made on the cell monolayer before treatments (0 h) to serve as a reference for observing cell migration. Photographs were taken at 24 and 48 h time points. ^{AA} $p < 0.01$ vs. control (24 h). ^{BB} $p < 0.01$ vs. control (48 h).

It is well known that the proliferation of cells is closely related to their migration ability. The proliferation of the epithelial cells was shown to be the primary force that drives this cell migration along the villus. Passive mitotic pressure generated by cell division in the intestinal crypts as well as the subsequent gradual expansion in cell diameter along the crypt–villus axis provides a plausible explanation for the steady, continuous migration of epithelial cells [23,24]. Indeed, previous computational models have suggested that these forces alone are sufficient to explain observed rates of cell migration, at least within the crypt [25–30]. Matrix metalloproteinases (MMPs) are a family of zinc-dependent extracellular matrix (ECM) re-modelling endopeptidases which have the ability to degrade almost all components of an extracellular matrix and are implicated in various physiological as well as pathological processes [31]. Carcinogenesis is a multi-stage process in which alteration of the microenvironment is required for the conversion of normal tissue to a tumor [32]. Initially, MMPs were considered to be important only in invasion and metastasis. However, recent studies have shown that MMPs are involved in several steps, such as proliferation, apoptosis, and angiogenesis during carcinogenesis [33].

As shown in Figure 4A, ADC treatment for 24 h led to a decrease in MMP-9 expression when compared with the negative control. As is already known, ginkgolide C (GC) is an MMP-9 activator. This was used to explore the exact effect of MMP on proliferation and migration of SPCA-1. As shown in Figure 4B, the proliferation rate of SPCA-1 increased by the combination treatment with 75 µM ADC and GC (10 and 20 µM) when compared with treatment of 150 µM ADC alone. We also found that co-incubation of 75 µM ADC and 10 µM GC with SPCA-1 could attenuate the inhibition effect of ADC on migration ability (Figure 4C,D). Taken together, the above results suggest that MMP-9 played a key role on both the proliferation and migration abilities of SPCA-1.

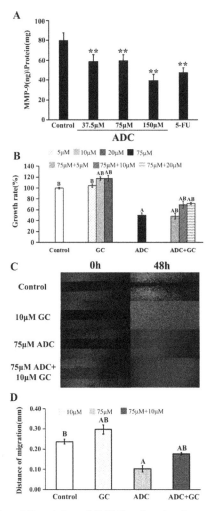

Figure 4. Effects of matrix metalloproteinase (MMP)-9 on the migration and proliferation of SPCA-1. (**A**) Expression of MMP-9 in SPCA-1 was detected by an ELISA kit after being treated with ADC for 72 h, and 5-FU was used as a positive control. ** $p < 0.01$ vs. control; (**B**) effects of MMP-9 activator (GC) on ADC-induced cell growth inhibition were detected by alamarBlue® assay, ^{A} $p < 0.01$ vs. control, ^{B} $p < 0.01$ vs. 75 µM ADC; (**C**) and (**D**) effects of the MMP-9 activator (GC) on ADC-induced cell migration inhibition were detected by microscope, ^{A} $p < 0.01$ vs. control, ^{B} $p < 0.01$ vs. 75 µM ADC.

2.4. Influence of ADC on Cell Cycle of SPCA-1 Cells

Analysis of the cell cycle based on PI staining and quantitative detection of cells were both performed using fluorescence-activated cell sorter (FCAS). As shown in Figure 5, following treatment with a gradient concentration of ADC (37.5, 75, and 150 µM), the percentage of cell cycles at the S phase rose from 36.87% to 43.66%, while at the G2/M phase, this decreased from 11.85% to 5.11%. This suggested that the S cell cycle arrest, which was induced by ADC, led to a further suppression of cell proliferation.

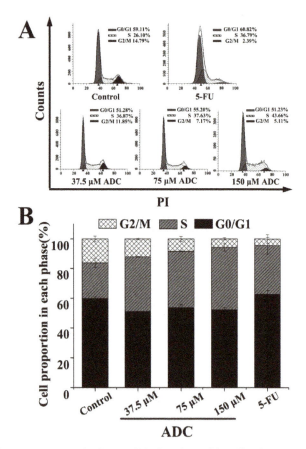

Figure 5. a fluorescence-activated cell sorter (FCAS) analysis of the cell cycle. (**A**) SPCA-1 cells were treated using 37.5, 75, and 150 µM ADC for 72 h. 100 µM 5-FU served as a positive control. The distribution of cell cycle, which was stained by PI, was analyzed using FCAS; (**B**) the percentage of cells in each phase of the cell cycles were calculated.

2.5. Effects of ADC on SPCA-1 Cell Apoptosis

In order to assess the effects of ADC on the apoptosis of lung cancer cells, SPCA-1 cells were exposed to 37.5, 150, and 240 µM ADC for 72 h. A FACS experiment was performed. As can been seen in Figure 6, ADC treatment resulted in an increased proportion of early apoptotic cells and late apoptotic/necrotic cells when compared with the negative control. The total impaired rate of cells was increased by 93% when using 240 µM ADC compared with untreated cells. This result suggests that ADC could have a significant effect on SPCA-1 apoptosis.

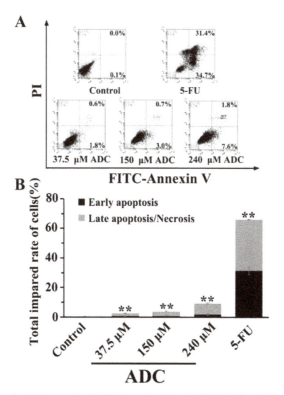

Figure 6. FCAS analysis of apoptosis. (**A**) Cells were incubated with the indicated doses (37.5, 150, and 240 μM) of ADC for 72 h and apoptotic cells were observed with FCAS following annexin V-FITC/PI double staining. (**B**) The total impaired rate of cells in each group was calculated, ** $p < 0.01$ vs. control.

2.6. Effects of ADC on ROS Released by SPCA-1 Cells

Although there is no clear evidence of its origin of reactive oxygen species (ROS), ROS was identified as being involved in cell apoptosis as a crucial intermediate stress response [34]. In this study, DCFH-DA was chosen to demonstrate whether apoptosis of SPCA-1 was induced by ADC through ROS generation. We found that ADC can increase ROS production at a concentration of 30 μM, 45 μM, and 60 μM (Figure 7A), and that ROS production increased 8.6%, 14.9%, and 23.1%, respectively (Figure 7B). This result suggested that ADC had the potential to induce proliferation inhibition and apoptosis of SPCA-1 through an ROS release. As we know, N-acetyl-L-cysteine (NAC) is an ROS scavenger. In this study, NAC was used to determine which ROS was involved in the ADC-induced proliferation inhibition as well as the apoptosis of SPCA-1. As can be seen in Figure 7C, the ADC-induced inhibition effect on cell proliferation was clearly attenuated by NAC. Results shown in Figure 7D,E demonstrate rates of early apoptosis, and total impaired cells were decreased following combination treatment with ADC and NAC when compared with just the ADC treatment.

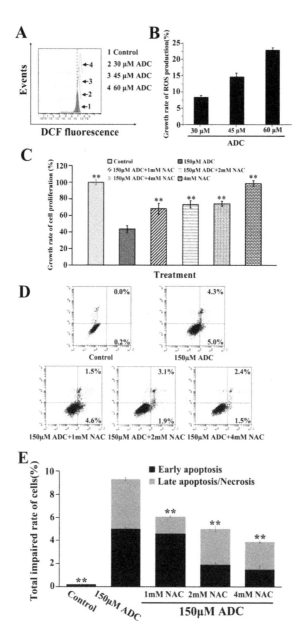

Figure 7. Reactive oxygen species (ROS) detection. (**A**) After being treated with ADC for 24 h, SPCA-1 cells were treated with redox-sensitive fluorescent dye DCFH-DA for 0.5 h, while the ROS generation was determined using FCAS. (**B**) The growth rate of ROS production was calculated. (**C**) Cells were pre-treated with N-acetyl-L-cysteine (NAC) for 2 h prior to exposure to ADC for 72 h, and finally the proliferation rate of cells was detected. ** $p < 0.01$ vs. 150 µM NAC. (**D**) Cells were pre-treated with NAC for 2 h prior exposure to ADC for 72 h, and apoptotic cells were observed with FCAS following annexin V-FITC/PI double staining. (**E**) The total impaired rate of cells in each group was calculated, ** $p < 0.01$ vs. 150 µM ADC.

2.7. Effects of ADC on Activation of Caspase-3, P53, and Bcl-2

Caspases are the important intermediate of apoptosis. Once the proteolytic is activated, procaspase forms will be converted to their enzymatically active forms. In addition, the consecutive activation of caspases is the crucial meditator of the mitochondrial apoptotic pathway [35]. Once SPCA-1 cells were treated with 100 µM ADC for 72 h, cleaved caspase-3 expression reached up to 15.33-fold. P53, known as a genes transregulator, is involved in DNA synthesis, repair, and apoptosis. When compared with untreated cells, the co-culture of SPCA-1 cells with 100 µM ADC caused P53 expression of up to 3.31-fold. In addition, Bcl-2 expression was down 1.45-fold after being treated with 100 µM ADC for 72 h compared with untreated cells. These results (Figure 8) suggest that caspases and the Bcl-2 family have taken part in ADC-induced SPCA-1 cell apoptosis by increasing the expression of pro-apoptotic members (cleaved caspase-3 and P53) and decreasing the expression of anti-apoptotic members (Bcl-2).

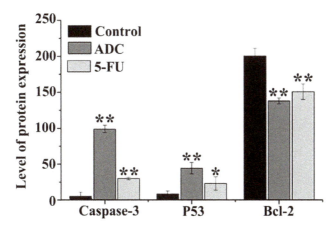

Figure 8. Expression of cleaved caspase-3, Bcl-2, and P53 in cells treated with ADC. Following treatment with 100 µM ADC for 72 h, cleaved caspase-3 and P53 were increased. The expression of Bcl-2 was reduced after ADC treatment for 72 h. * $p < 0.05$, ** $p < 0.01$ vs. control.

2.8. Effects of ADC on Autophagy Activation in SPCA-1 Cells

Autophagy has been of great interest for those developing novel anticancer treatments in recent years. In order to testify whether ADC induced autophagy activation of SPCA-1, TEM was employed to detect the formation of autophagic vacuoles. Autophagic vacuoles formation, which contain lamellar structures or residual digested material as well as empty vacuoles, is an indication of autophagy activation. Increased numbers of vacuoles and mature autophagosomes were frequently observed in cells when they were treated with 200 µM ADC, which indicated that ADC could activate autophagy of SPCA-1 (Figure 9A).

The conservation of LC3-I to LC3-II through lysis, lipofication, and proteolysis is a crucial character of autophagy activation, and so the LC3 protein serves as one of the most important markers of autophagy. Therefore, it is necessary to confirm whether ADC treatment could induce redistribution of LC3 (autophagosome) in SPCA-1. Results of the FCAS indicated that ADC could activate LC3-II formation (Figure 9B). Western blotting results also demonstrated that conversion of LC3-I to LC3-II increased 479.23%, 403.19%, and 382.27% following 50, 100, and 200 µM ADC treatment when compared with untreated cells (Figure 9C,D). The mRFP-GFP-LC3 detected by the live-cell imaging method was used to observe the flux rate of autophagy of SPCA-1 cells, while the reduction of GFP was used to indicate the process of autophagy. Results demonstrated that expressed GFP-LC3 was downregulated by ADC treatment, dependent on dose, while a combination of RFP and GFP was detected as yellow speckles in the untreated cells more than in the ADC treated cells, suggesting that autophagy was induced through activation of autophagic (Figure 9).

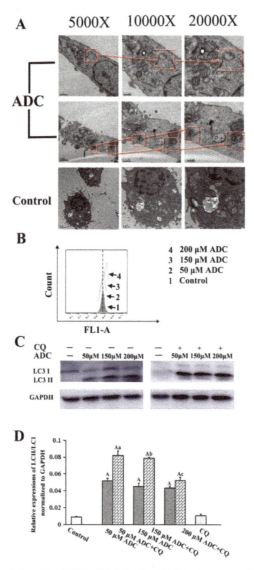

Figure 9. Autophagy induced by ADC in SPCA-1 cells. (A) Cytoplasm vacuolization enclosed in a double membrane in SPCA-1 cells was found using TEM. The arrow and rectangular frame indicate autophagic vacuoles; (B) conversion of LC3-I to LC3-II in ADC-treated SPCA-1 was found using FCAS; (C) effects of autophagy inhibitors (CQ) on ADC-induced cell growth inhibition was detected using Western blotting; (D) bar graphs demonstrated the expression of LC-II/LC-I. Data are the mean ± SEM of three independent experiments. [A] $p < 0.01$ vs. control; [a] $p < 0.01$ vs. 50 μM ADC; [b] $p < 0.01$ vs. 150 μM ADC; [c] $p < 0.01$ vs. 200 μM ADC.

As we know, both lysosome and autophagosome formation have the potential to induce an increased level of LC3B-II. However, in order to identify the reason for increased levels of LC3-II, the late-stage autophagy inhibitor chloroquine (CQ, 100 µM) must be used. Results indicated that LC3-II expression in SPCA-1 cells was increased 59.02%, 75.73%, and 21.15% following the combination treatment with 50, 100, and 200 µM ADC when compared with sole treatment of ADC, which suggests that increased levels of LC3B-II were mainly due to the increase in the number of vacuoles and mature autophagosomes (Figure 10A,B). In summary, these results demonstrate that autophagic activity (autophagic flux) was upregulated through the increased level of autophagosome formation in SPCA-1 cells which were treated with ADC.

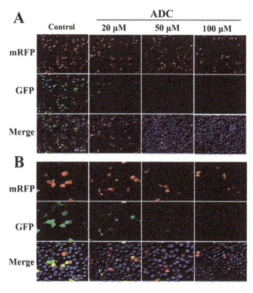

Figure 10. ADC induced autophagy flux. The living cell image was found using a laser scanning confocal microscope. The images were then magnified 30 times (**A**) and then 60 times (**B**).

2.9. Protective Effects of Autophagy on ADC-Induced Apoptotic SPCA-1 Cell Death

Autophagy has the potential to either protect cell survival by serving as an antagonist to block apoptotic cell death or by contributing to cell death [36]. Therefore, the exact role of autophagy in the anti-NSCLC activity of ADC was clarified in this study. A significant decrease (18.74%) or increase (17.08%) was found in the growth rate of SPCA-1, of which autophagy was inhibited by CQ or activated by RAPA following ADC treatment, in contrast with treatment by ADC alone (Figure 11A). Cell viability data was similar to results found in cell apoptosis. As shown in Figure 11C,D, apoptotic cells were increased 1-fold by combination treatment with 150 µM ADC and CQ when compared with treatment of 150 µM ADC alone. These results indicated that ADC induced protective autophagy in SPCA-1 cells. In this study, the RAPA was further used to explore the extract stage of protective autophagy, and the results showed that pre-treatment with RAPA did not restore the ADC-triggered cell death (Figure 11B). When combined with the results of RAPA in Figure 11A,B, this suggests that the survival effect only occurred when the cells were under stress.

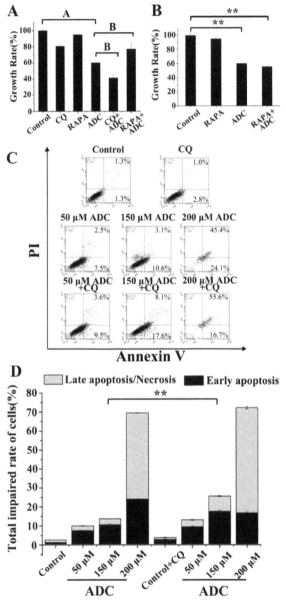

Figure 11. ADC induced protective autophagy. (**A**) Autophagy inhibition was able to protect cells from death. Cells which were pre-treated with ADC were treated with RAPA, and the proliferation rate of cells' death was detected. ^{A} $p < 0.01$ vs. control; ^{B} $p < 0.01$ vs. ADC. (**B**) Cells were pre-treated with RAPA for 3 h, then treated with ADC for 72 h, and finally the proliferation rate of cells was detected. ** $p < 0.01$ vs. control. (**C**) Cells were incubated with 50, 150, and 200 μM ADC for 72 h, and then treated with CQ. The early and late/necrosis apoptotic cells were observed with FCAS after annexin V-FITC/PI double staining. (**D**) The percentage of apoptosis in SPCA-1 was calculated. ** $p < 0.01$, 150 μM ADC vs. 150 μM ADC + CQ (total impaired rate of cells).

2.10. ADC Downregulated the AKT-mTOR Pathway and AMP-Activated Protein Kinase (AMPK) Pathway

As we know, AMPK and PI3K/Akt/mTOR are the two key signaling pathways which regulate autophagy and apoptosis. Recent reports have suggested that autophagy is negatively regulated by the PI3K/Akt/mTOR pathway [37]. To determine whether ADC-induced autophagy is involved in the Akt-mTOR pathway, FCAS were used for further detection. A phosflow analysis demonstrated that the phosphorylation of mTOR (Ser2448) and Akt (Ser473) was downregulated by ADC treatment in SPCA-1 cells, dependent on dose, which suggested that ADC-induced autophagy was activated through downregulation of the Akt-mTOR pathway (Figure 12A).

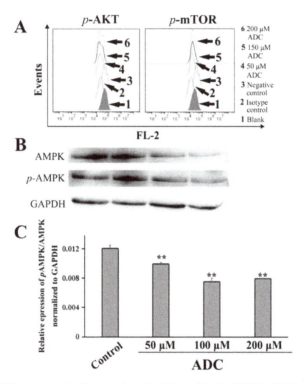

Figure 12. ADC-induced protective autophagy is attributed to the AMPK inhibition-independent blockade of the Akt/mTOR pathway. (**A**) ADC downregulated the AKT-mTOR pathway. Following treatment by ADC, cells were detected by FCAS. (**B**) ADC downregulated the AMPK pathway. Following treatment by ADC, cells underwent lysis and were detected by Western blotting. (**C**) Bar graphs demonstrate the relative expression of pAMPK/AMPK. Data are the mean ± SEM of the three independent experiments. ** $p < 0.01$ vs. control.

AMPK is a key signal factor involved in regulating the balance of cellular energy. When the cell is hungry, the AMPK signaling pathway will tell the body to increase glucose and fatty acid uptake. Recently, an increasing number of studies have shown that the AMPK signaling pathway is related to autophagy activation [37]. Western blotting results have indicated that the relative expression of pAMP/AMPK decreased by 17.67%, 37.63%, and 34.20% following 50, 100, and 200 µM ADC treatment, respectively, when compared with untreated cells (Figure 12B,C). Based on the results above, autophagy induced by ADC has a close relationship with the AMPK inhibition-independent blockade of the Akt/mTOR signaling pathway in SPCA-1 cells.

2.11. Metabolic Stability of ADC in SD Rat and Human Liver Microsomes

As shown in Table 1, following incubation with rat liver microsomes for 0, 5, 10, 20, 30, and 60 min, results were reported as percent remaining for ADC, which was 100%, 52.5%, 30.2%, 13.3%, 7.7%, and 4.6%, respectively (Figure 13). The value of $T_{1/2}$ (min) in the rat liver microsomes, which was calculated by the % of remaining data versus time, was 7.5 (Table 1). The values of $CL_{int(mic)}$ (μL/min/mg protein) and $CL_{int(liver)}$ (mL/min/kg) in the rat liver microsomes were 185.8 and 334.4, respectively (see Table 1, above).

Table 1. Metabolic stability of ADC in SD rat and human liver microsomes following incubation in the presence of NADPH.

Species	$T_{1/2}$ (min)	$CL_{int(mic)}$ (μL/min/mg protein)	$CL_{int(liver)}$ (mL/min/kg)
SD rat	7.5	185.8	334.4
Human	54.1	25.6	23.0

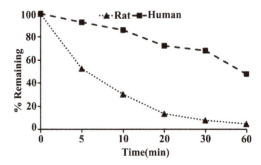

Figure 13. Percentage remaining versus time profile of ADC in SD rat and human liver microsomes.

As can be seen in Table 1, following incubation with human liver microsomes for 0, 5, 10, 20, 30, and 60 min, the percent remaining for the ADC was 100%, 92.5%, 85.8%, 72.2%, 68.0%, and 47.8% respectively (Figure 13). The value of $T_{1/2}$ (min) in the human liver microsomes was 54.1 (Table 1). The value of $CL_{int(mic)}$ (μL/min/mg protein) and $CL_{int(liver)}$ (mL/min/kg) in the human liver microsome were 25.6 and 23.0, respectively (Table 1).

3. Discussion

A pressing problem in the success of NSCLC therapy is tackling the diverse side effects and resistance of chemotherapy to current anticancer agents. Therefore, new strategies for drug development are urgently needed to solve the problem of drug resistance in anti-NSCLC treatment. In recent years, medicinal mushrooms have obtained global attention as a result of their strong therapeutic activity as both chemical inhibitors and immunomodulators.

Previous studies have reported that the ethanolic extract of *T. camphoratus* treatment could induce cell cycle arrest and proliferation inhibition of Hep3B and HepG5 [38]. Chen and colleagues [39] have found that both incubation amphotericin B and *T. camphoratus* could induce significant increases in proliferation inhibition and apoptosis rates of HT29 cells. The noteworthy anticancer-effect of *T. camphoratus* may contribute to the large amounts of polysaccharides, terpenoids, succinic acid, and maleic acid derivatives. According to existing research, antroquinonol has excellent anti-tumor effects on various malignant cells [40]. In this study, ADC exhibited anticancer effects on human NSCLC cell lines SPCA-1. These effects included cell growth inhibition, cell migration reduction, ROS release, cell cycle arrest, autophagy, and apoptosis. The effective G2 checkpoint mechanism is a key characteristic of many malignant cells. Based on these findings, abrogation of the cell cycle G2 checkpoint may become a new target for anti-NSCLC chemotherapy [41]. It has been found that the anti-tumor target of several lead compounds was the G2/M cycle arrest-dependent inhibition of cell proliferation [42,43]. In this study, the form of the cell proliferation inhibition induced by ADC has been shown to be dependent on the G2 cycle arrest, a finding which indicated that ADC can be used as a potential anti-NSCLC agent.

Inducing the apoptosis of malignant cells is one of the most important strategies for cancer chemotherapy. Therefore, the effect of ADC on NSCLC apoptosis was assessed in this study. The result of the FCAS experiment showed that ADC treatment could significantly increase the number of apoptotic cells, which is positively correlated with its concentration, and so suggested that ADC exhibited anticancer activity through apoptosis. Reactive oxygen species is a required metabolite of normal metabolism in cell growth which plays a critical role in the stress response as an important mediator. Some studies have shown that excessive ROS led to the occurrence of apoptosis [44–46]. The FCAS results showed that ADC induced ROS release, which could be a factor of the occurrence of apoptosis. The similar ROS-mediated cytotoxicity of other anti-NSCLC compounds has been found by other researchers. Zhou et al. [47] found that luteoloside could induce proliferation inhibition of NSCLC through ROS-mediated G_0/G_1 arrest. Song and colleagues [48] also discovered that the anti-tumor effect of SZC017, an oleanolic acid derivative, had a close relationship with ROS-dependent apoptosis. As we know, the loss of mitochondrial transmembrane potential will lead to ROS generation. The results of Zhang et al. [49] demonstrated that cedrol treatment could lead the loss of mitochondrial transmembrane potential and cause ROS-mediated apoptotic cell death in A549 NSCLC. The above results have indicated that the anti-tumor effects of these compounds were a close relationship with the AKT-related signaling pathway. Interestingly, Wang et al. [50] demonstrated that the role of ROS on endothelial dysfunction was opposite to its role on NSCLC. Their results showed that ADC could inhibit hyperglycemia-induced endothelial cell dysfunction via the inhibition of ROS generation, senescence, growth arrest, and apoptosis in cultured HUVECs [50]. In addition, Wang and colleagues [50] found that the Nrf2-related signaling pathway was involved in the protective effects of ADC on hyperglycemia-induced vascular endothelial cells' senescence and apoptosis. Therefore, the different signaling pathways in the two diseases may be the main reason for the inconsistency of ROS roles.

The Bcl-2-related protein family is the most representative apoptotic regulatory gene family [51]. Chaetoglobosin K25 [52] and buforin IIb 26 [53] have been found to activate apoptosis through a decrease of Bcl-2 expression. The Bcl-2-related protein family consists of a pro-apoptotic gene (Bcl-2, Bcl-xL, etc.) and an anti-apoptotic gene (Bax, Bak, etc.). The dialogue between the pro-apoptotic gene and the anti-apoptotic gene will determine the fate of a malignant cell [54].

Moreover, P53 plays a diverse role in Bcl-2-dependent apoptosis which substantially affects the ADC-mediated cell death. The proapoptotic protein Bax is activated by P53, while the anti-apoptotic protein Bcl-2 expression is decreased by P53 [55]. Once pro-apoptotic factors are released, the caspase family was further activated, resulting in caspase-3 cleavage. Our results indicated that ADC-induced apoptosis could be attributed to the decrease of Bcl-2 protein and increase of caspase-3 and P53 proteins in the SPCA-1 cells.

The role of autophagy in malignancy has attracted much attention due to the viable option it offers cancer therapies. Our results showed that ADC-induced autophagy promoted the formation of autophagosomes, activated the conversion of LC3B-I to LC3B-II, and induced autophagy flux.

The relationship between cancer and autophagy is complex and hard to unpick. According to the literature, autophagy activators are used clinically for anticancer treatment [56]. However, in contrast to the positive effects of autophagy activators on apoptosis-resistant malignancy treatment, some researchers have found that drug resistance in some malignant tumors might be the result of a relationship with autophagy activation [57]. Autophagy can inhibit tumor genesis by maintaining cell stability and reducing cell damage [58]. The apoptosis sensitivity of cancer cells can be increased by inhibiting autophagy, particularly in the late stage [59]. In this study, autophagy inhibitors CQ and autophagy activators RAPA were used to determine the exact effects of altered autophagy on NSCLC death, induced by ADC. In accord with these findings, the results of our study indicated that autophagy inhibition could accelerate ADC-induced cell death, and that autophagy activation could restore ADC-induced cell death. Furthermore, the combination of pro-apoptosis agents and anti-autophagy agents is perhaps a magic bullet in accelerating the anti-tumor effect on human NSCLC.

As we know, the PI3K/Akt/mTOR signaling pathway plays an important role on migration, invasion, and proliferation of malignant tumors. Duan and colleagues [60] found that the PI3K/ATK/mTOR pathway was related to invasion, migration, and proliferation of colorectal cancer induced by IMPDH2. Paul et al. [61] demonstrated that the binding between SDS22 (protein phosphatase 1 regulatory subunit) and AKT has the potential to lead to dephosphorylate ATK at Thr308 and Ser473 through PP1, and hence can inhibit proliferation, invasion, and migration of tumor cells. These findings indicate that the inhibition of the Akt/PI3K/mTOR pathway might lead to proliferation, migration, and invasion of SPCA-1. Studies have confirmed that many compounds suppress tumor growth and activate autophagy through the downregulation of this pathway. The autophagy activation of Rotten was found in prostate and pancreatic cancer stem cells, and results indicated this activation effect was dependent on the Akt/PI3K/mTOR pathway [62,63]. As shown in our results, the phosphorylation of mTOR and Akt was dose-dependently downregulated by the ADC treatment, thus initiating autophagy in SPCA-1 cells. Results implied that inhibition of the mTOR-Akt pathway was related to autophagy, which was induced by ADC autophagy.

As well Akt, AMPK, another important mTOR regulator, is activated in intracellular and external environmental pressures as a crucial energy sensor [64]. Studies have shown that AMPK participates in controlling the fate of cancer cells [65]. That is, AMPK activation inhibited mTOR both in the cell cycle arrest and apoptosis in vitro and in vivo. In contrast, cells were protected by the AMPK signaling pathway from chemotherapy-induced apoptosis and metabolic stress in certain circumstances [66]. The mTOR-mediated autophagy in cancer cells seems to be impacted both by AMPK-dependent cytotoxicity and cytoprotection [67–70]. Our results have shown that phosphorylation of AMPK and AMPK was dose-dependently downregulated by the ADC treatment, which indicates that ADC-induced autophagy was independent of AMPK inhibition while being mediated via the inhibition of the Akt/mTOR signaling pathway.

The microsomal stability assay is widely used in vitro model to characterize the metabolic conversion by phase I enzymes, in which the cytochrome P450 (CYP) family are the most important. Since metabolism is known to be highly variable in different species, microsome stability assay is commonly run in multiple species. The most commonly used species include humans for predicting clinical pharmacokinetics, rats for toxicity studies, mice for efficacy models, and dogs and monkeys for large animal toxicity. In this study, we tested the metabolism of ADC in rat and human liver microsomes. Results demonstrated that ADC was highly metabolized in SD rat liver microsomes, and moderately metabolized in human liver microsomes. This suggests the need for further modification of the metabolic stability of ADC and also contributes to predicting the in vivo PK performance [71,72].

4. Materials and Methods

4.1. Chemicals and Reagents

T. camphoratus (Access number: J1) was from the Preservation Center of Fungi, Institute of Edible Fungi, Shanghai Academy of Agricultural Sciences (Shanghai, China). Human lung adenocarcinoma cell line SPCA-1 (Access number: TCHu 53) was purchased from Cell Resource Center of Shanghai Institute of Life Sciences, Chinese Academy of Sciences (Shanghai, China). The BEAS-2B (Acess number: GDC0139) was purchased from China Center for Type Culture Collection (Wuhan, China). Ribonuclease A (RNase A, #EN053), Pierce™ BCA protein assay kit (#23225), Dulbecco's Modified Eagle Medium (DMEM, #11965-092), no phenol red DMEM (#31053-028) and fetal bovine serum (FBS, #10099-141) were from Thermo Fisher Scietific Inc. (Waltham, MA, USA). Annexin

V-fluoresceinisothiocyanate (FITC) apoptosis detection kit was from Nanjing KeyGen Biotech. Co. Ltd. (Nanjing, China). N-acetyl-L-cysteine (NAC), Chloroquine (CQ, #PHR1258), 5-Fluorouracil (5-Fu, #03738) and Propidium iodide (PI, #P4170) were from Sigma–Aldrich Co. LLC (St Louis, MO, USA). alamarBlue® (#BUF012B) was from Bio-Rad Laboratories, Inc. (Hercules, CA, USA). The Bcl-2 ELISA assay kit, P53 ELISA assay kit, and caspase-3 (Asp175) DuoSet IC ELISA were from R&D Systems Inc. (Minneapolis, MN, USA). Ginkgolide C(GC), rapamycin (RAPA) and chloroquine (CQ) were purchased from Sigma–Aldrich (St Louis, MO, USA). The Cyto-ID® Autophagy detection kit was obtained from Enzo Life Sciences Inc. (Farmingdale, NY, USA). The polyvinylidene difluoride (PVDF) membrane was purchased from EMD Millipore Inc. (Billerica, MA, USA). Phospho-PI3 kinase p85 (Tyr458)/p55 (Tyr199) Antibody was from Cell Signaling Technology (Beverly, MA, USA). The LC3 antibody was from Abcam Co. (Cambridge, UK). The Human liver microsomes (#452117) were from BD Gentest (Corning, NY, USA), and the rat liver microsomes (#R1000) were from Xenotech (Kansas, KS, USA). RayBio® Human MMP-9 ELISA kit (#P14780) was from RayBiotech, Inc. (Norcross, GA, USA).

4.2. Cell Lines and Cell Culture

The SPCA-1 and BEAS-2B were cultured in DMEM or DMEM/F12 supplemented with 10% FBS, 1% penicillin–streptomycin at 37 °C and 5% CO_2 in an incubator with a humidified atmosphere.

4.3. Cell Viability Assay

The SPCA-1 or BEAS-2B were seeded in 96-well plates at 2×10^4 cells/mL and cultured overnight. Cells were then treated with ADC (4.69, 9.38, 18.75, 37.5, 75, 150, and 300 µM) for 72 h. Medium containing 5‰ DMSO and 5-Fu (100 µM) served as negative and positive controls. After treatment, the medium was aspirated, and fresh phenol-red-free medium containing 5 µg/mL alamarBlue@ was added into all tested and control wells. After incubation at 37 °C for 4–6 h, when the medium color changed, the absorbance at 570 nm and 600 nm was measured using a spectrophotometric plate reader (Bio-Tek Instruments, Inc, Winooski, VT, USA). The proliferation rate was calculated according to the following formula:

$$\text{Proliferation rate (\%)} = 1 - \frac{117216 \times A_{570 \text{ (sample)}} - 80586 \times A_{600 \text{ (sample)}}}{117216 \times A_{570 \text{ (control)}} - 80586 \times A_{600 \text{ (control)}}} \times 100\%$$

4.4. Clone Formation Assay

Cells (2×10^5 cells/mL) were seeded in a 12-well plate and cultured overnight. The cells were then treated with 37.5, 150, and 240 µM ADC for 72 h. Medium containing 5‰ DMSO and 5-Fu (30, 60, and 75 µM) were served as the negative and positive controls, respectively. Cells were then seeded in 6-well culture plates at 100 cells/well. Each treatment had three repeated wells. After incubation for two weeks at 37 °C, cells were washed twice with PBS, and fixed with 40 µL/mL paraformaldehyde, and then stained with crystal violet. Groups of 50 or more cells were counted as colonies. Clone formation efficiency was calculated as (number of colonies/number of cells inoculated) × 100%.

4.5. Wound-Healing Assay

Cells were seeded in 12-well plates at 2×10^5 cells/mL and cultured overnight. A scrape was placed through the middle of the confluent cultures with a sterile pipette tip, and washed with PBS to remove debris, followed by treated with 37.5, 75, and 150 µM ADC. Medium containing 5‰ DMSO (5 µL/mL) and 200 µM 5-Fu were served as the negative and positive controls, respectively. The wound was observed under a phase-contrast microscope everyday (Olympus Corporation, Tokyo, Japan).

To further verify the role of MPP-9 on proliferation and migration of SPCA-1, an MMP-9 ELISA assay kit and MMP-9 activator GC were used. Cells (2×10^5 cells/mL) were seeded in 12-well plate and cultured overnight. The cells were then treated with 37.5, 75 and 150 µM ADC for 72 h. Medium containing 5‰ DMSO and 100 µM 5-Fu were served as the negative and positive controls, respectively.

After treatment, cells were collected and washed with cold PBS, and resuspended in RIPA lysis buffer, and then left on ice for 30 min. The lysate was centrifuged at 1×10^4 rpm at 4 °C for 30 min, and then total protein level was measured with a BCA assay kit. The absorbance was determined by a spectrophotometric plate reader at 450 nm. The samples (100 µL) were added to each well of the MMP-9 plate, and the plate was covered with a sealer. After incubation for 2.5 h at room temperature with a gentle shake, the plate was washed with 300 µL wash buffer four times. After the last wash, any remaining wash buffer was completely removed, and then the plate was inverted and blotted against clean paper towels. The 1× prepared biotinylated antibody was added to each well of the plate, and then the plate was incubated for 1 hr at room temperature with gentle shaking. The solution was dicarded, and the wash step was repeated. The prepared streptavidin solution was added to each well, and then plate was incubated for 45 min at room temperature with gentle shaking. The solution was dicarded, and the wash step was repeated. The TMB one-step substrate reagent was added to each well, and then the plate was incubated for 30 min at room temperature in the dark with gentle shaking. The stop solution (50 µL) was added to each well, and then the absorbance was determined by a spectrophotometric plate reader at 450 nm.

The SPCA-1 were seeded in 96-well plates at 2×10^4 cells/mL and cultured overnight. Cells were then treated with GC (5, 10, and 20 µM), ADC (75 µM) or ADC (75 µM) + GC (5, 10, and 20 µM) for 72 h, respectively. The proliferation was analyzed by alamarBlue® assay as shown in Section 4.3.

Cells were seeded in 12-well plates at 2×10^5 cells/mL and cultured overnight. A scrape was placed through the middle of the confluent cultures with a sterile pipette tip, and washed with PBS to remove debris, followed by treated with 10 µM GC, 75 µM ADC or 75 µM ADC + 10 µM GC for 48h, respectively. Medium containing 5‰ DMSO (5 µL/mL) was served as the negative control. The wound was observed under a phase-contrast microscope everyday (Olympus Corporation, Tokyo, Japan).

4.6. Cell Cycle Analysis

Cells were seeded in 12-well plates (2×10^5 cells/mL) and cultured overnight. The cells were treated with 37.5, 75, and 150 µM ADC for 72 h. Medium containing 5‰ DMSO and 5-Fu (100 µM) were served as the negative and positive controls, respectively. After treatments, the cells were harvested, washed with PBS, and fixed in 70% ethanol at 4 °C overnight. Cells were collected by centralization, washed with PBS twice, and added with 100 µL PBS containing 0.01 µg/mL PI and 0.005 mg/mL RNase A, then incubated at room temperature in dark for 15 min. Cells were analyzed using a fluorescence activated cell sorting (FCAS), and the results were analyzed using Modfit software.

4.7. Cell Apoptosis Detection

Cell apoptosis was determined by Annexin V-FITC apoptosis detection kit according to the manufacturer's instruction. Briefly, cells (2×10^5 cells/mL) were seeded in 12-well plate and cultured overnight. The cells were then treated with 37.5, 150, and 240 µM ADC for 72 h. Medium containing 5‰ DMSO and 5-Fu (100 and 400 µM) served as the negative and positive controls, respectively. After different treatments, cells were harvested and washed twice with cold PBS, and then resuspended in 100 µL 1× binding buffer solution containing 0.2 µg/mL Annexin V-FITC and 0.5 µg/mL PI, and then incubated at room temperature for 10 min in the dark. Apoptotic cells were analyzed using BD Accuri C6 flow cytometer (BD Biosciences, San Jose, CA, USA).

4.8. ROS Detection

2′,7′-Dichlorofluorescin diacetate (H_2DCFDA) is a cell permeable non-fluorescent probe. Upon oxidation by any hydroperoxide such as ROS, the non-fluorescent H_2DCFDA is converted to the highly fluorescent 7′-dichlorofluorescein (DCF). Fluorescent density reflected ROS level [73]. Therefore, intracellular ROS levels were determined using an H_2DCFDA probe as described. Briefly, cells were seeded in 24-well plates (2×10^4 cells/mL) and cultured overnight. Medium containing 5‰ DMSO served as the negative controls. After treatment with ADC (30, 45, and 60 µM) for 24 h in phenol-red-free DMEM, cells were incubated with 25 μM H$_2$DCFDA for 30 min at 37 °C. Then cells were harvested, washed, and resuspended in PBS. Cells were analyzed using FCAS, and the results were analyzed using Flowjo software.

To further verify the role of ROS in ADC-induced inhibition of SPCA-1 proliferation, a ROS scavenger was used to treat the cells before ADC treatment. The SPCA-1 cells were seeded in 96-well plates at 2×10^4 cells/mL and cultured overnight. Cells were pretreated with 1, 2, and 4 mM NAC for 2 h prior to exposure to 200 μM ADC for 72 h. The proliferation was analyzed by alamarBlue® assay as shown in Section 4.3.

To further verify the role of ROS in ADC-induced apoptosis of SPCA-1, NAC was used to treat the cells before ADC treatment. The SPCA-1 cells were seeded in 12-well plates at 2×10^5 cells/mL and cultured overnight. Cells were pre-treated with 1, 2, and 4 mM NAC for 2 h prior to exposure to 150 μM ADC for 72 h. The apoptotic cells were analyzed by FCAS after annexin V-FITC/PI double staining as shown in Section 4.7.

4.9. ELISA Assay

Cleaved caspase-3, Bcl-2, and P53 protein expression were measured using commercial cleaved caspase-3, Bcl-2, and P53 assay kits according to the manufacturer's instructions, respectively. Briefly, cells (2×10^5 cells/mL) were seeded in 12-well plates and cultured overnight. The cells were then treated with 150 μM ADC for 72 h. Medium containing 5‰ DMSO and 100 μM 5-Fu served as the negative and positive controls, respectively. After treatment, cells were collected and washed with cold PBS and resuspended in RIPA lysis buffer, and then left on ice for 30 min. The lysate was centrifuged at 1×10^4 rpm at 4 °C for 30 min, and then total protein level was measured with a BCA assay kit. The absorbance was determined by a spectrophotometric plate reader at 450 nm. The sample or standard (100 μL) were added to each well of the plate of cleaved caspase-3, Bcl-2 or P53, and the plates were covered with a sealer. After incubation for 2 h at room temperature, the plates were washed with wash butter three times. A diluted detection antibody (100 μL) was added to each well, and the plates were covered with a new plate sealer. After incubation for 2 h at room temperature, the plates were washed with wash butter three times again. The diluted Streptavidin-HRP A was added to each well of the plates, and the plates were incubated for 20 min in dark. After incubation for 20 min, the stop solution was added to each well, and the plates were gently tapped to ensure thorough mixing. The absorbance was determined by a spectrophotometric plate reader at 450 nm. The increase in protein expression was calculated as the ratio between values obtained from the treated samples versus those obtained in untreated controls.

4.10. Transmission Electron Microscopy Analysis

Transmission electron microscopy (TEM) was employed to identify autophagosomes in lung adenocarcinoma cell line SPCA-1. Cells were treated with 200 μM ADC for 72 h. Cells were harvested, washed and fixed in 2.5% glutaraldehyde in 0.1 M phosphate buffer, then post-fixed in 1% osmium tetroxide buffer. After dehydration in a graded series of ethanol, the cells were embedded in spur resin. Thin sections (70 nm) were cut on an ultramicrotome. The sectioned grids were stained with saturated solutions of uranyl acetate and lead citrate. The sections were examined by a JEM1230 electron microscope from JEOL (Tokyo, Japan).

4.11. Flow Cytometric Analysis of Autophagy with Cyto-ID Staining

The Cyto-ID autophagy detection kit (ENZ-51031-K200) from Enzo Life Sciences (New York, NY, USA) was used to stain live cells according to the manufacturer's instructions. This assay is dependent on a 488 nm-excitable green fluorescent detection reagent, which specifically fluoresces in autophagic vesicles [73]. Cells (2×10^5 cells/mL) were seeded in 12-well plates and cultured overnight. The cells were then treated with 50, 150, and 200 μM ADC for 72 h. Medium containing 5‰ DMSO and combination of 1 μM CQ or 500nM rapamycin served as the negative or positive controls, respectively.

After treatment, cells were washed with DPBS, and then incubated with Cyto-ID Green containing indicator free cell culture medium containing FBS (100 µL/mL) for 30 min at 37 °C, 5% CO_2 in the dark. At the end of staining procedure, the Cyto-ID Green containing medium were washed away, and then cells were trypsinized, washed, and resuspended in ice-cold DPBS containing FBS (40 µL/mL). Cells were analyzed using FCAS, and the results were analyzed using Flowjo software.

4.12. Western Blotting Analysis

Whole cell lysates were extracted using RIPA lysis buffer containing protease inhibitor and phosphatase inhibitor. Protein concentrations were measured by a Pierce BCA protein assay kit. Proteins were separated using SDS-PAGE and transferred to PVDF membranes and subsequently blocked with 5% BSA, and incubated with primary antibody (1:1000) from Cell Signaling Technology (Danvers, MA, USA) overnight at 4 °C. Immunoblots were followed by 1–2 h incubation in secondary antibody (1:300, BOSTER). After washing three times with TBS-T, the signal was detected using TanonTM High-sig ECL Western Blotting Substrate from Tanon Science & Technology Co., Ltd. (Shanghai, China) by the Amersham Imager 600 from GE Healthcare (Pittsburgh, PA, USA).

4.13. Live-Cell Imaging for Autophagic Flux

Laser scanning confocal microscopy were employed to identify autophagic flux. The SPCA-1 cells were transfected with mRFP-GFP-LC3B-expressing plasmid by Turbofectmin, and then cultured for 6 h. Cells were treated with 20, 150, and 200 µM ADC for 24 h, followed by incubation with DAPI for 5 min, cells were washed three times with PBS, and then cells were examined by an OLYMPUS FV1200 laser scanning confocal microscope.

4.14. FACS/Phosflow

A PE mouse anti-mTOR (pS2448), PE mouse anti-AKT (pS473), and PE mouse IgG1 kappa isotype control were measured by FCAS according to the manufacturer's instructions. Briefly, cells (2×10^5 cells/mL) were seeded in 12-well plates and cultured overnight. The cells were treated with 50, 150, and 200 µmol/L ADC for 24 h. Medium containing 5‰ DMSO served as the negative control. After treatments, cells were trypsinized and washed with cold stain buffer, and then fixed with cold fixation buffer for 30 min at 4 °C. After washing with staining buffer, cells were permeabilized with perm buffer III for 30 min at 4 °C. Cells were washed and resuspended in staining buffer, and 100 µL cell suspension to each test was continued with antibody staining for 1 h at room temperature. Then cells were washed with staining buffer and analyzed using FCAS, and the results were analyzed using Flowjo software.

4.15. Role of Autophagy

To explore the exact role of autophagy in the anticancer action of ADC in NSCLC, the autophagy inhibitor CQ and autophagy activator RAPA were used to treat the cells after ADC treatment. Cells were treated with 200 µM ADC for 72 h, and then were treated with 100 µM CQ and 5 µM RAPA for 4 h, respectively. The proliferation was analyzed by alamarBlue@ assay as shown in Section 4.3. To further confirm whether autophagy takes parts in ADC-induced apoptosis, cells were treated with 50, 150, and 200 µM ADC for 72 h, and then cells were treated with 100 µM CQ for 4 h. The apoptosis rate was analyzed by the Annexin V-FITC apoptosis detection kit and FCAS as shown in Section 4.7.

To determine at which stage (before or after ADC treatment) the protective autophagy occurs in NSCLC with ADC treatment, RAPA was used to treat the cells after ADC treatment. Cells were treated with the 5 µM RAPA for 4 h, and then cells were treated with 200 µM antrodin C for 72 h. The proliferation was analyzed by alamarBlue@ assay as shown in Section 4.3.

4.16. In Vitro ADC Metabolism

Many pharmacologically interesting molecules must be passed over because they are not sufficiently stable. Thus, it is necessary to focus on metabolic stability, which is widely considered one of the most significant challenges of drug discovery. The Phase I metabolism assay was performed by incubation of ADC (final concentrations, 1 µM) with an NADPH regenerating system (final concentration, 1 unit/mL) in SD rat and human liver microsomes (final concentration, 0.5 mg/mL). DMSO stock solution of ADC was prepared at a concentration of 10 mM. Working solution of ADC was prepared by adding methanol at 100 µM, and then by adding phosphate buffer at 10 µM. Liver microsome working solution was prepared by phosphate buffer at a concentration of 0.625 mg/mL. Eighty microliters of liver microsome working solution was added to the 10 µL ADC working solution directly, and 10 µL NADPH regenerating system solution were mixed, and then incubated at 37 °C for 0, 5, 10, 20, 30, and 60 min to initiate the reaction, followed by quenching of the reaction by adding 300 µL cold acetonitrile containing internal standards (100 ng/mL Tolbutamide) to each well. The Blank60 plate was referred to the no-test-compound treatment at 60 min incubation. The NCF60 plate was referred to the treatment of not adding the NADPH regenerating system at 60 min after incubation. All of the plates were centrifuged at 4000 rpm for 20 min, and the supernatants were transferred to new plates for analysis by LC/MS/MS.

4.17. Statistical Analyses

Results are presented as means ± standard deviation (SD). Inter-group comparisons were performed by one-way analysis of variance (ANOVA) and LSD's test. All of the variables were tested for normal and homogeneous variance using Levene's test. When necessary, Tamhane's T2 test was performed. A *p*-value of less than 0.05 or 0.01 was significant and very significant, respectively.

5. Conclusions

In summary, we provided in vitro evidence in human lung adenocarcinoma cancer cell lines which suggests that ADC retards cell growth, represses cell migration, disturbs cell cycle progression, and induces apoptotic death. We further demonstrated that protective autophagy could occur simultaneously in lung cancer cells exposed to ADC, and that these changes were partially mediated by the Akt-mTOR pathway. These findings may be helpful in the development of ADC as a chemotherapeutic lead compound for lung cancer, as well as to the rationale for enhancing its anti-lung cancer efficacy through the inhibition of protective autophagy.

Author Contributions: The manuscript was completed by the follow authors. The experiments were conducted by H.Y. The writing part was performed by H.Y., X.B. and Y.W., and W.W. and W.J. were regarded as co-corresponding author, as the manuscript was financially supported by their projects, which are the Natural Science Foundation of the Science and Technology Commission of Shanghai Municipality (Grant No. 16 ZR1431000) and the Shanghai Academy of Agricultural Science Program for Excellent Research Team (Grand No. 2017A06), respectively. The cell viability assay was directed by J.Z. and H.Z. The ADC metabolism assay was directed by C.T. The revised version of our manuscript was contributed to by Y.L., Y.Y., and Z.L.

Funding: The present study was supported by grants from the Natural Science Foundation of the Science and Technology Commission of Shanghai Municipality (Grant No. 16 ZR1431000) and the Shanghai Academy of Agricultural Science Program for Excellent Research Team (Grand No. 2017A06).

Conflicts of Interest: The authors declare no conflict of interest.

References

1. Siegel, R.L.; Miller, K.D.; Jemal, A. Cancer Statistics, 2017. *CA Cancer J. Clin.* **2010**, *60*, 277–300. [CrossRef] [PubMed]
2. Molina, J.R.; Yang, P.; Cassivi, S.D.; Schild, S.E.; Adjei, A.A. Non-small cell lung cancer: Epidemiology, risk factors, treatment, and survivorship. *Mayo Clin. Proc.* **2008**, *83*, 584–594. [CrossRef]

3. Ikari, A.; Sato, T.; Watanabe, R.; Yamazaki, Y.; Sugatani, J. Increase in claudin-2 expression by an EGFR/MEK/ERK/c-Fos pathway in lung adenocarcinoma A549 cells. *Biochim. Biophys. Acta* **2012**, *1823*, 1110–1118. [CrossRef] [PubMed]
4. Singh, R.P.; Deep, G.; Chittezhath, M.; Kaur, M.; Dwyer-Nield, L.D.; Malkinson, A.M.; Agarwal, R. Effect of silibinin on the growth and progression of primary lung tumors in mice. *J. Natl. Cancer Inst.* **2006**, *98*, 846–855. [CrossRef] [PubMed]
5. Li, Y.C.; He, S.M.; He, Z.X.; Li, M.; Yang, Y.; Pang, J.X.; Zhang, X.; Chow, K.; Zhou, Q.; Duan, W.; et al. Plumbagin induces apoptotic and autophagic cell death through inhibition of the PI3K/Akt/mTOR pathway in human non-small cell lung cancer cells. *Cancer Lett.* **2014**, *344*, 239–259. [CrossRef] [PubMed]
6. Zhu, X.L.; Lin, Z.B. Effects of Ganoderma lucidum polysaccharides on proliferation and cytotoxicity of cytokine-induced killer cells. *Acta Pharmacol. Sin.* **2005**, *26*, 1130–1137. [CrossRef] [PubMed]
7. Lu, M.C.; Du, YC.; Chuu, J.J.; Hwang, S.L.; Hsieh, P.C.; Hung, C.S.; Chang, F.R.; Wu, Y.C. Active extracts of wild fruiting bodies of Antrodia camphorata (EEAC) induce leukemia HL 60 cells apoptosis partially through histone hypoacetylation and synergistically promote anticancer effect of trichostatin A. *Arch. Toxicol.* **2009**, *83*, 121–129. [CrossRef] [PubMed]
8. Kong, Z.L.; Chang, J.S.; Chang, K.L.B. Antiproliferative effect of Antrodia camphorata polysaccharides encapsulated in chitosan–silica nanoparticles strongly depends on the metabolic activity type of the cell line. *J. Nanopart. Res.* **2013**, *15*, 1–13. [CrossRef]
9. Lee, J.; Lee, S.; Kim, S.L.; Choi, J.W.; Seo, J.Y.; Choi, D.J.; Park, Y.I. Corn silk maysin induces apoptotic cell death in PC-3 prostate cancer cells via mitochondria-dependent pathway. *Life Sci.* **2014**, *119*, 47–55. [CrossRef] [PubMed]
10. Zhang, Z.; Miao, L.; Lv, C.; Sun, H.; Wei, S.; Wang, B.; Huang, C.; Jiao, B. Wentilactone B induces G2/M phase arrest and apoptosis via the Ras/Raf/MAPK signaling pathway in human hepatoma SMMC-7721 cells. *Cell Death Dis.* **2013**, *4*, e657. [CrossRef]
11. Ryter, S.W.; Choi, A.M. Autophagy in lung disease pathogenesis and therapeutics. *Redox Biol.* **2015**, *4*, 215–225. [CrossRef]
12. Choi, K.S. Autophagy and cancer. *Exp. Mol. Med.* **2012**, *44*, 109–120. [CrossRef]
13. Maes, H.; Rubio, N.; Garg, A.D.; Agostinis, P. Autophagy: Shaping the tumor microenvironment and therapeutic response. *Trends Mol. Med.* **2013**, *19*, 428–446. [CrossRef] [PubMed]
14. Gozuacik, D.; Kimchi, A. Autophagy as a cell death and tumor suppressor mechanism. *Oncogene* **2004**, *23*, 2891–2906. [CrossRef]
15. Wu, S.H.; Ryvarden, L.; Chang, T.T. Antrodia camphorata ("niu-chang-chih"), new combination of a medicinal fungus in Taiwan. *Bot. Bull. Acad. Sin. Taipei* **1997**, *38*, 273–275.
16. Lee, K.H.; Morris-Natschke, S.L.; Yang, X.; Huang, R.; Zhou, T.; Wu, S.F.; Shi, Q.; Itokawa, H. Recent progress of research on medicinal mushrooms, foods, and other herbal products used in traditional Chinese medicine. *J. Tradit. Complement. Med.* **2012**, *2*, 84–95. [CrossRef]
17. Cha, W.S.; Ding, J.L.; Choi, D.B. Comparative evaluation of antioxidant, nitrite scavenging, and antitumor effects of Antrodia camphorata extract. *Biotechnol. Bioprocess Eng.* **2009**, *14*, 232–237. [CrossRef]
18. Phuong, D.T.; Ma, C.M.; Hattori, M.; Jin, J.S. Inhibitory effects of antrodins A-E from Antrodia cinnamomea and their metabolites on hepatitis C virus protease. *Phytother. Res.* **2010**, *23*, 582–584. [CrossRef] [PubMed]
19. Lu, M.C.; El-Shazly, M.; Wu, T.Y.; Du, Y.C.; Chang, T.T.; Chen, C.F.; Hsu, Y.M.; Lai, K.H.; Chiu, C.P.; Chang, F.R.; et al. Recent research and development of Antrodia cinnamomea. *Pharmacol. Ther.* **2013**, *139*, 124–156. [CrossRef] [PubMed]
20. Villaume, M.T.; Sella, E.; Saul, G.; Borzilleri, R.; Fargnoli, J.; Johnston, K.A.; Zhang, H.; Fereshteh, M.P.; Dhar, T.G.M.; Baran, P.S. Antroquinonol A: Scalable Synthesis and PreclinicalBiology of aPhase 2 Drug Candidate. *ACS Cent. Sci.* **2016**, *2*, 27–31. [CrossRef] [PubMed]
21. Baranczewski, P.; Stańczak, A.; Sundberg, K.; Svensson, R.; Wallin, A.; Jansson, J.; Garberg, P.; Postlind, H. Introduction to in vitro estimation of metabolic stability and drug interactions of new chemical entities in drug discovery and development. *Pharmacol. Rep.* **2006**, *58*, 453–472. [PubMed]
22. Di, L.; Kerns, E.H.; Carter, G.T. Drug-like property concepts in pharmaceutical design. *Curr. Pharm. Des.* **2009**, *15*, 2184–2194. [CrossRef] [PubMed]
23. Healt, J.P. Epithelial cell migration in the intestine. *Cell Biol. Int.* **1996**, *20*, 139–146.

24. Tsubouchi, S. Theoretical implications for cell migration through the crypt and the villus of labelling studies conducted at each position within the crypt. *Cell Tissue Kinet.* **1983**, *16*, 441–456. [PubMed]
25. Meineke, F.A.; Potten, C.S.; Loeffler, M. Cell migration and organization in the intestinal crypt using a lattice-free model. *Cell Prolif.* **2001**, *34*, 253–266. [CrossRef]
26. Pin, C.; Watson, A.J.; Carding, S.R. Modelling the Spatio-Temporal Cell Dynamics Reveals Novel Insights on Cell Differentiation and Proliferation in the Small Intestinal Crypt. *PLoS ONE* **2012**, *7*, e37115. [CrossRef] [PubMed]
27. Dunn, S.J.; Näthke, I.S.; Osborne, J.M. Computational models reveal a passive mechanism for cell migration in the crypt. *PLoS ONE* **2013**, *8*, e80516. [CrossRef] [PubMed]
28. Mirams, G.R.; Fletcher, A.G.; Maini, P.K.; Byrne, H.M. A theoretical investigation of the effect of proliferation and adhesion on monoclonal conversion in the colonic crypt. *J. Theor. Biol.* **2012**, *312*, 143–156. [CrossRef] [PubMed]
29. Osborne, J.M.; Walter, A.; Kershaw, S.K.; Mirams, G.R.; Fletcher, A.G.; Pathmanathan, P.; Gavaghan, D.; Jensen, O.E.; Maini, P.K.; Byrne, H.M. A hybrid approach to multi-scale modelling of cancer. *Philos. Trans. A Math. Phys. Eng Sci.* **2010**, *368*, 5013–5028. [CrossRef] [PubMed]
30. Van Leeuwen, I.M.; Mirams, G.R.; Walter, A.; Fletcher, A.; Murray, P.; Osborne, J.; Varma, S.; Young, S.J.; Cooper, J.; Doyle, B.; et al. An integrative computational model for intestinal tissue renewal. *Cell Prolif.* **2009**, *42*, 617–636. [CrossRef] [PubMed]
31. Vihinen, P.; Kähäri, V.M. Matrix metalloproteinases in cancer: Prognostic markers and therapeutic targets. *Int. J. Cancer.* **2002**, *99*, 157–166. [CrossRef]
32. Chaudhary, A.K.; Singh, M.; Bharti, A.C.; Asotra, K.; Sundaram, S.; Mehrotra, R. Genetic polymorphism of matrix metalloproteinases and their inhibitors in potentially malignant lesions of the head and neck. *J. Biomed. Sci.* **2010**, *17*, 10. [CrossRef] [PubMed]
33. Egeblad, M.; Werb, Z. New Functions for the Matrix Metalloproteinases in Cancer Progression. *Nat. Rev. Cancer* **2002**, *2*, 161–174. [CrossRef] [PubMed]
34. Kim, B.C.; Kim, H.G.; Lee, S.A.; Lim, S.; Park, E.H.; Kim, S.J.; Lim, C.J. Genipin-induced apoptosis in hepatoma cells is mediated by reactive oxygen species/c-Jun NH-terminal kinase-dependent activation of mitochondrial pathway. *Biochem. Pharmacol.* **2005**, *70*, 1398–1407. [CrossRef] [PubMed]
35. Rathore, S.; Datta, G.; Kaur, I.; Malhotra, P.; Mohmmed, A. Disruption of cellular homeostasis induces organelle stress and triggers apoptosis like cell-death pathways in malaria parasite. *Cell Death Dis.* **2015**, *6*, e1803. [CrossRef] [PubMed]
36. Zhang, P.; Sun, S.; Li, N.; Ho, A.S.; Kiang, K.M.; Zhang, X.; Cheng, Y.S.; Poon, M.W.; Lee, D.; Pu, J.K. Rutin increases the cytotoxicity of temozolomide in glioblastoma via autophagy inhibition. *J. Neurooncol.* **2015**, *132*, 1–8. [CrossRef]
37. Hu, Y.; Liu, J.; Wu, Y.F.; Lou, J.; Mao, Y.Y.; Shen, H.H.; Chen, Z.H. mTOR and autophagy in regulation of acute lung injury: A review and perspective. *Microb. Infect.* **2014**, *16*, 727–734. [CrossRef] [PubMed]
38. Lin, L.T.; Tai, C.J.; Su, C.H.; Chang, F.M.; Choong, C.Y.; Wang, C.K.; Tai, C.J. The Ethanolic Extract of Taiwanofungus camphoratus (Antrodia camphorata) Induces Cell Cycle Arrest and Enhances Cytotoxicity of Cisplatin and Doxorubicin on Human Hepatocellular Carcinoma Cells. *Biomed. Res. Int.* **2017**, *2015*, 415269.
39. Chen, L.Y.; Sheu, M.T.; Liu, D.Z.; Liao, C.K.; Ho, H.O.; Kao, W.Y.; Ho, Y.S.; Lee, W.S.; Su, C.H. Pretreatment with an ethanolic extract of Taiwanofungus camphoratus (Antrodia camphorata) enhances the cytotoxic effects of amphotericin B. *J. Agric. Food Chem.* **2011**, *59*, 11255–11263. [CrossRef]
40. Kumar, V.B.; Yuan, T.C.; Liou, J.W.; Yang, C.J.; Sung, P.J.; Weng, C.F. Antroquinonol inhibits NSCLC proliferation by altering PI3K/mTOR proteins and miRNA expression profiles. *Mutat. Res.* **2011**, *707*, 42–52. [CrossRef]
41. Kawabe, T. G2 checkpoint abrogators as anticancer drugs. *Mol. Cancer Ther.* **2004**, *3*, 513–519. [PubMed]
42. Ujiki, M.B.; Ding, X.Z.; Salabat, M.R.; Bentrem, D.J.; Golkar, L.; Milam, B.; Talamonti, M.S.; Bell, R.H.; Iwamura, T.; Adrian, T.E. Apigenin inhibits pancreatic cancer cell proliferation through G2/M cell cycle arrest. *Mol. Cancer* **2006**, *5*, 76. [CrossRef] [PubMed]
43. Zhou, X.; Zheng, M.; Chen, F.; Zhu, Y.; Yong, W.; Lin, H.; Sun, Y.; Han, X. Gefitinib Inhibits the Proliferation of Pancreatic Cancer Cells via Cell Cycle Arrest. *Anat. Rec.* **2010**, *292*, 1122–1127. [CrossRef] [PubMed]
44. Sun, S.Y.; Hail, N.; Lotan, R. Apoptosis as a novel target for cancer chemoprevention. *J. Natl. Cancer Inst.* **2004**, *96*, 662–672. [CrossRef] [PubMed]

45. Seo, B.R.; Min, K.J.; Woo, S.M.; Choe, M.; Choi, K.S.; Lee, Y.K.; Yoon, G.; Kwon, T.K. Inhibition of Cathepsin S Induces Mitochondrial ROS That Sensitizes TRAIL-Mediated Apoptosis Through p53-Mediated Downregulation of Bcl-2 and c-FLIP. *Antioxid. Redox Signal.* **2017**, *27*, 215–233. [CrossRef]
46. Guo, L.; Tan, K.; Wang, H.; Zhang, X. Pterostilbene inhibits hepatocellular carcinoma through p53/SOD2/ROS-mediated mitochondrial apoptosis. *Oncol. Rep.* **2016**, *36*, 3233–3240. [CrossRef]
47. Zhou, M.; Shen, S.; Zhao, X.; Gong, X. Luteoloside induces G0/G1 arrest and pro-death autophagy through the ROS-mediated AKT/mTOR/p70S6K signalling pathway in human non-small cell lung cancer cell lines. *Biochem. Biophys. Res. Commun.* **2017**, *494*, 263–269. [CrossRef]
48. Song, Y.; Kong, L.; Sun, B.; Gao, L.; Chu, P.; Ahsan, A.; Qaed, E.; Lin, Y.; Peng, J.; Ma, X.; et al. Induction of autophagy by an oleanolic acid derivative, SZC017, promotes ROS-dependent apoptosis through Akt and JAK2/STAT3 signaling pathway in human lung cancer cells. *Cell Biol. Int.* **2017**, *41*, 1367–1378. [CrossRef]
49. Zhang, S.Y.; Li, X.B.; Hou, S.G.; Sun, Y.; Shi, Y.R.; Lin, S.S. Cedrol induces autophagy and apoptotic cell death in A549 non-small cell lung carcinoma cells through the P13K/Akt signaling pathway, the loss of mitochondrial transmembrane potential and the generation of ROS. *Int. J. Mol. Med.* **2016**, *38*, 291–299. [CrossRef]
50. Senthil, K.K.J.; Gokila, V.M.; Wang, S.Y. Activation of Nrf2-mediated anti-oxidant genes by antrodin C prevents hyperglycemia-inducedsenescence and apoptosis in human endothelial cells. *Oncotarget* **2017**, *8*, 96568–96587. [CrossRef]
51. Tsujimoto, Y.; Shimizu, S. Role of the mitochondrial membrane permeability transition in cell death. *Apoptosis* **2007**, *12*, 835–840. [CrossRef] [PubMed]
52. Li, B.; Gao, Y.; Rankin, G.O.; Rojanasakul, Y.; Cutler, S.J.; Tu, Y.; Chen, Y.C. Chaetoglobosin K induces apoptosis and G2 cell cycle arrest through p53-dependent pathway in cisplatin-resistant ovarian cancer cells. *Cancer Lett.* **2015**, *356*, 418–433. [CrossRef]
53. Jang, J.H.; Kim, Y.J.; Kim, H.; Kim, S.C.; Cho, J.H. Buforin IIb induces endoplasmic reticulum stress-mediated apoptosis in HeLa cells. *Peptides* **2015**, *69*, 144–149. [CrossRef] [PubMed]
54. Brunelle, J.K.; Letai, A. Control of mitochondrial apoptosis by the Bcl-2 family. *J. Cell. Sci.* **2009**, *122*, 437–441. [CrossRef] [PubMed]
55. Okoshi, R.; Ozaki, T.; Yamamoto, H.; Ando, K.; Koida, N.; Ono, S.; Koda, T.; Kamijo, T.; Nakagawara, A.; Kizaki, H. Activation of AMP-activated protein kinase induces p53-dependent apoptotic cell death in response to energetic stress. *J. Biol. Chem.* **2008**, *283*, 3979–3987. [CrossRef] [PubMed]
56. Li, H.; Chen, C. Inhibition of autophagy enhances synergistic effects of *Salidroside* and anti-tumor agents against colorectal cancer. *Redox Biol.* **2015**, *4*, 215–225.
57. Komatsu, M.; Kominami, E.; Tanaka, K. Autophagy and neurodegeneration. *Autophagy* **2006**, *2*, 315–317. [CrossRef]
58. Jiang, X.; Overholtzer, M.; Thompson, C.B. Autophagy in cellular metabolism and cancer. *J. Clin. Investig.* **2015**, *125*, 47–54. [CrossRef]
59. Zhou, Y.; Liang, X.; Chang, H.; Shu, F.; Wu, Y.; Zhang, T.; Fu, Y.; Zhang, Q.; Zhu, J.D.; Mi, M. Ampelopsin-induced autophagy protects breast cancer cells from apoptosis through Akt-mTOR pathway via endoplasmic reticulum stress. *Cancer Sci.* **2014**, *105*, 1279–1287. [CrossRef]
60. Duan, S.; Huang, W.; Liu, X.; Liu, X.; Chen, N.; Xu, Q.; Hu, Y.; Song, W.; Zhou, J. IMPDH2 promotes colorectal cancer progression through activation of the PI3K/AKT/mTOR and PI3K/AKT/FOXO1 signaling pathways. *J. Exp. Clin. Cancer Res.* **2018**, *37*, 304. [CrossRef]
61. Paul, D.; Bargale, A.B.; Rapole, S.; Shetty, P.K.; Santra, M.K. Protein Phosphatase 1 Regulatory Subunit SDS22 Inhibits Breast Cancer Cell Tumorigenesis by Functioning as a Negative Regulator of the AKT Signaling Pathway. *Neoplasia* **2019**, *21*, 30–40. [CrossRef] [PubMed]
62. Kumar, D.; Shankar, S.; Srivastava, R.K. Rottlerin induces autophagy and apoptosis in prostate cancer stem cells via PI3K/Akt/mTOR signaling pathway. *Cancer Lett.* **2014**, *343*, 179–189. [CrossRef] [PubMed]
63. Singh, B.N.; Kumar, D.; Shankar, S.; Srivastava, R.K. Rottlerin induces autophagy which leads to apoptotic cell death through inhibition of PI3K/Akt/mTOR pathway in human pancreatic cancer stem cells. *Biochem. Pharmacol.* **2012**, *84*, 1154–1163. [CrossRef] [PubMed]
64. Mihaylova, M.M.; Shaw, R.J. The AMPK signalling pathway coordinates cell growth, autophagy and metabolism. *Nat. Cell Biol.* **2011**, *13*, 1016–1023. [CrossRef]

65. Kim, N.; Jeong, S.; Jing, K.; Shin, S.; Kim, S.; Heo, J.Y.; Kweon, G.R.; Park, S.K.; Wu, T.; Park, J.I.; et al. Docosahexaenoic Acid Induces Cell Death in Human Non-Small Cell Lung Cancer Cells by Repressing mTOR via AMPK Activation and PI3K/Akt Inhibition. *Biomed. Res. Int.* **2015**, *2015*, 239764. [CrossRef] [PubMed]
66. Vucicevic, L.; Misirkic, M.; Janjetovic, K.; Vilimanovich, U.; Sudar, E.; Isenovic, E.; Prica, M.; Harhaji-Trajkovic, L.; Kravic-Stevovic, T.; Bumbasirevic, V.; et al. Compound C induces protective autophagy in cancer cells through AMPK inhibition-independent blockade of Akt/mTOR pathway. *Autophagy* **2011**, *7*, 40–50. [CrossRef] [PubMed]
67. Papandreou, I.; Lim, A.L.; Laderoute, K.; Denko, N.C. Hypoxia signals autophagy in tumor cells via AMPK activity, independent of HIF-1, BNIP3, and BNIP3L. *Cell Death Differ.* **2008**, *15*, 1572–1581. [CrossRef]
68. Filomeni, G.; Desideri, E.; Cardaci, S.; Graziani, I.; Piccirillo, S.; Rotilio, G.; Ciriolo, M.R. Carcinoma cells activate AMP-activated protein kinase-dependent autophagy as survival response to kaempferol-mediated energetic impairment. *Autophagy* **2010**, *6*, 202–216. [CrossRef]
69. Law, B.Y.; Wang, M.; Ma, D.L.; Al-Mousa, F.; Michelangeli, F.; Cheng, S.H.; Ng, M.H.; To, K.F.; Mok, A.Y.; Ko, R.Y.; et al. Alisol B, a novel inhibitor of the sarcoplasmic/endoplasmic reticulum Ca (2+) ATPase pump, induces autophagy, endoplasmic reticulum stress, and apoptosis. *Mol. Cancer Ther.* **2010**, *9*, 718–730. [CrossRef]
70. Xu, Z.X.; Liang, J.; Haridas, V.; Gaikwad, A.; Connolly, F.P.; Mills, G.B.; Gutterman, J.U. A plant triterpenoid, avicin D, induces autophagy by activation of AMP-activated protein kinase. *Cell Death Differ.* **2007**, *14*, 1948–1957. [CrossRef]
71. Di, L.; Kerns, E.H.; Ma, X.J.; Huang, Y.; Carter, G.T. Applications of high throughput microsomal stability assay in drug discovery. *Comb. Chem. High Throughput Screen* **2008**, *11*, 469–476. [CrossRef] [PubMed]
72. Liu, R.; Schyman, P.; Wallqvist, A. Critically Assessing the Predictive Power of QSAR Models for Human Liver Microsomal Stability. *J. Chem. Inf. Model.* **2015**, *55*, 1566–1575. [CrossRef] [PubMed]
73. Nyhan, M.J.; O'Donovan, T.R.; Boersma, A.W.; Wiemer, E.A.; McKenna, S.L. MiR-193b promotes autophagy and non-apoptotic cell death in oesophageal cancer cells. *BMC Cancer* **2016**, *16*, 101. [CrossRef] [PubMed]

Sample Availability: Samples of the compounds are not available from the authors.

© 2019 by the authors. Licensee MDPI, Basel, Switzerland. This article is an open access article distributed under the terms and conditions of the Creative Commons Attribution (CC BY) license (http://creativecommons.org/licenses/by/4.0/).

Article

Bee Venom and Its Peptide Component Melittin Suppress Growth and Migration of Melanoma Cells via Inhibition of PI3K/AKT/mTOR and MAPK Pathways

Haet Nim Lim [1], Seung Bae Baek [2] and Hye Jin Jung [1],*

[1] Department of Pharmaceutical Engineering & Biotechnology, Sun Moon University, 70, Sunmoon-ro 221, Tangjeong-myeon, Asan-si, Chungnam 31460, Korea; gotsla9210@naver.com
[2] Eco system Lab., LOCORICO, Sun Moon University, 70, Sunmoon-ro 221, Tangjeong-myeon, Asan-si, Chungnam 31460, Korea; locorico@naver.com
* Correspondence: poka96@sunmoon.ac.kr; Tel.: +82-41-530-2354

Academic Editor: Roberto Fabiani
Received: 27 January 2019; Accepted: 3 March 2019; Published: 7 March 2019

Abstract: Malignant melanoma is the deadliest form of skin cancer and highly chemoresistant. Melittin, an amphiphilic peptide containing 26 amino acid residues, is the major active ingredient from bee venom (BV). Although melittin is known to have several biological activities such as anti-inflammatory, antibacterial and anticancer effects, its antimelanoma effect and underlying molecular mechanism have not been fully elucidated. In the current study, we investigated the inhibitory effect and action mechanism of BV and melittin against various melanoma cells including B16F10, A375SM and SK-MEL-28. BV and melittin potently suppressed the growth, clonogenic survival, migration and invasion of melanoma cells. They also reduced the melanin formation in α-melanocyte-stimulating hormone (MSH)-stimulated melanoma cells. Furthermore, BV and melittin induced the apoptosis of melanoma cells by enhancing the activities of caspase-3 and -9. In addition, we demonstrated that the antimelanoma effect of BV and melittin is associated with the downregulation of PI3K/AKT/mTOR and MAPK signaling pathways. We also found that the combination of melittin with the chemotherapeutic agent temozolomide (TMZ) significantly increases the inhibition of growth as well as invasion in melanoma cells compared to melittin or TMZ alone. Taken together, these results suggest that melittin could be potentially applied for the prevention and treatment of malignant melanoma.

Keywords: melanoma; bee venom; melittin; temozolomide; AKT; MAPK

---

1. Introduction

Melanoma, a malignant tumor of melanocytes, is the most rapid and fatal form of skin cancer [1,2]. The onset of malignant melanoma has been reported to be influenced by various environmental and genetic factors including ultraviolet (UV) light and oncogenic BRAF mutations, respectively [3,4]. Melanoma can easily spread to other parts of the body, making treatment more difficult and causing death. Over the past 30 years, numerous therapies have been developed, but the prognosis for patients with malignant melanoma has not been improved significantly, with an average survival time of 6–9 months. The primary treatment for melanoma is surgical removal of the tumor or chemotherapy, but the treatment is incomplete and the patient's prognosis is poor due to recurrence [5,6]. Therefore, the development of new therapies is still necessary.

The phosphatidylinositol 3-kinase (PI3K)/protein kinase B (AKT)/mammalian target of rapamycin (mTOR) and mitogen-activated protein kinase (MAPK) signaling pathways regulate cell survival, proliferation, migration and invasion, which are key functions of melanoma progression. Previous studies have shown that these pathways are constitutively activated through multiple mechanisms in malignant melanoma [7–9]. The sustained activation of these pathways not only protects melanoma cells from apoptosis, but also plays a decisive role in chemoresistance of melanoma. In pre-clinical models, the pharmacological inhibition of these pathways potently increases the sensitivity of melanoma cells to chemotherapeutic drugs such as cisplatin and temozolomide (TMZ). However, inhibiting only one of these pathways appears to be insufficient to effectively treat malignant melanoma [10]. Therefore, targeting PI3K/AKT/mTOR and MAPK pathways simultaneously can be a promising strategy to overcome chemoresistance of melanoma.

Bee venom (BV) is a natural toxin produced by honey bees (*Apis mellifera*) and has been widely used as a traditional medicine for a variety of diseases [11]. BV contains a variety of peptides including melittin, apamin, adolapin and enzymes, biological amines and non-peptide components [12]. Previous studies have revealed that BV possesses pain relief, anti-inflammatory, antimicrobial, antiviral and anticancer activities [13,14]. Most of all, BV's various cancer-controlling effects such as apoptosis, necrosis, cytotoxicity and growth arrest are being found in various types of cancer cells including prostate, breast, lung, liver, skin and bladder cancers [15,16]. BV triggered proliferation inhibition and apoptotic death of melanoma cells both in vitro and in vivo experiments, implying that BV may be useful in treating melanoma.

Melittin, the major bioactive component of BV (50% of its dry weight), is a small linear peptide composed of 26 amino acids. The pharmacological efficacies of BV appear to be due primarily to the synergistic effect of melittin [17–20]. Therefore, this amphiphilic anticancer peptide may be a better choice than BV in its original form for cancer management. Although BV is known to have the anticancer activity on melanoma, the antimelanoma effect of melittin and its mode of action at the molecular level remain largely unknown. In this study, we investigated the inhibitory effect and molecular action mechanism of BV and melittin against various melanoma cells including B16F10, A375SM and SK-MEL-28. Our results showed that BV and melittin potently suppress the growth and migration of melanoma cells via dual inhibition of PI3K/AKT/mTOR and MAPK pathways, suggesting that melittin could be a promising chemotherapeutic agent for malignant melanoma treatment.

2. Results

2.1. Inhibition of Melanoma Cell Growth by BV and Melittin

To determine whether BV and melittin affect melanoma cell growth, B16F10, A375SM and SK-MEL-28 cells were treated with various concentrations (0–100 µg/mL) of BV and melittin for 72 h, and the MTT assay was performed. BV and melittin remarkably inhibited the growth of melanoma cells, with IC_{50} values of 12.67 and 4.46 µg/mL in B16F10, 5.69 and 4.43 µg/mL in A375SM, and 3.05 and 3.06 µg/mL in SK-MEL-28, respectively (Figure 1).

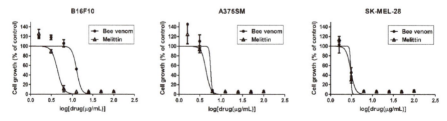

Figure 1. The effect of BV and melittin on the growth of melanoma cell lines. Cells were treated with various concentrations of BV and melittin for 72 h, and cell growth was measured using the MTT colorimetric assay. Data were presented as percentage relative to DMSO-treated control (% of control). Each value represents the mean ± SE from three independent experiments.

The growth inhibitory effect of BV and melittin against melanoma cells was further assessed using a colony formation assay. Treatment with BV and melittin resulted in a significant inhibition of the colony-forming ability in B16F10, A375SM and SK-MEL-28 melanoma cells (Figure 2). In particular, the inhibitory effect of melittin on the clonogenic growth was more potent than that of BV at the indicated concentrations. These results suggest that melittin may be considered a novel anticancer agent with potent antiproliferative activity against melanoma cells.

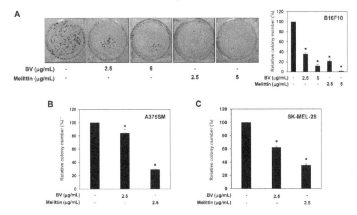

Figure 2. The effect of BV and melittin on the colony forming ability of melanoma cell lines. (**A**) B16F10, (**B**) A375SM and (**C**) SK-MEL-28 cells were treated with BV and melittin for 8–10 days. The cell colonies were detected by crystal violet staining and then counted. * $p < 0.05$ versus the control. Each value represents the mean ± SE from three independent experiments.

2.2. The Inhibitory Effect of BV and Melittin on Melanoma Cell Migration and Invasion

To assess the ability of BV and melittin to suppress the metastasis of melanoma cells, the effect of BV and melittin on the melanoma cell migration and invasion was evaluated by wound healing and Matrigel invasion assays, respectively. The wound scratch in untreated control cells was fully closed after 24 h of incubation, whereas treatment with BV and melittin led to the suppression of B16F10, A375SM and SK-MEL-28 melanoma cell migration in a dose-dependent manner (Figure 3).

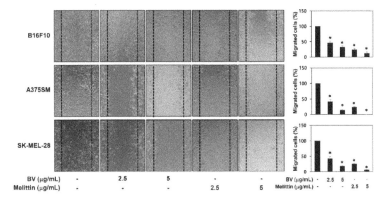

Figure 3. The effect of BV and melittin on the migration of melanoma cell lines. The migratory potential of melanoma cells was analyzed using wound healing assay. Cells were treated with BV and melittin for 24 h. Dotted black lines indicate the edge of the gap at 0 h. * $p < 0.05$ versus the control. Each value represents the mean ± SE from three independent experiments.

BV and melittin also decreased the invasion of A375SM and SK-MEL-28 cells when compared with controls (Figure 4). Notably, melittin more effectively inhibited the metastatic potential of melanoma cells than BV.

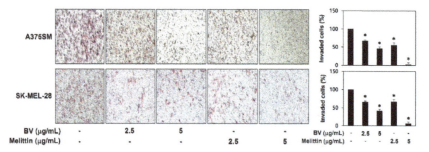

Figure 4. The effect of BV and melittin on the invasion of melanoma cell lines. The invasiveness of melanoma cells was analyzed using Matrigel-coated polycarbonate filters. Cells were treated with BV and melittin for 24 h. Cells penetrating the filters were stained and counted under an optical microscope. * $p < 0.05$ versus the control. Each value represents the mean ± SE from three independent experiments.

2.3. The Antimelanogenic Effect of BV and Melittin

Malignant melanocytes tend to exhibit upregulated melanogenesis and defective melanosomes [21,22]. To investigate the effect of BV and melittin on melanogenesis of B16F10 cells, we thus determined the melanin content. The cells were stimulated by α-MSH in the presence or absence of BV and melittin for 72 h. Treatment with BV and melittin dose-dependently downregulated the melanin formation of B16F10 cells induced by α-MSH, indicating that they inhibit the melanogenesis of melanoma cells (Figure 5).

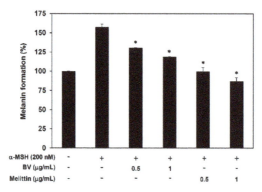

Figure 5. The effect of BV and melittin on the melanogenesis of α-MSH-stimulated B16F10 cells. Cells were treated with BV and melittin in the presence or absence of α-MSH for 72 h, and the cellular melanin contents were determined. * $p < 0.05$ versus the α-MSH control. Each value represents the mean ± SE from three independent experiments.

2.4. The Effect of BV and Melittin on Melanoma Cell Apoptosis

To further elucidate the anticancer effect of BV and melittin in melanoma cells, cellular apoptosis was quantitatively analyzed by flow cytometry following Annexin V-FITC and PI dual labeling. When melanoma cells were treated with BV and melittin for 24 h, the total amount of early and late apoptotic cells markedly increased in comparison with controls (from 0.99 to 7.80 and 46.45% for BV and melittin in B16F10 cells, from 1.18 to 35.45 and 98.60% in A375SM cells, and from 4.36 to 25.84 and 90.30% in

SK-MEL-28 cells, respectively) (Figure 6). In addition, the ability of melittin to induce apoptosis was superior to BV.

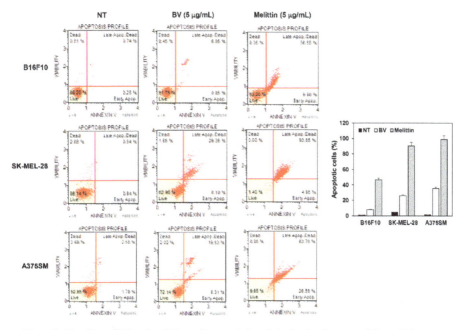

Figure 6. The effect of BV and melittin on the apoptotic cell death of melanoma cell lines. Cells were treated with BV and melittin for 24 h. Apoptotic cells were determined by flow cytometric analysis following annexin V-FITC and propidium iodide (PI) dual labeling. Each value represents the mean ± SE from three independent experiments.

Subsequently, the effect of BV and melittin on caspase activity was assessed to determine whether they induce caspase-dependent apoptosis in melanoma cells. Western blot analysis showed that treatment with BV and melittin resulted in the increased expression of cleaved caspase-3 and -9 in A375SM cells (Figure 7). Taken together, these data imply that the growth suppression of melanoma cells by BV and melittin is partly due to increased apoptosis in a caspase-dependent manner.

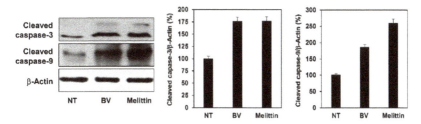

Figure 7. The effect of BV and melittin on the expression of apoptosis regulatory proteins in A375SM melanoma cells. Cells were treated with BV and melittin for 24 h, and the expression levels of cleaved caspase-3 and cleaved caspase-9 were detected by Western blotting. The levels of β-actin were used as an internal control. Each value represents the mean ± SE from three independent experiments.

2.5. The Effect of BV and Melittin on the Regulation of PI3K/AKT/mTOR and MAPK Pathways

The oncogenic PI3K/AKT/mTOR and MAPK signaling pathways regulate cell survival, proliferation, migration and invasion, which are key features of malignant melanoma progression [7]. The effect of BV and melittin on the PI3K/AKT/mTOR and MAPK activation was therefore investigated in melanoma cells. They significantly suppressed the phosphorylation of PI3K, AKT and mTOR as well as MAPKs such as extracellular signal-regulated kinase (ERK) and p38 in A375SM cells (Figure 8A).

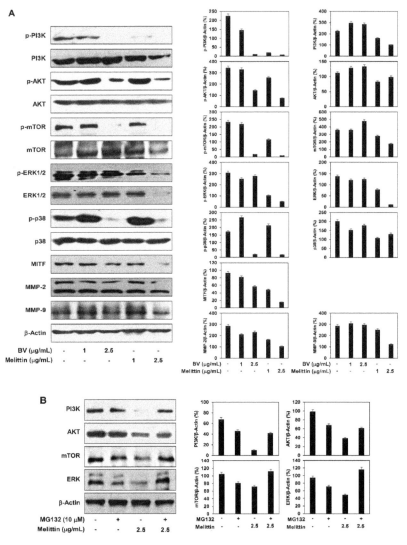

Figure 8. The effect of BV and melittin on the regulation of PI3K/AKT/mTOR and MAPK pathways. A375SM melanoma cells were treated with (**A**) BV, melittin and (**B**) MG132, and the protein levels were detected by Western blot analysis using specific antibodies. The levels of β-actin were used as an internal control. Each value represents the mean ± SE from three independent experiments.

Particularly, the inhibitory effect of melittin on the signaling pathways was better than that of BV. However, melittin also reduced the total protein levels of these signaling molecules. To further assess whether melittin affects their proteasomal degradation, we investigated the inhibitory effect of a proteasome inhibitor MG132 on the degradation of the proteins by the activity of melittin. Treatment with MG132 abolished the degradation of PI3K, AKT, mTOR and ERK proteins by melittin, indicating that melittin induces the proteolysis of these signaling effectors in melanoma cells (Figure 8B). Therefore, the suppressive effect of melittin on the PI3K/AKT/mTOR and MAPK signaling may be partly associated with its inhibitory effect on the protein stability of such signaling molecules. Taken together, these results suggest that BV and melittin inhibit the growth and metastatic potential of melanoma cells by downregulating both PI3K/AKT/mTOR and MAPK signaling pathways. As a result, the inhibition of these signaling pathways led to a reduction in the expression of microphthalmia-associated transcription factor (MITF), which is an important regulator of melanogenesis and malignant melanoma development, as well as matrix metalloproteinase (MMP)-2 and MMP-9, which play a critical role in melanoma metastasis (Figure 8A) [23,24].

2.6. The Effect of a Combination of Melittin with TMZ on Melanoma Cell Growth and Invasion

Temozolomide (TMZ) is used as a chemotherapeutic agent in many carcinomas including melanoma, but its anticancer activity is insufficient due to tumor chemoresistance [25,26]. In order to further develop a promising combination therapy of melittin for malignant melanoma, we investigated the effect of melittin on chemosensitivity of melanoma cells to TMZ.

As shown in Figure 9A, combination of melittin with TMZ markedly increased growth inhibition in both A375SM and SK-MEL-28 melanoma cell lines compared with melittin or TMZ alone (inhibition rates of 42.51, 48.62 and 96.02% with 2.5 µg/mL melittin, 50 µM TMZ and melittin/TMZ combination in A375SM cells and 44.90, 39.38 and 90.32% in SK-MEL-28 cells, respectively). In addition, the combination treatment of melittin with TMZ greatly increased invasion inhibition compared with single agent treatment (inhibition rates of 51.90, 30.51 and 89.48% with 2.5 µg/mL melittin, 50 µM TMZ and melittin/TMZ combination in A375SM cells and 54.76, 36.50 and 94.95% in SK-MEL-28 cells, respectively) (Figure 9B). These findings suggest that melittin might be used to treat melanoma in combination with TMZ.

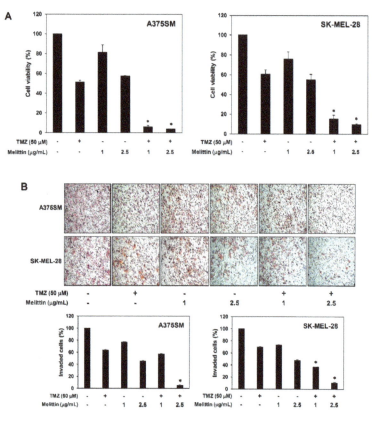

Figure 9. The effect of the combination of melittin with TMZ on the growth and invasion of melanoma cell lines. (**A**) Cells were treated with melittin and TMZ for 72 h, and cell growth was measured using the MTT colorimetric assay. (**B**) Cells were treated with melittin and TMZ for 24 h. Cells penetrating the Matrigel-coated polycarbonate filters were stained and counted under an optical microscope. * $p < 0.05$ versus the single agent treatment. Each value represents the mean ± SE from three independent experiments.

3. Discussion

Melanoma is one of the malignant tumors whose incidence and mortality are increasing worldwide and is notably resistant to conventional chemotherapeutic agents [1,2]. Through our continuing efforts in searching potential anticancer agents from natural products, the antimelanoma effect and underlying molecular mechanism of melittin, a main ingredient of bee venom (BV), were newly found in this study.

Melittin is an amphiphilic peptide of linear structure composed of 26 amino acids (GIGAVLKVLTTGLPALISWIKRKRQQ) in which the N-terminal part is hydrophobic, whereas the C-terminal part is hydrophilic and strongly basic [27]. Recent studies have revealed that melittin exhibits potent antitumor effect against several human cancer cell lines including ovarian, liver, cervical, bladder, gastric and breast cancers. Melittin induced apoptosis of human ovarian cancer cells via increase of death receptor expression and inhibition of signal transducer and activator of transcription 3 (STAT3) pathway [28]. On the other hand, the apoptotic effect of melittin on human gastric cancer cells was associated with activation of mitochondrial signaling pathway but not death receptor-mediated pathway [29]. Besides, melittin inhibited growth of human hepatoma cells through HDAC2-mediated

PTEN upregulation and inhibition of PI3K/AKT signaling pathway [30]. Furthermore, melittin showed anticancer and antiangiogenic effects by suppressing expression of hypoxia-inducible factor-1α (HIF-1α) and vascular endothelial growth factor (VEGF) through inhibition of ERK and mTOR pathways in human cervical cancer cells [31]. However, even though BV has been reported to inhibit the proliferation of malignant melanoma cells via cell cycle arrest at G1 stage and calcium-dependent apoptotic cell death [32,33], the cellular mechanisms of the antimelanoma effect of melittin remain fully unexplained.

In the present study, we found that melittin effectively inhibits the growth of melanoma cells by inducing a caspase-dependent apoptosis. Apoptosis is caused by the organic reactions of various proteins regulated by internal and external cellular pathways [34,35]. Caspase plays an important role in apoptosis and is involved in the common pathway of various apoptotic signals. Caspases are present as inactive pro-enzymes that are activated by proteolytic cleavage. Our results indicated that melittin induces apoptosis in melanoma cells through upregulation of cleaved caspase-3 and -9.

Malignant melanoma begins with a progressive disease in the local area, but as it progresses, the tumor cells begin to penetrate into the surrounding tissues. The metastatic ability of melanoma cells determines the severity of this disease and is therefore considered an important goal in malignant melanoma management [36]. We confirmed the antimetastatic potential of melittin through in vitro migration and invasion assays using melanoma cells. In particular, melittin showed stronger anticancer efficacy than BV in both proliferative and metastatic abilities of melanoma cells.

We also observed that BV and melittin have antimelanogenic activity in α-MSH-stimulating condition. The synthesis of melanin represents a major differentiated function of melanocytes. Although the main function of melanin is to protect against UV-induced damage, melanin pigment can also attenuate effectiveness of radiation or chemotherapy. In addition, melanogenesis can enhance melanoma progression owing to immunosuppressive, genotoxic and mutagenic properties [21,22,37]. In this study, treatment with melittin efficiently blocked the melanin formation of melanoma cells stimulated by α-MSH, indicating that melittin downregulates the differentiation of melanoma cells associated with melanogenesis.

Previous studies have shown that melittin downregulates PI3K/AKT/mTOR and MAPK signaling pathways, thereby leading to apoptosis in several tumor cells such as liver and cervical cancers and inhibits the metastatic ability of cancer cells [30,31]. These pathways have been also demonstrated to be important in development of malignant melanoma by promoting cell proliferation and metastasis [7]. To elucidate the molecular mechanism responsible for the malignant melanoma inhibitory effect of melittin, we confirmed whether melittin affects PI3K/AKT/mTOR and MAPK signaling. Treatment with melittin caused a reduction in the phosphorylation of PI3K, AKT and mTOR as well as MAPKs including ERK and p38 in melanoma cells. Notably, the inhibitory effect of melittin on the PI3K/AKT/mTOR and MAPK signaling was better than that of BV. Meanwhile, melittin also decreased the total protein levels of these signaling molecules, implying that the suppressive effect of melittin on the PI3K/AKT/mTOR and MAPK signaling is partly due to its inhibitory effect on the protein expression of such signaling molecules. Taken together, melittin could be a more efficient anticancer agent against malignant melanoma by simultaneously targeting the PI3K/AKT/mTOR and MAPK pathways.

Current standard therapy for malignant melanoma includes a combination of resectional surgery with chemotherapy such as temozolomide (TMZ), an alkylating agent that induces apoptosis through DNA strand breaks [25,26]. However, metastatic melanoma is highly chemoresistant and the chemotherapeutic drug yielded very limited survival benefit. Recent studies have shown that inhibition of PI3K/AKT/mTOR signaling sensitizes melanoma cells to TMZ [38,39]. We thus evaluated whether melittin ameliorates the anticancer effect of TMZ. Combined treatment with melittin and TMZ more effectively inhibited growth and invasiveness of melanoma cells compared with melittin or TMZ alone. These data suggest that melittin may have a promising therapeutic potential to overcome the chemoresistance to TMZ of melanoma patients.

Nevertheless, melittin is known to have a strong hemolytic activity [19,27]. Even though we demonstrated significant anticancer activity of melittin against malignant melanoma, further studies to reduce the nonspecific cell lysis and toxicity of melittin will be needed.

4. Materials and Methods

4.1. Materials

Melittin, temozolomide, MG132 and alpha-melanocyte stimulating hormone (α-MSH) were obtained from Sigma-Aldrich (St. Louis, MO, USA). Bee venom, Matrigel and Transwell chamber system were obtained from Chung Jin Biotech (Ansan, Korea), BD Biosciences (San Jose, CA, USA) and Corning Costar (Acton, MA, USA), respectively. Dulbecco's modified Eagle's medium (DMEM) and fetal bovine serum (FBS) were purchased from Invitrogen (Grand Island, NY, USA). Anti-phospho-PI3K, anti-PI3K, anti-phospho-AKT, anti-AKT, anti-phospho-mTOR, anti-mTOR, anti-phospho-ERK1/2, anti-ERK1/2, anti-phospho-p38, anti-p38, anti-cleaved caspase-3, anti-cleaved capase-9, anti-MITF, anti-MMP-2, anti-MMP-9 and anti-β-actin antibodies were purchased from Cell Signaling Technology (Beverly, MA, USA).

4.2. Cell Culture and Cell Growth Assay

B16F10, A375SM and SK-MEL-28 melanoma cell lines were obtained from the Korean Cell Line Bank (KCLB). All cells were grown in DMEM supplemented with 10% FBS and maintained at 37 °C in a humidified 5% CO_2 incubator. Cell growth was examined using the 3-(4,5-dimethylthiazol-2-yl)-2,5-diphenyltetrazolium bromide (MTT) colorimetric assay. The cells were seeded in 96-well culture plates at a density of 2×10^3 cells/well. After 24 h incubation, various concentrations of BV and melittin were added to each well. After 72 h, 50 µL of MTT solution (2 mg/mL; Sigma-Aldrich) was added to each well, and the cells were incubated for 3 h. To dissolve formazan crystals, the culture medium was removed and an equal volume of DMSO was added to each well. The absorbance of each well was determined at a wavelength of 540 nm using a microplate reader (Thermo Fisher Scientific, Vantaa, Finland).

4.3. Colony Formation Assay

Melanoma cells were seeded in 6-well culture plates at a density of 5×10^2 cells/well. After 24 h incubation, the cells were treated with BV and melittin for 8–10 days. Following this, the cell colonies were fixed with 4% formaldehyde and stained with 0.5% crystal violet solution.

4.4. Chemoinvasion Assay

Cell invasion was assayed using Transwell chamber inserts. The lower side of the polycarbonate filter was coated with 10 µL of gelatin (1 mg/mL), and the upper side was coated with 10 µL of Matrigel (3 mg/mL). Melanoma cells (1×10^5) were seeded in the upper chamber of the filter, and BV and melittin were added to the lower chamber filled with medium. The chamber was incubated for 24 h, and the cells were fixed with methanol and stained with H & E. The total number of cells that invaded the lower chamber of the filter was counted using an optical microscope (Olympus, Center Valley, PA, USA).

4.5. Cell Migration Assay

Melanoma cells were seeded in 24-well culture plates at a density of 2×10^5 cells/well and grown to 90% confluence. The confluent monolayer cells were scratched using a pipette tip and each well was washed with phosphate-buffered saline (PBS) to remove non-adherent cells. The cells were treated with BV and melittin and then incubated for up to 24 h. The perimeter of the central cell-free zone was confirmed under an optical microscope (Olympus).

4.6. Apoptosis Analysis

The apoptotic cell distribution was determined using the MUSE Annexin V & Dead Cell Kit (Merck KGaA, Darmstadt, Germany) according to the manufacturer's instructions. Briefly, after treatment with BV and melittin, all cells were collected and diluted with PBS containing 1% bovine serum albumin (BSA) as a dilution buffer to a concentration of 5×10^5 cells/mL. 100 µL of Annexin V/Dead Cell reagent and 100 µL of a single cell suspension were mixed in a microtube and in the dark for 20 min at room temperature. The cells were then analyzed using the Muse Cell Analyzer (Millipore Corporation, Hayward, CA, USA).

4.7. Western Blot Analysis

Cell lysates were separated by 10% sodium dodecyl sulfate-polyacrylamide gel electrophoresis (SDS-PAGE), and the separated proteins were transferred to polyvinylidene difluoride (PVDF) membranes (Millipore, Billerica, MA, USA) using standard electroblotting procedures. The blots were blocked and immunolabeled with primary antibodies against phospho-PI3K, PI3K, phospho-AKT, AKT, phospho-mTOR, mTOR, phospho-ERK1/2, ERK1/2, phospho-p38, p38, cleaved capase-3, cleaved capase-9, MITF, MMP-2, MMP-9 and β-actin overnight at 4 °C. Immunolabeling was detected with an enhanced chemiluminescence (ECL) kit (Bio-Rad Laboratories, Hercules, CA, USA), according to the manufacturer's instructions.

4.8. Measurement of Melanin Content

B16F10 cells (15×10^4 cells/well) were plated in 12-well culture plates and then treated with BV and melittin in the presence or absence of α-MSH (200 nM) for 72 h. The cells were then washed with PBS and lysed in 150 µL of 1 M NaOH at 95 °C. The lysate was added in 96-well microplate, and the absorbance was measured at 405 nm using a microplate spectrophotometer (Thermo Fisher Scientific).

4.9. Statistical Analysis

The results are expressed as the mean ± standard error (SE). Student's *t*-test was used to determine statistical significance between the control and the test groups. A *p*-value of <0.05 was considered to indicate a statistically significant difference.

5. Conclusions

In this study, we demonstrated that BV and its main component melittin potently suppress multiple oncogenic processes including growth, clonogenicity, migration, invasion and melanogenesis in malignant melanoma cells. Particularly, melittin showed stronger anticancer effects than BV. We also found that BV and melittin induce apoptosis in a caspase-dependent manner. Moreover, their antimelanoma effects are involved in the downregulation of PI3K/AKT/mTOR and MAPK signaling pathways. Noticeably, the combination of melittin with the chemotherapeutic agent TMZ increases sensitivity of melanoma cells towards TMZ. Based on these findings, we suggest that melittin could be an attractive candidate to treat malignant melanoma.

Author Contributions: H.J.J. conceived and designed the experiments; H.N.L. performed the experiments and analyzed the data; H.J.J. and S.B.B. contributed reagents/materials/analysis tools; H.N.L. and H.J.J. wrote the paper.

Funding: This research was funded by the Ministry of Education (NRF-2016R1D1A1B03932956) and the Ministry of Science and ICT (No. 2019R1A2C1009033), Republic of Korea.

Acknowledgments: This research was supported by Basic Science Research Program through the National Research Foundation of Korea (NRF) funded by the Ministry of Education (NRF-2016R1D1A1B03932956) and the NRF grant funded by the Ministry of Science and ICT (No. 2019R1A2C1009033). This work was also supported by the Brain Korea 21 Plus Project, Republic of Korea.

Conflicts of Interest: The authors declare no conflict of interest.

References

1. Miller, K.D.; Siegel, R.L.; Lin, C.C.; Mariotto, A.B.; Kramer, J.L.; Rowland, J.H.; Stein, K.D.; Alteri, R.; Jemal, A. Cancer treatment and survivorship statistics. *CA Cancer J. Clin.* **2016**, *66*, 271–289. [CrossRef] [PubMed]
2. Cummins, D.L.; Cummins, J.M.; Pantle, H.; Silverman, M.A.; Leonard, A.L.; Chanmugam, A. Cutaneous malignant melanoma. *Mayo Clin. Proc.* **2006**, *81*, 500–507. [CrossRef] [PubMed]
3. Kato, M.; Liu, W.; Akhand, A.A.; Hossain, K.; Takeda, K.; Takahashi, M.; Nakashima, I. Ultraviolet radiation induces both full activation of ret kinase and malignant melanocytic tumor promotion in RFP-RET-transgenic mice. *J. Investig. Dermatol.* **2000**, *115*, 1157–1158. [CrossRef] [PubMed]
4. Libra, M.; Malaponte, G.; Navolanic, P.M.; Gangemi, P.; Bevelacqua, V.; Proietti, L.; Bruni, B.; Stivala, F.; Mazzarino, M.C.; Travali, S.; et al. Analysis of BRAF mutation in primary and metastatic melanoma. *Cell Cycle* **2005**, *4*, 1382–1384. [CrossRef] [PubMed]
5. Hocker, T.L.; Singh, M.K.; Tsao, H. Melanoma genetics and therapeutic approaches in the 21st century: Moving from the benchside to the bedside. *J. Investig. Dermatol.* **2008**, *128*, 2575–2595. [CrossRef] [PubMed]
6. Tsao, H.; Chin, L.; Garraway, L.A.; Fisher, D.E. Melanoma: From mutations to medicine. *Genes Dev.* **2012**, *26*, 1131–1155. [CrossRef] [PubMed]
7. Meier, F.; Schittek, B.; Busch, S.; Garbe, C.; Smalley, K.; Satyamoorthy, K.; Li, G.; Herlyn, M. The RAS/RAF/MEK/ERK and PI3K/AKT signaling pathways present molecular targets for the effective treatment of advanced melanoma. *Front. Biosci.* **2005**, *10*, 2986–3001. [CrossRef] [PubMed]
8. Kolch, W. Ras/Raf signalling and emerging pharmacotherapeutic targets. *Expert Opin. Pharmacother.* **2002**, *3*, 709–718. [CrossRef] [PubMed]
9. Vivanco, I.; Sawyers, C.L. The phosphatidylinositol 3-Kinase AKT pathway in human cancer. *Nat. Rev. Cancer* **2002**, *2*, 489–501. [CrossRef] [PubMed]
10. Flaherty, K.T. Chemotherapy and targeted therapy combinations in advanced melanoma. *Clin. Cancer Res.* **2006**, *12*, 2366s–2370s. [CrossRef] [PubMed]
11. Ali, M.A.A.S.M. Studies on bee venom and its medical uses. *Int. J. Adv. Res. Technol.* **2012**, *1*, 69–83.
12. Bogdanov, S. Bee venom: Composition, health, medicine: A review. *Peptides* **2015**, *1*, 1–20.
13. Son, D.J.; Lee, J.W.; Lee, Y.H.; Song, H.S.; Lee, C.K.; Hong, J.T. Therapeutic application of anti-arthritis, pain-releasing, and anti-cancer effects of bee venom and its constituent compounds. *Pharmacol. Ther.* **2007**, *115*, 246–270. [CrossRef] [PubMed]
14. Uddin, M.B.; Lee, B.H.; Nikapitiya, C.; Kim, J.H.; Kim, T.H.; Lee, H.C.; Kim, C.G.; Lee, J.S.; Kim, C.J. Inhibitory effects of bee venom and its components against viruses in vitro and in vivo. *J. Microbiol.* **2016**, *54*, 853–866. [CrossRef] [PubMed]
15. Oršolić, N. Bee venom in cancer therapy. *Cancer Metastasis Rev.* **2012**, *31*, 173–194. [CrossRef] [PubMed]
16. Liu, C.C.; Hao, D.J.; Zhang, Q.; An, J.; Zhao, J.J.; Chen, B.; Zhang, L.L.; Yang, H. Application of bee venom and its main constituent melittin for cancer treatment. *Cancer Chemother. Pharmacol.* **2016**, *78*, 1113–1130. [CrossRef] [PubMed]
17. Rady, I.; Siddiqui, I.A.; Rady, M.; Mukhtar, H. Melittin, a major peptide component of bee venom, and its conjugates in cancer therapy. *Cancer Lett.* **2017**, *402*, 16–31. [CrossRef] [PubMed]
18. Kim, W.H.; An, H.J.; Kim, J.Y.; Gwon, M.G.; Gu, H.; Jeon, M.; Kim, M.K.; Han, S.M.; Park, K.K. Anti-inflammatory effect of melittin on *Porphyromonas gingivalis* LPS-stimulated human keratinocytes. *Molecules* **2018**, *23*, 332. [CrossRef] [PubMed]
19. Lee, G.; Bae, H. Anti-inflammatory applications of melittin, a major component of bee venom: Detailed mechanism of action and adverse effects. *Molecules* **2016**, *21*, 616. [CrossRef] [PubMed]
20. Hossen, M.S.; Gan, S.H.; Khalil, M.I. Melittin, a potential natural toxin of crude bee venom: Probable future arsenal in the treatment of diabetes mellitus. *J. Chem.* **2017**, *2017*, 4035626. [CrossRef]
21. Slominski, R.M.; Zmijewski, M.A.; Slominski, A.T. The role of melanin pigment in melanoma. *Exp. Dermatol.* **2015**, *24*, 258–259. [CrossRef] [PubMed]
22. Sim, D.Y.; Sohng, J.K.; Jung, H.J. Anticancer activity of 7,8-dihydroxyflavone in melanoma cells via downregulation of α-MSH/cAMP/MITF pathway. *Oncol. Rep.* **2016**, *36*, 528–534. [PubMed]
23. Levy, C.; Khaled, M.; Fisher, D.E. MITF: Master regulator of melanocyte development and melanoma oncogene. *Trends Mol. Med.* **2006**, *12*, 406–414. [CrossRef] [PubMed]

24. Hofmann, U.B.; Westphal, J.R.; Van Muijen, G.N.; Ruiter, D.J. Matrix metalloproteinases in human melanoma. *J. Investig. Dermatol.* **2000**, *115*, 337–344. [CrossRef] [PubMed]
25. Middleton, M.R.; Grob, J.J.; Aaronson, N.; Fierlbeck, G.; Tilgen, W.; Seiter, S.; Gore, M.; Aamdal, S.; Cebon, J.; Coates, A.; et al. Randomized phase III study of temozolomide versus dacarbazine in the treatment of patients with advanced metastatic malignant melanoma. *J. Clin. Oncol.* **2000**, *18*, 158–166. [CrossRef] [PubMed]
26. Quirt, I.; Verma, S.; Petrella, T.; Bak, K.; Charette, M. Temozolomide for the treatment of metastatic melanoma: A systematic review. *Oncologist* **2007**, *12*, 1114–1123. [CrossRef] [PubMed]
27. Raghuraman, H.; Chattopadhyay, A. Melittin: A membrane-active peptide with diverse functions. *Biosci. Rep.* **2007**, *27*, 189–223. [CrossRef] [PubMed]
28. Jo, M.; Park, M.H.; Kollipara, P.S.; An, B.J.; Song, H.S.; Han, S.B.; Kim, J.H.; Song, M.J.; Hong, J.T. Anti-cancer effect of bee venom toxin and melittin in ovarian cancer cells through induction of death receptors and inhibition of JAK2/STAT3 pathway. *Toxicol. Appl. Pharmacol.* **2012**, *258*, 72–81. [CrossRef] [PubMed]
29. Kong, G.M.; Tao, W.H.; Diao, Y.L.; Fang, P.H.; Wang, J.J.; Bo, P.; Qian, F. Melittin induces human gastric cancer cell apoptosis via activation of mitochondrial pathway. *World J. Gastroenterol.* **2016**, *22*, 3186–3195. [CrossRef] [PubMed]
30. Zhang, H.; Zhao, B.; Huang, C.; Meng, X.M.; Bian, E.B.; Li, J. Melittin restores PTEN expression by down-regulating HDAC2 in human hepatocelluar carcinoma HepG2 cells. *PLoS ONE* **2014**, *9*, e95520. [CrossRef] [PubMed]
31. Shin, J.M.; Jeong, Y.J.; Cho, H.J.; Park, K.K.; Chung, I.K.; Lee, I.K.; Kwak, J.Y.; Chang, H.W.; Kim, C.H.; Moon, S.K.; et al. Melittin suppresses HIF-1α/VEGF expression through inhibition of ERK and mTOR/p70S6K pathway in human cervical carcinoma cells. *PLoS ONE* **2013**, *8*, e69380.
32. Tu, W.C.; Wu, C.C.; Hsieh, H.L.; Chen, C.Y.; Hsu, S.L. Honeybee venom induces calcium-dependent but caspase-independent apoptotic cell death in human melanoma A2058 cells. *Toxicon* **2008**, *52*, 318–329. [CrossRef] [PubMed]
33. Liu, X.; Chen, D.; Xie, L.; Zhang, R. Effect of honey bee venom on proliferation of K1735M2 mouse melanoma cells in-vitro and growth of murine B16 melanomas in-vivo. *J. Pharm. Pharmacol.* **2002**, *54*, 1083–1089. [CrossRef] [PubMed]
34. Green, D.R. Apoptotic pathways: The roads to ruin. *Cell* **1998**, *94*, 695–698. [CrossRef]
35. Elmore, S. Apoptosis: A review of programmed cell death. *Toxicol. Pathol.* **2007**, *35*, 495–516. [CrossRef] [PubMed]
36. Luther, C.; Swami, U.; Zhang, J.; Milhem, M.; Zakharia, Y. Advanced stage melanoma therapies: Detailing the present and exploring the future. *Crit. Rev. Oncol. Hematol.* **2019**, *133*, 99–111. [CrossRef] [PubMed]
37. Brożyna, A.A.; Jóźwicki, W.; Carlson, J.A.; Slominski, A.T. Melanogenesis affects overall and disease-free survival in patients with stage III and IV melanoma. *Hum. Pathol.* **2013**, *44*, 2071–2074. [CrossRef] [PubMed]
38. Niessner, H.; Kosnopfel, C.; Sinnberg, T.; Beck, D.; Krieg, K.; Wanke, I.; Lasithiotakis, K.; Bonin, M.; Garbe, C.; Meier, F. Combined activity of temozolomide and the mTOR inhibitor temsirolimus in metastatic melanoma involves DKK1. *Exp. Dermatol.* **2017**, *26*, 598–606. [CrossRef] [PubMed]
39. Sinnberg, T.; Lasithiotakis, K.; Niessner, H.; Schittek, B.; Flaherty, K.T.; Kulms, D.; Maczey, E.; Campos, M.; Gogel, J.; Garbe, C.; et al. Inhibition of PI3K-AKT-mTOR signaling sensitizes melanoma cells to cisplatin and temozolomide. *J. Investig. Dermatol.* **2009**, *129*, 1500–1515. [CrossRef] [PubMed]

Sample Availability: Samples of the compounds are not available from the authors.

© 2019 by the authors. Licensee MDPI, Basel, Switzerland. This article is an open access article distributed under the terms and conditions of the Creative Commons Attribution (CC BY) license (http://creativecommons.org/licenses/by/4.0/).

Article

The Extracts of *Artemisia absinthium* L. Suppress the Growth of Hepatocellular Carcinoma Cells through Induction of Apoptosis via Endoplasmic Reticulum Stress and Mitochondrial-Dependent Pathway

Xianxian Wei [†], Lijie Xia [†], Dilinigeer Ziyayiding, Qiuyan Chen, Runqing Liu, Xiaoyu Xu and Jinyao Li *

Xinjiang Key Laboratory of Biological Resources and Genetic Engineering, College of Life Science and Technology, Xinjiang University, Urumqi, Xinjiang 830046, China; 15099141611@189.cn (X.W.); xialijie1219@163.com (L.X.); dilnigar9696@sina.com (D.Z.); m15276567620_1@163.com (Q.C.); Lucyducy@163.com (R.L.); 15276654427@139.com (X.X.)
* Correspondence: ljyxju@xju.edu.cn; Tel.: +86-991-8583259; Fax: +86-991-8583517
† These two authors contributed equally.

Academic Editor: Roberto Fabiani
Received: 27 January 2019; Accepted: 28 February 2019; Published: 5 March 2019

Abstract: *Artemisia absinthium* L. has pharmaceutical and medicinal effects such as antimicrobial, antiparasitic, hepatoprotective, and antioxidant activities. Here, we prepared *A. absinthium* ethanol extract (AAEE) and its subfractions including petroleum ether (AAEE-Pe) and ethyl acetate (AAEE-Ea) and investigated their antitumor effect on human hepatoma BEL-7404 cells and mouse hepatoma H22 cells. The cell viability of hepatoma cells was measured by 3-(4,5-dimethylthiazol-2-yl)-2,5-diphenyltetrazolium bromide (MTT) assay. The apoptosis, cell cycle, mitochondrial membrane potential (Δψm), and reactive oxygen species (ROS) were analyzed by flow cytometry. The levels of proteins in the cell cycle and apoptotic pathways were detected by Western blot. AAEE, AAEE-Pe, and AAEE-Ea exhibited potent cytotoxicity for both BEL-7404 cells and H22 cells through the induction of cell apoptosis and cell cycle arrest. Moreover, AAEE, AAEE-Pe, and AAEE-Ea significantly reduced Δψm, increased the release of cytochrome c, and promoted the cleavage of caspase-3, caspase-9, and poly(ADP-ribose) polymerase (PARP) in BEL-7404 and H22 cells. AAEE, AAEE-Pe, and AAEE-Ea significantly upregulated the levels of ROS and C/EBP-homologous protein (CHOP). Further, AAEE, AAEE-Pe, and AAEE-Ea significantly inhibited tumor growth in the H22 tumor mouse model and improved the survival of tumor mice without side effects. These results suggest that AAEE, AAEE-Pe, and AAEE-Ea inhibited the growth of hepatoma cells through induction of apoptosis, which might be mediated by the endoplasmic reticulum stress and mitochondrial-dependent pathway.

Keywords: *Artemisia absinthium*; apoptosis; endoplasmic reticulum stress; mitochondrial-dependent pathway

1. Introduction

Hepatocellular carcinoma (HCC) is one of the most common malignant tumors and was the sixth most commonly diagnosed cancer and the fourth leading cause of cancer death worldwide in 2018, with about 841,000 new cases and 782,000 deaths annually [1,2]. The main risk factors for HCC are chronic infection with hepatitis B virus (HBV) or hepatitis C virus (HCV), aflatoxin-contaminated foodstuffs, heavy alcohol intake, obesity, smoking, and type 2 diabetes [3]. Currently, therapeutic options for the treatment of HCC include liver resection, transplantation, palliative intra-arterial therapies, immunotherapy strategies, and so on [4,5]. However, the prognosis of most patients with

HCC is poor [6]. For HCC treatment, the main drugs, including oxaliplatin and sorafenib, remain unsatisfactory because of their side effects and multidrug resistance [7,8]. Therefore, it is urgent to develop novel therapeutic agents to treat HCC.

Artemisia absinthium L. belongs to the Asteraceae family and is commonly known as wormwood. The chemical components of *A. absinthium* include sesquiterpene lactone, sesquiterpene lactone-pinene, β-thujone, α-thujone, sabinyl acetate, and β-thujone [9]. *A. absinthium* has pharmaceutical and medicinal effects such as antimicrobial [9], insecticidal [10], antiparasitic [11], antitumor [12], antipyretic [13], hepatoprotective [14,15], and antioxidant activities [16,17]. In the present study, *A. absinthium* ethanol extract (AAEE) and its subfractions including petroleum ether (AAEE-Pe) and ethyl acetate (AAEE-Ea) were prepared, and their antitumor effects on HCC were investigated both in vitro and in vivo. AAEE, AAEE-Pe, and AAEE-Ea selectively inhibited the growth of hepatoma cells both in vitro and in vivo without cytotoxic effects on normal hepatic cells. Moreover, these extracts could arrest the cell cycle at the G2/M phase and induce apoptosis through endoplasmic reticulum (ER) stress and the mitochondrial-dependent pathway in human hepatoma BEL-7404 cells and mouse hepatoma H22 cells, and they might be used as safe and effective agents for the treatment of HCC.

2. Results

2.1. AAEE, AAEE-Pe, and AAEE-Ea Suppress the Growth of BEL-7404 and H22 Cells In Vitro

To investigate the anti-proliferative effects of AAEE, AAEE-Pe, and AAEE-Ea, BEL-7404 and H22 cells were treated with 25, 75, and 150 µg/mL of AAEE, AAEE-Pe, and AAEE-Ea. After 24 h, the morphology of BEL-7404 and H22 cells was observed with an inverted microscope. Compared to untreated cells, BEL-7404 and H22 cells treated with AAEE, AAEE-Pe, and AAEE-Ea became shrunk and round, and cell numbers were reduced in a dose-dependent manner (Figure 1A). Cell viability was detected by MTT assay after treatment for 24, 48, and 72 h. The viability of BEL-7404 and H22 cells was dose- and time-dependently decreased after treatment with AAEE, AAEE-Pe, or AAEE-Ea (Figure 1B). The IC_{50} (50% inhibitory concentration) values of AAEE, AAEE-Pe, and AAEE-Ea for BEL-7404 and H22 cells at 24, 48, and 72 h are shown in Table 1. The IC_{50} values of H22 cells followed the order AAEE-Pe ≤ AAEE < AAEE-Ea. The IC_{50} values of BEL-7404 cells followed the order AAEE ≤ AAEE-Pe < AAEE-Ea. The effect of AAEE, AAEE-Pe, and AAEE-Ea was also detected on normal liver cells NCTC1469. AAEE and AAEE-Pe showed some cytotoxicity on NCTC1469 cells, but it was much lower than that of BEL-7404 and H22 cells. AAEE-Ea has no cytotoxicity on NCTC1469 cells (Figure 1C). These results suggest that AAEE, AAEE-Pe, and AAEE-Ea selectively inhibited the growth of hepatoma cells in vitro.

Table 1. IC_{50} values of AAEE, AAEE-Pe, and AAEE-Ea for BEL-7404 and H22 cells.

		IC_{50}		
		24 h	48 h	72 h
H22 cells	AAEE	59.16	33.40	14.71
	AAEE-Pe	32.91	29.64	23.38
	AAEE-Ea	97.76	43.57	25.69
BEL-7404 cells	AAEE	89.86	56.86	60.39
	AAEE-Pe	98.71	75.88	89.51
	AAEE-Ea	171.7	82.04	71.97

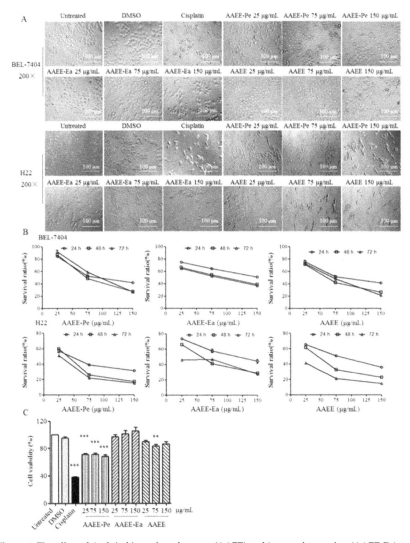

Figure 1. The effect of *A. absinthium* ethanol extract (AAEE) and its petroleum ether (AAEE-Pe) and ethyl acetate (AAEE-Ea) subfractions on the growth of BEL-7404, H22, and NCTC1469 cells. Cells were treated with different concentrations of AAEE, AAEE-Pe, and AAEE-Ea. (**A**) After 24 h, the morphology of BEL-7404 and H22 cells was observed by inverted microscope. (**B**) After 24, 48, and 72 h, the viability of BEL-7404 and H22 cells was detected by MTT assay. (**C**) After 24 h, the viability of NCTC1469 cells was detected by MTT assay. ** $p < 0.01$, *** $p < 0.001$ compared to Untreated.

2.2. AAEE, AAEE-Pe, and AAEE-Ea Induce Apoptosis in BEL-7404 and H22 Cells

To study whether AAEE, AAEE-Pe, and AAEE-Ea induce apoptosis, BEL-7404 and H22 cells were stained with Annexin V and propidium iodide (PI) after treatment and analyzed by flow cytometry. The frequencies of apoptotic BEL-7404 and B22 cells were significantly increased by each of AAEE, AAEE-Pe, and AAEE-Ea in a dose-dependent manner (Figure 2A,B). AAEE, AAEE-Pe, and AAEE-Ea did not induce the necrosis of BEL-7404 cells but significantly induced necrosis of H22 cells (Figure 2A,B). The pro- and anti-apoptotic members of the B-cell lymphoma-2 (BCL-2) protein family serve important roles in the regulation of cell apoptosis. After treatment with AAEE, AAEE-Pe, and AAEE-Ea, the levels of pro-apoptotic Bax and anti-apoptotic Bcl-2 in BEL-7404 and H22 cells were upregulated and downregulated, respectively. The ratios of Bax/Bcl2 significantly increased upon AAEE, AAEE-Pe, and AAEE-Ea treatment (Figure 2C). The results indicate that AAEE, AAEE-Pe, and AAEE-Ea inhibited the growth of BEL-7404 and H22 cells through the induction of apoptosis.

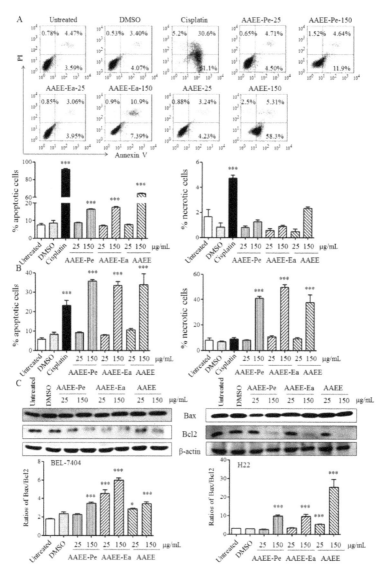

Figure 2. AAEE, AAEE-Pe, and AAEE-Ea induced apoptosis in BEL-7404 and H22 cells. Cells were treated with different concentrations of AAEE, AAEE-Pe, and AAEE-Ea for 24 h. After staining with Annexin V and PI, BEL-7404 (**A**) and H22 (**B**) cells were analyzed by flow cytometry. (**C**) Proteins were isolated, and the levels of Bax and Bcl-2 were analyzed by Western blot. * $p < 0.05$; *** $p < 0.001$ compared to Untreated.

2.3. AAEE, AAEE-Pe, and AAEE-Ea Induce Cell Cycle Arrest in BEL-7404 and H22 Cells

The morphological characteristics of apoptosis include chromatin condensation and DNA fragmentation. After treatment with AAEE, AAEE-Pe, and AAEE-Ea, BEL-7404 and H22 cells were stained with Hoechst 33258 and observed by inverted fluorescence microscopy. The nuclei of untreated cells showed homogeneous staining, while the nuclei of cells treated with AAEE, AAEE-Pe, and AAEE-Ea showed condensed chromatin (Figure 3A), suggesting that AAEE, AAEE-Pe, and AAEE-Ea induce apoptosis in BEL-7404 and H22 cells.

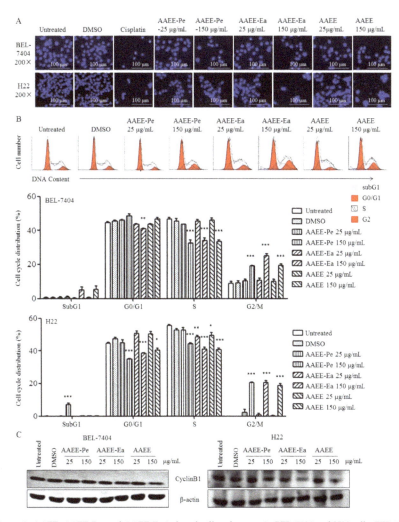

Figure 3. AAEE, AAEE-Pe, and AAEE-Ea induced cell cycle arrest in BEL-7404 and H22 cells. BEL-7404 and H22 cells were treated with different concentrations of AAEE, AAEE-Pe, and AAEE-Ea for 24 h. (**A**) BEL-7404 and H22 cells were stained with Hoechst 33258 and observed by inverted fluorescence microscope. (**B**) DNA contents in BEL-7404 cells were analyzed by flow cytometry and are shown in the upper panels. The summaries of the cell cycle distributions in BEL-7404 and H22 cells are shown in the middle and bottom panels, respectively. (**C**) Expression of Cyclin B1 was analyzed by Western blot. * $p < 0.05$; ** $p < 0.01$; *** $p < 0.001$ compared to Untreated.

Next, the distribution of the cell cycle in BEL-7404 and H22 cells was detected by PI staining after treatment with AAEE, AAEE-Pe, and AAEE-Ea for 24 h. We observed that cells in the G2/M phase were significantly increased and cells in the S phases were significantly decreased upon AAEE, AAEE-Pe, and AAEE-Ea treatment (Figure 3B), indicating that AAEE, AAEE-Pe, and AAEE-Ea arrested the cell cycle of BEL-7404 and H22 cells at the G2/M phase. Consistently, AAEE, AAEE-Pe, and AAEE-Ea reduced the expression of cyclin B1 in BEL-7404 and H22 cells (Figure 3C). The results suggest that AAEE, AAEE-Pe, and AAEE-Ea induced apoptosis and cell cycle arrest in hepatoma cells.

2.4. AAEE, AAEE-Pe, and AAEE-Ea Reduce Mitochondrial Membrane Potential ($\Delta\psi m$)

Mitochondrial membrane integrity is strictly regulated by the pro- and anti-apoptotic members of the BCL-2 protein family. After treatment with AAEE, AAEE-Pe, and AAEE-Ea for 24 h, the $\Delta\psi m$ was measured by flow cytometry using JC-1 as a fluorescent dye. As shown in Figure 4A, AAEE, AAEE-Pe, and AAEE-Ea significantly reduced the $\Delta\psi m$ values of BEL-7404 and H22 cells in a dose-dependent manner. Consistently, AAEE, AAEE-Pe, and AAEE-Ea promoted the release of cytochrome c in the cytosol of BEL-7404 and H22 cells. The activation of caspases is generally considered to be a key hallmark of apoptosis. We also observed that the levels of cleaved caspase-3 and caspase-9 were increased upon AAEE, AAEE-Pe, and AAEE-Ea treatment, which promoted the cleavage of poly(ADP-ribose) polymerase (PARP) in both BEL-7404 and H22 cells (Figure 4B). The results indicate that AAEE, AAEE-Pe, and AAEE-Ea induced apoptosis in hepatoma cells through the mitochondria-dependent pathway.

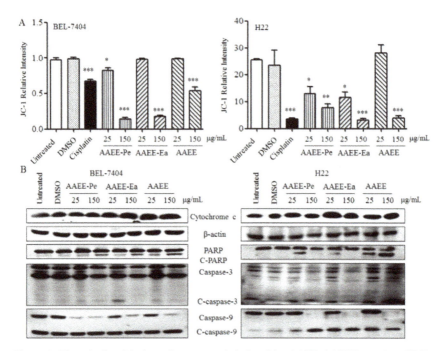

Figure 4. The mitochondria-dependent apoptosis induced by AAEE, AAEE-Pe, and AAEE-Ea. BEL-7404 and H22 cells were treated with different concentrations of AAEE, AAEE-Pe and AAEE-Ea for 24 h. (**A**) Cells were stained with JC-1 dye and analyzed by flow cytometry. (**B**) Total protein was isolated to detect the release of cytochrome c and the levels of poly(ADP-ribose) polymerase (PARP), cleaved-PARP (C-PARP), cleaved-caspase-3, caspase-3, cleaved-caspase-9, and caspase-9 by Western blot. * $p < 0.05$; ** $p < 0.01$; *** $p < 0.001$ compared to Untreated.

2.5. AAEE, AAEE-Pe, and AAEE-Ea Promote Reactive Oxygen Species (ROS) Generation and ER Stress

To examine the effects of AAEE, AAEE-Pe, and AAEE-Ea on oxidative stress, ROS generation was detected by flow cytometry at the indicated time points. After treatment with AAEE, AAEE-Pe, and AAEE-Ea, ROS levels were significantly increased at 6 h, reached a peak at 12 h, and then were maintained to 24 h in BEL-7404 cells (Figure 5A). Similarly, AAEE, AAEE-Pe, and AAEE-Ea dose-dependently increased ROS levels in H22 cells after treatment for 24 h (Figure 5B).

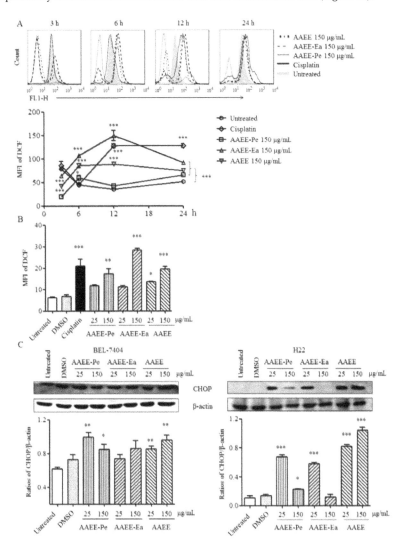

Figure 5. AAEE, AAEE-Pe, and AAEE-Ea induced ROS generation and ER stress. (**A**) BEL-7404 cells were treated with 150 μg/mL of AAEE, AAEE-Pe, and AAEE-Ea and the levels of ROS were analyzed by flow cytometry at indicated time points. (**B**) H22 cells were treated with different concentrations of AAEE, AAEE-Pe, and AAEE-Ea for 24 h and the levels of ROS were analyzed by flow cytometry. (**C**) BEL-7404 and H22 cells were treated with different concentrations of AAEE, AAEE-Pe, and AAEE-Ea for 24 h. Cell lysates were used to analyze the levels of CHOP by Western blot. * $p < 0.05$; ** $p < 0.01$; *** $p < 0.001$ compared to Untreated.

ER stress exacerbates mitochondrial dysfunction by activating caspase-9 and increasing the release of cytochrome c [18]. CHOP, DNA damage inducible gene 153 (GADD153), is the main apoptotic factor activated by ER stress, and its overexpression promotes apoptosis in cancer [19]. After treatment with AAEE, AAEE-Pe, and AAEE-Ea for 24 h, the levels of CHOP were detected by Western blot. The results showed that AAEE, AAEE-Pe, and AAEE-Ea significantly increased the levels of CHOP in BEL-7404 and H22 cells (Figure 5C), suggesting that ER stress might be involved in the induction of apoptosis by AAEE, AAEE-Pe, and AAEE-Ea.

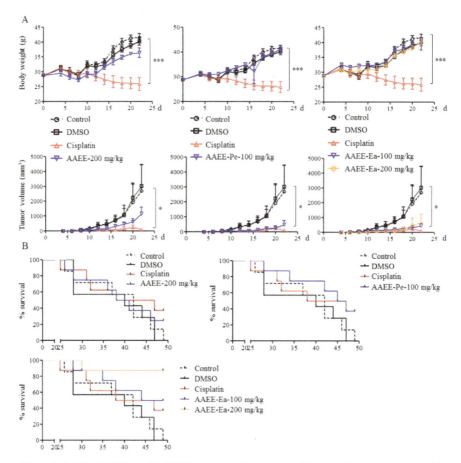

Figure 6. AAEE, AAEE-Pe, and AAEE-Ea suppressed tumor growth in vivo. A tumor mouse model was established by injection of H22 cells. After 3 days, tumor mice were treated with DMSO, cisplatin, AAEE, AAEE-Pe, and AAEE-Ea. Body weight of mice, tumor sizes (**A**), and survival rates (**B**) were monitored at the indicated time points. * $p < 0.05$; *** $p < 0.001$ compared to control.

2.6. AAEE, AAEE-Pe, and AAEE-Ea Inhibit the Growth of H22 Cells In Vivo

To further confirm the inhibitory effect of AAEE, AAEE-Pe, and AAEE-Ea on tumor growth in vivo, an H22 tumor mouse model was established in Kunming mice. Tumor mice were treated with different doses of AAEE, AAEE-Pe, and AAEE-Ea after 3 days of H22 cell injection. Cisplatin was used as a positive control and DMSO was used as a solvent control. We observed that cisplatin significantly reduced the weight of mice, but AAEE, AAEE-Pe, and AAEE-Ea did not affect the weight of mice. The tumor growth was significantly inhibited in the cisplatin, 100 mg/kg AAEE-Pe, and 100 and 200 mg/kg AAEE-Ea groups. AAEE at a 200 mg/kg dose also suppressed tumor growth to a certain degree (Figure 6A). At the end of the tumor study, the survival rates of tumor mice in each group were calculated. All mice were dead in the control (7 out of 7) and DMSO (7 out of 7) groups. The survival rates were 37.5%, 25%, 37.5%, 50%, and 87.5% in the cisplatin (3 out of 8), 200 mg/kg AAEE (2 out of 8), 100 mg/kg AAEE-Pe (3 out of 8), and 100 (4 out of 8) and 200 mg/kg (7 out of 8) AAEE-Ea groups, respectively (Figure 6B). These data suggest that AAEE, AAEE-Pe, and AAEE-Ea effectively inhibited the growth of H22 cells in vivo and improved the survival of tumor mice without obvious toxicity.

3. Discussion

Chinese herbal medicine (CHM) has a long history of use in treating cancers and provides potential antitumor remedies. *A. absinthium*, a kind of CHM, has been used as an antipyretic, antiseptic, and anti-parasitic agent for the treatment of chronic fevers and inflammation of the liver [20]. In this study, we prepared AAEE, AAEE-Pe, and AAEE-Ea and investigated their antitumor effects on hepatoma cells. We found that AAEE, AAEE-Pe, and AAEE-Ea significantly suppressed the growth of BEL-7404 and H22 cells, induced apoptosis and cell cycle arrest, reduced $\Delta\psi m$, increased the release of cytochrome c, activated caspases, and promoted ROS production and ER stress.

It has been reported that a number of components of CHM can inhibit the growth of tumor cells both in vitro and in vivo, such as polysaccharides, flavones, terpenoids, and phenols [21–26]. The components of polysaccharides, flavones, and triterpenes in AAEE, AAEE-Pe, and AAEE-Ea were quantified. Although the three extracts contained different concentrations of polysaccharides, flavones, and triterpenes, they had similar antitumor effects. The results suggest that polysaccharides, flavones, and triterpenes might not be the major antitumor components in AAEE, AAEE-Pe, and AAEE-Ea. We will further identify the major antitumor components in the extracts of *A. absinthium* in future study.

The BCL-2 protein family strictly controls the apoptosis of cells [27,28]. An imbalance of proteins in the BCL-2 family triggers the intrinsic apoptosis pathway that increases mitochondrial permeability and the release of cytochrome c and activates caspase-9/caspase-3 [29]. We found that the levels of Bax/Bcl2 significantly increased in BEL-7404 and H22 cells after AAEE, AAEE-Pe, and AAEE-Ea treatment, which might cause the reduction of $\Delta\psi m$ and the release of cytochrome c to activate caspases and apoptosis. Similarly, Shafi et al. [30] reported that the methanol extract of *A. absinthium* induced the apoptosis of human breast cancer cells through the modulation of BCL-2 family proteins.

Accumulating evidence points to the role of ER stress in the induction of apoptosis in various cancer cells [31–33]. ER stress exacerbates mitochondrial dysfunction by activating caspase-9 and increasing the release of cytochrome c [34,35]. Our results showed that AAEE, AAEE-Pe, and AAEE-Ea treatment significantly increased the expression of the ER stress-related protein CHOP, indicating that AAEE, AAEE-Pe, and AAEE-Ea might induce apoptosis of BEL-7404 and H22 cells through ER stress and the mitochondria-dependent pathway.

We observed that the ROS levels in hepatoma cells treated with AAEE and AAEE-Pe were lower than that in untreated cells at the beginning of 3 h, although they were significantly upregulated after 6 h, suggesting that the extracts might have antioxidant activities. This is similar to results found in other studies [16,17]. The increased ROS might be due to the reduction of $\Delta\psi m$ induced by AAEE and AAEE-Pe treatment.

Finally, the antitumor and side effects of AAEE, AAEE-Pe, and AAEE-Ea were evaluated in an H22 tumor mouse model. All three extracts did not affect the weight of mouse but cisplatin significantly reduced the weight of mouse, suggesting that AAEE, AAEE-Pe, and AAEE-Ea might have no side effects in vivo. Moreover, AAEE-Pe and AAEE-Ea significantly inhibited tumor growth, which was similar with cisplatin. AAEE, AAEE-Pe, and AAEE-Ea further improved the survival of H22 tumor mice, and the high dose of AAEE-Ea showed the highest survival rate.

In conclusion, AAEE, AAEE-Pe, and AAEE-Ea inhibited the growth of hepatoma cells through the induction of apoptosis that might be mediated by ER stress and the mitochondria-dependent pathway.

AAEE, AAEE-Pe, and AAEE-Ea suppressed H22 tumor growth and improved the survival of H22 tumor mice without obvious side effects, indicating that the three extracts might be used to develop safe and effective antitumor agents.

4. Materials and Methods

4.1. Preparation of AAEE, AAEE-Pe, and AAEE-Ea

The AAEE, AAEE-Pe, and AAEE-Ea were prepared according to the following procedure. Briefly, the powder was made using the aerial parts of *A. absinthium* including stems, leaves, flowers, and seeds (Alikang Uygur medicine technology co., Ltd., Urumqi, Xinjiang, China) and extracted overnight using 10 volumes of distilled water at 4 °C. After centrifugation at 8000 rpm for 20 min, the pellet was collected and extracted with 10 volumes of distilled water for 2 h at 60 °C. After centrifugation, the pellet was extracted using 10 volumes of 85% ethanol at 60 °C three times (2 h/time). The supernatant was collected after filtration and concentrated by a rotary evaporator to obtain the extractum. Some extractum was dried by a vacuum freeze-dryer to obtain AAEE. The remaining extractum was dissolved in distilled water and extracted by an equal volume of petroleum ether eight times, then the upper layer was collected and dried by a vacuum freeze–dryer to obtain AAEE-Pe. The bottom layer was extracted by an equal volume of ethyl acetate eight times, then the supernatant was collected and dried by a vacuum freeze–dryer to obtain AAEE-Ea. The AAEE, AAEE-Pe, and AAEE-Ea were dissolved in dimethyl sulfoxide (DMSO) (Sigma, St. Louis, MO, USA) and filtered with a 0.22 μm filter. The contents of flavonoids, terpenoids, and polysaccharides were determined by $AlCl_3$–KA_C, vanillin–glacial acetic acid, and anthrone–sulfuric acid colorimetry, respectively, which were shown in Table 2.

Table 2. The contents of polysaccharides, flavonoids, and triterpenoids in AAEE, AAEE-Pe, and AAEE-Ea.

	AAEE	AAEE-Pe	AAEE-Ea
Polysaccharides	31.15%	0.68%	9.62%
Flavonoids	16.89%	10.26%	24.97%
Triterpenes	22.57%	31.59%	29.89%

4.2. Cell Culture

The NCTC1469, H22, and BEL-7404 cells were obtained from the Xinjiang Key Laboratory of Biological Resources and Genetic Engineering, Xinjiang University (Urumqi, Xinjiang, China) and cultured in Roswell Park Memorial Institute (RPMI) 1640 medium (Gibco, Thermo Fisher Scientific, Waltham, MA, USA) supplemented with 10% fetal bovine serum (MRC, Changzhou, China) and 1% L-glutamine (100 mM), 100 U/mL penicillin, and 100 μg/mL streptomycin (MRC, Changzhou, China) at 37 °C in humidified air with 5% CO_2.

4.3. Cell Viability Assay

The proliferation of NCTC1469, H22, and BEL-7404 cells was analyzed by 3-(4,5-dimethylthiazol-2-yl)-2,5-diphenyltetrazolium bromide (MTT) (Sigma, Louis, MO, USA) assay. Briefly, cells (5000 cells/well) were seeded in 96-well plates and treated with various doses of AAEE, AAEE-Pe, and AAEE-Ea for 24 h, 48 h, or 72 h. DMSO (0.3%) was used as a solvent control and cisplatin (35 μg/mL) was used as a positive control. The supernatant was discarded after centrifugation at 1200 rpm for 5 min and 100 μL of MTT solution (0.5 mg/mL in PBS) was added to each well and incubated at 37 °C for 3 h. The formed formazan crystals were dissolved in 200 μL DMSO. The OD_{490} values were measured by a 96-well microplate reader (Bio-Rad Laboratories, Hercules, CA, USA). The relative cell viability was calculated according to the formula Cell viability (%) = ($OD_{treated}$/$OD_{untreated}$) × 100%. This experiment was conducted three times independently.

4.4. Observation of Cell Morphology

H22 and BEL-7404 cells were seeded in 96-well plates and were treated with different concentrations of AAEE, AAEE-Pe, and AAEE-Ea for 24 h. After treatment, the morphology of H22 and BEL-7404 cells was observed by inverted fluorescence microscope (Nikon Eclipse Ti-E, Tokyo, Japan).

4.5. Analysis of Apoptosis

H22 and BEL-7404 cells were treated with different concentrations of AAEE, AAEE-Pe, and AAEE-Ea for 24 h, and then stained with an Annexin V-FITC/propidium iodide (PI) Apoptosis Detection Kit (YEASEN, Shanghai, China) according to the manufacturer's instructions. Cisplatin and DMSO were used as positive and negative controls, respectively. Samples were analyzed by flow cytometry (BD FACSCalibur, San Jose, CA, USA). This experiment was conducted three times independently.

4.6. Hoechst 33258 Staining

H22 and BEL-7404 cells were seeded in 6-well plates at the concentration of 1×10^5 cells/well in 2 mL medium. After 60%~70% confluence, the cells were treated with AAEE, AAEE-Pe, and AAEE-Ea for 24 h. The cells were collected and fixed with 4% ice-cold Paraformaldehyde at 4 °C for 10 min. After washing with PBS, cells were stained with Hoechst 33258 (Beyotime, Shanghai, China) at 4 °C for 10 min. Samples were observed using an inverted fluorescence microscope.

4.7. Analysis of the Cell Cycle

H22 and BEL-7404 cells were inoculated in 60 mm culture dishes and treated with different concentrations of AAEE, AAEE-Pe, and AAEE-Ea for 24 h. All cells were collected and washed twice with PBS, then fixed in 70% ice-cold ethanol overnight at 4 °C. After washing twice with PBS, cells were re-suspended in 250 µL propidium iodide/RNase staining buffer (BD Biosciences, San Jose, CA, USA). After 10 min at room temperature, samples were collected by flow cytometry and the cell cycle distribution was analyzed using ModFit LT 3.0 software (BD FACS Calibur, San Jose, AC, USA). This experiment was conducted three times independently.

4.8. Analysis of $\Delta\psi m$

H22 and BEL-7404 cells were treated with different concentrations of AAEE, AAEE-Pe, and AAEE-Ea for 24 h and then stained with the membrane-permeable JC-1 dye (Beyotime, Shanghai, China) for 30 min at 37 °C. Samples were analyzed by flow cytometry. This experiment was conducted three times independently.

4.9. Analysis of ROS

BEL-7404 cells were treated with AAEE, AAEE-Pe, and AAEE-Ea for 2, 4, 6, 12, and 24 h. H22 cells were treated with AAEE, AAEE-Pe, and AAEE-Ea for 24 h. Cells were stained by 10 mM of fluorescent probe 2′,7′-dichlorodihydrofluorescein diacetate (DCFH-DA) (Beyotime, Shanghai, China) for 20 min at 37 °C. After washing three times with ice-cold PBS, samples were analyzed by flow cytometry. This experiment was conducted two times independently.

4.10. Western Blot

The antibodies against caspase-9, Bax and Bcl-2, and anti-mouse IgG-HRP and anti-rabbit IgG-HRP were purchased from BBI Life Sciences (Shanghai, China). The antibodies against caspase-3, PARP, cytochrome c, and β-actin were obtained from Cell Signaling Technology (Danvers, MA, USA). The antibodies against CHOP and CyclinB1 were bought from Beyotime (Shanghai, China).

H22 and BEL-7404 cells were treated with AAEE, AAEE-Pe, and AAEE-Ea for 24 h. After washing twice with PBS, cell lysates were prepared with RIPA Lysis Buffer (Beijing ComWin Biotech Co.,

Ltd., Beijing, China) and protein concentrations were detected using a BCA Kit (Thermo Fisher Scientific, Waltham, MA, USA) according to the manufacturer's instructions. Proteins were separated on 12% sodium dodecyl sulfate polyacrylamide gel electrophoresis (SDS-PAGE) and transferred to a polyvinylidene difluoride (PVDF) membrane. After incubation with primary and secondary antibodies, target proteins were detected by chemiluminescence (Beyotime, Shaghai, China). Signals were quantified using ImageJ digitizing software (ImageJ 1.50, National Institutes of Health, Bethesda, MD, USA). This experiment was conducted three times independently.

4.11. Animals and Ethics Statement

Six- to eight-week-old male Kunming mice were purchased from Animal Laboratory Center, Xinjiang Medical University (Urumqi, Xinjiang, China). Mice were kept in a standard temperature-controlled, light-cycled animal facility at Xinjiang University. All animal experiments were approved by the Committee on the Ethics of Animal Experiments of Xinjiang Key Laboratory of Biological Resources and Genetic Engineering (BRGE-AE001) and performed under the guidelines of the Animal Care and Use Committee of College of Life Science and Technology, Xinjiang University.

4.12. Tumor Mouse Study

For establishment of a tumor mouse model, male Kunming mice were injected with 1×10^6 H22 cells in 100 μL PBS subcutaneously. After 3 days, mice were randomly divided into seven groups (7 mice/group for Control and DMSO, 8 mice/group for the other five groups). The solvent control group intraperitoneally received 0.1 mL DMSO daily. The positive group was intraperitoneally injected with 5 mg/kg cisplatin at intervals of five days. The experimental groups were intraperitoneally injected with 200 mg/kg AAEE, 100 mg/kg AAEE-Pe, or 100 mg/kg or 200 mg/kg AAEE-Ea in 0.1 mL DMSO every two days. Tumor sizes were measured using calipers and tumor volume was calculated according to the formula tumor volume (mm^3) = (length × width^2)/2.

4.13. Statistical Analysis

The data are expressed as mean ± standard error of the mean (SEM). Statistical significance was analyzed using one-way analysis of variance (ANOVA) by Tukey's Multiple Comparison Test. $p < 0.05$ was considered statistically significant.

Author Contributions: X.W., D.Z., Q.C., R.L. and X.X. performed the experiments; L.X. and J.L. designed the experiments, analyzed the data, and wrote the paper.

Funding: This work was supported by the Chinese National Natural Science Foundation Grant [U1803381 to Jinyao Li and 31860258 to Lijie Xia], the 1000 Young Talents Program of China to Jinyao Li and the "Tianshan Youth Project" Young Ph.D. Science and Technology talents Project [no. 2017Q077] to Lijie Xia.

Conflicts of Interest: The authors declare no conflict of interest.

References

1. Wang, D.; Sun, Q.; Wu, J.; Wang, W.; Yao, G.; Li, T.; Li, X.; Li, L.; Zhang, Y.; Cui, W.; et al. A new Prenylated Flavonoid induces G0/G1 arrest and apoptosis through p38/JNK MAPKpathways in Human Hepatocellular Carcinoma cells. *Sci. Rep.* **2017**, *7*, 5736. [CrossRef] [PubMed]
2. Bray, F.; Ferlay, J.; Soerjomataram, I.; Siegel, R.L.; Torre, L.A.; Jemal, A. Global cancer statistics 2018: GLOBOCAN estimates of incidence and mortality worldwide for 36 cancers in 185 countries. *CA-A Cancer J. Clin.* **2018**, *68*, 394–424. [CrossRef] [PubMed]
3. London, W.T.; Petrick, J.L.; McGlynn, K.A.; Thun, M.J.; Linet, M.S.; Cerhan, J.R.; Haiman, C.A.; Schottenfeld, D. (Eds.) *Cancer Epidemiology and Prevention*, 4th ed.; Oxford University Press: New York, NY, USA, 2018; pp. 635–660.
4. Mauer, K.; O'Kelley, R.; Podda, N.; Flanagan, S.; Gadani, S. New treatment modalities for hepatocellular cancer. *Curr. Gastroenterol. Rep.* **2015**, *17*, 19. [CrossRef] [PubMed]

5. Wang, Y.; Deng, T.; Zeng, L.; Chen, W. Efficacy and safety of radiofrequency ablation and transcatheter arterial chemoembolization for treatment of hepatocellular carcinoma: A meta-analysis. *Hepatology* **2016**, *46*, 58–71. [CrossRef] [PubMed]
6. Liu, G.; Fan, X.; Tang, M.; Chen, R.; Wang, H.; Jia, R.; Zhou, X.; Jing, W.; Wang, H.J.; Yang, Y.; et al. Osteopontin induces autophagy to promote chemo-resistance in human hepatocellular carcinoma cells. *Cancer Lett.* **2016**, *383*, 171–182. [CrossRef] [PubMed]
7. Horgan, A.M.; Dawson, L.A.; Swaminath, A.; Knox, J.J. Sorafenib and radiation therapy for the treatment of advanced hepatocellular carcinoma. *J. Gastrointest. Cancer* **2012**, *43*, 344–348. [CrossRef] [PubMed]
8. Tabernero, J.; Garcia-Carbonero, R.; Cassidy, J.; Sobrero, A.; Van, C.E.; Kohne, C.H.; Tejpar, S.; Gladkov, O.; Davidenko, I.; Salazar, R.; et al. Sorafenib in combination with oxaliplatin, leucovorin, and fluorouracil (modified FOLFOX6) as first-line treatment of metastatic colorectal cancer: The respect trial. *Clin. Cancer Res.* **2013**, *19*, 2541–2550. [CrossRef] [PubMed]
9. Kamel, M.; Nidhal, S.; Olfa, B.; Slim, D.; Sonia, T.; Abdulkhaleg, A.; Khaldoun, A.S.; Wided, B.A.; Sana, A.; Adel, H.B.; et al. Chemical composition and antioxidant and antimicrobial activities of wormwood (*Artemisia absinthium* L.) essential oils and phenolics. *J. Chem.* **2015**, *2015*, 1–12.
10. Azizi, K.; Shahidi-Hakak, F.; Asgari, Q.; Hatam, G.R.; Fakoorziba, M.R.; Miri, R.; Moemenbellah-Fard, M.D. In vitro efficacy of ethanolic extract of *Artemisia absinthium* (*Asteraceae*) against *Leishmania major* L. using cell sensitivity and flow cytometry assays. *J. Parasit. Dis.* **2016**, *40*, 735–740. [CrossRef] [PubMed]
11. Tamargo, B.; Monzote, L.; Piñón, A.; Machín, L.; García, M.; Scull, R.; Setzer, W.N. In vitro and in vivo evaluation of essential oil from *Artemisia absinthium* L. formulated in nanocochleates against cutaneous Leishmaniasis. *Medicines* **2017**, *4*, 38. [CrossRef] [PubMed]
12. Turak, A.; Shi, S.P.; Jiang, Y.; Tu, P.F. Dimeric guaianolides from *Artemisia absinthium*. *Phytochemistry* **2014**, *105*, 109–114. [CrossRef] [PubMed]
13. Caner, A.; Doskaya, M.; Degirmenci, A.; Can, H.; Baykan, S.; Uner, A.; Basdemir, G.; Zeybek, U.; Guruz, Y. Comparison of the effects of *Artemisia vulgaris* and *Artemisia absinthium* growing in western Anatolia against trichinellosis (*Trichinella spiralis*) in rats. *Exp. Parasitol.* **2008**, *119*, 173–179. [CrossRef] [PubMed]
14. Amat, N.; Upur, H.; Blazeković, B. In vivo hepatoprotective activity of the aqueous extract of *Artemisia absinthium* L. againstchemically and immunologically induced liver injuries in mice. *J. Ethnopharmacol.* **2010**, *131*, 478–484. [CrossRef] [PubMed]
15. Gilani, A.H.; Janbaz, K.H. Preventive and curative effects of *Artemisia absinthium* on acetaminophen and CCl4-induced hepatotoxicity. *Gen. Pharmacol.* **1995**, *26*, 309. [CrossRef]
16. Craciunescu, O.; Constantin, D.; Gaspar, A.; Toma, L.; Utoiu, E.; Moldovan, L. Evaluation of antioxidant and cytoprotective activities of *Arnica montana* L. and *Artemisia absinthium* L. ethanolic extracts. *Chem. Cent. J.* **2012**, *6*, 1–11. [CrossRef] [PubMed]
17. Bora, K.S.; Sharma, A. Evaluation of antioxidant and free-radical scavenging potential of *Artemisia absinthium*. *Pharm. Biol.* **2011**, *49*, 1216. [CrossRef] [PubMed]
18. Zhang, M.Z.; Du, H.X.; Huang, Z.X.; Zhang, P.; Yue, Y.Y.; Wang, W.Y.; Liu, W.; Zeng, J.; Ma, J.B.; Chen, G.Q.; et al. Thymoquinone induces apoptosis in bladder cancer cell via endoplasmic reticulum stress-dependent mitochondrial pathway. *Chem.-Biol. Interact.* **2018**, *292*, 65–75. [CrossRef]
19. Vuyolwethu, S.; Georgia, S.; Roger, H.; Grafov, A.; Grafova, I.; Nieger, M.; Katz, A.A.; Parker, M.I.; Kaschula, C.H. The cytotoxicity of the ajoene analogue BisPMB in WHCO1 oesophageal cancer cells is mediated by CHOP/GADD153. *Molecules* **2017**, *22*, 892. [CrossRef]
20. Zhang, H.P. *Drug Standard of Ministry of Public Health of the People's Republic of CHINA Xingjiang Technological and Health Publishing House, Xingjiang, Uighur Medicine Part*; Xinjiang Publishing House of Science: Urumqi, China, 1999; p. 53.
21. Song, G.C.; Yu, Y.J.; Wang, X.J. Experiments on antitumor activity of *Scutellaria barbata* polysaccharides and its immunological mechanisms. *Open Access Libr. J.* **2011**, *13*, 641–643.
22. Wang, D.D.; Wu, Q.X.; Pan, W.J.; Hussain, S.; Mehmood, S.; Chen, Y. A novel polysaccharide from the *Sarcodon aspratus* triggers apoptosis in Hela cells via induction of mitochondrial dysfunction. *Food Nutr. Res.* **2018**, *62*, 1285. [CrossRef]
23. Ashokkumar, R.; Jamuna, S.; Sakeena Sadullah, M.S.; Niranjali, D.S. Vitexin protects isoproterenol induced post myocardial injury by modulating hipposignaling and ER stress responses. *Biochem. Biophys. Res. Commun.* **2018**, *496*, 731–737. [CrossRef] [PubMed]

24. Fan, C.; Yang, Y.; Liu, Y.; Jiang, S.; Di, S.Y.; Hu, W.; Ma, Z.Q.; Li, T.; Zhu, Y.F.; Xin, Z.L.; et al. Icariin displays anticancer activity against human esophageal cancer cells via regulating endoplasmic reticulum stress-mediated apoptotic signaling. *Sci. Rep.* **2016**, *6*, 21145. [CrossRef] [PubMed]
25. Cai, Y.L.; Zheng, Y.F.; Gu, J.Y.; Wang, S.Q.; Wang, N.; Yang, B.W.; Zhang, F.X.; Wang, D.M.; Fu, W.J.; Wang, Z.Y. Betulinic acid chemosensitizes breast cancer by triggering ER stress-mediated apoptosis by directly targeting GRP78. *Cell Death Dis.* **2018**, *9*, 636. [CrossRef] [PubMed]
26. Wang, B.; Zhou, T.Y.; Nie, C.H.; Wang, D.L.; Zheng, S.S. Bigelovin, a sesquiterpene lactone, suppresses tumor growth through inducing apoptosis and autophagy via the inhibition of mTOR pathway regulated by ROS generation in liver cancer. *Biochem. Biophys. Res. Commun.* **2018**, *499*, 156–163. [CrossRef] [PubMed]
27. Ramos, S. Effects of dietary flavonoids on apoptotic pathways related to cancer chemoprevention. *J. Nutr. Biochem.* **2007**, *18*, 427–442. [CrossRef]
28. Marzo, I.; Naval, J. Bcl-2 family members as molecular targets in cancer therapy. *Biochem. Pharmacol.* **2008**, *76*, 939–946. [CrossRef]
29. Garner, T.P.; Lopez, A.; Reyna, D.E.; Spitz, A.Z.; Gavathiotis, E. Progress in targeting the BCL-2 family of proteins. *Curr. Opin. Chem. Biol.* **2017**, *39*, 133–142. [CrossRef]
30. Shafi, G.; Hasan, T.N.; Syed, N.A.; Al-Hazzani, A.A.; Alshatwi, A.A.; Jyothi, A.; Munshi, A. *Artemisia absinthium(AA)*: A novel potential complementary and alternative medicine for breast cancer. *Mol. Biol. Rep.* **2012**, *39*, 7373–7379. [CrossRef]
31. Banerjee, A.; Banerjee, V.; Czinn, S.; Blanchard, T. Increased reactive oxygen species levels cause ER stress and cytotoxicity in andrographolide treated colon cancer cells. *Oncotarget* **2017**, *8*, 26142–26153. [CrossRef]
32. Yu, X.S.; Du, J.; Fan, Y.J.; Liu, F.J.; Cao, L.L.; Liang, N.; Xu, D.G.; Zhang, J.D. Activation of endoplasmic reticulum stress promotes autophagy and apoptosis and reverses chemoresistance of human small cell lung cancer cells by inhibiting the PI3K/AKT/mTOR signaling pathway. *Oncotarget* **2016**, *7*, 76827–76839. [CrossRef]
33. Merlot, A.; Shafie, N.; Yu, Y.; Richardson, V.; Jansson, P.J.; Sahni, S.; Lane, D.J.; Kovacevic, D.S.; Richardson, D.R. Mechanism of the induction of endoplasmic reticulum stress by the anti-cancer agent, di-2-pyridylketone 4,4-dimethyl-3-thiosemicarbazone (Dp44mT): Activation of PERK/eIF2a, IRE1a, ATF6 and calmodulin kinase. *Biochem. Pharmacol.* **2016**, *109*, 27–47. [CrossRef] [PubMed]
34. Wu, M.H.; Chiou, H.L.; Lin, C.L.; Lin, C.Y.; Yang, S.F.; Hsieh, Y.H. Induction of endoplasmic reticulum stress and mitochondrial dysfunction dependent apoptosis signaling pathway in human renal cancer cells by Norcantharidin. *Oncotarget* **2018**, *9*, 4787–4797. [CrossRef] [PubMed]
35. Zhang, D.; Gao, C.; Li, R.; Zhang, L.; Tian, J.K. TEOA, a triterpenoid from *Actinidia eriantha*, induces autophagy in SW620 cells via endoplasmic reticulum stress and ROS-dependent mitophagy. *Arch. Pharm. Res.* **2017**, *40*, 1–13. [CrossRef] [PubMed]

Sample Availability: Samples of the compounds are available from the authors.

© 2019 by the authors. Licensee MDPI, Basel, Switzerland. This article is an open access article distributed under the terms and conditions of the Creative Commons Attribution (CC BY) license (http://creativecommons.org/licenses/by/4.0/).

Article

Cytotoxic Properties of Damiana (*Turnera diffusa*) Extracts and Constituents and A Validated Quantitative UHPLC-DAD Assay

Johanna Willer [1], Karin Jöhrer [2], Richard Greil [2,3], Christian Zidorn [1] and Serhat Sezai Çiçek [1,*]

1. Department of Pharmaceutical Biology, Kiel University, Gutenbergstraße 76, 24118 Kiel, Germany; jwiller@pharmazie.uni-kiel.de (J.W.); czidorn@pharmazie.uni-kiel.de (C.Z.)
2. Tyrolean Cancer Research Institute, Innrain 66, 6020 Innsbruck, Austria; karin.joehrer@tkfi.at (K.J.); r.greil@salk.at (R.G.)
3. Department of Internal Medicine III, Oncologic Center, Salzburg Cancer Research Institute–Laboratory for Immunological and Molecular Cancer Research (SCRI-LIMCR), Paracelsus Medical University Salzburg, Cancer Cluster Salzburg, Müllner Hauptstraße 48, 5020 Salzburg, Austria
* Correspondence: scicek@pharmazie.uni-kiel.de; Tel.: +49-431-880-1077

Academic Editor: Roberto Fabiani
Received: 30 January 2019; Accepted: 25 February 2019; Published: 28 February 2019

Abstract: In our continuing search for new cytotoxic agents, we assayed extracts, fractions, and pure compounds from damiana (*Turnera diffusa*) against multiple myeloma (NCI-H929, U266, and MM1S) cell lines. After a first liquid-liquid solvent extraction, the ethyl acetate layer of an acetone (70%) crude extract was identified as the most active fraction. Further separation of the active fraction led to the isolation of naringenin (**1**), three apigenin coumaroyl glucosides **2–4**, and five flavone aglycones **5–9**. Naringenin (**1**) and apigenin 7-*O*-(4″-*O*-*p*-*E*-coumaroyl)-glucoside (**4**) showed significant cytotoxic effects against the tested myeloma cell lines. Additionally, we established a validated ultra-high performance liquid chromatography diode array detector (UHPLC-DAD) method for the quantification of the isolated components in the herb and in traditional preparations of *T. diffusa*.

Keywords: multiple myeloma; quality control; naringenin; flavonoids; traditional preparation

1. Introduction

Turnera diffusa Willd., Passifloraceae, commonly referred to as damiana, is a shrub occurring in north-eastern Brazil, Mesoamerica, the Caribbean, Mexico, and Texas [1]. The traditional use of *T. diffusa* in Latin America encompasses usage as an aphrodisiac, a tonic, and for the treatment of diabetes [2,3]. Damiana extracts with tequila are allegedly used as love potions [3]. Due to the long history of *T. diffusa* as an aphrodisiac, both the stimulating effect as well as the underlying mechanisms are relatively well investigated. In animal testing, the aqueous extract of *T. diffusa* was found to increase the sexual activity of rats [4]. The flavonoids obtained by percolation with methanol, and here especially the flavanone pinocembrin, were identified as aromatase inhibitors resulting in increased testosterone levels and improved libido [5]. Regarding the antidiabetic activity, conflicting results were obtained. Whereas Alarcon-Aguilar et al. could not find any hypoglycemic effect of an ethanol-water extract [6], Parra-Naranjo et al. demonstrated hypoglycemic effects of a methanol extract and identified teuhetenone A, a nor-sesquiterpene, as the active principle [7].

Apart from traditional usage, an aqueous extract of *T. diffusa* was found to inhibit the monoamine oxidase A with IC_{50} values of 130 mg/mL as well as the acetyl- and butyrylcholinesterase with IC_{25} values of 0.352 and 0.370 mg/mL, respectively [8]. Additionally, cytotoxic effects of a methanolic extract against breast carcinoma cell line MDA-MD-231 were demonstrated [9]. Though the mechanism behind the observed cytotoxicity remains unsolved, the compounds responsible for cell death induced by the methanolic extract were identified as arbutin and apigenin.

The antioxidative effects of flavonoids are well documented and linked to certain structural features such as a dihydroxylated B-ring, a double bond located between C2 and C3, and a 4-oxo function at ring C [10]. Moreover, flavonoids are known to influence the metabolism, e.g., by inhibiting oxidases or activating antioxidative enzymes [11,12]. However, for flavonoids isolated from *T. diffusa* a variety of additional activities were reported. Velutin (**7**), a dimethoxylated hydroxyflavone known from the açaí berry (*Euterpe oleracea* Mart., Arecaceae), has been shown to possess strong anti-inflammatory effects by inhibiting NF-κB activation as well as p38 and JNK phosphorylation and hence by down-modulation of the expression of TNF-α and IL-6 [13]. At low doses, velutin (**7**) is more potent than established anti-inflammatory agents such as apigenin. Brito et al. found that the anti-inflammatory effect of velutin (**7**) against periodontitis is caused by an inhibition of HIF-1α expression [14]. Besides, the compound was reported to possess cytotoxic effects against human nasopharynx carcinoma (KB) cells with an IC_{50} value of 4.8 µM [15]. Acacetin 7-*O*-methyl ether (**9**), another methoxylated flavonoid, showed moderate cytotoxic effects on HeLa cells. Subsequent testing focused on the influence of aminoalkylation of acacetin 7-*O*-methyl ether (**9**) and the resulting antiproliferative activity against three human cancer cell lines (HeLa, HCC1954, SK-OV-3) [16]. The Mixe Indians (Oaxaca, Mexico) use an aqueous extract of *Calea zacatechichi* Schltdl. (Asteraceae) as a remedy for malaria. In subsequent experiments, the flavone genkwanin (**6**) was identified as active principle [17]. Boege et al. reported acacetin (**5**) to inhibit the topoisomerase I [18], though the compound lacks the structural feature (3-hydroxy group) supposed necessary for topoisomerase activity [19].

In the present study, the effect of extracts, fractions, and pure compounds from *T. diffusa* on multiple myeloma (MM) cell lines was investigated. Flavones, e.g., apigenin, chrysin, and luteolin, have been shown to block proteasome catalytic activities in tumor cells [20] and to induce cell death in myeloma cells [21]. Proteasome inhibitors are state-of-the-art in the therapy of multiple myeloma, but most patients develop resistance over time and new drugs are urgently needed. In addition, compounds of damiana are expected to induce reactive oxygen species [22], presumably adding to the expected cytotoxicity in myeloma cells. Our bioactivity-guided approach led to the isolation of seven flavonoids and a mixture of acacetin and genkwanin (Figure 1, **1–9**). In a second step, a validated ultra-high performance liquid chromatography (UHPLC) diode-array detector (DAD) method for the quantification of phenolic constituents in extracts and preparations of *T. diffusa* has been established.

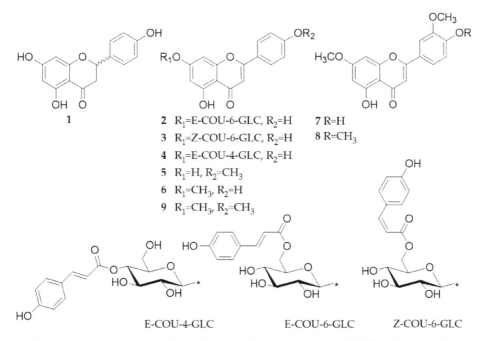

Figure 1. Chemical structures of flavonoids isolated from the aerial parts of *T. diffusa*. Naringenin (**1**), apigenin 7-*O*-(6″-*O*-*p*-*E*-coumaroyl)-glucoside (**2**), apigenin 7-*O*-(6″-*O*-*p*-*Z*-coumaroyl)-glucoside (**3**), apigenin 7-*O*-(4″-*O*-*p*-*E*-coumaroyl)-glucoside (**4**), acacetin (**5**), genkwanin (**6**), velutin (**7**), gonzalitosin I (**8**), and acacetin 7-*O*-methyl ether (**9**).

2. Results

2.1. Bioactivity of Tested Fractions

After an acetone extract (70%) of *T. diffusa* showed cytotoxic potential in an initial screening against MM cell lines (Figure 2a,b (positive control)), the crude extract was fractionated with organic solvents of increasing polarity. Subsequently, the obtained fractions TD-1 (ethyl acetate), TD-2 (*n*-butanol), TD-3 (acidified *n*-butanol), and TD-4 (aqueous layer) were evaluated for their cytotoxicity. As shown in Figure 2c,d, the cytotoxic activity was mainly retained in the ethyl acetate fraction (TD-1), resulting in decreased viability in all tested concentrations after both 24 and 48 h of treatment, respectively. In contrast, treatment with TD-2 decreased the amount of viable cells only moderately after 48 h of treatment, whereas no effects were observed for fractions TD-3 and TD-4.

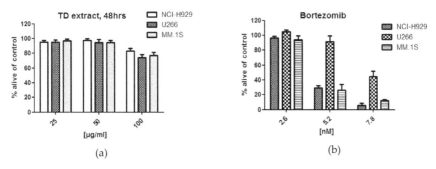

Figure 2. *Cont.*

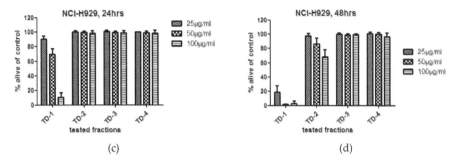

Figure 2. Viability of multiple myeloma (MM) cell lines NCI-H929, U266, and MM1S (**a**) 48 h after treatment with an acetone 70% extract of *T. diffusa* and after treatment with bortezomib (**b**). Viability of NCI-H929 cells 24 and 48 h after treatment with ethyl acetate (TD-1), *n*-butanol (TD-2), acidified *n*-butanol (TD-3) fractions and the remaining water layer (TD-4) derived from an acetone (70%) extract of *T. diffusa* (**c**,**d**). Three concentration levels (25, 50, 100 µg/mL) are shown.

In a second approach, the crude acetone extract was fractionated in smaller polarity steps using *n*-hexane (TD-1a), diethyl ether (TD-1b), and ethyl acetate (TD-1c). The resulting fractions as well as TD-1 were then evaluated for their cytotoxic potential against MM cell lines MM1S, U266, and NCI-H929 after 24 h of incubation (Figure 3). As expected, TD-1 decreased viable cells significantly at a concentration of 100 µg/mL, but not at lower concentrations. Unlike before, partitioning in smaller polarity steps did not result in a single active fraction but revealed different activities for TD-1a to TD-1c. At a concentration of 100 µg/mL, TD-1a and TD-1b showed pronounced effects against MM1S cell lines, which were higher in TD-1b and lower in TD-1a, compared to TD-1. TD-1b moreover showed higher activity against NCI-H929 cell lines than TD-1 and slightly stronger effects against U266 cell lines. In contrast, TD-1a was less active against these two cell lines. TD-1c was only moderately active against all three tested cell lines.

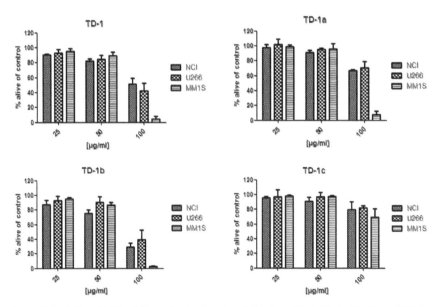

Figure 3. Viability of MM cell lines after treatment with fractions TD-1, TD-1a, TD-1b, and TD-1c. Three concentration levels (25, 50, 100 µg/mL) and 24 h incubation times are shown.

2.2. Chromatographic Analyses of Tested Fractions

The tested fractions were analyzed by UHPLC in order to eventually attribute the observed effects to specific peaks (Figure 4). Comparison of fractions TD-1 and TD-2, which were obtained in the first separation step, clearly shows that the active fraction TD-1 contains a number of peaks that are missing in fraction TD-2. These peaks, which are eluting after 25 min comprise, amongst other, the later isolated compounds **2, 3** and **5–9**. UHPLC analysis of the three fractions of the second partitioning step (TD-1a to TD-1c) resulted in similar findings. Here, fraction TD-1c, which was significantly less active than TD-1a and TD-1b, was lacking compounds **5–7** and showed clearly lower amounts of compounds **8** and **9**. However, compound **4** was present in more or less the same concentration and compounds **2, 3** and various polar constituents were present in higher amounts. The difference between fraction TD-1a and the slightly more active fraction TD-1b was less significant. Apart from the presence of a few peaks in the more polar region, TD-1b showed higher amounts of compounds **5–8** and lower concentrations of compounds **2** and **9**. Thus, the isolation of peaks eluting after 25 min was considered most promising for the identification of the cytotoxic principle(s) present in *T. diffusa*. As the respective compounds were quantitatively extracted in the first separation step, but not in the second one, fraction TD-1 was chosen for isolation of the desired substances.

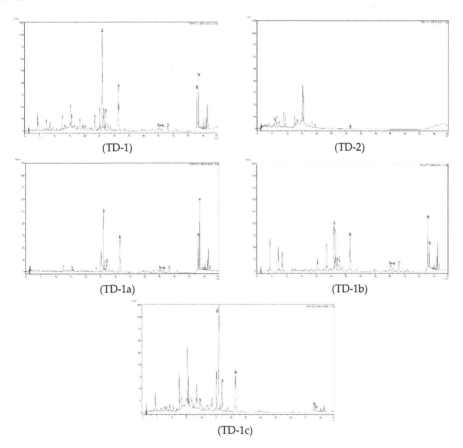

Figure 4. Ultra-high performance liquid chromatography (UHPLC) chromatograms of TD-1, TD-2, and TD-1a to TD-1c.

2.3. Isolation of Natural Compounds

Separation of fraction TD-1 was accomplished by silica gel flash chromatography, Sephadex LH-20 gel permeation chromatography, as well as preparative medium- and semi-preparative high-performance liquid chromatography and yielded seven flavonoids and a mixture of two compounds. Using analytical reference standards, mass spectrometry (MS) and nuclear magnetic resonance (NMR) spectroscopy, the isolated compounds were identified as naringenin (**1**), apigenin 7-O-(6″-O-p-E-coumaroyl)-glucoside (**2**), apigenin 7-O-(6″-O-p-Z-coumaroyl)-glucoside (**3**), apigenin 7-O-(4″-O-p-E-coumaroyl)-glucoside (**4**), acacetin (**5**), genkwanin (**6**), velutin (**7**), gonzalitosin I (**8**), and acacetin 7-O-methyl ether (**9**).

2.4. Bioactivity of Isolated Compounds

Cytotoxic assays of the pure compounds resulted in the identification of two compounds (**1** and **4**) with significant impact on cell viability (Figure 5). Of these two compounds, naringenin (**1**) was found cytotoxic even at low concentrations, showing a decreased viability of NCI-H929 and U266 cell lines of 25.5 ± 12.5 and 79.6 ± 15.2%, respectively, after 24 h incubation time. At the same incubation time and at a concentration of 50 µM, compound **4** decreased the viability of the same cell lines (NCI-H929 and U266) to 66.1 ± 17.4 and 84.4 ± 3.7%, respectively. Peripheral blood mononuclear cells (PBMC) from healthy donor (HD) were affected by the treatment with compounds **1** and **4**, if however, in different extent than the cancer cell lines.

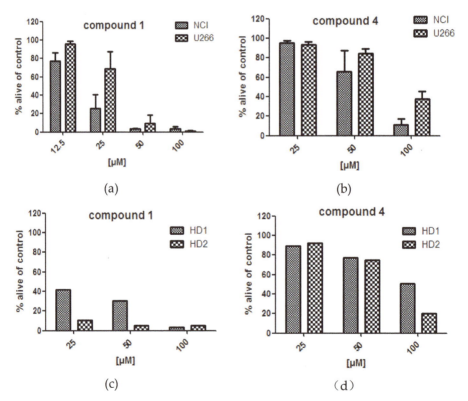

Figure 5. Viability of MM cell lines (U266, NCI-H929) (a,b) and healthy donor (HD) cells (c,d) 24 h after treatment with compounds **1** and **4**, respectively.

2.5. Validation of UHPLC-DAD Assay

Of the isolated flavonoids, compounds **4** to **8** were chosen as calibration standards, using compound **4** to co-quantify compounds **2** and **3** and compound **8** to co-quantify compound **9**. Linearity was achieved by five-point calibration with a coefficient of determination of >0.99 (Table 1). Repeatability and precision measurements were performed using a dimethyl sulfoxide (DMSO) extract (Figure 6). Repeatability experiments revealed relative standard deviations (RSD) below 2.3% (Table 2). Intra-day precision showed RSD values below 6.9% for all compounds on day 1. The relative standard deviation on day 2 showed similar results for most compounds except for compounds **5–7**, which had higher deviations. Also, inter-day precision was higher for these compounds, with values of 6.9 and 7.8%, whereas the other compounds showed RSD values below 6%. Spiking experiments over four calibration levels conducted for compounds **4** to **8** showed recovery rates ranging from 98.0 to 127% (Table 3).

Table 1. Regression curves, coefficients of determination, limit of detection (LOD), and limit of quantitation (LOQ) of the UHPLC method.

Compound	Regression Curve	R^2	LOD [1]	LOQ [1]
4	$6.971 \times 10^9 \times (-3.027 \times 10^5)$	0.9985	1.39	4.60
5 + 6	$8.170 \times 10^9 \times (-1.455 \times 10^6)$	0.9955	0.162	0.535
7	$3.605 \times 10^9 \times (-5.394 \times 10^5)$	0.9956	0.236	0.780
8	$9.720 \times 10^9 \times (-1.862 \times 10^6)$	0.9924	0.120	0.396

[1] Concentrations are given in µg/mL.

Table 2. Repeatability and precision of the UHPLC method.

Compound	Repeatability [1]	Intra-day 1 [1]	Intra-day 2 [1]	Inter-day [1]
2	12.9 (0.138)	10.9 (4.95)	10.6 (4.90)	10.8 (4.82)
3	2.16 (2.29)	1.73 (5.52)	1.68 (4.20)	1.71 (4.97)
4	8.38 (1.19)	7.73 (4.18)	7.29 (4.83)	7.51 (5.25)
5 + 6	1.96 (1.62)	1.72 (3.02)	1.55 (5.48)	1.64 (6.89)
7	2.12 (1.05)	1.63 (6.89)	1.71 (9.01)	1.67 (7.82)
8	1.08 (0.389)	0.907 (3.37)	0.897 (3.66)	0.902 (3.40)
9	1.73 (0.169)	1.39 (4.87)	1.39 (5.34)	1.39 (4.88)

[1] Concentrations are given in µg/mL, relative standard deviations are given in parentheses and are stated in percent.

Table 3. Accuracy of the UHPLC method.

Compound	Sample Concentration [1]	Spiked Amount [2]	Total Concentration [1]	Recovery [3]
4	6.56	3.08	4.72	105
		7.69	12.6	113
		15.4	18.7	112
		23.1	24.7	111
5 + 6	1.38	0.401	0.746	102
		1.07	2.10	99.7
		2.14	2.83	106
		3.21	3.55	107
7	1.52	0.585	0.964	102
		1.04	2.18	100
		2.08	2.84	102
		3.12	3.50	99.2
8	0.798	0.297	0.497	109
		0.793	1.39	109
		1.59	1.99	109
		2.09	2.29	127

[1] Concentrations are given in µg/mL. [2] Amounts are given in µg. [3] Recovery is stated in percent.

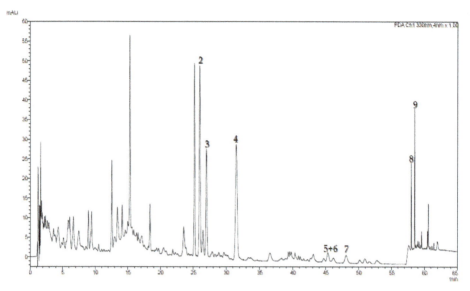

Figure 6. Chromatogram of a dimethyl sulfoxide (DMSO) extract prepared from *T. diffusa*. UHPLC was performed using a solvent mixture of 0.1% formic acid in water (solvent A) and acetonitrile (solvent B) with the following gradient: 15% B to 25% B in 15 min, to 29% B in 9 min, to 29% B in 11 min, to 36% B in 1 min, to 36% B in 19 min, to 95% B in 0.1 min, to 95% B in 9.9 min. Post-run was set to 10 min, temperature to 32 °C. The injection volume was 5 µL. The UV trace was recorded at 330 nm. The flow was 0.2 mL/min.

2.6. Chromatographic Analyses of a Traditional Preparation

The established method was additionally used for the analysis of a traditional preparation of *T. diffusa* (Figure 7). The sample of a traditional preparation of *T. diffusa* revealed minor compounds in the medium to low polarity range and a high content of polar compounds. By comparison with MS data and retention times of isolated compounds from *T. diffusa* compounds **4** to **9** were identified and their yield quantified using the method described above. Ultraviolet (UV) spectroscopy experiments indicated the presence of phenolic compounds in the polar range. Apigenin (**e**) was identified using reference standards. Previous investigations of an aqueous extract of *T. diffusa* by Bernardo et al. revealed a variety of apigenin, luteolin, and quercetin glycosides [8]. By comparison with ultraviolet (UV) and MS data larycitrin-3-*O*-(6-glucosyl)glucoside (**a**), apigenin 8-*C*-(2-rhamnosyl)glucoside (**b**), (luteolin 8-*C*-(2-rhamnosyl)ketodeoxihexoside (**c**), apigenin 7-*O*-(2-rhamnosyl)ketodeoxihexoside (**d**) were detected.

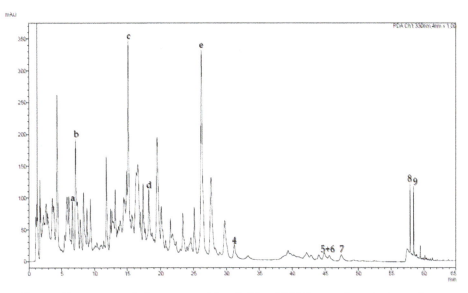

Figure 7. Chromatogram of a liquor prepared from *T. diffusa*. UHPLC was performed using a solvent mixture of 0.1% formic acid in water (solvent A) and acetonitrile (solvent B) with the following gradient: 15% B to 25% B in 15 min, to 29% B in 9 min, to 29% B in 11 min, to 36% B in 1 min, to 36% B in 19 min, to 95% B in 0.1 min, to 95% B in 9.9 min. Post-run was set to 10 min, temperature to 32 °C. The injection volume was 5 µL. The ultraviolet (UV) trace was recorded at 330 nm. The flow was 0.2 mL/min. **4** to **9** refer to the isolated compounds (see Figure 1), compounds **a** to **d** were identified by UV spectroscopy and MS spectrometry experiments as larycitrin-3-O-(6-glucosyl)glucoside (**a**), apigenin 8-C-(2-rhamnosyl)glucoside (**b**), (luteolin 8-C-(2-rhamnosyl)ketodeoxihexoside (**c**), apigenin 7-O-(2-rhamnosyl)ketodeoxihexoside (**d**), and **e** was identified as apigenin by an authentic standard.

3. Discussion

3.1. Bioactivity of Tested Fractions

A crude acetone (70%) extract of *T. diffusa* displayed promising cytotoxicity against MM cell lines. Partitioning of the acetone extract located the cytotoxic compounds in the ethyl acetate (TD-1) fraction. In a second approach, using smaller polarity steps, the activity was found in the *n*-hexane (TD-1a) and the diethyl ether (TD-1b) fraction, while the remaining ethyl acetate fraction (TD-1c) was only slightly active. UHPLC experiments of the active fractions revealed similar peak patterns in the low polarity range. This peak pattern was absent in the inactive fraction TD-2, and different in TD-1c. Thus, further purification steps with fraction TD-1 focused on this polarity range and afforded the isolation of naringenin (**1**), apigenin 7-O-(6″-O-p-E-coumaroyl)-glucoside (**2**), apigenin 7-O-(6″-O-p-Z-coumaroyl)-glucoside (**3**), apigenin 7-O-(4″-O-p-E-coumaroyl)-glucoside (**4**), acacetin (**5**), genkwanin (**6**), velutin (**7**), gonzalitosin I (**8**), and acacetin 7-O-methyl ether (**9**).

Regarding the isolated compounds, compound **4** was abundant in all active fractions (TD-1, TD-1a to TD-1c) but was absent in TD-2. Treatment with compound **4** (50, 100 µM) led to decreased viability of NCI-H929 and in a lesser extent of U266 after incubation for 24 h. This indicates that compound **4** contributes to the cytotoxic effect observed for the tested fractions. The corresponding aglycone, the flavone apigenin, was reported repeatedly for its cytotoxic activity [9,20,21,23,24]. However, the moderately active fraction TD-1c contained less of lipophilic compounds **5** to **9**, suggesting that these compounds might at least contribute to the observed cytotoxic effects.

The most active compound of our study, the flavanone naringenin (**1**), is well investigated for its ability to induce apoptosis against HL-60 cells via the activation of caspase-3, a member of the caspase-cascade that plays an important role in apoptosis. Interestingly, the cytotoxic effect of naringenin (**1**) was found to be weakened if C7 is substituted with a sugar moiety (rutinoside) [25,26]. In the present work, naringenin (**1**) was found to possess the highest activity of all tested substances and showed a pronounced decrease of viability especially in NCI-H929 cells. Nevertheless, the active principle resulting in the lowered viability in this study requires further testing.

The experiments performed under the same conditions with HD cells indicate a negative influence on cells of the immune system after treatment with compound **1** or **4** (Figure 5). Interestingly, Chen et al. (2003) did not observe apoptosis in polymorphonuclear neutrophils (PMN) after their treatment with naringenin (**1**). However, testing conditions differed from the ones used in this study. Nevertheless, though flavonoids are generally assumed safe due to regular uptake with fruits and vegetables, they were found to evoke cytotoxic effects at higher doses [27,28].

3.2. Validation

By testing different columns (Synergi 4μm Polar-RP 80 Å 150 × 2.00 mm, Luna Omega 1.6 μm C18 100 Å 100 × 2.1 mm, Kinetex 1.7 μm XB-C18 100 Å 100 × 2.1 mm) the best separation was achieved for the C18 Luna Omega column. Consequently, it was chosen as starting point for method development. Different compositions of pure water or formic acid in water as well as methanol or acetonitrile were tested for their influence on the separation of the compounds of the DMSO extract. Thereby, 0.1% formic acid and acetonitrile were identified as suitable eluents and a column temperature of 32 °C was found to give a good resolution of the peaks. Due to the complex mixture and a variety of similar compounds within the crude extract, isocratic steps were performed at different solvent concentrations as part of the gradient. Because of the absorption maxima around 330 nm of compounds **4** to **8**, this wavelength was chosen for detection. An injection volume of 5 μL was found to provide a good repeatability with an acceptable peak resolution.

The calibration curves for quantified constituents **4** to **8** were obtained on five levels each regarding the concentration of the corresponding analyte in the extract. The established calibration curves had determination coefficients of more than 0.99 and were thus accepted for quantification purposes. Spiking experiments showed acceptable recovery rates over a broad calibration range (Table 3).

Interday precision revealed good relative standard deviation (RSD) values for peaks **5** to **8**; compound **7** however varies in a bigger extent. These findings also account for the co-quantified peaks with acacetin 7-O-methylether (**9**) and the compounds **2** and **3** having similar deviations. Displaying structural isomers, the latter two compounds were co-quantified by compound **4**, while acacetin and genkwanin (**5** + **6**) were chosen to co-quantify acacetin 7-O-methylether (**9**) since the molecules only differ in the amount of one methyl group.

Limit of quantification (LOQ) for all quantified compounds **4**–**8** was set to the lower limit of the calibration curve. Those values provide acceptable standard deviation (data not shown). Commonly, limit of detection (LOD) is defined as one third of LOQ. Therefore, LOD was calculated from the LOQ (Table 2).

3.3. Traditional Preparation

T. diffusa is used as remedy for various diseases in the traditional medicine of Latin America. Additionally, a liquor of *T. diffusa* is used to increase sexual activity [3]. In literature, tequila is described as extraction agent while in this study Kornbrand (Bauerndank/Edeka/Hamburg, Germany) was used. Both liquors contain 38% ethanol and thus were considered comparable. Due to the contained water, the liquor was expected to contain a high number of polar compounds. This assumption was verified by UHPLC analysis, which revealed most of the peaks eluting in the polar range. Compounds **4** to **9** were present, however, at low concentrations. In the present study, compound **4** showed moderate cytotoxicity in an in vitro assay against myeloma cell lines and healthy donor (HD) cells at higher concentrations of 50 and 100 μM. This corresponds with 28.9 mg/mL or 57.8 mg/mL, a concentration unlikely to be met in traditional preparations.

4. Materials and Methods

4.1. General Experimental Procedures

Solvents and reagents for isolation were of analytical quality. Solvents used for UHPLC were of LC-MS grade quality. Column chromatography was performed with silica gel (40–63 μm particle size) (Merck, Darmstadt, Germany) or with Sephadex LH-20 (GE Healthcare, Chicago, IL, USA). Thin layer chromatography (TLC) was performed on silica gel 60 F_{254} plates (Merck) using ethyl acetate-methanol-water (10:1:1) or n-hexane-ethyl acetate-methanol-formic acid (7:4:1.5:0.1) as mobile phase and vanillin-sulphuric acid for detection. Preparative medium-performance liquid chromatography (MPLC) was accomplished using a Büchi PrepChrom C-700 equipped with a Büchi PrepChrom MPLC column C18 (250 × 30 mm, 15 μm) (Büchi, Flawil, Switzerland). Semi-preparative high-performance liquid chromatography (HPLC) was carried out on Waters a Alliance e2695 Separations Module with Alliance 2998 photodiode array (PDA), 2410 RI, and WFC III fraction collector (Waters, Milford, MA, USA), and either an Aqua 5 μ C18 column (250 × 10 mm, 5 μm particle size, Phenomenex, Aschaffenburg, Germany) or a VP Nucleodur C18 (250 × 10 mm, 5 μm particle size, Macherey-Nagel, Düren, Germany). Extracts, fractions, and pure compounds were analyzed by a Hitachi ChromasterUltra R_S System (VWR, Darmstadt, Germany) connected to an autosampler, column heater, PDA and a Sederé Sedex 100 evaporative light scattering detector (ELSD), using a Phenomenex Synergi Polar-RP column (150 × 2 mm, 4 μm particle size). Pure compounds were additionally analyzed by Nexera X2 system (Shimadzu, Kyoto, Japan) connected to an autosampler, column heater, PDA and a Shimadzu LCMS 8030 Triple Quadrupole Mass Spectrometer with electron spray ionization. Quantification was performed on the same instrument using a Phenomenex Luna Omega C18 column (100 × 2.1 mm, 1.6 μm particle size) and NMR spectra were recorded on an Avance III 300 NMR spectrometer (Bruker, Billerica, MA, USA) connected to a BACs-autosampler (Bruker). Centrifugation was performed on a Heraeus Megafuge 16 (Thermo Fisher Scientific Inc., Waltham, MA, USA). The authentic standard of Apigenin >98% HPLC was purchased from TransMIT (Gießen, Germany).

4.2. Plant Material

Dried and cut aerial parts of *T. diffusa* (HAB-2014 quality) were obtained from Caesar & Loretz GmbH (Caelo), Hilden, Germany (art.-No.: 257a, lot number: 15294206).

4.3. Extraction and Fractionation

Dried herb (1.00 kg) was ground and extracted five times with 2 liters of acetone 70% undergoing sonification. The solvent was evaporated under reduced pressure to afford 85 g of crude extract. For isolation the crude extract was repeatedly partitioned between ethyl acetate and water and the ethyl acetate layer was evaporated to dryness, yielding 32.6 g (TD-1). Subsequently, this procedure was repeated with butanol (14.6 g, TD-2). After acidification of the water layer with 2.5 mL formic acid, the solution was again extracted with butanol (5.40 g, TD-3) and the aqueous layer was then evaporated to dryness, yielding 32.4 g (TD-4).

In a second approach, 5.0 g of crude extract were suspended in water and extracted with n-hexane. The fraction was evaporated to dryness, yielding 0.53 g (TD-1a). The procedure was repeated with diethyl ether (0.21 g, TD-1b) and ethyl acetate (0.22 g, TD-1c).

For the validated UHPLC-DAD assay, sieved plant material (800 μm mesh width) was extracted threefold with DMSO, centrifuged, and the supernatants were collected and diluted in a 20 mL volumetric flask.

The traditional liquor from *T. diffusa* was prepared by maceration of 35 g of ground drug material with 0.7 liters of 38% alcohol (Bauerndank) for four weeks. Subsequently, the extract was filtered.

4.4. Isolation

TD-1 was subjected to silica gel column chromatography (40 × 8 cm) and eluted in a gradient manner with petroleum ether-ethyl acetate-methanol-water (10:0:0:0 to 0:0:19:10) yielding 160 fractions. After characterization by TLC the obtained subfractions were combined to 20 fractions (TD-1_1 to TD-1_20).

From TD-1_2 compound **1**, a two to one mixture of **5** and **6**, and compound **8** were obtained after column chromatography with Sephadex LH20 and acetone–dichloromethane (15:85) as eluting solvent. The thereby eluted subfraction TD-1_2_D was subjected to semi-preparative HPLC using 0.025% formic acid in water and methanol in a gradient matter, yielding 14.2 mg of velutin (**7**) and 27.8 mg of linoleic acid. 27 mg of acacetin 7-O-methylether (**9**) and another 58 mg of linoleic acid were obtained from TD-1_4 by preparative column chromatography using 0.025% formic acid in water and methanol in a gradient matter. TD-1_8 (196 mg) was submitted to further separation by Sephadex LH20 (using acetone as eluent) to give 11.7 mg of compound **4**. TD-1_12 was subjected to semi-preparative HPLC using 0.025% formic acid in water and acetonitrile in an isocratic matter, yielding 6.3 mg of compound **3**. TD-1_14 yielded 10.0 mg of sitosterol and TD-1_ 20 9.0 mg of compound **2**. Structure elucidation of the isolated compounds was accomplished by comparison of MS- and NMR-spectra with literature data [29–36]. MS and NMR spectra (^{1}H, HSQC, HMBC) are provided in the Supplementary Materials (**S1** to **S15**).

4.5. Cytotoxic Assays

Cytotoxicity was assessed for TD-1 to TD-4, TD-1a to TD-1c as well as compounds **1** and **3** to **9**. Induction of apoptosis was measured in myeloma cell lines NCI-H929, MM1S, and U266 as well as in PBMCs of healthy donors by flow cytometry using established protocols [37] thereby staining the cells with Annexin-fluorescein isothiocyanate and propidium iodide. Bortezomib (Eubio, Vienna, Austria) was used as positive control. Cell lines were purchased from DSMZ (Braunschweig, Germany) and routinely fingerprinted and tested for mycoplasma negativity. All cells (cell lines and PBMC) were grown in RPMI-1640 medium (Life Technologies, Paisley, UK) supplemented with 10% fetal calf serum (FCS; PAA, Linz, Austria), L-glutamine 100 µg/ml, and penicillin-streptomycin 100 U/ml. PBMCs from healthy donors were utilized after obtaining written informed consent at the University Hospital Salzburg (ethics committee approval 415-E/1287/6-2011). Cells were subjected to Ficoll separation (Ficoll Paque^{TM}, VWR, Darmstadt, Germany), and incubated in RPMI-1640 Media with supplements as above. In brief, 0.5×10^6 myeloma cells/mL or similar numbers of PMBCs were incubated for 24 h and 48 h with or without the tested compounds dissolved in DMSO in different concentrations. At least three analysis in triplicates were performed for each cell line and a solvent control was included. The extent of non-apoptotic cells (AnnexinV/propidium iodide negative) was calculated as percentage of viable cells in respect to the untreated control. Data are shown as mean percentage of viable cells and standard deviation (error bars).

4.6. Chromatographic Analyses

Solvents used for UHPLC analyses during isolation steps were 0.1% formic acid in water and methanol using a gradient from 40% of methanol to 95% in 80 min with a flow of 0.2 mL/min. Post-run was 10 min, injection volume 5 µL, temperature 30 °C. UV traces were detected at 210 nm, 254 nm, and 280 nm. Additionally, an evaporative light scattering signal was recorded.

For quantification a solvent mixture of 0.1% formic acid in water (solvent A) and acetonitrile (solvent B) was used with the following gradient: 15% B to 25% B in 15 min, to 29% B in 9 min, to 29% B in 11 min, to 36% B in 1 min, to 36% B in 19 min, to 95% B in 0.1 min, to 95% B in 9.9 min. Post-run was set to 10 min, temperature to 32 °C. The injection volume was 5 µL. The UV trace was recorded by at 330 nm. The flow was 0.2 mL/min.

4.7. Method Validation

The UHPLC-DAD method was validated for linearity, LOD and LOQ, accuracy, precision, and repeatability. For the determination of linearity calibration curves were established by serial dilution of compounds **4** to **8**. Thus, calibration ranges of 5 µg/mL to 50 µg/mL (**4**), 0.75 µg/mL to 7.5 µg/mL (**7**), and 0.5 µg/mL to 5 µg/mL (**5 + 6** and **8**) were obtained. Corresponding regressions curves, coefficients of determination as well as LOD and LOQ are given in Table 1.

Repeatability was determined by measuring one sample six-fold while intra-day precision was studied by measuring six different samples once. Interday precision of the method was verified by assessing six samples on two different days. Consistency of compound concentrations was thereby investigated (Table 2).

Spiking experiments were performed on four levels for each quantified compound. For compound **5 + 6**, **7**, and **8** stock solutions of 0.04 mg/mL were prepared. From these, 0.75 mL, 0.5 mL or 0.25 mL were taken and mixed with 0.25 mL, 0.5 mL or 0.75 mL of the plant extract. Additionally, for all quantified constituents 0.25 mL of the lowest level of the calibration curve was added to 0.75 mL of the extract. Of these resulting solutions 5 µL were injected three-fold. Results are given in Table 3.

Acacetin 7-*O*-methylether (**9**) was co-quantified by acacetin and genkwanin (**5 + 6**), compounds **2** and **3** by compound **4**.

5. Conclusions

Investigation of the cytotoxic properties of *T. diffusa* revealed significant effects for different apolar extracts against the myeloma cell lines MM1S, U266 and NCI-H929. Systematic evaluation of the active extracts by UHPLC led to the reduction of the complex metabolite to a range of possible candidates, which were subsequently isolated. Of these compounds, naringenin (**1**) and apigenin 7-*O*-(4″-*O*-*p*-*E*-coumaroyl)-glucoside (**4**) were identified as two components responsible for the observed activity. The cytotoxicity of naringenin is in line with previous findings, if however, observed for other cell lines [25,26]. Up to the best of our knowledge, compound **4** is described as cytotoxic for the first time. Nevertheless, its aglycone apigenin has been found active against cancer cell lines before [9,20,21,23,24]. Interestingly, only one of the two tested apigenin coumaroyl glucosides (compounds **3** and **4**) showed activity in the conducted assays, indicating steric effects to play a pivotal role for the cytotoxicity of these compounds.

Furthermore, the present study describes the first validated UHPLC-DAD method for the quantification of phenolic constituents in *T. diffusa*. The established assay allows the quantitation of eight flavonoids in both, the herb and the traditional preparation of *T. diffusa*, and coupled to mass spectrometry gives information on the abundance of another five flavonoids occurring in hydroethanolic damiana extracts.

Supplementary Materials: The following are available online, Figures S1–S15.

Author Contributions: Conceptualization, S.S.Ç. and C.Z.; validation, J.W.; investigation, J.W. and K.J.; resources, R.G. and C.Z.; writing—original draft preparation, J.W. and K.J.; writing—review and editing, C.Z. and S.S.Ç.; supervision, S.S.Ç.; project administration, S.S.Ç.

Funding: We acknowledge financial funding support by Land Schleswig-Holstein within the funding programme Open Access Publikationsfonds.

Acknowledgments: The authors thank Matthias Mayr for conducting the cytotoxicity assays.

Conflicts of Interest: The authors declare no conflict of interest.

References

1. Arbo, M.M.; Espert, S.M. Morphology, phylogeny and biogeography of *Turnera* L. (Turneraceae). *Taxon* **2009**, *58*, 457–467. [CrossRef]
2. Andrade-Cetto, A.; Heinrich, M. Mexican plants with hypoglycaemic effect used in the treatment of diabetes. *J. Ethnopharmacol.* **2005**, *99*, 325–348. [CrossRef] [PubMed]

3. Szewczyk, K.; Zidorn, C. Ethnobotany, phytochemistry, and bioactivity of the genus *Turnera* (Passifloraceae) with a focus on damiana—*Turnera diffusa*. *J. Ethnopharmacol.* **2014**, *152*, 424–443. [CrossRef] [PubMed]
4. Estrada-Reyes, R.; Ortiz-López, P.; Gutiérrez-Ortíz, J.; Martínez-Mota, L. *Turnera diffusa* Wild (Turneraceae) recovers sexual behavior in sexually exhausted males. *J. Ethnopharmacol.* **2009**, *123*, 423–429. [CrossRef] [PubMed]
5. Zhao, J.; Dasmahapatra, A.K.; Khan, S.I.; Khan, I.A. Anti-aromatase activity of the constituents from damiana (*Turnera diffusa*). *J. Ethnopharmacol.* **2008**, *120*, 387–393. [CrossRef] [PubMed]
6. Alarcon-Aguilar, F.J.; Roman-Ramos, R.; Flores-Saenz, J.L.; Aguirre-Garcia, F. Investigation on the hypoglycaemic effects of extracts of four Mexican medicinal plants in normal and Alloxan-diabetic mice. *Phytother. Res.* **2002**, *16*, 383–386. [CrossRef] [PubMed]
7. Parra-Naranjo, A.; Delgado-Montemayor, C.; Fraga-López, A.; Castañeda-Corral, G.; Salazar-Aranda, R.; Acevedo-Frenández, J.J.; Waksman, N. Acute Hypoglycemic and Antidiabetic Effect of Teuhetenone A Isolated from *Turnera diffusa*. *Molecules* **2017**, *22*, 599. [CrossRef] [PubMed]
8. Bernardo, J.; Ferreres, F.; Gil-Izquierdo, Á.; Valentão, P.; Andrade, P.B. Medicinal species as MTDLs: *Turnera diffusa* Willd. Ex Schult inhibits CNS enzymes and delays glutamate excitotoxicity in SH-SY$_5$Y cells *via* oxidative damage. *Food. Chem. Toxicol.* **2017**, *106*, 466–476. [CrossRef] [PubMed]
9. Avelino-Flores, M.D.C.; Cruz-López, M.D.C.; Jiménez-Montejo, F.E.; Reyes-Leyva, J. Cytotoxic Activity of the Methanolic Extract of *Turnera diffusa* Willd on Breast Cancer Cells. *J. Med. Food* **2015**, *18*, 299–305. [CrossRef] [PubMed]
10. Williams, R.J.; Spencer, J.P.E.; Rice-Evans, C. Flavonoids: Antioxidants or signalling molecules? *Free Radic. Biol. Med.* **2004**, *36*, 838–849. [CrossRef] [PubMed]
11. Procházková, D.; Boušová, I.; Wilhelmová, N. Antioxidant and prooxidant properties of flavonoids. *Fitoterapia* **2011**, *82*, 513–523.
12. Nijveldt, R.J.; Van Nood, E.; Van Hoorn, D.E.C.; Boelens, P.G.; Van Norren, K.; van Leeuwen, P.A.M. Flavonoids: A review of probable mechanisms of action and potential applications. *Am. J. Clin. Nutr.* **2001**, *74*, 418–425. [CrossRef] [PubMed]
13. Xie, C.; Kang, J.; Li, Z.; Schauss, A.G.; Badger, T.M.; Nagarajan, S.; Wu, T.; Wu, X. The açai flavonoid velutin is a potent anti-inflammatory agent: Blockade of LPS-mediated TNF-α and IL-6 production through inhibiting NF-κB activation and MAPK pathway. *J. Nutr. Biochem.* **2012**, *23*, 1184–1191. [CrossRef] [PubMed]
14. Brito, C.; Stavroullakis, A.; Oliveira, T.; Prakki, A. Cytotoxicity and potential anti-inflammatory activity of velutin on RAW 264.7 cell line differentiation: Implications in periodontal bone loss. *Arch. Oral Biol.* **2017**, *83*, 348–356. [CrossRef] [PubMed]
15. Zahir, A.; Jossang, A.; Bodo, B. DNA Topoisomerase I Inhibitors: Cytotoxic Flavones from *Lethedon tannaensis*. *J. Nat. Prod.* **1996**, *59*, 701–703. [CrossRef] [PubMed]
16. Yan, L.; Liu, H.; Wang, Q.; Wang, G. Synthesis and antiproliferative activity of novel aminoalkylated flavones. *Chem. Heterocycl. Com.* **2017**, *53*, 871–875. [CrossRef]
17. Köhler, I.; Jenett-Siems, K.; Siems, K.; Hernández, M.A.; Ibarra, R.A.; Berendsohn, W.G.; Bienzle, U.; Eich, E. *In vitro* Antiplasmodial Investigation of Medicinal Plants from El Salvador§. *Z. Naturforsch.* **2002**, *57*, 277–281. [CrossRef]
18. Boege, F.; Straub, T.; Kehr, A.; Boesenberg, C.; Christiansen, K.; Andersen, A.; Jakob, F.; Köhrle, J. Selected Novel Flavones Inhibit the DNA Binding or the DNA Religation Step of Eukaryotic Topoisomerase I. *J. Biol. Chem.* **1996**, *271*, 2262–2270. [CrossRef] [PubMed]
19. Constantinou, A.; Mehta, R.; Runyan, C.; Rao, K.; Vaughan, A.; Moon, R. Flavonoids as DNA Topoisomerase Antagonists and Poisons: Structure-Activity Relationships. *J. Nat. Prod.* **1995**, *58*, 217–225. [CrossRef] [PubMed]
20. Wu, Y.X.; Fang, X. Apigenin, Chrysin, and Luteolin Selectively Inhibit Chymotrypsin-Like and Trypsin-Like Proteasome Catalytic Activities in Tumor Cells. *Planta Med.* **2010**, *76*, 128–132. [CrossRef] [PubMed]
21. Zhao, M.; Ma, J.; Zhu, H.Y.; Zhang, X.H.; Du, Z.Y.; Xu, Y.J.; Yu, X.D. Apigenin Inhibits Proliferation and Induces Apoptosis in Human Multiple Myeloma Cells through Targeting the Trinity of CK2, Cdc37 and Hsp90. *Mol. Cancer* **2011**, *10*, 104. [CrossRef] [PubMed]

22. Perez-Meseguer, J.; Garza-Juarez, A.; Salazar-Aranda, R.; Salazar-Cavazos, M.L.; Rodriguez, Y.C.D.L.T.; Rivas-Galindo, V.; de Torres, N.W. Development and Validation of an HPLC-DAD Analytical Procedure for Quality Control of Damiana (*Turnera diffusa*), Using an Antioxidant Marker Isolated from the Plant. *J. AOAC Int.* **2010**, *93*, 1161–1168. [PubMed]
23. Wang, I.-K.; Lin-Shiau, S.-Y.; Lin, J.-K. Induction of apoptosis by apigenin and related flavonoids through cytochrome c release and activation of caspase-9 and caspase-3 in leukaemia HL-60 cells. *Eur. J. Cancer* **1999**, *35*, 1517–1525. [CrossRef]
24. Ruela-de-Sousa, R.R.; Fuhler, G.M.; Blom, N.; Ferreira, C.V.; Aoyama, H.; Peppelenbosch, M.P. Cytotoxicity of apigenin on leukemia cell lines: Implications for prevention and therapy. *Cell Death Dis.* **2010**, *1*, 1–12. [CrossRef] [PubMed]
25. Kanno, S.-I.; Tomizawa, A.; Hiura, T.; Osanai, Y.; Shouji, A.; Ujibe, M.; Ohtake, T.; Kimura, K.; Ishikawa, M. Inhibitory Effects of Naringenin on Tumor Growth in Human Cancer Cell Lines and Sarcoma S-180-Implanted Mice. *Biol. Pharm. Bull.* **2005**, *28*, 527–530. [CrossRef] [PubMed]
26. Chen, Y.-C.; Shen, S.-C.; Lin, H.-Y. Rutinoside at C7 attenuates the apoptosis-inducing activity of flavonoids. *Biochem. Pharm.* **2003**, *66*, 1139–1150. [CrossRef]
27. Chen, R.; Hollborn, M.; Grosche, A.; Reichenbach, A.; Wiedemann, P.; Bringmann, A.; Kohen, L. Effects of the vegetable polyphenols epigallocatechin-3-gallate, luteolin, apigenin, myricetin, quercetin, and cyanidin in primary cultures of human retinal pigment epithelial cells. *Mol. Vis.* **2014**, *20*, 242–258. [PubMed]
28. Matsuo, M.; Sasaki, N.; Saga, K.; Kaneko, T. Cytotoxicity of Flavonoids toward Cultured Normal Human Cells. *Biol. Pharm. Bull.* **2005**, *28*, 253–259. [CrossRef] [PubMed]
29. Zhao, J.; Pawar, R.S.; Ali, Z.; Khan, I.A. Phytochemical Investigation of *Turnera diffusa*. *J. Nat. Prod.* **2007**, *70*, 289–292. [CrossRef] [PubMed]
30. Nawwar, M.A.M.; El-Mousallamy, A.M.D.; Barakat, H.H.; Buddrus, J.; Linscheid, M. Flavonoid lactates from leaves of *Marrubium vulgare*. *Phytochemistry* **1989**, *28*, 3201–3206. [CrossRef]
31. Venturella, P.; Bellino, A.; Marino, M.L. Three acylated flavone glycosides from *Sideritis syriaca*. *Phytochemistry* **1995**, *38*, 527–530. [CrossRef]
32. Nath, L.R.; Gorantla, J.N.; Joseph, S.M.; Antony, J.; Thankachan, S.; Menon, D.B.; Sankar, S.; Lankalapalli, R.S.; Anto, R.J. Kaempferide, the most active among the four flavonoids isolated and characterized from *Chromolaena odorata*, induces apoptosis in cervical cancer cells while being pharmacologically safe. *RSC Adv.* **2015**, *5*, 100912–100922. [CrossRef]
33. Suleimenov, E.M.; Jose, R.A.; Rakhmadieva, S.B.; De Borggraeve, W.; Dehaen, W. Flavonoids from *Senecio viscosus*. *Chem. Nat. Compd.* **2009**, *45*, 731–732. [CrossRef]
34. Tang, J.; Li, H.-L.; Li, Y.-L.; Zhang, W.-D. Flavonoids from rhizomes of *Veratrum dahuricum*. *Chem. Nat. Compd.* **2007**, *43*, 696–697. [CrossRef]
35. Kou, L.-Q.; Cheng, X.-L.; Zhang, Z.-T. Syntheses and Crystal Structures of Two Apigenin Alkylation Derivatives. *J. Chem. Crystallogr.* **2008**, *38*, 21–25. [CrossRef]
36. Nakasugi, T.; Komai, K. Antimutagens in the Brazilian Folk Medicinal Plant Carqueja (*Baccharis trimera* Less.). *J. Agric. Food Chem.* **1998**, *46*, 2560–2564. [CrossRef]
37. Koopman, G.; Reutelingsperger, C.P.M.; Kuijten, G.A.M.; Keehnen, R.M.J.; Pals, S.T.; van Oers, M.H.J. Annexin V for Flow Cytometric Detection of Phosphatidylserine Expression on B Cells Undergoing Apoptosis. *Blood* **1994**, *84*, 1415–1420. [PubMed]

Sample Availability: Samples of the compounds **5** + **6** and **7** are available from the authors.

© 2019 by the authors. Licensee MDPI, Basel, Switzerland. This article is an open access article distributed under the terms and conditions of the Creative Commons Attribution (CC BY) license (http://creativecommons.org/licenses/by/4.0/).

Article

Cucurbitacin B Induces the Lysosomal Degradation of EGFR and Suppresses the CIP2A/PP2A/Akt Signaling Axis in Gefitinib-Resistant Non-Small Cell Lung Cancer

Pengfei Liu [1,2,†], Yuchen Xiang [1,2,†], Xuewen Liu [1,2], Te Zhang [2], Rui Yang [1,2], Sen Chen [1,2], Li Xu [2], Qingqing Yu [2], Huzi Zhao [1], Liang Zhang [1], Ying Liu [1,2,3,*] and Yuan Si [1,2,*]

[1] Laboratory of Molecular Target Therapy of Cancer, Institute of Basic Medical Sciences, Hubei University of Medicine, Shiyan 442000, China; LiuPF_001@163.com (P.L.); xiangyc4026@163.com (Y.X.); liuxw1110@163.com (X.L.); yangr0512@163.com (R.Y.); sener_chen@163.com (S.C.); zhaohz07@163.com (H.Z.); zl_19820321@163.com (L.Z.)
[2] Laboratory of Molecular Target Therapy of Cancer, Biomedical Research Institute, Hubei University of Medicine, Shiyan 442000, China; zhangte519@163.com (T.Z.); xuli103028@sina.com (L.X.); 18064081841@163.com (Q.Y.)
[3] Hubei Key Laboratory of Wudang Local Chinese Medicine Research and Institute of Medicinal Chemistry, Hubei University of Medicine, Shiyan 442000, China
* Correspondence: ying_liu1002@163.com (Y.L.); siyuan138@126.com (Y.S.)
† These authors contributed equally.

Received: 10 December 2018; Accepted: 1 February 2019; Published: 12 February 2019

Abstract: Non-small cell lung cancer (NSCLC) patients carrying an epidermal growth factor receptor (EGFR) mutation are initially sensitive to EGFR-tyrosine kinase inhibitors (TKIs) treatment, but soon develop an acquired resistance. The treatment effect of EGFR-TKIs-resistant NSCLC patients still faces challenges. Cucurbitacin B (CuB), a triterpene hydrocarbon compound isolated from plants of various families and genera, elicits anticancer effects in a variety of cancer types. However, whether CuB is a viable treatment option for gefitinib-resistant (GR) NSCLC remains unclear. Here, we investigated the anticancer effects and underlying mechanisms of CuB. We report that CuB inhibited the growth and invasion of GR NSCLC cells and induced apoptosis. The inhibitory effect of CuB occurred through its promotion of the lysosomal degradation of EGFR and the downregulation of the cancerous inhibitor of protein phosphatase 2A/protein phosphatase 2A/Akt (CIP2A/PP2A/Akt) signaling axis. CuB and cisplatin synergistically inhibited tumor growth. A xenograft tumor model indicated that CuB inhibited tumor growth in vivo. Immunohistochemistry results further demonstrated that CuB decreased EGFR and CIP2A levels in vivo. These findings suggested that CuB could suppress the growth and invasion of GR NSCLC cells by inducing the lysosomal degradation of EGFR and by downregulating the CIP2A/PP2A/Akt signaling axis. Thus, CuB may be a new drug candidate for the treatment of GR NSCLC.

Keywords: Cucurbitacin B; gefitinib-resistant NSCLC; EGFR; lysosomal degradation; CIP2A

1. Introduction

Lung cancer is the most commonly diagnosed cancer and the leading cause of cancer-related death. An estimated two million new lung cancer cases were recorded in 2018, and these cases account for approximately 11.6% of the total number of cancer cases [1]. Non-small cell lung cancer (NSCLC) accounts for the majority (80%) of lung cancer cases [2]. Although most NSCLC patients initially respond to chemotherapy, they gradually become drug-resistant, which in turn leads to cancer recurrence and poor prognosis [3]. Gefitinib and erlotinib are epidermal growth factor receptor-tyrosine kinase inhibitors (EGFR-TKIs). The treatment effect of EGFR-TKIs is significant for NSCLC patients with EGFR activating mutations (such as exon 19 deletion and the L858R point mutation). However, cancer cells often develop TKI resistance, which in turn causes tumor recurrence [4]. Therefore, acquired EGFR-TKI resistance is a clinical problem that needs to be solved. Patients with acquired resistance to gefitinib or erlotinib have acquired a second mutation in exon 20 of the EGFR gene, resulting in the replacement of threonine at position 790 in the protein kinase domain with methionine (T790M). Threonine 790 is an important amino acid residue in EGFR that occupies the adenosine triphosphate (ATP)-binding pocket adjacent to the ATP-binding cleft, and it determines the binding specificity of the inhibitor. The replacement of Thr790 by Met increases the affinity for ATP and reduces the binding of any ATP-competitive kinase inhibitors. [5]. Thus, treatment strategies for secondary mutations of EGFR (T790M) should be developed to overcome EGFR-TKI resistance, which would benefit NSCLC patients.

In the past 10 years, the cancerous inhibitor of protein phosphatase 2A (CIP2A) has been increasingly recognized as a key oncoprotein in several human malignancies, including myeloma [6], breast cancer [7], gastric cancer [8], glioma [9], and colorectal cancer [10]. Previous independent studies have shown that abnormal CIP2A overexpression is associated with tumor growth, anti-apoptotic effects, drug resistance, metastasis, and poor prognosis of the malignant tumors mentioned above. Additionally, CIP2A is involved in the occurrence of NSCLC, and the overexpression of CIP2A is associated with cigarette smoking and poor prognosis [11,12]. CIP2A is an endogenous inhibitor of the key tumor suppressor protein phosphatase 2A (PP2A) [13]. A previous review proposed an interactive regulatory network (carcinogenic nexus) involving CIP2A [14]. In this network, CIP2A interacts with various key cellular protein/transcription factors or components of key oncogenic signaling pathways through direct interaction or through indirect CIP2A-PP2A interactions. The primary role of CIP2A in the "carcinogenic nexus" is the inhibition of another important associated component, PP2A. PP2A is a tumor suppressor that regulates homeostasis by inhibiting intracellular signaling pathways that are driven by the constitutive activation of multiple kinases [15]. Mutations leading to the abnormal expression of PP2A scaffolds and regulatory subunits are common in many human cancers [16]. Therefore, based on its tumor suppressive properties, the reactivation of PP2A is a potential strategy for cancer treatment [17,18]. Targeting the oncoprotein CIP2A is an important strategy to reactivate PP2A to treat cancer.

Cucurbitacin is a natural tetracyclic triterpenoid compound mainly found in *Cucurbitaceae* [9]. In China and India, the use of *Cucurbitaceae* as an herbal medicine is based on its different biological activities, such as its anti-diabetic, anti-inflammatory, and anti-cancerous activities against different cancer types [19,20]. Cucurbitacin B (CuB), one of the most important members of the cucurbitacin family, has been shown to have antiplasmodial, immunomodulatory, hepatoprotective, antioxidant, cardiovascular, anthelmintic, anti-inflammatory, and anti-fertility activities [21]. Recently, several studies have reported that CuB-mediated anti-cancer activities are mainly mediated through the activation of apoptosis, cell cycle arrest, and autophagy, as well as through the suppression of the STAT3 and Raf/MEK/ERK pathways [22]. However, no study has examined the efficacy of CuB in gefitinib-resistant (GR) NSCLC. This study is the first to report that CuB induces EGFR degradation and has CIP2A/PP2A/Akt inhibitory activities in GR NSCLC cells.

2. Materials and Methods

2.1. Reagents

Cucurbitacin B (CuB) with a purity of up to 98% was purchased from Shanghai Yuanye Bio-Technology Co., Ltd. (Shanghai, China). CuB was dissolved in DMSO, (Sigma-Aldrich; Merck Millipore, Darmstadt, Germany) at a stock solution of 40 mM and stored at −20 °C.

2.2. Cell Culture

Human gefitinib-resistant NSCLC cell lines A549, NCI-H1299 (H1299), NCI-H1975 (H1975), and NCI-H820 (H820), and human normal lung epithelial cell line (16-HBE) were obtained from American Type Culture Collection (ATCC, Manassas, VA, USA). A549 and H1299 harbor wild-type EGFR. H1975 harbors L858R and T790M double mutation on EGFR, and H820 harbors exon 19 in frame deletion and T790M double mutation on EGFR. A549, H1299, and 16-HBE cells were cultured in Dulbecco modified Eagle medium (DMEM, Gibco; Thermo Fisher Scientific, Inc., Waltham, MA, USA). H1975 and H820 cells were cultured in Roswell Park Memorial Institute (RPMI) 1640 medium (Gibco; Thermo Fisher Scientific, Inc.). DMEM and RPMI 1640 medium were supplemented with 10% fetal bovine serum (FBS; HyClone, Logan, UT, USA), 100 U/mL penicillin, and 100 µg/mL streptomycin (both from Gibco; Thermo Fisher Scientific, Inc.), and cultured in a humidified atmosphere with 5% CO_2 at 37 °C.

2.3. Cytotoxic Assay and Cell Viability

Cells were seeded into a 96-well plate and pre-cultured for 24 h, and then treated with CuB or geftinib for 24 h. Cell cytotoxicity was determined by an 3-(4,5-dimethylthiazol-2-yl)-2,5-diphenyltetrazolium bromide (MTT) assay. The absorbance was measured at 570 nm by an automated microplated reader (BioTek Instruments, Inc., Winooski, VT, USA), and the cell death rate was calculated as follows: inhibition rate (%) = (average A_{570} of the control group − average A_{570} of the experimental group)/(average A_{570} of the control group − average A_{570} of the blank group) × 100%. Cell viability was estimated by trypan blue dye exclusion.

2.4. Soft-Agar Colony Formation Assay

Cells were suspended in 1 ml of RPMI 1640 containing 0.3% low-melting-point agarose (Amresco, Cleveland, OH, USA) and 10% FBS, and plated on a bottom layer containing 0.6% agarose and 10% FBS in a six-well plate in triplicate. After two weeks, plates were stained with 0.2% gentian violet and the colonies were counted under a light microscope (IX70; Olympus Corporation, Tokyo, Japan) after two weeks.

2.5. Invasion Assay

An invasion assay was carried out using a 24-well plate (Corning, Inc., Corning, NY, USA). A polyvinyl-pyrrolidone-free polycarbonate filter (8 µm pore size) (Corning) was coated with matrigel (BD Biosciences, Franklin Lakes, NJ, USA). The lower chamber was filled with medium containing 20% FBS as a chemoattractant. The coated filter and upper chamber were laid over the lower chamber. Cells (1×10^4 cells/well) were seeded onto the upper chamber wells. After incubation for 20 h at 37 °C, the filter was fixed and stained with 2% ethanol containing 0.2% crystal violet (15 min). After being dried, the stained cells were enumerated under a light microscope at 10× objective. For quantification, the invaded stained cells on the other side of the membrane were extracted with 33% acetic acid. The absorbance of the eluted stain was determined at 570 nm.

2.6. Wound Healing Assay

Cells (4×10^5 cells/2 mL) were seeded in a six-well plate and incubated at 37 °C until 90% to 100% confluence. After this, the confluent cells were scratched with a 200 µL pipet tip, followed by washing with PBS, and then treated with serum free medium. After 24 h of incubation, the cells were fixed and stained with 2% ethanol containing 0.2% crystal violet powder (15 min), and randomly chosen fields were photographed under a light microscope at 4× objective. The number of cells that had migrated into the scratched area was calculated.

2.7. Western Blot

Cell pellets were lysed in radioimmunoprecipitation assay (RIPA) buffer containing 50 mM Tris at pH 8.0, 150 mM NaCl, 0.1% sodium lauryl sulfate (SDS), 0.5% deoxycholate, 1% nonidet P-40 (NP-40), 1 mM DL-dithiothreitol (DTT), 1 mM NaF, 1 mMNaVO$_3$, 1 mM phenylmethanesulfonyl fluoride (PMSF, Sigma-Aldrich; Merck Millipore, Darmstadt, Germany), and 1% protease inhibitors cocktail (Merck, Millipore). Lysates were normalized for total protein (25 μg) and loaded on 8% to 12% sodium dodecyl sulfate polyacrylamide gel, electrophoresed, and transferred to a PVDF membrane (Millipore, Kenilworth, NJ, USA), followed by blocking with 5% skimmed milk at room temperature for 1 h. The membrane was incubated with primary antibodies overnight at 4 °C and rinsed with Tris-buffered saline with Tween 20. The primary antibodies used were anti-caspase-3 (1:1000 dilution; catalog no. 9662), anti-caspase-8 (1:1000 dilution; catalog no. 9746), anti-poly(adenosine diphosphate (ADP) ribose) polymerase (PARP; 1:1000 dilution; catalog no. 9542), anti-EGFR (1:1000 dilution; catalog no. 4267), anti-ERK1/2 (1:1000 dilution; catalog no. 9102), anti-phospho-ERK1/2 (Thr202/Tyr204) (1:1000 dilution; catalog no. 9101), anti-PP2A (1:1000 dilution; catalog no. 2038) (all Cell Signaling Technology, Inc., Danvers, MA, USA), anti-CIP2A (1:500 dilution; catalog no. sc-80662), anti-phospho-Akt (S473) (1:500 dilution; catalog no. sc-7985), anti-Akt (1:500 dilution; catalog no. sc-8312) (all Santa Cruz Biotechnology, Inc., Dallas, TX, USA), and anti-glyceraldehyde-3-phosphate dehydrogenase (GAPDH, 1:5000 dilution; catalog no. M20006; Abmart, Shanghai, China). The blots were then washed and incubated with horseradish peroxidase (HRP)-conjugated secondary antibody (1:10,000 dilution; catalog no. E030120-01 and E030110-01; EarthOx, LLC, San Francisco, CA, USA) for 1.5 h at room temperature. Detection was performed by using a SuperSignal® West Pico Trial kit (catalog no. QA210131; Pierce Biotechnology, Inc., Rockford, IL, USA) [23]. The defined sections of the film were scanned for image capture and quantification using Adobe Photoshop software (CS4, Adobe Systems Incorporated, California, USA) and Image J software (National Institutes of Health, Bethesda, MD, USA).

2.8. Quantitative Polymerase Chain Reaction

The expression level of the *EGFR* gene was examined by quantitative polymerase chain reaction (QPCR). GAPDH was used as an endogenous control for each sample. Total RNA from SW620 or HT29 cells or patients' tissues was extracted using TRIzol reagent (Invitrogen; Thermo Fisher Scientifc, Inc.,) according to the manufacturer's protocols. Total RNA (2 μg) and the ReverTra Ace qPCR real time kit (Toyobo Life Science, Osaka, Japan) were used for the QPCR analysis of *CIP2A*. Reverse transcription occurred at 37 °C for 15 min and 98 °C for 5 min, with storage at −20 °C. RNA (2 μg), 4 μL 5 RT Buffer, 1 μL RT Enzyme mix, 1 μL Primer mix, and Nuclease-free Water were mixed to a 20 μL total volume. The primers used in this study were as follows: *EGFR* forward primer: 5′- TTGTTCCTCACTGCTGTTCAC-3′ and *EGFR* reverse primer: 5′-GTCCATCATCTGTCTCCTTTC-3′; and *GAPDH* forward, 5′-TGTTGCCATCAATGACCCCTT-3′ and reverse, 5′-CTCCACGACGTACTCAGCG-3′. QPCR was performed using an ABI StepOnePlus™ Real-Time PCR system (Applied Biosystems; Thermo Fisher Scientific, Inc.) with the Power SYBR® Green PCR Master mix (Toyobo Life Science). SYBR Green PCR Master Mix (10 μL), forward and reverse primers (200 nM), a cDNA template (100 ng), and doubly-distilled H$_2$O were mixed to a 20 μL total volume. PCR conditions consisted of the following: 95 °C for 3 min, 95 °C for 15 s, and 60 °C for 1 min, for 40 cycles. The threshold cycle for each sample was selected from the linear range and converted to a starting quantity by interpolation from a standard curve generated on the same plate for each set of primers. The *CIP2A* mRNA levels were evaluated using the $2^{-\Delta\Delta Cq}$ method, standardized to levels of GAPDH amplification [24]. Each test was performed in triplicate.

2.9. Immunofluorescence Staining

H1975 cells were incubated in the presence or absence of CuB for 24 h. Cells were then fixed and penetrated. Primary antibodies were added at a dilution of 1:50 and incubated with cells at 4 °C overnight. Dylight 488 or Dylight 594-conjugated secondary antibodies (EarthOx, LLC, San Francisco, CA, USA) were diluted 1:500 in 3% BSA in PBS for 1.5 h at room temperature. For visualization of the cell nucleus, 4′,6-diamidino-2-phenylindole (DAPI) was used. Sections were observed using an Olympus laser scanning confocal microscope with imaging software (Olympus Fluoview FV-1000, Tokyo, Japan).

2.10. PP2A Activity Assay

PP2A phosphatase activity was tested using a PP2A immunoprecipitation phosphatase assay kit (Upstate Biotechnology, Inc., Lake Placid, NY, USA). According to the manufacturer's instructions, 100 µg protein isolated from the cells and 4 µg anti-PP2A monoclonal antibody (1:100 dilution; catalog no. 2038; Cell Signaling Technology, Inc.) were incubated together at 4 °C overnight. Protein A agarose beads (40 µL) were added to the mixture and incubated at 4 °C for 2 h, and the beads were then collected and washed three times with 700 µL ice-cold TBS and once with 500 µL Ser/Thr Assay Buffer (Upstate Biotechnology, Inc.). The beads were further incubated with 750 mM phosphopeptide in assay buffer at 30 °C for 10 min with continuous agitation. Malachite Green Phosphate Detection Solution (100 µL) was added and the absorbance at 650 nm was measured, as described previously [25].

2.11. Transfection of DNA

The pOTENT-1-CIP2A expression plasmid was purchased from Youbio Co. (Changcha, China). The pOTENT-1-CIP2A plasmid (1 µg/µL) was transfected into GR NSCLC cells using Lipofectamine®3000 transfection reagent (Invitrogen; Thermo Fisher Scientific, Inc.) following the manufacturer's protocols.

2.12. Drug Combination Assay

Drug combination is widely used in cancer treatment to achieve a synergistic therapeutic effect and overcome drug resistance in clinics. To estimate the effect of CuB and DDP combination, the combination index (CI) was calculated by the Chou-Talalay equation [23]. H1975 or H820 cells were seeded in 96-well plates. Drugs were added alone or together at an indicated concentration. The inhibition effect was measured by an MTT assay, as mentioned above. The formula of CI = (D)CuB/(Dx)CuB + (D)DDP/(Dx)DDP. (D: the doses of compounds CuB or DDP, respectively, necessary to produce the same effect in combination. Dx: the dose of one compound alone required producing an effect). With this formula and the assistance of CalcuSyn software (Version 2.1, Biosoft, Cambridge, UK), the combined effects of the two compounds could be assessed as follows: CI < 1 indicates synergism; CI = 1 indicates additive effect; and CI > 1 indicates Antagonism.

2.13. Human NSCLC Xenograft Experiments

Equal numbers of female and male ($n = 24$), five-week-old, nude immunodeficient mice (nu/nu) (weighing ~16 g) were purchased from Hunan SJA Laboratory Animal Co., Ltd. (Changsha, China), and maintained and monitored in a specific pathogen-free environment (temperature 22~24 °C, barrier environment, 12 h/12 h, sterile water, full nutritive feed). All animal studies were conducted according to protocols approved by the Hubei University of Medicine Animal Care and Use Committee, complying with the rules of Regulations for the Administration of Affairs Concerning Experimental Animals (Approved by the State Council of China, No. SYXK (Hubei) 2016-0031). The mice were injected subcutaneously with GR NSCLC H1975 cells (2.5×10^6) suspended in 100 µL RPMI 1640 medium into the right flank of each mouse. Treatments were started when the tumors reached a palpable size. Caliper measurements of the longest perpendicular tumor diameters were performed twice a week to estimate the tumor volume, using the following formula: $4\pi/3 \times (\text{width}/2)^2 \times (\text{length}/2)$, representing the three-dimensional volume of an ellipse. Animals were sacrificed when tumors reached 1.5 cm or if the mice appeared moribund to prevent unnecessary morbidity to the mice. At the time of the animals' death, tumors were excised for immunohistochemistry.

2.14. Immunohistochemistry of Tissues

Formalin-fixed, paraffin-embedded tissues from mice were selected for immunohistochemical examination by using an indirect immunoperoxidase method. The antibodies used for immunohistochemical staining were EGFR and CIP2A.

2.15. Statistical Analysis

All statistical analyses were conducted using GraphPad Prism 5 (GraphPad Software, Inc., La Jolla, CA, USA) and SPSS 22.0 software for Windows (IBM Corp., Armonk, NY, USA). Results from three independent experiments were presented as the mean ± standard deviation, unless otherwise noted. Statistically significant values were compared using Student's t-test of unpaired data or one-way analysis of variance and Bonferroni's post hoc test, $P < 0.05$ was used to indicate a statistically significant difference.

3. Results

3.1. CuB Induces Cytotoxicity in Gefitinib-Resistant Non-Small Cell Lung Cancer Cells

The effect of CuB on cell proliferation was determined using four GR NSCLC cell lines, namely, H1975, H820, A549, and H1299, and one normal lung epithelial cell line, 16-HBE. These four GR NSCLC cell lines have different EGFR gene mutations. The H1975 cell line has a double mutation of L858R and T790M in EGFR, and the H820 cell line has a frameshift deletion of exon 19 and a T790M mutation in EGFR. Both the A549 and H1299 cell lines express the wild-type EGFR protein. MTT assays suggested that CuB was moderately cytotoxic to all four cell lines, with an IC_{50} value between 4.23 µM and 0.19 µM (Table 1). As shown in Figure 1B–D, CuB was effective in suppressing the proliferation of GR NSCLC (H1975 and H820) cells. Interestingly, CuB had the weakest inhibitory effect on normal lung epithelial cells (16-HBE). Trypan blue exclusion assays suggested that CuB decreased the viability of H1975 (Figure 1E) and H820 (Figure 1F) cells in a dose- and time-dependent manner. We next determined the effect of CuB on cell colony formation activity, and we found that CuB markedly inhibited the clonogenic ability of H1975 (Figure 1G) and H820 (Figure 1H) cells. These data indicated that CuB suppressed the anchorage-dependent (growth) and anchorage-independent (clonogenic ability) proliferation of GR NSCLC cells. In the remainder of the study, the CuB dose that was selected for inhibition was less than 30% to ensure cellular integrity.

Table 1. IC_{50} of CuB on GR NSCLC cell lines [a].

Cell Lines	16-HBE	H1299	A549	H1975	H820
IC_{50} (µM)	4.23 ± 0.81	0.77 ± 0.04	0.76 ± 0.06	0.63 ± 0.06	0.19 ± 0.04

[a] The cells were treated with CuB at various concentrations for 24 h, the cell cytotoxicity was analyzed by MTT assay, and the IC_{50} was calculated using CalcuSyn. Values shown are means plus or minus SD of quadruplicate determinations.

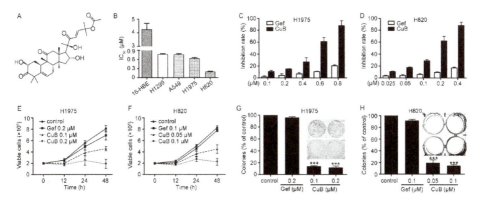

Figure 1. Cucurbitacin (CuB) inhibits gefitinib-resistant non-small cell lung cancer cells (GR-NSCLC) cells. (**A**): Chemical structure of CuB. (**B**): The IC_{50} of CuB for indicated cell lines. (**C–D**): H1975 and H820 cells were treated with increasing concentration of CuB or gefitinib for 24 h, and analyzed by MTT assay. Gef: gefitinib. (**E–F**): Inhibitory effects of CuB on cell viability of H1975 and H820 cells assayed by trypan blue exclusion assay. (**G–H**): The colony formation assays of H1975 and H820 cells treated with CuB at indicated concentration.

3.2. CuB Inhibits Invasion and Migration and Induces Caspase-Dependent Apoptosis of Gefitinib-Resistant Non-Small Cell Lung Cancer Cells

We investigated whether CuB suppressed the invasive behavior of H1975 cells. An invasion assay suggested that low doses of CuB (0–0.1 μM) inhibited the invasion of H1975 cells (Figure 2A,C). Furthermore, the wound healing assay suggested that CuB markedly decreased H1975 cell migration in a dosage-dependent manner (Figure 2B,C). These data indicated that CuB inhibited the invasive behavior of GR NSCLC cells at relatively lowly cytotoxic concentrations.

We next determined the effect of CuB on apoptosis in GR NSCLC cells. Western blot analysis suggested that CuB induced a marked increase in the active form of both caspase-8 (casp-8) and caspase-3 (casp-3) and induced the cleavage of poly(ADP-ribose) polymerase (PARP) in H1975 cells and H820 cells in a dose-dependent manner (Figure 2D). These data suggested that CuB induced caspase-dependent apoptosis in GR NSCLC cells.

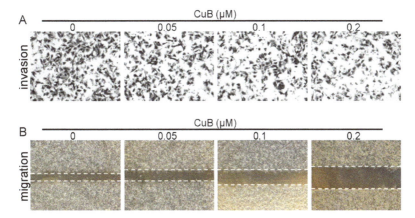

Figure 2. *Cont.*

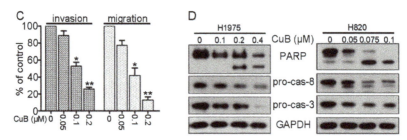

Figure 2. CuB reduces invasive behavior and induces apoptosis of GR NSCLC cells. (**A**) Invasion assay was carried out using modified 24-well microchemotaxis chambers. H1975 cells were pretreated with CuB for 30 min. (**B**) Confluent H1975 cells were scratched and then treated with CuB in a basic medium for 24 h. (**C**) Statistical results of Figure 2A,B. Data are shown as the mean ± SD of three independent experiments.* $P < 0.05$; ** $P < 0.01$ vs. 0 µM. (**D**) H1975 and H820 cells were treated with increasing concentrations of CuB for 24 h. Western blot was performed using antibodies indicated. GAPDH was used as the loading control.

3.3. CuB Induces the Lysosomal Degradation of EGFR and, thus, Inhibits ERK Signaling

Since mutated EGFR plays a critical role in the growth and invasion of NSCLC cells, we next determined the effect of CuB on EGFR expression in H1975 and H820 cells. Interestingly, we found that treatment with CuB at 0.1 µM in H1975 cells and 0.05 µM in H820 cells caused the downregulation of EGFR expression at the protein level (Figure 3A). We further showed that CuB caused the downregulation of EGFR in a time-dependent manner (Figure 3B). We next determined whether CuB affected *EGFR* gene transcription by QPCR. The results suggested that CuB had no significant effect on *EGFR* mRNA expression (Figure 3C). These data indicated that CuB may affect EGFR protein stability. Next, we blocked protein synthesis by the protein synthesis inhibitor cycloheximide (CHX) and found that EGFR remained stable after more than 12 h of CHX treatment. However, it was downregulated at 6 h in cells treated with CHX plus CuB (Figure 3D). These data indicated that CuB induced EGFR proteolysis. Previous work has reported that EGFR degradation is mediated by the lysosomal pathway [26]. Immunofluorescence analysis showed an increased colocalization of EGFR and the lysosomal marker lysosomal-associated membrane protein 1 (LAMP-1) (Figure 3E) in CuB-treated H1975 cells, suggesting that CuB promoted EGFR trafficking to lysosomes. Extracellular regulated protein kinases (ERK) are proteins in major downstream signaling of EGFR that promote cell proliferation. Activated ERK translocates to the nucleus and transactivates transcription factors, altering gene expression to promote cell cycle progression and invasion [27,28]. We measured ERK activity in CuB-treated GR NSCLC cells and found that CuB can decrease phosphorylated ERK (pERK) in a dose-dependent manner, without causing clear changes in the total ERK expression in H1975 and H820 cells (Figure 3F). In the presence of the lysosome inhibitor chloroquine (Chl), EGFR accumulation (Figure 3H) and colocalization of EGFR with lysosomes was reduced (Figure 3G), suggesting that CuB promoted the lysosomal degradation of EGFR. In addition, Chl antagonized the inhibitory effect of CuB on cell proliferation (Figure 3I). Furthermore, we compared pERK in cells treated with CuB in the presence and absence of Chl. The data indicated that Chl partially reversed the inhibitory effect of CuB on ERK phosphorylation (Figure 3H) and cell invasion (Figure 3J). These results further suggested that CuB reduced invasion and pERK levels via EGFR degradation.

3.4. CuB Downregulates the CIP2A/PP2A/Akt Signaling Axis in Gefitinib-Resistant Non-Small Cell Lung Cancer Cells

In our previous work, we reported that CIP2A plays an important role in the proliferation and aggressiveness of NSCLC and the natural compound oridonin could downregulate CIP2A levels in GR NSCLC cells [20]. Here, we reported that treatment with CuB at 0.1–0.4 µM for 24 h could downregulate CIP2A expression in H1975 cells (0.05–0.1 µM in H820 cells) (Figure 4A). As shown in the left panel, CIP2A protein expression was decreased in H1975 cells exposed to 0.2 µM of CuB. Furthermore, treatment of H820 cells with 0.075 µM of CuB caused an apparent downregulation of CIP2A. We also demonstrated that CuB induced the downregulation of CIP2A in a time-dependent manner (Figure 4B). CIP2A is an endogenous inhibitor of the tumor suppressor protein phosphatase 2A (PP2A) and is highly expressed in a variety of tumors [29]. Next, we examined the activity of PP2A and found that PP2A activity was significantly increased in H1975 and H820 cells after CuB treatment (Figure 4C,D). The expression and activation of Akt downstream of PP2A was further examined, and we found that CuB downregulated Akt phosphorylation (pAkt) in H1975 and H820 cells (Figure 4E,F), and the total Akt level did not clearly change. These results suggest that CuB downregulated the CIP2A/PP2A/Akt pathway in GR NSCLC cells.

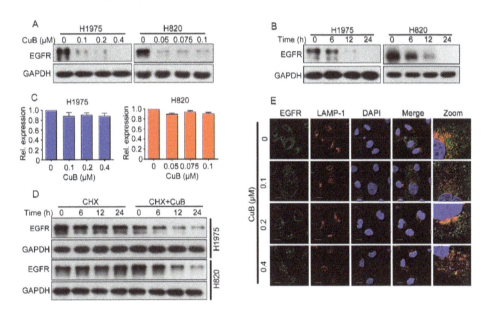

Figure 3. Cont.

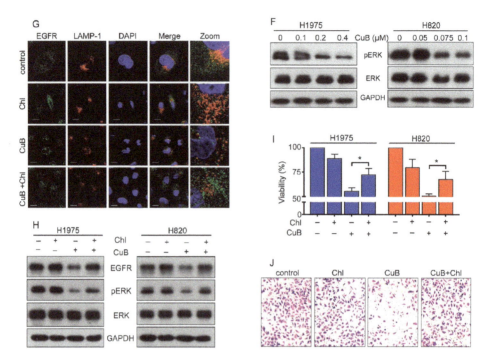

Figure 3. CuB induces lysosomal degradation of EGFR and thus inhibits ERK signaling in GR NSCLC cells. (**A**) H1975 and H820 cells were treated with increasing concentrations of CuB for 24 h. Western blot was performed using antibodies indicated. (**B**) H1975 (or H820) cells were treated with 0.2 μM (or 0.075 μM) CuB for the indicated times, and cell lysates were subjected to western blot assay. (**C**) The mRNA level of *EGFR* in H1975 and H820 cells treated with CuB for 24 h was analyzed by QPCR. (**D**) H1975 (or H820) cells were treated with 40 μg/ml cycloheximide (CHX) in the absence or presence of 0.2 μM (or 0.075 μM) CuB for the indicated times, and cell lysates were harvested for western blot assay. (**E**) H1975 cells were exposed to increasing concentrations of CuB for 24 h. For immunofluorescence analysis, cells was stained with an anti-EGFR, anti-LAMP-1 antibodies, and DAPI and observed by confocal microscopy. Scale bar = 20 μm. (**F**) H1975 and H820 cells were treated with increasing concentrations of CuB for 24 h. Western blot was performed using antibodies indicated. (**G**) H1975 cells were pretreated with chloroquine (Chl; 10 μM) for 2 h, followed by addition of CuB (0.4 μM) for 22 h. For immunofluorescence analysis, cells was stained with an anti-EGFR, anti-LAMP-1 antibodies, and DAPI and observed by confocal microscopy. Scale bar = 20 μm. (**H**) H1975 (or H820) cells were pretreated with Chl (10 μM) for 2 h, followed by addition of 0.2 μM (or 0.075 μM) CuB for 22 h. Western blot was performed using antibodies indicated. (**I**) H1975 (or H820) cells were pretreated with Chl (10 μM) for 2 h, followed by addition of 0.5 μM (or 0.15 μM) CuB for 22 h and then analyzed by MTT assay. (**J**) H1975 cells were pretreated with Chl (10 μM) for 2 h, followed by addition of 0.2 μM CuB for 2 h. Invasion assay was carried out using modified 24-well microchemotaxis chambers. * $P < 0.05$.

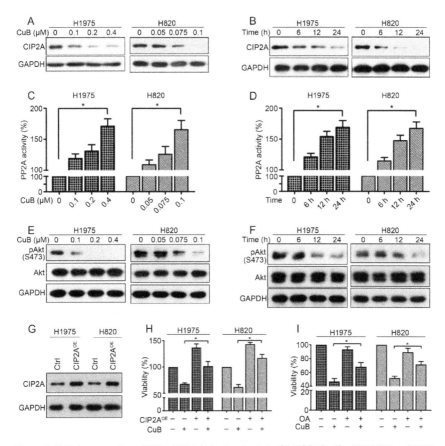

Figure 4. CuB down-regulates CIP2A/PP2A/Akt signal axis in GR NSCLC cells. (**A**) H1975 and H820 cells were treated with increasing concentrations of CuB for 24 h. Western blot was performed using antibodies indicated. (**B**) H1975 (or H820) cells were treated with 0.2 µM (or 0.075 µM) CuB for the indicated times, and cell lysates were subjected to western blot assay. (**C**) H1975 and H820 cells were treated with increasing concentrations of CuB for 24 h. Cell lysates were prepared for detecting PP2A activity, as mentioned before. (**D**) H1975 (or H820) cells were treated with 0.2 µM (or 0.075 µM) CuB for the indicated times, and cell lysates were prepared for detecting PP2A activity, as mentioned before. (**E**) H1975 and H820 cells were treated with increasing concentrations of CuB for 24 h. Western blot was performed using antibodies indicated. (**F**): H1975 (or H820) cells were treated with 0.2 µM (or 0.075 µM) CuB for the indicated times, and cell lysates were subjected to western blot assay. (**G**) H1975 (or H820) cells were transfected with a CIP2A expression plasmid (CIP2A^{OE}), and total protein was isolated and then subjected to western blot analysis. (**H**) H1975 (or H820) cells were transfected with the CIP2A^{OE}, and then treated with CuB (H1975: 0.4 µM; H820: 0.1 µM), MTT assay was used to detect growth 48 h after transfection. (**I**) H1975 (or H820) cells were treated with CuB (H1975: 0.45 µM; H820: 0.15 µM) and/or OA (50 nM) for 24 h and analyzed by MTT assay. * $P < 0.05$.

To further confirm the role of the CIP2A pathway in mediating the growth inhibition of GR NSCLC cells by CuB, we generated H1975 and H820 cells that overexpressed a CIP2A (CIP2A^{OE}) plasmid by transient transfection (Figure 4G). Compared with that of wild-type cells, the proliferation of CIP2A^{OE} cells significantly increased (Figure 4H). Notably, CIP2A overexpression significantly antagonized CuB-induced growth inhibition (Figure 4H). These findings demonstrated that CIP2A may play a critical role in CuB-triggered GR NSCLC growth. We subsequently examined whether

PP2A inhibition would alter cellular sensitivity to CuB. Okadaic acid (OA), a PP2A inhibitor, was applied to H1975 and H820 cells with or without CuB treatment. Pretreatment with OA antagonized the effects of CuB on growth in H1975 and H820 cells (Figure 4I). Thus, we confirmed that CuB induced cell growth inhibition, at least in part, by downregulating the CIP2A/PP2A/Akt pathway.

Based on the results presented above, we concluded that CuB inhibited GR NSCLC growth and induced apoptosis by inducing the lysosomal degradation of EGFR and by downregulating the CIP2A/PP2A/Akt signaling axis (Figure 5).

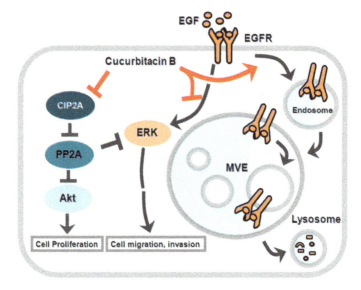

Figure 5. Diagram of CuB blockage possible mechanism in GR NSCLC cells.

3.5. CuB and Cisplatin Synergistically Inhibit the Proliferation and Apoptosis of Gefitinib-Resistant Non-Small Cell Lung Cancer Cells

It has been reported that the combination of chemotherapy and natural compounds can exhibit a synergistic effect to decrease breast cancer cell viability [30]. To explore the inhibitory capacity of a combination of CuB and cisplatin (DDP), we examined cell viability after combined CuB and DDP treatment. CuB plus DDP showed synergistic effects against H1975 and H820 cells (Figure 6A,B). The results from the analysis using CalcuSyn software (version 2.1) showed that the CI value was less than 1 (Table 2). These data suggested that CuB and DDP synergistically suppressed the viability of GR NSCLC cells. DDP further enhanced the CuB-dependent induction of cell apoptosis and its inhibitory effects on EGFR and the CIP2A/Akt pathway (Figure 6C,D).

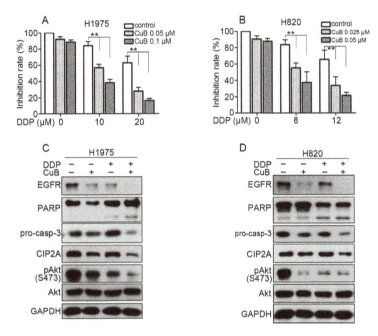

Figure 6. CuB and DDP synergistically inhibit GR NSCLC cells. (**A,B**): H1975 and H820 cells were treated for 24 h with DDP in the presence of CuB. MTT assay was used to test the proliferation of cells. * $P < 0.05$, ** $P < 0.001$. (**C**): H1975 cells were cultured with control media, CuB (0.1 µM), DDP (10 µM), or CuB (0.1 µM) plus DDP (10 µM) for 24 h. Cells were then lysed and subjected to Western blot using indicated antibodies. (**D**): H820 cells were cultured with control media, CuB (0.05 µM), DDP (8 µM), or CuB (0.05 µM) plus DDP (8 µM) for 24 h. Cells were then lysed and subjected to Western blot using indicated antibodies.

Table 2. CuB and DDP combination index (CI) values [a].

H1975				H820			
CuB (µM)	DDP (µM)	Effect	CI (CuB+DDP)	CuB (µM)	DDP (µM)	Effect	CI (CuB+DDP)
0.05	10	0.43	0.44	0.025	8	0.44	0.51
0.05	20	0.72	0.29	0.025	12	0.66	0.53
0.1	10	0.61	0.28	0.05	8	0.63	0.36
0.1	20	0.84	0.18	0.05	12	0.79	0.39

[a] H1975 or H820 cells were treated with CuB and DDP combinedly or alone with indicated concentrations for 24 h, the cytotoxicity was analyzed by MTT assay, and the CI values were calculated using CalcuSyn software.

3.6. CuB Inhibits Tumor Growth In Vivo

To test the in vivo anti-tumor effect of CuB on NSCLC, we implanted 5×10^6 H1975 cells that had been resuspended in 100 µL of RPMI 1640 medium on the right side of nude mice to construct a xenograft mouse model. Treatment began once the tumors reached a palpable size (0.5 cm in diameter). Each of the three groups was administered the vehicle control, gefitinib (30 mg/kg), or CuB (0.5 mg/kg) five times per week for 24 days. The tumor-bearing mice were sacrificed when the tumor reached 1.5 cm in diameter or severe pain diminished their quality of life. We showed that CuB significantly suppressed tumor growth compared to the vehicle control or gefitinib ($P < 0.01$; Figure 7A,B). CuB treatment also markedly decreased tumor weight in the mice (Figure 7C). Importantly, CuB treatment did not significantly decrease the body weight of the mice, suggesting that CuB did not cause evident side effects (Figure 7D). When all of the mice were sacrificed, the tumor specimens were isolated and examined using immunohistochemistry, and the data suggested that the expression levels of CIP2A and EGFR were downregulated in the CuB-treated groups (Figure 7E). Therefore, CuB is predicted to be a potential therapy for GR NSCLC.

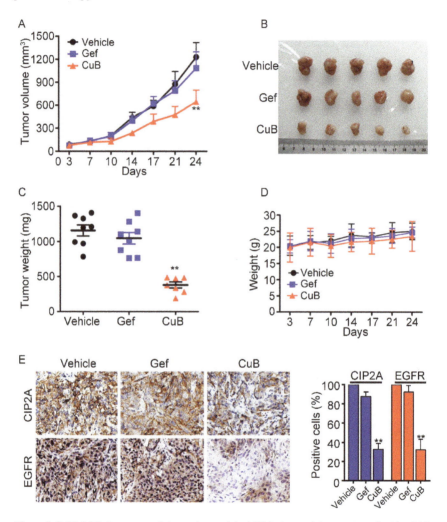

Figure 7. CuB inhibits tumor growth in murine models. (**A**) Murine models were treated with vehicle, gefitinib (Gef, 30 mg/kg), or CuB (0.5 mg/kg) and the tumor volumes were calculated twice a week. ** $P < 0.01$ vs. vehicle or Gef. (**B**) Images of xenograft tumors obtained from mice with different treatment after 24 days. (**C**) Weight of the tumor from each group taken out from the sacrificed mice at the end of the study ** $P < 0.01$ vs. vehicle or Gef group. (**D**) CuB treatment did not affect the murine model body weight. (**E**) The expressions of EGFR and CIP2A in xenograft tumors were analyzed by immunohistochemistry (original magnification 400×), and their expression levels were quantified in percentages of positive cells within five medium-power fields under microscope and shown in histograms; * $P < 0.05$, ** $P < 0.01$ compared with the vehicle group.

4. Discussion

CuB has anti-proliferative effects on various lung cancer cell lines in vitro and in vivo [31–33]. However, the cytotoxic effects of CuB on EGFR-mutant GR NSCLC cells remain poorly understood. This study reports, for the first time, that CuB suppressed the proliferation of GR NSCLC cells in vitro and in vivo by inducing lysosome-mediated EGFR degradation and by downregulating the CIP2A/PP2A/Akt signaling axis. These results strongly suggest the possible therapeutic value of CuB in patients with GR NSCLC that carry EGFR mutations.

More than 90% of solid tumor deaths are due to tumor metastasis [34]. Thus, inhibiting or preventing cancer metastasis is an important means to improve the survival rate of cancer patients. Our results indicated that CuB significantly suppressed the invasion (Figure 2A) and migration (Figure 2B) of H1975 cells. Escape from apoptosis is an important feature of cancer progression and drug resistance, and the activation of apoptosis has become another important strategy for cancer treatment [35]. Casp-3 is an effector of extrinsic and intrinsic apoptotic signaling [10]. We showed that CuB induced a reduction in the levels of pro-casp-8 and pro-casp-3 and induced the proteolysis of PARP (Figure 2D), which suggested that casp-8 and casp-3 were activated. Thus, CuB may promote apoptosis by activating extrinsic apoptosis signaling, indicated by the activation of casp-8.

The abnormal expression or activation of EGFR and its downstream signaling pathways can promote malignant processes, invasion, and drug tolerance in many human cancers [36]. It is well-known that GR NSCLC cells are largely dependent on the constitutive activity of EGFR kinase signaling. Activation of EGFR, in turn, activates its downstream kinases, such as Akt and ERK, thereby promoting the proliferation and invasion of cancer cells [27]. Given the cytotoxic effects of CuB on GR NSCLC cells, we investigated whether CuB affects EGFR. The results showed that with increased CuB dose and exposure time, the level of EGFR protein was significantly decreased, and the mRNA expression level was not affected (Figure 3A–E). These results suggested that CuB may affect the protein stability of EGFR. A key mechanism for downregulating EGFR signaling is lysosomal-mediated trafficking and degradation [36]. Next, we showed a decrease in EGFR–LAMP-1 colocalization in the absence of CuB (Figure 3F). To further assess whether the CuB-induced degradation of EGFR was mediated by lysosomes, the effects of the lysosome inhibitor chloroquine on CuB-induced EGFR degradation were examined (Figure 3G). The results demonstrated that CuB induced the lysosomal-mediated degradation of EGFR. We next assessed which downstream signaling pathways may have mediated EGFR signaling. ERK is an important downstream signaling protein of EGFR and is involved in the regulation of biological processes such as cell proliferation, invasion, and apoptosis [37]. We found that CuB inhibited ERK phosphorylation (Figure 3H). Therefore, CuB is a novel anti-tumor drug that treats GR NSCLC by inhibiting the EGFR/ERK pathway.

Previous studies have shown that high CIP2A expression was highly correlated with cancer invasiveness and poor prognosis in lung cancer; thus CIP2A is used as a potential molecular marker and therapeutic target for the treatment of lung cancer [38]. We found that CuB was also able to induce a marked dose- and time-dependent reduction of CIP2A at the protein levels in GR NSCLC (Figure 4A,B). Recent studies have implicated CIP2A-mediated increases in Akt activity in the inactivation of PP2A phosphatase activity. Some natural compounds that target the CIP2A protein have shown potential effects on a variety of tumors in vivo and in vitro [39,40]. We next examined the PP2A activity and pAkt levels in GR NSCLC cells after CuB was administered. The data suggested that CuB reactivated PP2A activity and inactivated Akt (Figure 4C–F), indicating that the CIP2A/PP2A/Akt pathway may serve as an alternative mechanism that underlies the effects of CuB.

When CuB was combined with the conventional drug DDP, they had a synergistic cytotoxic effect on GR NSCLC cells (Figure 6). In the H1975 xenograft mouse model, CuB significantly inhibited tumor growth and had little effect on body weight of the mice (Figure 7A–D). Tumor tissues isolated from mice also showed that CuB inhibited EGFR and CIP2A expression in vivo (Figure 7E). Our data suggest that CuB not only directly affects EGFR degradation, but also affects the CIP2A/PP2A signaling pathway. Furthermore, the growth and invasion of GR NSCLC cells were inhibited by CuB activity via the CIP2A/PP2A signaling pathway. Therefore, CuB may become a novel anti-tumor drug for the prevention and treatment of GR NSCLC.

5. Conclusions

In conclusion, we reported that CuB significantly suppressed tumor growth and invasion and activated apoptosis in GR NSCLC in vitro and in vivo. Our data further revealed that CuB inhibited ERK and Akt phosphorylation by inducing the lysosomal degradation of EGFR and that CuB inhibits the CIP2A/PP2A signaling axis. These observations indicate that CuB could be a promising therapeutic agent for treating GR NSCLC, and additional toxicological experiments are necessary to verify this conclusion.

Author Contributions: Y.S. and Y.L. conceived and designed the experiments, wrote the paper; P.L. and Y.X. performed most of the experiments; X.L., R.Y., T.Z., S.C., L.X. and Q.Y. performed the experiments; H.Z. and L.Z. took part in the analysis of the data.

Funding: This work was supported by grants from the National Natural Science Foundation of China (no. 81802387), the Scientific and Technological Project of Shiyan City of Hubei Province (nos. 17Y01, 18Y10, and 18Y13), the Faculty Development grants from Hubei University of Medicine (nos. 2018QDJZR03, 2018QDJZR27, 2016QDJZR22), the Principal Investigator Grant of Hubei University of Medicine (HBMUPI201806), and the National Training Program of Innovation and Entrepreneurship for Undergraduates (grant nos. 201810929012 and 201810929056).

Conflicts of Interest: The authors have no conflicts of interest to declare.

References

1. Bray, F.; Ferlay, J.; Soerjomataram, I.; Siegel, R.L.; Torre, L.A.; Jemal, A. Global cancer statistics 2018: GLOBOCAN estimates of incidence and mortality worldwide for 36 cancers in 185 countries. *CA Cancer J. Clin.* **2018**, *68*, 394–424. [CrossRef]
2. Xiao, X.; He, Z.; Cao, W.; Cai, F.; Zhang, L.; Huang, Q.; Fan, C.; Duan, C.; Wang, X.; Wang, J.; et al. Oridonin inhibits gefitinib-resistant lung cancer cells by suppressing EGFR/ERK/MMP-12 and CIP2A/Akt signaling pathways. *Int. J. Oncol.* **2016**, *48*, 2608–2618. [CrossRef] [PubMed]
3. Lopez Sambrooks, C.; Baro, M.; Quijano, A.; Narayan, A.; Cui, W.; Greninger, P.; Egan, R.; Patel, A.; Benes, C.H.; Saltzman, W.M.; et al. Oligosaccharyltransferase Inhibition Overcomes Therapeutic Resistance to EGFR Tyrosine Kinase Inhibitors. *Cancer Res.* **2018**, *78*, 5094–5106. [CrossRef] [PubMed]
4. Kim, J.H.; Nam, B.; Choi, Y.J.; Kim, S.Y.; Lee, J.E.; Sung, K.J.; Kim, W.S.; Choi, C.M.; Chang, E.J.; Koh, J.S.; et al. Enhanced Glycolysis Supports Cell Survival in EGFR-Mutant Lung Adenocarcinoma by Inhibiting Autophagy-Mediated EGFR Degradation. *Cancer Res.* **2018**, *78*, 4482–4496. [CrossRef] [PubMed]
5. Yun, C.H.; Mengwasser, K.E.; Toms, A.V.; Woo, M.S.; Greulich, H.; Wong, K.K.; Meyerson, M.; Eck, M.J. The T790M mutation in EGFR kinase causes drug resistance by increasing the affinity for ATP. *Proc. Natl. Acad. Sci. USA* **2008**, *105*, 2070–2705. [CrossRef] [PubMed]
6. Liu, X.; Cao, W.; Qin, S.; Zhang, T.; Zheng, J.; Dong, Y.; Ming, P.; Cheng, Q.; Lu, Z.; Guo, Y.; et al. Overexpression of CIP2A is associated with poor prognosis in multiple myeloma. *Signal Transduct. Target* **2017**, *2*, 17013. [CrossRef]
7. Li, S.; Feng, T.-T.; Guo, Y.; Yu, X.; Huang, Q.; Zhang, L.; Tang, W.; Liu, Y. Expression of cancerous inhibitor of protein phosphatase 2A in human triple negative breast cancer correlates with tumor survival, invasion and autophagy. *Oncol. Lett.* **2016**, *12*, 5370–5376. [CrossRef]
8. Ji, J.; Zhen, W.; Si, Y.; Ma, W.; Zheng, L.; Li, C.; Zhang, Y.; Qin, S.; Zhang, T.; Liu, P.; et al. Increase in CIP2A expression is associated with cisplatin chemoresistance in gastric cancer. *Cancer Biomark.* **2018**, *21*, 307–316. [CrossRef]
9. Qin, S.; Li, J.; Si, Y.; He, Z.; Zhang, T.; Wang, D.; Liu, X.; Guo, Y.; Zhang, L.; Li, S.; et al. Cucurbitacin B induces inhibitory effects via CIP2A/PP2A/Akt pathway in glioblastoma multiforme. *Mol. Carcinog.* **2018**, *57*, 687–699. [CrossRef]
10. Jin, L.; Si, Y.; Hong, X.; Liu, P.; Zhu, B.; Yu, H.; Zhao, X.; Qin, S.; Xiong, M.; Liu, Y.; et al. Ethoxysanguinarine inhibits viability and induces apoptosis of colorectal cancer cells by inhibiting CIP2A. *Int. J. Oncol.* **2018**, *52*, 1569–1578. [CrossRef]

11. Liu, Z.; Ma, L.; Wen, Z.S.; Hu, Z.; Wu, F.Q.; Li, W.; Liu, J.; Zhou, G.B. Cancerous inhibitor of PP2A is targeted by natural compound celastrol for degradation in non-small-cell lung cancer. *Carcinogenesis* **2014**, *35*, 905–914. [CrossRef] [PubMed]
12. Ma, L.; Wen, Z.S.; Liu, Z.; Hu, Z.; Ma, J.; Chen, X.Q.; Liu, Y.Q.; Pu, J.X.; Xiao, W.L.; Sun, H.D.; et al. Overexpression and small molecule-triggered downregulation of CIP2A in lung cancer. *PLoS ONE* **2011**, *6*, e20159. [CrossRef] [PubMed]
13. Junttila, M.R.; Puustinen, P.; Niemela, M.; Ahola, R.; Arnold, H.; Bottzauw, T.; Ala-aho, R.; Nielsen, C.; Ivaska, J.; Taya, Y.; et al. CIP2A inhibits PP2A in human malignancies. *Cell* **2007**, *130*, 51–62. [CrossRef] [PubMed]
14. De, P.; Carlson, J.; Leyland-Jones, B.; Dey, N. Oncogenic nexus of cancerous inhibitor of protein phosphatase 2A (CIP2A): An oncoprotein with many hands. *Oncotarget* **2014**, *5*, 4581–4602. [CrossRef]
15. Perrotti, D.; Neviani, P. Protein phosphatase 2A: A target for anticancer therapy. *Lancet. Oncol.* **2013**, *14*, e229–e238. [CrossRef]
16. Seshacharyulu, P.; Pandey, P.; Datta, K.; Batra, S.K. Phosphatase: PP2A structural importance, regulation and its aberrant expression in cancer. *Cancer Lett.* **2013**, *335*, 9–18. [CrossRef] [PubMed]
17. Huang, C.Y.; Hung, M.H.; Shih, C.T.; Hsieh, F.S.; Kuo, C.W.; Tsai, M.H.; Chang, S.S.; Hsiao, Y.J.; Chen, L.J.; Chao, T.I.; et al. Antagonizing SET Augments the Effects of Radiation Therapy in Hepatocellular Carcinoma through Reactivation of PP2A-Mediated Akt Downregulation. *J. Pharmacol. Exp. Ther.* **2018**, *366*, 410–421. [CrossRef] [PubMed]
18. Sangodkar, J.; Farrington, C.C.; McClinch, K.; Galsky, M.D.; Kastrinsky, D.B.; Narla, G. All roads lead to PP2A: Exploiting the therapeutic potential of this phosphatase. *FEBS J.* **2016**, *283*, 1004–1024. [CrossRef]
19. Chan, K.T.; Meng, F.Y.; Li, Q.; Ho, C.Y.; Lam, T.S.; To, Y.; Lee, W.H.; Li, M.; Chu, K.H.; Toh, M. Cucurbitacin B induces apoptosis and S phase cell cycle arrest in BEL-7402 human hepatocellular carcinoma cells and is effective via oral administration. *Cancer Lett.* **2010**, *294*, 118–124. [CrossRef]
20. Cai, F.; Zhang, L.; Xiao, X.; Duan, C.; Huang, Q.; Fan, C.; Li, J.; Liu, X.; Li, S.; Liu, Y. Cucurbitacin B reverses multidrug resistance by targeting CIP2A to reactivate protein phosphatase 2A in MCF-7/adriamycin cells. *Oncol. Rep.* **2016**, *36*, 1180–1186. [CrossRef]
21. Shukla, S.; Khan, S.; Kumar, S.; Sinha, S.; Farhan, M.; Bora, H.K.; Maurya, R.; Meeran, S.M. Cucurbitacin B Alters the Expression of Tumor-Related Genes by Epigenetic Modifications in NSCLC and Inhibits NNK-Induced Lung Tumorigenesis. *Cancer Prev. Res.* **2015**, *8*, 552–562. [CrossRef] [PubMed]
22. Guo, J.; Zhao, W.; Hao, W.; Ren, G.; Lu, J.; Chen, X. Cucurbitacin B induces DNA damage, G2/M phase arrest, and apoptosis mediated by reactive oxygen species (ROS) in leukemia K562 cells. *Anti-Cancer Agents Med. Chem.* **2014**, *14*, 1146–1153. [CrossRef]
23. Cao, W.; Liu, Y.; Zhang, R.; Zhang, B.; Wang, T.; Zhu, X.; Mei, L.; Chen, H.; Zhang, H.; Ming, P.; et al. Homoharringtonine induces apoptosis and inhibits STAT3 via IL-6/JAK1/STAT3 signal pathway in Gefitinib-resistant lung cancer cells. *Sci. Rep.* **2015**, *5*, 8477. [CrossRef] [PubMed]
24. Livak, K.J.; Schmittgen, T.D. Analysis of relative gene expression data using real-time quantitative PCR and the $2^{-\Delta\Delta C_T}$ Method. *Methods* **2001**, *25*, 402–408. [CrossRef] [PubMed]
25. Liu, H.; Gu, Y.; Wang, H.; Yin, J.; Zheng, G.; Zhang, Z.; Lu, M.; Wang, C.; He, Z. Overexpression of PP2A inhibitor SET oncoprotein is associated with tumor progression and poor prognosis in human non-small cell lung cancer. *Oncotarget* **2015**, *6*, 14913–14925. [CrossRef] [PubMed]
26. Wen, J.; Fu, J.; Ling, Y.; Zhang, W. MIIP accelerates epidermal growth factor receptor protein turnover and attenuates proliferation in non-small cell lung cancer. *Oncotarget* **2016**, *7*, 9118–9134. [CrossRef] [PubMed]
27. Gao, J.; Liu, X.; Yang, F.; Liu, T.; Yan, Q.; Yang, X. By inhibiting Ras/Raf/ERK and MMP-9, knockdown of EpCAM inhibits breast cancer cell growth and metastasis. *Oncotarget* **2015**, *6*, 27187–27198. [CrossRef] [PubMed]
28. Yang, X.S.; Liu, S.A.; Liu, J.W.; Yan, Q. Fucosyltransferase IV enhances expression of MMP-12 stimulated by EGF via the ERK1/2, p38 and NF-κB pathways in A431 cells. *Asian Pac. J. Cancer Prev.* **2012**, *13*, 1657–1662. [CrossRef]
29. Rincon, R.; Cristobal, I.; Zazo, S.; Arpi, O.; Menendez, S.; Manso, R.; Lluch, A.; Eroles, P.; Rovira, A.; Albanell, J.; et al. PP2A inhibition determines poor outcome and doxorubicin resistance in early breast cancer and its activation shows promising therapeutic effects. *Oncotarget* **2015**, *6*, 4299–4314. [CrossRef]

30. Feng, T.; Cao, W.; Shen, W.; Zhang, L.; Gu, X.; Guo, Y.; Tsai, H.-I.; Liu, X.; Li, J.; Zhang, J.; et al. Arctigenin inhibits STAT3 and exhibits anticancer potential in human triple-negative breast cancer therapy. *Oncotarget* **2017**, *8*, 329–344. [CrossRef]
31. Kausar, H.; Munagala, R.; Bansal, S.S.; Aqil, F.; Vadhanam, M.V.; Gupta, R.C. Cucurbitacin B potently suppresses non-small-cell lung cancer growth: Identification of intracellular thiols as critical targets. *Cancer Lett.* **2013**, *332*, 35–45. [CrossRef] [PubMed]
32. Shukla, S.; Sinha, S.; Khan, S.; Kumar, S.; Singh, K.; Mitra, K.; Maurya, R.; Meeran, S.M. Cucurbitacin B inhibits the stemness and metastatic abilities of NSCLC via downregulation of canonical Wnt/β-catenin signaling axis. *Sci. Rep.* **2016**, *6*, 21860. [CrossRef] [PubMed]
33. Khan, N.; Jajeh, F.; Khan, M.I.; Mukhtar, E.; Shabana, S.M.; Mukhtar, H. Sestrin-3 modulation is essential for therapeutic efficacy of cucurbitacin B in lung cancer cells. *Carcinogenesis* **2017**, *38*, 184–195. [CrossRef] [PubMed]
34. Pritchard, V.L.; Dimond, L.; Harrison, J.S.; Velázquez, C.C.S.; Zieba, J.T.; Burton, R.S.; Edmands, S. Interpopulation hybridization results in widespread viability selection across the genome in *Tigriopus californicus*. *BMC Genet.* **2011**, *12*, 54. [CrossRef] [PubMed]
35. Hanahan, D.; Weinberg, R.A. Hallmarks of cancer: The next generation. *Cell* **2011**, *144*, 646–674. [CrossRef]
36. Mao, J.; Ma, L.; Shen, Y.; Zhu, K.; Zhang, R.; Xi, W.; Ruan, Z.; Luo, C.; Chen, Z.; Xi, X.; et al. Arsenic circumvents the gefitinib resistance by binding to P62 and mediating autophagic degradation of EGFR in non-small cell lung cancer. *Cell Death Dis.* **2018**, *9*, 963. [CrossRef] [PubMed]
37. Binshtok, U.; Sprinzak, D. The Domino Effect in EGFR-ERK Signaling. *Dev. Cell* **2018**, *46*, 128–130. [CrossRef] [PubMed]
38. Cha, G.; Xu, J.; Xu, X.; Li, B.; Lu, S.; Nanding, A.; Hu, S.; Liu, S. High expression of CIP2A protein is associated with tumor aggressiveness in stage I-III NSCLC and correlates with poor prognosis. *Oncotargets Ther.* **2017**, *10*, 5907–5914. [CrossRef]
39. Huang, Q.; Qin, S.; Yuan, X.; Zhang, L.; Ji, J.; Liu, X.; Mai, W.; Zhang, Y.; Liu, P.; Sun, Z.; et al. Arctigenin inhibits triple-negative breast cancers by targeting CIP2A to reactivate protein phosphatase 2A. *Oncol. Rep.* **2017**, *38*, 598–606. [CrossRef]
40. Liu, X.; Sun, Z.; Deng, J.; Liu, J.; Ma, K.; Si, Y.; Zhang, T.; Feng, T.; Liu, Y.; Tan, Y. Polyphyllin I inhibits invasion and epithelial-mesenchymal transition via CIP2A/PP2A/ERK signaling in prostate cancer. *Int. J. Oncol.* **2018**, *53*, 1279–1288. [CrossRef]

Sample Availability: Not available.

© 2019 by the authors. Licensee MDPI, Basel, Switzerland. This article is an open access article distributed under the terms and conditions of the Creative Commons Attribution (CC BY) license (http://creativecommons.org/licenses/by/4.0/).

Article

Total Phenols from Grape Leaves Counteract Cell Proliferation and Modulate Apoptosis-Related Gene Expression in MCF-7 and HepG2 Human Cancer Cell Lines

Selma Ferhi [1], Sara Santaniello [2], Sakina Zerizer [3], Sara Cruciani [2], Angela Fadda [4], Daniele Sanna [5], Antonio Dore [4], Margherita Maioli [2,6,7,8,*] and Guy D'hallewin [4]

1. Department of Applied Biology, University Larbi Tébessi Tebessa, 12000 Tebessa, Algeria; s_ferhi@yahoo.fr
2. Department of Biomedical Sciences, University of Sassari, Viale San Pietro 43/B, 07100 Sassari, Italy; sara.santaniello@gmail.com (S.S.); sara.cruciani@outlook.com (S.C.)
3. Laboratoire d'Obtention de Substances Thérapeutiques (L.O.S.T.), Département de Biologie Animale, Université Des Frères Mentouri-Constantine, 25000 Constantine, Algeria; zerizer.sakina@umc.edu.dz
4. Institute of Sciences of Food Production, National Research Council, Traversa la Crucca, 3. Loc Baldinca Li Punti, 07100 Sassari, Italy; angela.fadda@cnr.it (A.F.); antonio.dore@cnr.it (A.D.); guy.dhallewin@cnr.it (G.D.)
5. Institute of Biomolecular Chemistry, National Research Council, Traversa La Crucca, 3. Località Baldinca Li Punti, 07100 Sassari, Italy; Daniele.Sanna@cnr.it
6. Istituto di Ricerca Genetica e Biomedica, Consiglio Nazionale delle Ricerche (CNR), Monserrato, 09042 Cagliari, Italy
7. National Laboratory of Molecular Biology and Stem Cell Engineering-National Institute of Biostructures and Biosystems-Eldor Lab, at Innovation Accelerators, CNR, 40129 Bologna, Italy
8. Center for Developmental Biology and Reprogramming-CEDEBIOR, Department of Biomedical Sciences, University of Sassari, Viale San Pietro 43/B, 07100 Sassari, Italy
* Correspondence: mmaioli@uniss.it; Tel.: +39-079-228-277

Academic Editor: Roberto Fabiani
Received: 18 December 2018; Accepted: 5 February 2019; Published: 10 February 2019

Abstract: Grape leaves influence several biological activities in the cardiovascular system, acting as antioxidants. In this study, we aimed at evaluating the effect of ethanolic and water extracts from grape leaves grown in Algeria, obtained by accelerator solvent extraction (ASE), on cell proliferation. The amount of total phenols was determined using the modified Folin-Ciocalteu method, antioxidant activities were evaluated by the 2,2-diphenyl-l-picrylhydrazyl free radical (DPPH*) method and ˙OH radical scavenging using electron paramagnetic resonance (EPR) spectroscopy methods. Cell proliferation of HepG2 hepatocarcinoma, MCF-7 human breast cancer cells and vein human umbilical (HUVEC) cells, as control for normal cell growth, was assessed by 3-(4,5-dimethylthiazol-2-yl)-2,5-diphenyltetrazolium bromide reduction assay (MTT). Apoptosis-related genes were determined by measuring Bax and Bcl-2 mRNA expression levels. Accelerator solvent extractor yield did not show significant difference between the two solvents (ethanol and water) ($p > 0.05$). Total phenolic content of water and ethanolic extracts was 55.41 ± 0.11 and 155.73 ± 1.20 mg of gallic acid equivalents/g of dry weight, respectively. Ethanolic extracts showed larger amounts of total phenols as compared to water extracts and interesting antioxidant activity. HepG2 and MCF-7 cell proliferation decreased with increasing concentration of extracts (0.5, 1, and 2 mg/mL) added to the culture during a period of 1–72 h. In addition, the expression of the pro-apoptotic gene Bax was increased and that of the anti-apoptotic gene Bcl-2 was decreased in a dose-dependent manner, when both MCF-7 and HepG2 cells were cultured with one of the two extracts for 72 h. None of the extracts elicited toxic effects on vein umbilical HUVEC cells, highlighting the high specificity of the antiproliferative effect, targeting only cancer cells. Finally, our results suggested that ASE crude extract from grape leaves represents a source of bioactive compounds such as phenols, with potential antioxidants activity, disclosing a novel antiproliferative effect affecting only HepG2 and MCF-7 tumor cells.

Keywords: grape leaves; ASE; TP; Antioxidant activities; Antiproliferative; pro-apoptotic effects; Gene expression; Nutraceuticals

1. Introduction

Oxidative stress is a pathogenetic mechanism associated with several diseases, including atherosclerosis, neurodegenerative diseases, such as Alzheimer's and Parkinson's disease, cancer, diabetes mellitus, inflammatory diseases, as well as psychological diseases or aging processes [1]. Indeed, increased formation of free radicals (FR) can promote the development of malignancy, and "normal" rates of FR generation may account for the increased risk of cancer development in the elderly [2]. Cancer is the major cause of morbidity and mortality in modern society. The number of deaths by cancer in 2008 was estimated to be 7.6 million, a number predicted to double by 2030 [3]. In developed countries, cancer is the main cause of death after cardiac disease [4]. Many treatments against cancer are possible, such as surgical removal, chemotherapy, radiation therapy and immunotherapy.

Apoptosis, or programmed cell death, is a normal and fundamental event that occurs in a highly regulated and precise manner. This process plays a key role in normal tissue development and maturation, maintaining the homeostasis in the body by controlling the immune system. Apoptosis is the most potent defense against cancer since it is the mechanism used by metazoans to eliminate deleterious cells. Furthermore, a large number of chemo preventive agents exert their effectiveness by inducing apoptosis in transformed cells, as shown both in vitro and in vivo [5,6]. Since apoptosis provides a physiologic mechanism to eliminate abnormal cells, dietary factors affecting apoptosis can elicit an important effect on carcinogenesis. For these reasons, activation of apoptosis by dietary factors in pre-cancerous cells may represent a preventive mechanism (chemoprevention) [6,7].

Nearly 90 out of 121 drugs prescribed to treat cancer originate from plants [8]. The term "nutraceutical" was coined in 1989 by Stephen De Felice to define "food, or parts of a food, that provide medical or health benefits, including the prevention and treatment of disease" [9–11]. Many studies demonstrate that grapes are rich in anthocyanins, flavanols, flavonoids, terpenes, organic acids, vitamins, carbohydrates, lipids and enzymes [12,13]. These findings have created considerable interest in grape leaves as a promising source of compounds with nutritional properties and biological potential. Moreover, the use of grape leaves provides a way of solving the disposal problems arising from the large amounts of industrial residues generated by the wine and juice industries [14,15].

Extraction is the most important step to recover and isolate bioactive molecules from plant materials. Various extraction techniques have been developed to obtain nutraceuticals from plants in order to shorten extraction time, reduce solvent consumption, increase extraction yield, improve the quality of extracts and increase pollution prevention [16]. Among those, accelerated solvent extraction (ASE) is a solid-liquid extraction process performed at elevated temperature and under pressure to maintain the solvent in its liquid state. The solvent remains below its critical condition during ASE. The increased temperature accelerates the extraction kinetics and the elevated pressure keeps the solvent in the liquid state, thus achieving a safe and rapid extraction. The only disadvantage of ASE is the high cost of the needed equipment [17].

The aim of the present study was to analyze the polyphenol anti-oxidative and anti-proliferative properties of water and ethanol ASE crude extracts from grape leaves grown in Medea (Algeria).

2. Results

2.1. Yield and Total Phenolic Content

Table 1 shows the yield and total phenolic content of ethanolic and aqueous ASE crude extracts obtained from grape leaves. The aqueous extract gave a higher total phenolic yield (22.8 ± 3.21%) as compared to ethanol (18.87 ± 0.6%), despite not being statistically significant ($p = 0.116$). However, the ethanolic extracts exhibited larger amounts of TP (around 2.8 times) as compared to the water extract ($p = 0.001$). The ethanol polarity might be responsible for the observed TP content difference.

Table 1. Yield extraction (%), total phenols and EPR-spin trapping and DPPH-radical scavenging activity (IC50) of ethanolic and water crude extracts obtained by accelerator solvent extraction (ASE).

Type of Extracts	Total Phenols (mg GAE/gr DW ± SD) [y]	Yield (% ± SD)	IC50·OH (mg/mL ± SD)	IC50 DPPH (mg/mL ± SD)
WACE	55.41 ± 0.11 [a]	22.8 ± 3.21 [a]	0.67 ± 0.53 [a] $R^2 = 0.9791$	0.15 ± 0.41 [a] $R^2 = 0.9711$
EACE	155.73 ± 1.20 [b]	18.87 ± 0.6 [a]	0.64 ± 0.71 [a] $R^2 = 0.9989$	0.09 ± 0.32 [b] $R^2 = 09922$

[y] GAE: gallic acid equivalent; DW: Dry weight; SD: standard deviation; IC50: sample concentration at which 50% of the free radical activity was inhibited. a: ASE water crude extract; b: ASE ethanolic crude extract. The unlike letters represent values significantly different at $p < 0.05$

2.2. DPPH and EPR Radical-Scavenging Activity

The antioxidant capability was expressed as the quantity of antioxidant inducing a 50% decrease in DPPH concentration or a 50% inhibition of the hydroxyl radical production (IC50). The quenching efficiency of DPPH or hydroxyl radical is inversely proportional to the IC50. Table 2 shows the IC50 of grape leaves ethanolic and aqueous crude extracts. The ethanolic extract of grape leaves showed higher activity of the scavenging DPPH radical (0.09 mg/mL) as compared to the aqueous extract (0.15 mg/mL) ($p = 0.035$). Ethanolic and water extracts provided IC50 of 0.67 (±0.53) and 0.64 (±0.71) mg/mL respectively. The trapping of hydroxyl radical did not show any significant difference between the two extracts ($p = 0.181$).

Table 2. IC50* of grape leaves ethanolic and water ASE crude extracts on MCF-7, HepG2 and HUVEC cells.

Extract \ Cells	MCF-7	HepG2	HUVEC
WACE [y] IC50* (mg/mL)	0.71	1.1	>>2
EACE [x] IC50* (mg/mL)	0.43	0.7	>>2

* IC50: sample concentration at which 50% of cell proliferation was inhibited; [y] WACE: ASE aqueous crude extract; [x] EACE: ASE ethanolic crude extract.

2.3. Effect of Grape Leaves EACE and WACE Extract on HUVEC Cell Proliferation

Both the ethanolic (EACE) and aqueous (WACE) extracts were not toxic for HUVEC cells, with the IC50 being higher that 2 mg/mL. Ethanolic and water extracts inhibited HUVEC cells proliferation in a dose-dependent manner ($p = 0.01$ and HUVEC cells induced an inhibition of cell growth (96%) at 10 µM (Figure 1)).

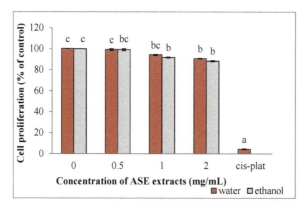

Figure 1. Effect of ASE crude extracts on HUVEC cell proliferation (untreated group: concentration = 0). Data are expressed as mean ± SD, n = 3. Bars marked by unlike letters within a group are significantly different at $p < 0.05$, according to Duncan's Multiple Range Test (DMRT).

2.4. EACE and WACE Extract Counteract HepG2 Proliferation

The survival of HepG2 cells was significantly reduced following incubation with ethanol ($p = 0.001$) and water extracts ($p = 0.001$) (cell proliferation is expressed as the mean percentages of viable cells relative to untreated cells) (Figure 2). In addition, inhibition of HepG2 cell proliferation by both extracts were dose-dependent. In particular, IC50 was obtained when 0.7 mg/mL or 1.1 mg/mL of ethanolic or water extracts, respectively, were added to the culture medium. In all cases, ethanolic extracts were significantly more active than water extracts ($p = 0.001$). The maximum growth inhibition was obtained using Cisplatin (93.52%), representing the positive control, followed by 2 mg/mL ethanolic extracts (82.5%) and 2 mg/mL water extracts (68.63%).

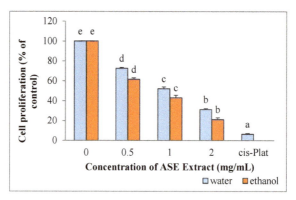

Figure 2. Effect of ASE crude extracts on of HepG2 cell proliferation. Each value is expressed as mean ± SD, n = 3 (untreated group: concentration = 0). Bars marked by unlike letters within a group are significantly different at $p < 0.05$, according to Duncan's Multiple Range Test (DMRT).

2.5. EACE and WACE Extracts Influence the Expression of Apoptosis-Related Genes in HepG2 Cells

HepG2 cultured in the presence of EACE or WACE exhibited a significant increase in Bax mRNA levels in a concentration-dependent manner, as compared to untreated control ($p < 0.05$) (Figure 3). Moreover, Bcl-2 gene expression was down-regulated in a concentration-dependent manner ($p < 0.05$) (Figure 4). The effect of ethanolic extracts was more prominent on HepG2 cells than water extracts ($p = 0.002$). In particular, the maximum effect on both Bax and Bcl-2 genes was observed using the highest concentration (2 mg/mL) of ethanolic extracts.

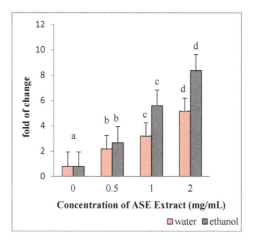

Figure 3. Effect of ASE crude extracts on Bax gene expression in HepG2 cells. The mRNA levels for each gene are expressed as fold of change ($2^{-\Delta\Delta Ct}$) relative to the untreated control (defined as 1) (mean ± SD; n = 3) and normalized to the Glyceraldehyde-3-Phosphate-Dehidrogenase (GAPDH). Data are expressed as mean± SD referred to the control. Bars marked by unlike letters within a group are significantly different at $p < 0.05$, according to Duncan's Multiple Range Test (DMRT).

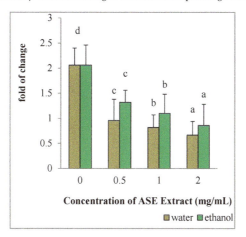

Figure 4. Effect of ASE crude extracts on Bcl-2 gene expression in HepG2 cells. The mRNA levels for each gene are expressed as fold of change ($2^{-\Delta\Delta Ct}$) relative to the untreated control (CTRL-), defined as 1 (mean ± SD; n = 3), and normalized to the Glyceraldehyde-3-Phosphate-Dehidrogenase (GAPDH). Data are expressed as mean ± SD referred to the control. Bars marked by unlike letters within a group are significantly different at $p < 0.05$, according to Duncan's Multiple Range Test (DMRT).

2.6. EACE and WACE Extracts Influence MCF-7 Proliferation

Similar to what was observed in HepG2 cells, both crude extracts significantly inhibited MCF-7 cell proliferation (Figure 5). In particular, the IC50 for EACE and WACE of grape leaves was 0.43 mg/mL and 0.71 mg/mL, respectively. The ethanolic extracts were significantly more active than the water extracts ($p = 0.002$). The largest percentage of growth inhibition was obtained by Cisplatin (99.34%), followed by ethanolic (88.56%) and water extracts (79.31%) (Figure 5). Results revealed that MCF-7 cells were more sensitive to extracts than HepG2 cells.

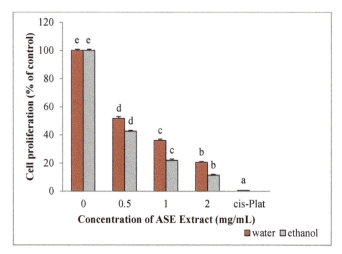

Figure 5. Effects of ASE crude extracts on MCF-7 cell proliferation (concentration 0 corresponding to the untreated group). Data are expressed as mean ± SD, n = 3. Bars marked by unlike letters within a group are significantly different at $p < 0.05$, according to Duncan's Multiple Range Test (DMRT).

2.7. EACE and WACE Extracts Influenced the Expression Of Apoptosis-Related Genes in MCF-7 Cells

Ethanol and water extracts significantly modulated Bax and Bcl-2 mRNA expression levels in MCF-7 cells in a concentration-dependent manner, with Bax expression being significantly upregulated ($p = 0.001$) and Bcl-2 significantly down-regulated ($p = 0.002$). The maximum effect was observed at the highest concentration (2 mg/mL) of ethanolic or water extracts (Figures 6 and 7).

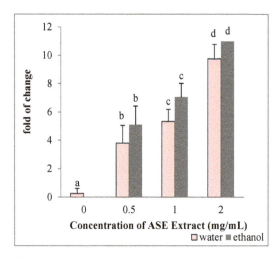

Figure 6. Effect of ASE crude extracts on Bax gene expression in MCF-7 cells. The mRNA levels are expressed as fold of change ($2^{-\Delta\Delta Ct}$) as compared to untreated HepG2 cells (defined as 1) (mean ± SD; n = 3) and normalized to GAPDH. Bars marked by unlike letters within a group are significantly different at $p < 0.05$, according to Duncan's Multiple Range Test (DMRT).

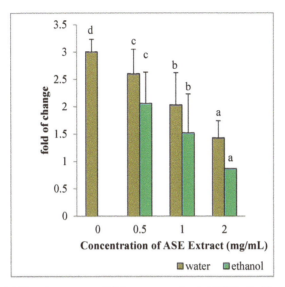

Figure 7. Effect of ASE crude extracts on Bcl-2 gene expression in MCF-7 cells. The mRNA levels are expressed as fold of change ($2^{-\Delta\Delta Ct}$) as compared to untreated HepG2 cells (defined as 1) (mean ± SD; n = 6) and normalized to GAPDH. Bars marked by unlike letters within a group are significantly different at $p < 0.05$, according to Duncan's Multiple Range Test (DMRT).

3. Discussion

Plant bioactive compounds have drawn increasing attention due to their potent antioxidant properties and their marked effects in the prevention of various oxidative-stress-associated diseases, such as cancer. In the last few years, the identification and development of these compounds or extracts from different plants has become a major area of health- and medical-related research [18]. Phenolic compounds are considered as bioactive compounds, widely present in all parts of plant and crude extracts [19].

In this study, we utilized the accelerator solvent extraction method to prepare crude extracts of grape leaves, grown in Algeria, in ultrapure water and 60% ethanol. We aimed at evaluating the anti-proliferative effects of these extracts on HepG2 hepatocarcinoma cells and MCF-7 breast cancer cells. The amount of total phenols and the antioxidant activity were evaluated by scavenging DPPH* and trapping of hydroxyl radical using EPR-spin trapping technique. Then, cell viability was analyzed by using different concentrations of the extracts.

This study showed for the first time, the extraction of bioactive compounds such as phenolic compounds from grape leaves by ASE. ASE provided fast (10 min), easy (automated technique), safe (no direct exposure to the solvent) and inexpensive (in 34 mL of solvent) extraction, leading to high yields and high phenolic contents. Leelavinothan and Arumugam (2008) found that grape leaves contain 99 mg of gallic acid equivalents (mg GAE)/g of phenolic compounds in 70% hydroalcoholic solvent after 72 h of extraction [20], a value lower than the one obtained with the extraction methods described in the present study and requiring a longer extraction time and more solvent. Orhan et al., (2007) describe a phenolic compound yield of 16,07% by extracting 500 g of *Vitis vinifera* dried powder leaves with 80% ethanol at room temperature (5 L * 6 times) [21]. The pressure exerted by ASE allows the extraction cell to be filled faster and helps to force liquid into the solid matrix. Elevated temperatures enhance the diffusivity of the solvent, resulting in an increased extraction kinetic [22–24]. Consequently, ASE may be used to obtain a higher yield in an extremely short time as compared to all previously described methods. Indeed, in recent years, ASE has been successfully applied to the extraction of phenolic compounds from different plant materials, such as grape seeds and skin [25–27] apples [28], spinach [29], eggplants [30] and barley flours [31].

Electron paramagnetic resonance (EPR) spin trapping has become an indispensable tool for the specific detection of reactive oxygen free radicals in biological systems [32]. The EPR spin-trapping technique was used to study the ability of ASE grape leaves extracts to quench OH radicals, which are common reactive oxygen species associated with oxidative cell damage [33]. The hydroxyl radical reacts unselectively and very quickly with any chemical compound able to lose a hydrogen atom [34]. Our results indicate that water and ethanol grape leaf extracts possessed similar ˙OH radicals quenching activity. In water extract, the content of TP, despite being lower than that of ethanol, was high enough to react with the hydroxyl radicals produced, thus excluding any dose-dependent mechanism in the reaction between antioxidants and ˙OH.

DPPH* free radical was used to evaluate the ability of phenolic compounds to transfer labile hydrogen atoms to radicals [32]. Our extracts showed high capability to scavenge DPPH*, due to the presence of different polyphenols, including flavonoids, which can be found in grape leaves [35,36]. Generally, the chemical structure of flavan-3-ol family grants a good antioxidant response towards DPPH. The hydrogen-donating substituents (hydroxyl groups), attached to the aromatic ring structures of flavonoids, allow for a redox reaction able to scavenge free radicals [21,37].

Apoptosis can be activated through two major pathways, the mitochondria-dependent pathway and the death-receptor-dependent pathway. In the mitochondria-dependent signaling pathway, the Bcl-2 family of proteins is divided into two groups: suppressors of apoptosis (e.g., Bcl-2, Bcl-XL, Mcl-1) and activators of apoptosis (e.g., Bax, Bok, Hrk, Bad). The Bax/Bcl-2 ratio might represent a critical factor influencing cell behavior. Suppression of Bcl-2 promotes apoptosis in response to several stimuli, including anticancer drugs [38]. Bax is a pro-apoptotic protein residing in the cytosol in an inactive form and translocating, after activation, to the mitochondria, where it plays an important role in mitochondria-mediated apoptosis. Activated Bax, either in homo-oligomeric form or as complex with other proteins, creates pores in the outer mitochondrial membrane, which leads to the leakage of ions, essential metabolites and cytochrome c from mitochondria to cytosol, thus promoting cell death [39]. Our results demonstrated that grape leaves have an anti-proliferative effect on HepG2 and MCF-7 cells. EACE and WACE markedly inhibited HepG2 and MCF-7 cell viability.

In cells cultured with these extracts, the mRNA levels of the anti-apoptotic factor, Bcl-2, were downregulated, while the expression of the pro-apoptotic gene Bax, was significantly induced. Within this context, other authors have demonstrated that molecules as Diazaphenothiaznes exert an antiproliferative activity in MCF7 cells and C32 human amelanotic melanoma, by regulating BAX and BCL2 gene expression [40,41].

Deepak et al. (2015) show that desert plant extracts are able to induce apoptosis in HepG2 cells. They also describe an upregulation of Bax, Bad, cytochrome c, caspase-3, caspase-7, caspase-9 and poly (adenosine diphosphate-ribose) polymerase [42]. Furthermore, the *Allium atroviolaceum* flower extracts was found to inhibit HepG2 cell growth, revealing a sub-G_0 cell cycle arrest, changes in morphological features and annexin-V and propidium iodide positive staining, which correlates with Bcl-2 down-regulation and caspase-3 activity [43]. Lu Y et al. (2011) report that injectable seed extracts from *Coix lacryma-jobi* L. induce apoptosis in HepG2 cells, with elevated and prolonged expression of caspase-8, which do not significantly influence the expression of Bcl-2 [44]. Moreover, Jun et al. (2009) report that quercetin can inhibit proliferation and induce apoptosis in HepG2 cells by decreasing the levels of surviving cells and Bcl-2 protein expression, and significantly increasing the protein levels of p53 [38].

We found that the ethanolic crude extracts were able to induce a larger anti-proliferative effect as compared to the aqueous crude extracts, which may be due to the different amount of phenols detected in the two different extracts. Nevertheless, further experiments are needed in order to understand if apoptosis could definitely explain the antiproliferative effects induced by the extracts tested in the present study.

Our extracts showed growth inhibition in MCF-7 cells, confirming what has been previously described by other authors using different plant extracts. Blassan et al. (2016) report that *Rubus fairholmianus* root extracts inhibit MCF-7 cells growth via caspase 3/7-induced apoptosis [45]. Reis et al. (2013) report that *Leccinum vulpinum* induces DNA damage, decreases cell proliferation and induces apoptosis in MCF-7 cells [46]. Miris et al. (2011) report that pomegranate (*Punica granatum* L.), at certain concentration, inhibits MCF-7 cell proliferation and induces increased expression of the pro-apoptotic gene Bax and decreased the expression of the anti-apoptotic gene Bcl-2 [47]. ASE extracts of grape leaves grown in Algeria were not cytotoxic for HUVEC cells. Atmaca et al. (2016) report that *Salvia triloba* L. extract has pro-apoptotic and anti-angiogenic effect in prostate cancer cell lines while being not cytotoxic for normal cells [48]. Aghbali et al. (2013) describe the pro-apoptotic potential of grape seeds extracts, confirmed by a significant inhibition of cell growth and viability in a dose- and time-dependent manner without inducing damage to HUVEC non-cancerous cells [49]. Indeed, the bioactive phytochemicals, Honokiol and Magnolol contained in *Magnolia officinalis* and their derivatives show an antiproliferative effect on HepG2 cell proliferation while being unable to elicit any effect on fibroblasts [50].

Finally, the literature strongly suggests that grape is a potential source of antioxidant, anticancer and cancer chemo-preventive phytochemicals. The other parts of the grapes, the skin and seeds, the whole grape by itself, grape-derived raisins and phytochemicals within the grapes have also been found to bear potential anticancer properties in various preclinical and clinical studies [51].

4. Materials and Methods

4.1. Chemicals and Cells

All solvents used were HPLC (High Performance Liquid Chromatography) grade purified (Merck, Darmstadt, Germany); water was purified using a milli-Qplus system from Millipore (Milford, MA, USA). Reagents employed were of analytical grade; Folin-Ciocalteu reagent and Sodium Carbonate (Na_2CO_3) were purchased from Carlo Erba (Milan, Italy); DPPH (2,2-diphenyl-1-picrylhydrazyl) and gallic acid (3,4,5-trihydroxybenzoic acid) were purchased from Sigma-Aldrich, 5,5-dimethyl-1-pyrroline-*N*-oxide (DMPO) spin trap and dimethyl sulfoxide (DMSO) were purchased from Sigma-Aldrich (Milan, Italy).

HepG2 and MCF-7 cells were obtained from the Hospital of Cagliari, 09121, Cagliari, Italy. HUVECs cells were obtained from Gibco™ (Grand Island, NY, USA). Cisplatin was obtained from the Oncological Services Hospital of Sassari, Italy.

Dulbecco's phosphate buffered saline (DPBS) was purchased from Euroclone (Milano, Italy); Dulbecco's modified Eagle's Medium with phenol red (DMEM) and fetal bovine serum (FBS) from Life Technologies (Grand Island, NY, USA); Medium 200 and LSGS (5-003-10) from Gibco™. TRIzol reagent, SuperScript® VILO™ cDNA Synthesis Kit, Platinum Quantitative PCR Supermix UDG Kit, SybrGreen I, primer and fluorescein from Life Technologies (Grand Island, NY, USA). L-glutamine, Penicillin, Streptomycin, nonessential amino acids from Euroclone (Milano, Italy). (3-(4,5-dimethylthiazol-2-yl)-2,5-diphenyltetrazolium bromide) tetrazolium reduction MTT Cell Proliferation Assay ATCC® 30-1010K kit was purchased from Invitrogen Co. All the primer sequences are represent in Table 3.

Table 3. Primers sequences used for real-time PCR reactions.

Primers	Forward	Reverse
hGAPDH	GAGTCAACGGATTTGGTCGT	GACAAGCTTCCCGTTCTCAG
BAX	TCTGACGGCAACTTCAACTG	TTGAGGAGTCTCACCCAACC
BCL-2	AGGATTGTGGCCTTCTTTGA	ACAGTTCCACAAAGGCATCC

4.2. Plant Material

Mature leaves from the *Vitis vinifera* L. apical portion were collected in Medea, Algeria in August. Leaves were rinsed with tap water and dried at room temperature (25 ± 3 °C). Finally, they were ground into a fine powder and kept in the dark at 5 °C in a sterile bag and under vacuum for further use.

4.3. Extraction Procedure

ASE was performed on a Dionex ASE 350 (Dionex Thermo FisherScientific Inc., Waltham, MA, USA). Powdered leaves (1 g) were weighed into a 22 mL Dionex (ASE 350) stainless-steel cell. The cells were equipped with a stainless-steel fit and a cellulose filter. The optimized operating conditions for ASE extraction are indicated in Table 4.

Table 4. Conditions of ASE extraction procedure.

Temperature (°C)	40
Pressure (PSI)	1500
Number of Cycle	2
Extraction time of one cycle (min)	5
Concentration of Ethanol (%)	60 Ethanol/40 water
Type of water used	Ultrapure

Two solvents were tested: ethanol 60% (v/v) and water. The extraction was performed in quadruplicate. After the extraction process, water extracts were immediately freeze-dried whereas the ethanolic ones were first evaporated under a nitrogen flow to remove ethanol, then freeze dried. The freeze-dried extracts were weighed and stored at −80 °C until analysis. Accelerated solvent extraction was performed with the lowest extraction temperature to avoid the maximum degradation of thermolabile compounds.

The extraction yield was calculated as follows:

$$\text{Yield\%} = \frac{\text{(the weight of freeze} - \text{dried recover)}}{1 \text{ gram (initial weight of leaf powder used)}} \times 100 \quad (1)$$

4.4. Total Phenolic (TP) Content

The total phenolic content was measured using the modified Folin-Ciocalteu method [52–54]. 1 mg of each lyophilized extract was mixed with 9 mL of cold ethanol (80%) (1:10 w/v), vortexed (Stuart, U.K. model SA8.) at 1600 rpm for 2 min and centrifuged (ALC-Centrifuge 4227R, Milan, Italy) at 16,000× g for 15 min at 4 °C. 200 µL of each extract were mixed with the Folin-Ciocalteu reagent (1 mL) and allowed to react for 8 min before adding 800 µL of sodium carbonate solution (0.075 mL^{-1}). The mixture was incubated in the dark for one hour at room temperature (20 ± 3 °C) followed by an additional hour at 0 °C. The absorbance was read at 760 nm with a spectrophotometer (8453 Agilent Technologies, Santa Clara, CA, USA).

Results were expressed as milligrams of gallic acid equivalent/g of dry weight on the basis of a gallic acid calibration curve (50 to 500 mg/L with R^2 = 0.996).

4.5. Antioxidant Activity

4.5.1. Spin Trapping Assay of the •OH Radical

The hydroxyl radical scavenging activity was determined with the spin trapping method coupled with electron paramagnetic resonance spectroscopy according to Fadda et al. [34]. The hydroxyl radicals were generated by the Fenton reaction and trapped with a nitrone spin trap 5,5-dimethyl-pyrroline N-oxide (DMPO) [55]. 20 mg of the freeze-dried extract mixed with 1 mL of ultrapure water degassed under nitrogen flow was prepared as stock solution. Serial dilutions were prepared from the stock solution, and depending on the results, the correct concentration for each extract was established. 100 µL of the diluted samples were mixed with Fe(II) sulfate 0.1 mM (100 µL), 112 µL DMPO 26 mM (112 µL) and H_2O_2 1 mM (100 µL).

The DMPO-OH adduct was detected with a Bruker EMX EPR spectrometer operating at the X-band (9.4 GHz) using a Bruker Aqua-X capillary cell. The EPR instrument was set under the following conditions: modulation frequency, 100 kHz; modulation amplitude, 1 G; receiver gain, 1×10^5; microwave power, 20 mW. EPR spectra were recorded at room temperature immediately after the preparation of the reaction mixture. The concentration of the spin adduct DMPO-OH was estimated from the double integration of spectra. The hydroxyl radical scavenging activity was expressed as IC50 on the basis of the percentage of inhibition calculated as follows:

$$\% \text{ inhibition} = \frac{(A_0 - A_s)}{A_0} \times 100 \qquad (2)$$

where A_0 is the intensity of the spin adducts without extract and A_s is the absorbance of the adduct after the reaction with the extract. Different sample's concentrations were used to calculate the IC50, that is, the extract concentration that halves the concentration of hydroxyl radical adduct of the blank. Three replications were performed for each dilution.

4.5.2. DPPH

The radical scavenging activity of ethanolic and water extracts of grape leaves was determined spectrophotometrically with the DPPH test [56].

30 µL ASE ethanolic and water crude extract at different concentrations (0.05, 0.1, 0.2 mg mL^{-1}) were mixed with 3 mL of a DPPH methanol solution (0.3 mM). A blank solution was prepared using methanol instead of the extract.

Solutions were stored in the dark at room temperature for 30 minutes. The absorbance was measured at 518 nm and converted into the percentage of inhibition using the following equation:

$$\% \text{ inhibition} = \frac{A_0 - A_s}{A_0} \times 100 \qquad (3)$$

where A_0 is the absorbance of the sample without extract and A_s is the absorbance of the sample after the reaction with the extract. The DPPH radical scavenging activity was expressed as IC50. Three replications were performed for each dilution.

4.6. Cell Culture

HepG2 and MCF-7 cells were maintained in Dulbecco's modified Eagle's Medium with phenol red (DMEM), supplemented with 10% heat-inactivated fetal bovine serum (FBS), 200 of µM L-glutamine, 200 U/mL of penicillin, 10 µg/mL of streptomycin and 0.1 mM of non-essential amino acids. HUVEC cells were cultured in Medium 200 (Gibco™), containing LSGS (5-003-10; Gibco™), 200 U/mL of penicillin and 10 µg/mL of streptomycin. Cells were grown in 75 cm^2 tissue culture flasks in the culture incubator at 37 °C with 5% CO_2 and saturated humidity.

4.7. MTT Viability Assay

The anti-proliferative activity of ethanolic and aqueous ASE extracts of *Vitis vinifera* L. leaves on HepG2, MCF-7 and HUVEC cells was determined using a cell viability test.

The MTT [3-(4,5-dimethylthiazol-2-yl)-2,5-diphenyltetrazolium bromide] tetrazolium reduction assay is a colorimetric assay based on the ability of functional mitochondria to reduce by succinate dehydrogenase enzyme an insoluble formazan crystal, which displays a purple color [57].

Then, the effects of the treatments on the overall growth of a particular cell population were assessed by determining the number of living cells remaining in the analyzed cell culture. After counting, HepG2, MCF-7 and HUVEC cells were seeded on a 96-well plate at concentration of 10,000/well in 200 µL and incubated at 37 °C in a 5% CO_2 incubator (Thermo Fisher Scientific, Waltham, MA, USA).

After 24 h, the medium was replaced with fresh medium containing compounds tested (ethanol and aqueous ASE crude extract) at concentration of 0.5 mg/mL, 1 mg/mL and 2 mg/mL. The negative control is performed in growing medium but positive control is prepared in medium supplemented with cosplatin 10 µM. Every test was repeated three times. After one day, we again substituted medium with or without compounds and repeated the same treatment (treatment 2). The MTT substrate was prepared in a sterile Dulbecco's phosphate buffered saline (DPBS), then added to cells in culture at a final concentration of 650 µg/mL and incubated for 3 h in the culture incubator at 37 °C with 5% CO_2 and saturated humidity. After incubation, the medium was removed by aspiration and 200 µL/well Dimethylsufoxide DMSO (Sigma Aldrich) was added to each well. Absorbance was read at 570 nm in a Gemini EMMicroplate Reader (Molecular devices). The percentage of cell proliferation was calculated relative to control wells designated as 100% viable cells using the following formula:

$$\frac{(At - Ab)}{(Ac - Ab)} \times 100 = \% \; cell \; proliferation \qquad (4)$$

where At = absorbance value of test compound (ASE extract), Ab = absorbance value of blank (medium alone), Ac = absorbance value of control.

4.8. Gene Expression

HepG2 and MCF-7 cells were plated into 24-well cell culture plates (60,000 cells/500 µL for each well) in culture medium with ethanolic and aqueous ASE extracts of grape leaves to evaluate the expression levels of apoptotic-related genes. Extracts were prepared fresh just before each experiment and dissolved in DMEM.

After treatment, the total RNA was isolated using TRIzol reagent and quantified by measuring the absorbance at 260/280 nm (NanoDrop 2000, spectrophotometer Thermoscientific ND8008, Thermo Fisher Scientific, Waltham, MA, USA). Approximately 1 µg of total RNA was reverse-transcribed to cDNA by SuperScript® VILO™ cDNA Synthesis Kit (Life Technologies, Grand Island, NY, USA).

Quantitative polymerase chain reaction was run in triplicate using a CFX Thermal Cycler (Bio-Rad, Hercules, CA, USA). 2 µL of cDNA were amplified in 25 µL reactions using Platinum Quantitative PCR Supermix UDG Kit. A Supermix 2X was mixed with Sybr Green I, 0.1 µM of primer and 10 nM fluorescein (Life Technologies, Grand Island, NY, USA). Relative target Ct (the threshold cycle) values of Bcl-2 and Bax were normalized to GAPDH, as housekeeping gene. The mRNA levels of cells treated with ethanolic and aqueous ASE extract were expressed using the $2^{-\Delta\Delta Ct}$ method [58], relative to the mRNA level of the untreated sample for each experiment.

4.9. Statistical Analysis

Results are expressed as mean ± standard deviation (SD) and were analyzed by ANOVA with Duncan's multiple range tests procedure (DMRT) and Student's t-test using 25.0 SPSS Windows software. Differences were considered significant for $p < 0.05$.

5. Conclusions

In the present study, we revealed for the first time that accelerator solvent extraction yielded a higher extraction rate of total phenols and antioxidant activity in an extremely short time. The extract obtained from grape leaves grown in the Medea region (Algeria) exhibited an antiproliferative effect on MCF-7 breast cancer cells and HepG2 hepatocarcinoma cells. Moreover, considering previous reports by other authors and the present results that provide evidence for the modulation of Bax/Bcl2 mRNA levels by leaf extracts, which affects the balance between apoptosis and cell survival, it may be concluded that these extracts could be used as an easily accessible source of natural antioxidants, and as a matrix to prepare drugs counteracting distinctive cancer cells' proliferation.

Author Contributions: S.F. is the leading author, who developed most of the idea and wrote the article. S.S., S.C., A.F. and D.S. provided technical guidance during simulations and experiments; M.M. provided the necessary technical tools for the realization of this work and followed the writing of the paper and its revision. S.Z. is the supervisor of this work. A.D. and G.D. supervised the work and are the coordinators of the guesting Institute.

Funding: This research received no external funding.

Acknowledgments: The authors are grateful to: ATRSS–DGRSDT (MESRS, Algeria), the Institute of Science of Food Production from the Italian National Research Council. Angela Fadda, Antonio Dore and Daniele Sanna for the precious help provided during the realization of this work. Margherita Maioli, Sara Santaniello and Sara Cruciani for accepting me in their lab and for providing the interesting tracking.

Conflicts of Interest: The authors declare no conflict of interest.

References

1. Ďurackova, Z. Some Current Insights into Oxidative Stress. *Physiol. Res.* **2010**, *59*, 459–469. [PubMed]
2. Barry, H. Oxidative stress and cancer: Have we moved forward? *Biochem. J.* **2007**, *401*, 1–11. [CrossRef]
3. Ferlay, J.; Shin, H.R.; Bray, F.; Forman, D.; Mathers, C.; Parkin, D.M. Estimates of worldwide burden of cancer in 2008: GLOBOCAN 2008. *Int. J. Cancer* **2010**, *127*, 2893–2917. [CrossRef] [PubMed]
4. You, J.S.; Jones, P.A. Cancer genetics and epigenetics: Two sides of the same coin? *Cancer Cell* **2012**, *22*, 9–20. [CrossRef] [PubMed]
5. Sun, S.Y.; Hail, N., Jr.; Lotan, R. Apoptosis as a novel target for cancer chemoprevention. *J. Natl. Cancer Inst.* **2004**, *96*, 662–672. [CrossRef] [PubMed]
6. Fresco, P.; Borges, F.; Marques, M.P.M.; Diniz, C. The Anticancer Properties of Dietary Polyphenols and its Relation with Apoptosis. *Curr. Pharm. Des.* **2010**, *16*, 114–134. [CrossRef] [PubMed]
7. Martin, K.R. Targeting apoptosis with dietary bioactive agents. *J. Med. Food* **2006**, *231*, 117–129. [CrossRef]
8. Chandrasekara Reddy, G.; Shiva Prakash, S.; Diwakar, L. Stilbene heterocycles: Synthesis, antimicrobial, antioxidant and anticancer activities. *J. Pharm. Innov.* **2015**, *3*, 24–30.
9. Kalra Ekta, K. Nutraceutical-Definition and Introduction. *AAPS Pharm. Sci.* **2003**, *5*. [CrossRef]
10. Basoli, V.; Santaniello, S.; Cruciani, S.; Ginesu, G.C.; Cossu, M.L.; Delitala, A.P.; Serra, P.A.; Ventura, C.; Maioli, M. Melatonin and Vitamin D Interfere with the Adipogenic Fate of Adipose-Derived Stem Cells. *Int. J. Mol. Sci.* **2017**, *18*, 981. [CrossRef]
11. Santaniello, S.; Cruciani, S.; Basoli, V.; Balzano, F.; Bellu, E.; Garroni, G.; Ginesu, G.C.; Cossu, M.L.; Facchin, F.; Delitala, A.P.; et al. Melatonin and Vitamin D Orchestrate Adipose Derived Stem Cell Fate by Modulating Epigenetic Regulatory Genes. *Int. J. Med. Sci.* **2018**, *15*, 1631–1639. [CrossRef] [PubMed]
12. Mazza, G.; Miniati, E. *Anthocyanins in Fruits, Vegetables and Grains*; CRC Press: Boca Raton, FL, USA, 1993.
13. Felicio, J.D.; Santos, R.S.; Gonzalez, E. Chemical constituents from Vitis Vinifera (Vitaceae). *Arg. Inst. Biol.* **2001**, *68*, 47–50.
14. Monagas, M.; Garrido, I.; Bartolomé, B.; Gomez-Cordovés, B. Chemical characterization of commercial dietary ingredients from *Vitis vinifera* L. *Anal. Chim. Acta* **2006**, *563*, 401–410. [CrossRef]
15. Monagas, M.; Hernandez-Ledesma, B.; Gomez-Cordovés, C.; Bartolomé, B. Commercial dietary ingredients from Vitisvinifera L. leaves and grape skins: Antioxidant and chemical characterization. *J. Agric. Food Chem.* **2006**, *54*, 319–327. [CrossRef] [PubMed]
16. Wang, L.J.; Weller, C.L. Recent advances in extraction of nutraceuticals from plants. *Trends Food Sci. Technol.* **2006**, *17*, 300–312. [CrossRef]
17. Romanik, G.; Gilgenast, E.; Przyjazny, A.; Kamiński, M. Techniques of preparing plant material for chromatographic separation and analysis. *J. Biochem. Biophys. Methods* **2007**, *70*, 253–261. [CrossRef] [PubMed]
18. Jin, D.; Russell, J.M. Plant Phenolics: Extraction, Analysis and Their Antioxidant and Anticancer Properties. *Moeclues* **2010**, *15*, 7313–7352. [CrossRef]

19. Asma, H.A.-S.; Mohammad, A.H. Total phenols, total flavonoids contents and free radical scavenging activity of seeds crude extracts of pigeon pea traditionally used in Oman for the treatment of several chronic diseases. *Asian Pac. J. Trop. Dis.* **2015**, *5*, 316–321. [CrossRef]
20. Pari, L.; Suresh, A. Effect of grape (Vitisvinifera L.) leaf extract on alcohol induced oxidative stress in rats. *Food Chem. Toxic.* **2008**, *46*, 1627–1634. [CrossRef]
21. Deliorman, O.D.; Orhan, N.; Ergun, E.; Ergun, F. Hepatoprotective effect of Vitisvinifera L. leaves on carbon tetrachloride-induced acute liver damage in rats. *J. Ethnopharmacol.* **2007**, *112*, 145–151. [CrossRef]
22. Richter, B.E.; Jones, B.A.; Ezzell, J.L.; Porter, N.L.; Avdalovic, N.; Pohl, C. Accelerated solvent extraction: A technology for sample preparation. *Anal. Chem.* **1996**, *68*, 1033–1039. [CrossRef]
23. Brachet, A.; Rudaz, S.; Mateus, L.; Christen, P.; Veuthey, J. Optimisation of accelerated solvent extraction of cocaine and benzoylecgonine from coca leaves. *J. Sep. Sci.* **2001**, *24*, 865–873. [CrossRef]
24. Kaufmann, B.; Christen, P.; Veuthey, J.L. Study of factors influencing pressurized solvent extraction of polar steroids from plant material Application to the Recovery of Withanolides. *Chromatographia* **2001**, *54*, 394–398. [CrossRef]
25. Ju, Z.Y.; Howard, L.R. Effects of solvent and temperature on pressurized liquid extraction of anthocyanins and total phenolics from dried red grape skin. *J. Agric. Food Chem.* **2003**, *51*, 5207–5213. [CrossRef] [PubMed]
26. Pineiro, Z.; Palma, M.; Barroso, C.G. Determination of trans-resveratrol in grapes by pressurised liquid extraction and fast high-performance liquid chromatography. *J. Chromatogr. A* **2006**, *1110*, 61–65. [CrossRef]
27. Luque-Rodriguez, J.M.; Luque de Castro, M.D.; Perez-Juan, P. Dynamic superheated liquid extraction of anthocyanins and other phenolics from red grape skins of winemaking residues. *Bioresour. Technol.* **2007**, *98*, 2705–2713. [CrossRef] [PubMed]
28. Alonso-Salces, R.M.; Korta, E.; Barranco, A.; Berrueta, L.A.; Gallo, B.; Vicente, F. Pressurized liquid extraction for the determination of polyphenols in apple. *J. Chromatogr. A* **2001**, *933*, 37–43. [CrossRef]
29. Howard, L.; Pandjaitan, N. Pressurized liquid extraction of flavonoids from spinach. *J. Food Sci.* **2008**, *73*, 151–157. [CrossRef]
30. Luthria, D.L.; Mukhopadhyay, S. Influence of sample preparation on assay of phenolic acids from eggplant. *J. Agric. Food Chem.* **2006**, *54*, 41–47. [CrossRef]
31. Bonoli, M.; Marconi, E.; Caboni, M.F. Free and bound phenolic compounds in barley (Hordeumvulgare L.) flours. *J. Chromatogr. A* **2004**, *1057*, 1–12. [CrossRef]
32. Hawkins, C.L.; Davies, M.J. Detection and characterization of radicals in biological materials using EPR methodology. *Biochim. Biophys. Acta* **2014**, *1840*, 708–721. [CrossRef] [PubMed]
33. Goupy, P.; Dufour, C.; Loonis, M.; Dangles, O. Quantitative kinetic analysis of hydrogen transfer reactions from dietary polyphenols to the DPPH radical. *J. Agric. Food Chem.* **2003**, *51*, 615–622. [CrossRef] [PubMed]
34. Fadda, A.; Barberis, A.; Sanna, D. Influence of pH, buffers and role of quinolinic acid, a novel iron chelating agent, in the determination of hydroxyl radical scavenging activity of plant extracts by Electron Paramagnetic Resonance (EPR). *Food Chem.* **2018**, *240*, 174–182. [CrossRef] [PubMed]
35. Samoticha, J.; Wojdyło, A.; Golis, T. Phenolic composition, physicochemical properties and antioxidant activity of interspecific hybrids of grapes growing in Poland. *Food Chem.* **2017**, *215*, 263–273. [CrossRef] [PubMed]
36. Farhadi, K.; Esmaeilzadeh, F.; Hatami, M.; Forough, M.; Molaie, R. Determination of phenolic compounds content and antioxidant activity in skin, pulp, seed, cane and leaf of five native grape cultivars in West Azerbaijan province. *Iran. Food Chem.* **2016**, *15*, 847–855. [CrossRef] [PubMed]
37. Brand-Williams, W.; Cuvelier, M.E.; Berset, C. Use of a free-radical method to evaluate antioxidant activity. *Food Sci. Technol.* **1995**, *28*, 25–30. [CrossRef]
38. Tan, J.; Wang, B.C.; Zhu, L.C. Regulation of Survivin and Bcl-2 in HepG2 Cell Apoptosis Induced by Quercetin. *Chem. Biodevers.* **2009**, *6*, 1101–1110. [CrossRef]
39. Farahmandzad, F.; Ahmadi, R.; Riyahi, N.; Mahdavi, E. The Effects of Scrophulariastriata extract on Apoptosis in Glioblastoma Cells in Cell Culture. *Int. J. Adv. Chem. Eng. Bio. Sci.* **2015**, *2*, 62–63. [CrossRef]
40. Morak-Młodawska, B.; Pluta, K.; Latocha, M.; Suwińska, K.; Jeleń, M.; Kuśmierz, D. 3,6-Diazaphenothiazines as potential lead molecules – synthesis, characterization and anticancer activity. *J. Enzyme Inhib. Med. Chem.* **2016**, *31*, 1512–1519. [CrossRef]
41. Morak-Młodawska, B.; Pluta, K.; Latocha, M.; Jeleń, M.; Kùsmierz, D. Synthesis, Anticancer Activity, and Apoptosis Induction of Novel 3,6-Diazaphenothiazines. *Molecules* **2019**, *24*, 267. [CrossRef]

42. Deepak, B.; Animesh, M.; Eviatar, N.; Anupam, B. Apoptosis-inducing effects of extracts from desert plants in HepG2 human hepatocarcinoma cells. *Asian Pac. J. Trop. Biomed.* **2015**, *5*, 87–92. [CrossRef]
43. Khazaei, S.; Abdul Hamid, R.; MohdEsa, N.; Ramachandran, V.; GhomiTabatabaee, F.; Aalam, A.; Ismail, P. Promotion of HepG2 cell apoptosis by flower of Allium atroviolaceum and the mechanism of action. *BMC Complement. Altern. Med.* **2017**, *17*, 1594–1596. [CrossRef] [PubMed]
44. Lu, Y.; Zhang, B.Y.; Jia, Z.X.; Wu, W.J.; Lu, Z.Q. Hepatocellular carcinoma HepG2 cell apoptosis and caspase-8 and Bcl-2 expression induced by injectable seed extract of Coixlacryma-jobi. *Hepatob. Pancreat. Dis. Int.* **2011**, *10*, 303–307. [CrossRef]
45. Blassan, P.G.; Abrahamse, H.; Parimelazhagan, T. Caspase dependent apoptotic activity of *Rubus fairholmianus* Gard.on MCF-7 human breast cancer cell lines. *J. Appl. Biomed.* **2016**, *89*, 1–9. [CrossRef]
46. Filipa, S.; Reis, D.S.; Lillian, B.; Anabela, M.; Patricia, M.; Isabel, C.F.R.; Vasconcelos, M.H. Leccinumvulpinum Watling induces DNA damage, decreases cell proliferation and induces apoptosis on the human MCF-7 breast cancer cell line. *Food Chem. Toxicol.* **2013**, *90*, 45–54. [CrossRef]
47. Dikmen, M.; Ozturk, N.; Ozturk, Y. The Antioxidant Potency of Punicagranatum, L. Fruit Peel Reduces Cell Proliferation and Induces Apoptosis on Breast Cancer. *J. Med. Food.* **2011**, *14*, 1638–1646. [CrossRef] [PubMed]
48. Atmaca, H.; Bozkurt, E. Apoptotic and anti-angiogenic effects of *Salvia triloba* extract in prostate cancer cell lines. *Tumor Biol.* **2015**, *37*, 3639–4657. [CrossRef] [PubMed]
49. Amirala, A.; Speidel, V.H.; Abbas, D.; Nader, K.; Gharavi, F.; Zare, S.; Mona, O.; Ali, B.; Behzad, B. Induction of apoptosis by grape seed extract (*Vitisvinifera*) in oral squamous cell carcinoma. *Bosn. J. Basic Med. Sci.* **2013**, *13*, 186–191. [CrossRef]
50. Maioli, M.; Basoli, V.; Carta, P.; Fabbri, D.; Antonietta, D.M.; Cruciani, S.; Andrea, S.P.; Delogu, G. Synthesis of magnolol and honokiol derivatives and their effect against hepatocarcinoma cells. *PLoS ONE* **2018**, *13*, e0192178. [CrossRef]
51. Zhou, K.Q.; Julian, J.R. Potential Anticancer Properties of Grape Antioxidants. *J. Oncol.* **2012**. [CrossRef]
52. Folin, D.; Denis, W. On phosphotungstic-phosphomolybdic compounds as color reagents. *J. Bio. Chem.* **1912**, *12*, 239–243.
53. Folin, D.; Ciocalteu, V. On tyrosine and tryptophan determinations in proteins. *J. Bio. Chem.* **1927**, *73*, 627–650.
54. Singleton, V.; Rossi, J.A., Jr. Colorimetry of total phenolic with phosphotungstic-phosphomolybdic acid reagents. *Am. J. Enol. Viticult.* **1965**, *16*, 144–157.
55. Fadda, A.; Bernardo, P.; Alberto, A.; Antonio, B.; Maria, C. Suitability for ready-to-eat processing and preservation of six green and red baby leaves cultivars and evaluation of their antioxidant value during storage and after the expiration date. *J. Food Process Preserv.* **2015**, *40*, 550–558. [CrossRef]
56. Choi, C.W.; Kim, S.C.; Hwang, S.S.; Choi, B.K. Antioxidant activity and free radical scavenging capacity between Korean medicinal plants and flavonoids by assay guided comparison. *Plant Sci.* **2002**, *163*, 1161–1168. [CrossRef]
57. Mosmann, T. Rapid colorimetric assay for cellular growth and survival: Application to proliferation and cytotoxicity assays. *J. Immunol. Methods* **1983**, *65*, 55–63. [CrossRef]
58. Kenneth, J.L.; Schmittgen, T.D. Analysis of Relative Gene Expression Data Using Real-Time Quantitative PCR and the $2^{-\Delta\Delta Ct}$ Method. *Methods* **2001**, *25*, 402–408. [CrossRef]

Sample Availability: Samples of the compounds from grape leaves are available from the authors.

© 2019 by the authors. Licensee MDPI, Basel, Switzerland. This article is an open access article distributed under the terms and conditions of the Creative Commons Attribution (CC BY) license (http://creativecommons.org/licenses/by/4.0/).

MDPI
St. Alban-Anlage 66
4052 Basel
Switzerland
Tel. +41 61 683 77 34
Fax +41 61 302 89 18
www.mdpi.com

Molecules Editorial Office
E-mail: molecules@mdpi.com
www.mdpi.com/journal/molecules

Lightning Source UK Ltd.
Milton Keynes UK
UKHW051952090821
388566UK00003B/139